Lorenzo Cantoni and James A. Danowski (Eds.)
Communication and Technology

Handbooks of Communication Science

Edited by
Peter J. Schulz and Paul Cobley

Volume 5

Communication and Technology

Edited by
Lorenzo Cantoni and James A. Danowski

DE GRUYTER
MOUTON

Università della Svizzera italiana | Faculty of Communication Sciences

The publication of this series has been partly funded by the Università della Svizzera italiana – University of Lugano.

ISBN 978-3-11-026653-5
e-ISBN (PDF) 978-3-11-027135-5
e-ISBN (EPUB) 978-3-11-039344-6
ISSN 2199-6288

Library of Congress Cataloging-in-Publication Data
A CIP catalog record for this book has been applied for at the Library of Congress.

Bibliographic information published by the Deutsche Nationalbibliothek
The Deutsche Nationalbibliothek lists this publication in the Deutsche Nationalbibliografie; detailed bibliographic data are available on the Internet at http://dnb.dnb.de.

Cover image: Oliver Rossi/Photographer's Choice RF/Gettyimages
Typesetting: Meta Systems Publishing & Printservices GmbH, Wustermark
Printing and binding: CPI books GmbH, Leck
♾ Printed on acid-free paper
Printed in Germany

MIX
Papier aus verantwortungsvollen Quellen
FSC® C083411

www.degruyter.com

Preface to *Handbooks of Communication Science* series

This volume is part of the series *Handbooks of Communication Science*, published from 2012 onwards by de Gruyter Mouton. When our generation of scholars was in their undergraduate years, and one happened to be studying communication, a series like this one was hard to imagine. There was, in fact, such a dearth of basic and reference literature that trying to make one's way in communication studies as our generation did would be unimaginable to today's undergraduates in the field. In truth, there was simply nothing much to turn to when you needed to cast a first glance at the key objects in the field of communication. The situation in the United States was slightly different; nevertheless, it is only within the last generation that the basic literature has really proliferated there.

What one did when looking for an overview or just a quick reference was to turn to social science books in general, or to the handbooks or textbooks from the neighbouring disciplines such as psychology, sociology, political science, linguistics, and probably other fields. That situation has changed dramatically. There are more textbooks available on some subjects than even the most industrious undergraduate can read. The representative key multi-volume *International Encyclopedia of Communication* has now been available for some years. Overviews of subfields of communication exist in abundance. There is no longer a dearth for the curious undergraduate, who might nevertheless overlook the abundance of printed material and Google whatever he or she wants to know, to find a suitable Wikipedia entry within seconds.

'Overview literature' in an academic discipline serves to draw a balance. There has been a demand and a necessity to draw that balance in the field of communication and it is an indicator of the maturing of the discipline. Our project of a multi-volume series of *Handbooks of Communication Science* is a part of this coming-of-age movement of the field. It is certainly one of the largest endeavours of its kind within communication sciences, with almost two dozen volumes already planned. But it is also unique in its combination of several things.

The series is a major publishing venture which aims to offer a portrait of the current state of the art in the study of communication. But it seeks to do more than just assemble our knowledge of communication structures and processes; it seeks to *integrate* this knowledge. It does so by offering comprehensive articles in all the volumes instead of small entries in the style of an encyclopedia. An extensive index in each *Handbook* in the series, serves the encyclopedic task of find relevant specific pieces of information. There are already several handbooks in sub-disciplines of communication sciences such as political communication, methodology, organisational communication – but none so far has tried to comprehensively cover the discipline as a whole.

For all that it is maturing, communication as a discipline is still young and one of its benefits is that it derives its theories and methods from a great variety of work in other, and often older, disciplines. One consequence of this is that there is a variety of approaches and traditions in the field. For the *Handbooks* in this series, this has created two necessities: commitment to a pluralism of approaches, and a commitment to honour the scholarly traditions of current work and its intellectual roots in the knowledge in earlier times.

There is really no single object of communication sciences. However, if one were to posit one possible object it might be the human communicative act – often conceived as "someone communicates something to someone else." This is the departure point for much study of communication and, in consonance with such study, it is also the departure point for this series of *Handbooks*. As such, the series does not attempt to adopt the untenable position of understanding communication sciences as the study of everything that can be conceived as communicating. Rather, while acknowledging that the study of communication must be multifaceted or fragmented, it also recognizes two very general approaches to communication which can be distinguished as: a) the semiotic or linguistic approach associated particularly with the humanities and developed especially where the Romance languages have been dominant and b) a quantitative approach associated with the hard and the social sciences and developed, especially, within an Anglo-German tradition. Although the relationship between these two approaches and between theory and research has not always been straightforward, the series does not privilege one above the other. In being committed to a plurality of approaches it assumes that different camps have something to tell each other. In this way, the *Handbooks* aspire to be relevant for all approaches to communication. The specific designation "communication science" for the *Handbooks* should be taken to indicate this commitment to plurality; like "the study of communication", it merely designates the disciplined, methodologically informed, institutionalized study of (human) communication.

On an operational level, the series aims at meeting the needs of undergraduates, postgraduates, academics and researchers across the area of communication studies. Integrating knowledge of communication structures and processes, it is dedicated to cultural and epistemological diversity, covering work originating from around the globe and applying very different scholarly approaches. To this end, the series is divided into 6 sections: "Theories and Models of Communication", "Messages, Codes and Channels", "Mode of Address, Communicative Situations and Contexts", "Methodologies", "Application areas" and "Futures". As readers will see, the first four sections are fixed; yet it is in the nature of our field that the "Application areas" will expand. It is inevitable that the futures for the field promise to be intriguing with their proximity to the key concerns of human existence on this planet (and even beyond), with the continuing prospect in communication sciences that that future is increasingly susceptible of prediction.

Note: administration on this series has been funded by the Università della Svizzera italiana – University of Lugano. Thanks go to the president of the university, Professor Piero Martinoli, as well as to the administration director, Albino Zgraggen.

Peter J. Schulz, Università della Svizzera italiana, Lugano
Paul Cobley, Middlesex University, London

Contents

Introduction

Lorenzo Cantoni and James A. Danowski

Communication technologies: An itinerary

Abstract: This introductory chapter provides an overview on the entire volume. In particular, it explains the very structure of the book, organized in three sections: (I) “Mediavolution”: communication media between evolution and revolution; (II) Communication technologies and their environment; and (III) Communication technologies and new practices of communication in the information and communication society.

Every chapter is then briefly outlined, so to better understand the overall architecture and the richness of the collected papers. A paragraph especially devoted to diffusion models and communication technologies, and some textual analyses do complement the introduction.

Keywords: ICT: Information and Communication Technologies, media history, online communication, media research

When Peter J. Schulz and Paul Cobley invited us to be part of this extraordinary enterprise of the *Handbooks of Communication Science*, that seemed to us a challenge you cannot refuse, and at the same time an (almost) impossible mission.

On the one hand, we are fully convinced that one of our goals – one might even say missions – as academics is not only to move further and further research in specific areas, but also to try to make sense of all new advances, so to integrate them in a meaningful – even if always incomplete – picture. As per the famous suggestion by Aristotle, understanding is always a unifying process, which finds commonalities, analogies, rules, even theories under endless changes of our experience. And this is, in fact, what is suggested by the name itself of University: finding centers, polarities, which explain, and cast light onto countless individual elements. Which attract, order, and explain our changing experiences.

On the other hand, we must acknowledge that, while the research enterprise is always challenging and a work-in-progress, the topic itself of this volume makes it even more challenging: in the end it is not without reason that Communication Technologies are also referred to as *New Media*: to emphasize their ever-changing nature, their fast running towards new communication technologies, practices, configurations ... The topic itself is clearly discouraging any attempt to provide any established (enough) knowledge: any encyclopedic overview appears condemned from the very beginning to fail both because it won’t be able to reach a too fast-moving target, and because won’t be able to reach enough depth while presenting (b)leading innovations, processes started just a few years ago, and far from being clearly understood.

Nonetheless, we accepted. First, because we knew that we could share such challenge – and eventually succeed – with the help and patient collaboration of so many good colleagues and friends; second, because communication processes (and technologies) are always rooted in basic human needs, which might find unforeseen embodiments, but can't escape a thorough analysis, able to go beyond what appears at first sight; third, because this is anyway our mission as academic researchers: now more than ever students and young researchers need to be provided with good maps and compasses to navigate in the very perilous sea of Information and Communication Technologies (ICTs). And maybe (fourth) because we are not only hungry of knowledge, but also fools ...

This introductory chapter is organized as follows: it provides first a paragraph on Changing Diffusion Models and Communication Technologies, followed by an helicopter view on the volume and on its structure, and then by a detailed presentation of the various contributions. A very last paragraph acknowledges the many people who have made this mission possible (if successful, will be judged by the readers).

1 Changing diffusion models and communication technologies

As mentioned above, new media – which in turn are going to become old and to be substituted by newer ones – are anyway distributed on a timeline, which runs in parallel with the very history of human beings: our need to communicate between and among us, to share one of our richest treasures: what we think (love or fear, like or dislike ...), a richness that requires the full communication toolbox human beings have received and developed across centuries and continents. A good/responsibility, which in Latin was said "munus", and where the very term "communication" is rooted, to suggest such sharing of meanings.

While several attempts have been made to uncover the development from pure orality to various communication media or "technologies of the word" (see for instance Ong, 1982), and their *mediamorphosis* (Fidler, 1997), new media are presenting a somehow accelerated adoption/diffusion pattern, if compared with previous technological innovations. Such peculiarity has to be briefly introduced here, so to stress one more time not only the challenge of editing such a book, but the new context brought about by (currently) new media.

In the early 1970s, the modern convergence of communication technologies began. The two initial technologies were known as 'telecommunications' (mainly telephony systems) and 'computers.' Oettinger (1971), a professor of Information Resources Policy at Harvard, and frequent participant in congressional hearings about new communication technologies in the U. S., coined a new term to describe this convergence of computers and telecommunications: "Compunications." This heralded a new age of communication and technology that required changes in public policy regarding what would later become better known as the "Information Society" (Porat, 1977).

Prior to this time, particularly before the mid-1960s, diffusion of communication media, and other innovations – concepts, products, and processes – followed an S-shaped diffusion curve (Rogers and Shoemaker, 1971). This cumulative normal distribution over time was because of the primary importance to adoption of interpersonal communication networks. Because of the constraints on how many individuals a person could maintain close relationships with, talking with others to persuade them adopt was quite limited. Media were not significant in these classic diffusion processes. Rather, the process was one of contagion, with information moving virally from one person to another.

At the same time, industrial societies had been producing age-graded cohorts (e.g. generations) primarily based on being born near the same time and experiencing the same historical contexts and events. When peak industrialization occurred in the mid-1960s in countries such as the U. S., age-grading as the basis for cohort formation began to change. Evidence shows that after 1966 "post-industrial societies" (Bell, 1967), in which manufacturing comprised less than 50 % of GDP, cohort formation also changed. Rather than based on birth, cohort formation was increasingly based on shared interest in common communication media content (Danowski & Ruchinskas, 1985).

This dynamic of developing communication cohorts impacted as well on the diffusion process for new communication and technology innovations. Adoption became increasingly based on messages people obtained from media and less from direct interpersonal network communication. The previously dominant S-shaped diffusion curve, a cumulative normal distribution of adoption over time, became increasingly supplanted by the convex-shaped diffusion curve, which Bass (1969) called the R-curve. Figure 1 shows the differences in shape of these two diffusion curves. As the bulge of the curve becomes more convex, media messages are having more influence on adoption than are interpersonal communication networks.

As these fundamental changes in processes of diffusion occurred, the adoption of new communication technologies increasingly followed r-shaped rather than s-shaped diffusion curves. New theories were needed to explain these processes (Danowski, Riopelle, and Gluesing, 2011). R-shaped diffusion has been explained as "herd effects" (Choi, 1997), taken from animal behavior. Once lead herd members begin to move in a particular direction, the other members quickly follow suit with no knowledge of environmental conditions. This results in an adoption curve with very rapid growth over time followed by a tapering off as laggards adopt.

In communication terms, these increasingly convex r-shaped diffusion curves are largely the result of changes in communication and technology. Rosenkopf and Abramson (1999) refer to these r-curves as "information cascades," theorizing that their occurrence becomes more likely as there are: 1) a greater number of mediated

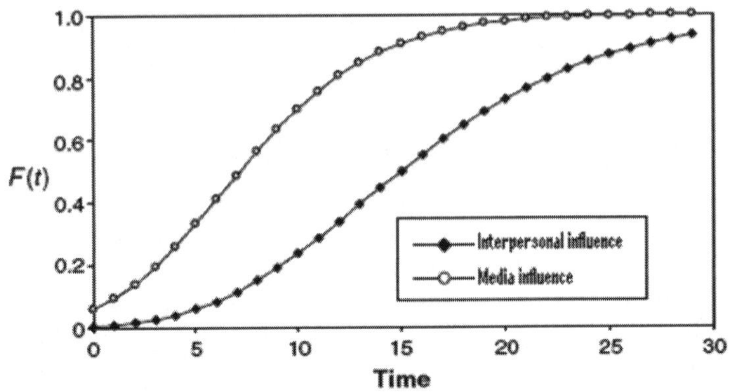

Fig. 1: Convex media influenced adoption curve compared to interpersonal influence S-shaped curve.

messages about the innovation, 2) more ambiguity about its efficiency/effectiveness, 3) more mediated messages about the number of adopters, and 4) more mediated messages about the social status of adopters.

We would add that the smaller the time interval of the communication media news cycle, the greater the likelihood of information cascade processes. These intervals have shrunk from daily, during the broadcast media dominance of the industrial era, to a virtually real-time cycle with a time interval of seconds in contemporary post-industrial information societies.

As well, there are the effects of the post-industrial communication cohorts based on shared information interests and media channels, rather than on age. Communication cohort members identify with their cohort and make inferences that their own interests in communication content are shared by other members without having to talk with them about this. When they see in social media evidence of other communication cohort members adopting an innovation, they are more likely to adopt without exchanging messages with others. Communication technology diffusion, as well as diffusion of concepts, products, and processes through these new media, is increasingly likely to show bulging convex diffusion curves rather than the s-shaped curves of the industrial era.

2 An helicopter view

The volume presents thirty chapters, organized into three sections: *"Mediavolution": communication media between evolution and revolution*; *Communication technologies and their environment*; and *Communication technologies and new practices of communication in the information and communication society*. The first two sections feature nine chapters each, while the third one features twelve chapters. In

Tab. 1: The 50 most common words present in the book.

Item	#	Item	#
new	1'057	computer	281
communication	1'038	http	272
information	1'035	design	271
social	1'017	public	270
technology	1'006	www	259
media	797	recording	251
internet	726	systems	247
research	655	user	247
digital	610	health	246
virtual	526	members	229
online	514	networks	228
community	493	users	225
world	477	experience	223
web	455	international	223
mobile	434	society	222
development	398	news	221
data	392	reality	214
network	361	access	212
learning	324	university	211
press	318	work	211
human	303	studies	210
people	299	theory	209
journal	298	music	205
example	292	content	202
time	288	government	201

general, while the first two sections include longer contributions, the last one presents shorter papers.

The first section approaches the topic moving from semiotic codes and technologies themselves, and explores their changes and innovations in recent time, as well as models to interpret them. It deals with newspapers wires services, telecommunication networks, cinema, music, video-games, hypermedia and the web, virtual reality, social networks, up to so-called web2.0 and 3.0 ...

The second section deals with specific topics, particularly "hot" when approaching communication technologies, from user experience and usability to information quality, from digital divide to the social impact of mobile technologies, from legal to ethical issues, from the impact of new media onto old ones to so-called "digital natives", up to research methods on the internet.

The third and last section is aimed at mapping specific areas where Communication Technologies do play a major role, and have promoted/allowed significant changes in practices. It explores areas like commerce, workplace, marketing and PR, government, terrorism, religion, learning, health, tourism, journalism, libraries, science.

Tab. 2: Top 46 Directed Word Pairs within 3 words on either side.

Word Pair		#	Word Pair		#
social	media	310	social	networks	78
communication	technologies	217	health	information	76
virtual	communities	210	online	games	76
new	communication	183	world	wide	74
user	experience	176	new	forms	73
virtual	community	157	world	web	73
new	technologies	145	mobile	phone	72
social	network	139	wide	web	72
information	quality	131	landline	telephone	72
information	overload	116	network	theory	68
mobile	communication	111	open	access	68
virtual	reality	99	community	members	68
information	communication	98	electric	telegraph	68
united	states	97	digital	media	65
virtual	snss	96	virtual	worlds	65
new	media	92	online	communication	65
social	theory	92	electronic	governance	64
social	networking	92	health	care	64
communities	snss	92	media	communication	63
mobile	telephone	90	icts	development	63
information	technologies	89	player	communities	63
quality	information	84	no	longer	60
record	companies	80	information	technology	59

As the reader can notice, the editors have tried to be as inclusive as possible, nonetheless, the available space (and time) required to make many (difficult) choices, so to leave outside several other possible topics/perspectives.

Chapter authors have been selected for their respective prominence and leadership in the various areas, as well as with the intention to provide several voices and perspectives, and to invite scholars with different and complementary backgrounds, both in terms of geography, of disciplines, and of current affiliations.

Once all manuscripts have been collected, we have run a few statistical analyses on such corpus, so to explore some peculiarities and recurrences. Having stopped generic/irrelevant terms and other linguistic elements, and after having combined similar/identical items (e.g.: technology and technologies), the fifty most frequent items are presented in Table 1 (elaboration done with WordSmith Tools 6).

In addition to individual word frequencies as shown in Table 1, it is informative to examine words in their immediate local context in the corpus, such as frequency counts of what pairs of words occur within 3 words on either side of each word in the text (Danowski, 1993). Furthermore, by retaining the order of words in the pairs as they occurred in the text, these directed word pairs enable

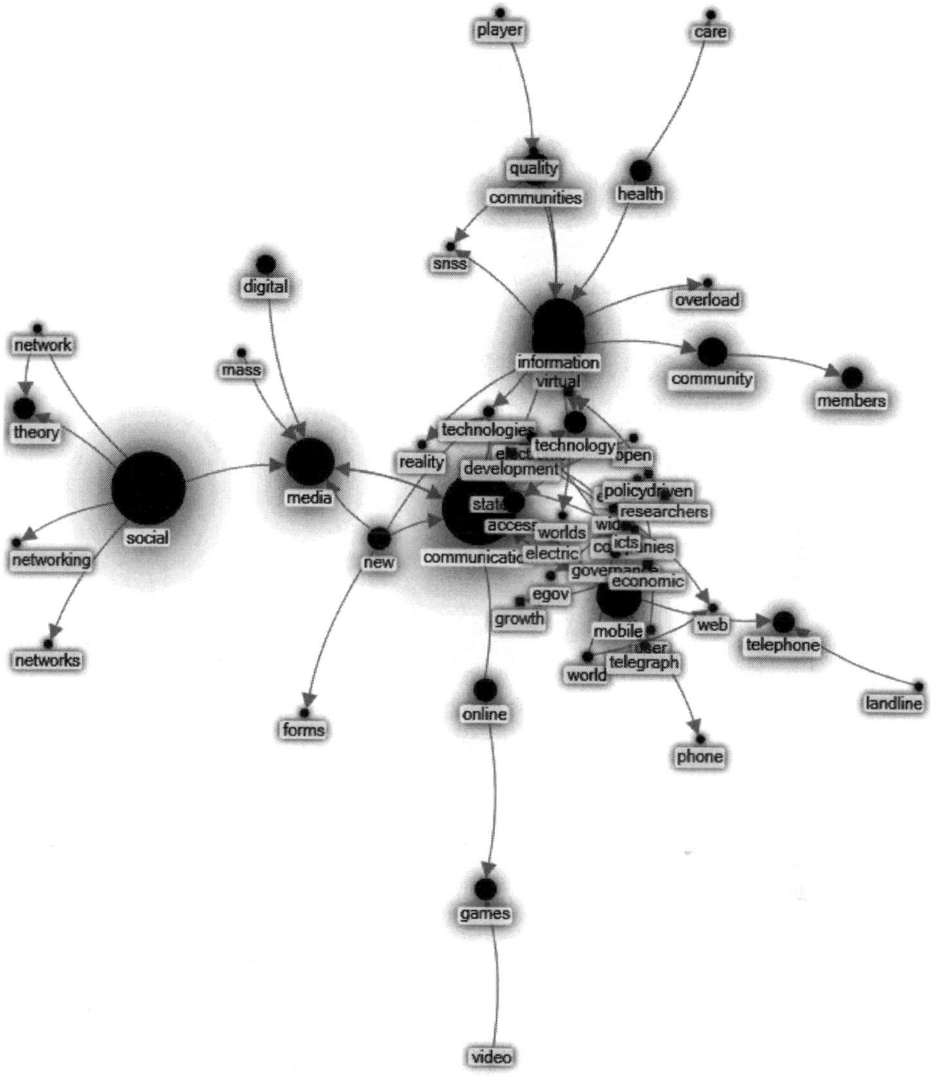

Fig. 2: Network among the top word pairs.

one to gain more detailed representations of meanings. The local context and proxemics among words provided by word-pair analysis adds value to the interpretation of the statistical analysis of the text. Moreover, when the overlaps of these words paired in the word window is graphed, it adds additional interpretative information. Strings of highly-related pairs longer than 2 words create groups of words that are more related within the group to one another than to words outside the group. For the word-pair analysis we used WORDij (Danowski, 2013), here we did not perform stemming of words to their simple roots for of significant information, for reasons discussed in (Danowski, 1993).

Fig. 3: Word-cloud of most frequent words present in the book.

Fig. 4: Word-cloud of chapters’ keywords.

Once WORDij has processed the text, we used NodeXL (Smith, Milic-Frayling, Shneiderman, Mendes Rodrigues, Leskovec, Dunne, 2010) to create the optimal graph layout. The size of the nodes reflects their betweenness centrality: the lager the circle the more central the word in the overall network.

A different kind of representation is word clouds, produced by Wordle software (www.wordle.net) which are seen in Figures 3 and 4. Although one gets a more simplified representation than word networks as shown in Figure 2, the clouds are useful for a quick picture of the overall contents, ignoring specific word strings and groups. Clouds are effective for showing the overall word domain as well as relative word frequency as this is represented in the size of the word.

Tab. 3: List of words appearing at least three times among chapters' keywords.

Item	#	Item	#
social	14	online	4
communication	12	communities	3
digital	12	economy	3
media	11	games	3
web	8	learning	3
internet	7	market	3
mobile	7	mediated	3
technology	7	methods	3
information	6	new	3
electronic	5	overload	3
research	5	quality	3
virtual	5	semantic	3
data	4	systems	3
governance	4	telegraph	3
learners	4	theory	3

Hereafter, it is represented a word-cloud of the whole book, which includes more terms than the fifty above listed ones.

As it can be noted, the most prominent terms provide the very core topical area of this book: Information and Communication Technology, new (media), the internet and social (media). Also its research orientation is clearly visible, as well as different perspectives, focusing on individuals and/or on their communities.

A similar result can be achieved through a slightly different strategy. All contributors have been requested not only to provide a short summary of their chapter, but also to list some relevant keywords. If we just consider such keywords (221 in total, which might consist of more than one word), 30 words appear at least three times, as presented in the table above.

If a word-cloud is done using only such keywords, which are particularly relevant because provided by the authors themselves, the following image appears.

It's now time to introduce the three sections and each individual contribution, providing a short summary of them, adapted from their abstract section.

3 Three sections, thirty chapters

The first section is devoted to analyze the *"Mediavolution": communication media between evolution and revolution*; it approaches this task studying different semiotic codes and media.

Brett Oppegaard opens this section with a chapter titled *From orality to newspaper wire services: Conceptualizing a medium*. The very concept of "medium" enables researchers to connect cave paintings to wearable computing, a concept in which the media form creates a setting, or environment, in which communication takes place. That environment shapes, encourages, promotes, constricts, and restricts the messages in ways that affect cultural and social behaviors, meaning the medium – through which communication takes place – also is an important part of the message, to quote a very successful suggestion by Marshall McLuhan.

Gabriele Balbi and Richard R. John present *Point-to-point: telecommunications networks from the optical telegraph to the mobile telephone*. Their chapter defines "telecommunications" and sketches the main dimensions of four telecommunications networks over a two-hundred-year period – the optical telegraph, the electric telegraph, the landline telephone, and the mobile telephone (and its predecessor, the wireless telegraph). Then, it shows how historical scholarship on topics in the history of telecommunications has been shaped by three intellectual traditions: the Large Technical Systems approach, Political Economy, and the Social Construction of Technology.

Alejandro Pardo presents a chapter on *Cinema and technology: from painting to photography and cinema, up to digital motion pictures in theatres and on the net*. After having noted that the relation between cinema and technology has been present since the very inception of motion pictures, this chapter presents a comprehensive compilation of key scholarly literature and identifies major theoretical issues and emerging concepts. It is divided into three main sections: the first one focuses on the relationship between the arts and technology, and specifically between cinema and the arts, and between cinema and technology; the second section draws a brief historical summary on the technological development of (audio)visual media, moving from the primitive canvas to the first photographic plates and from the birth of cinema to the digital image; the third part is a synthesis of some of the most relevant theoretical and critical issues regarding the imbrication of art, technology and cinema.

Tom McCourt's chapter on *Recorded music* surveys the history of recording identifying three eras: acoustic, electric, and digital. It explores also some common characteristics: first, a shifting oligopoly of record companies has controlled this process; second, each era claimed to more accurately capture sound through greater technological intervention; third, changes in recording and distribution have repurposed and decentralized music, affecting its creation and reception.

Marko Siitonen explores *Communication in video games: from players to player communities*. Digital games research and communication studies intertwine at several points: gaming, and play in general, is a social activity, and digital gaming and online game worlds offer near endless ways for self-expression and socializing. This chapter looks at questions of social interaction within the realm of online multiplayer games; the topics introduced proceed from motivations of individual players to the social dynamics of player groups and communities, to exploring games as communication systems and platforms.

Stefano Tardini and Lorenzo Cantoni present a chapter on *Hypermedia, internet and the web*. In this chapter, two of the most important instances of ICTs are introduced: internet and the web, together with the concept of hypermedia/hypertext, which played a pivotal role in the theoretical discussions about ICT-mediated communication as well as in the widespread diffusion of the internet and the web. The concept and the history of hypertext are presented, and some relevant interpretations of it are provided, borrowed from the field of communication sciences: a linguistic and semiotic approach, a rhetorical one, and a literary one. The internet, its history, diffusion and different layers are then presented, to introduce the result of the application of hypertext to the internet: the world wide web, which decreed the success of the internet as the most widespread and powerful communication technology at the beginning of the third millennium. A model to design and interpret websites is then explained: OCM – Online Communication Model.

Rita M. Lauria and Jacquelyn Ford Morie have co-authored a chapter on *Virtuality: VR as metamedia and herald of our future realities*. They examine the concept of virtual reality (VR) as an advanced telecommunications medium that transcends all that has gone before, forming, in essence, a new and advanced metamedium. By acknowledging the porous boundaries between the simulated and the "real," virtuality constitutes a phenomenological structure of "seeming," where the computer-constructed reality feels experientially authentic. After presenting a brief history of developments in VR, from both technological and more philosophical viewpoints, they explore the complementary concepts of the computer system as an active participant, and the embodiment of the human actor within the simulated reality, and discuss how the concept of virtuality serves to fuse these potential dichotomies.

Constance Elise Porter's chapter on *Virtual communities and social networks* reviews findings from previous research and identifies key scholarly issues from historical, contemporary and forward-looking perspectives. Ultimately, she calls for scholars to go beyond descriptive accounts of human behavior in virtual communities and social networking sites by developing theoretical explanations for such behavior. In doing so, the author lays a foundation upon which marketing and communications scholars might build programmatic research.

Ulrike Gretzel presents a chapter titled *Web 2.0 and 3.0*, which are summary terms widely used to describe emerging trends in the development of internet technologies. This chapter describes the technological foundations of Web 2.0 and 3.0, and discusses the economic factors and cultural/social practices that are typically associated with the two phenomena.

* * *

The second section, titled *Communication technologies and their environment*, approaches specific relevant issues linked with communication technologies.

Tim Unwin's chapter opens the section dealing with *ICTs and the dialectics of development*. It provides an overview of some of the challenges that need to be considered in defining the notions of both ICTs and 'development', arguing that both must be seen as contested terms that serve specific interests. The chapter adopts a dialectical approach that first seeks to identify the main grounds for a thesis of the 'good' in the use of ICTs in development practices, it then develops an antithesis that proposes that the use of ICTs has actually increased inequality at a range of scales, and has thus worked against a definition of 'development' based on social equity. After that, it seeks to explore what a synthesis of these two diametrically opposed positions might look like.

Martin J. Eppler's chapter focuses on *Information quality and information overload: the promises and perils of the information age*. In this contribution, he presents two key concepts in the realm of modern day communication infrastructures: the prescriptive notion of information quality, and the descriptive concept of information overload, i.e.: not being able to process the provided quantity of information adequately.

Davide Bolchini's chapter discusses *User experience and usability*. It reviews key concepts related to both usability and user experience of interactive communication, coming from different disciplinary areas: computer-mediated communication, computer science, information science, software engineering, and human-computer interaction. Moreover, it seeks to synthesize a practical, integrated perspective across knowledge domains, which mainly stems from usability engineering, the growing area of user experience, and interaction design.

Brian Winston's chapter *Impact of new media: a corrective* is aimed at rebalancing the narratives about old and new media, and at demystifying the technicist hyperbolic rhetoric of the "information revolution."

Claire Hewson's chapter deals with *Research methods on the internet*. She discusses how internet technologies can be used to support primary research on the internet, or internet-mediated research (IMR). After having depicted a brief history of IMR, key methods that have been implemented in IMR are outlined: surveys and questionnaires, interviews and focus groups, experiments, observation and document analysis. Key issues and debates which have emerged in IMR are discussed, including issues of data quality, sampling and sample bias, and ethics.

Wenhong Chen and Rich Ling discuss the very emerging issue of *Mobile media and communication*. In fact, mobile telephony is the most widespread mediation technology in the world: with access to the wireless internet, mobile devices have expanded from a tool of voice or text-based communication to devices and services for multimedia communication, consumption and production. The chapter provides a brief history of mobile media and communication, highlighting the technological affordances of the mobile internet; it reviews theoretical issues and reports major streams of empirical research.

Emanuele Rapetti and Francesc Pedró present a chapter titled *Digital Natives, New Millennium Learners and Generation Y, does age matter? Data and reflection from the higher education context*. Along the same line of B. Winston, they aim at providing a balanced answer to the very controversial question whether it exists a generation of digital(ized) learners. From the analysis of relevant literature, three different views do emerge: the "enthusiasts", the "concerned ones", and the "critics".

Joanna Kulesza's chapter deals with *Legal issues in a networked world*. She sketches the evolution of international internet governance as the background for all ongoing discussions on the appropriate legal framework for the global network. Among the discussed issues: local notions of privacy confronted with the global need for cybersecurity, national perceptions of decency and online expression exercised across territorial borders, intellectual property rights ... She points to two crucial criteria in confronting all legal challenges online: the need to reinterpret the notions of jurisdiction, and due diligence applied to online communications.

Adriano Fabris' chapter on *Ethical issues in internet communication* concludes the second section. He addresses the specific ethical problems that emerge in connection with the world of the internet, and examines different strategies used to tackle them. To this end, he distinguishes two different perspectives: "Ethics of the Internet" and "Ethics in the Internet".

* * *

The third and last section of the volume presents twelve specific areas, in which the impact of ICTs has been particularly relevant, if not disruptive.

Rolf T. Wigand explores the domain of *Commerce*. He discusses the efficiency of electronic markets, especially their gain in efficiency over traditional markets. According to the author, added value of eCommerce comes from a proper alignment of information and communication technologies, business strategy/goals, and business processes.

Kevin B. Wright's chapter deals with *Workplace relationships: telework, work-life balance, social support, negative features, and individual/organizational outcomes*. The chapter explores several theoretical frameworks that have been applied to the study of new communication technologies and workplace relationships: telecommuting, work-life balance, and negative behaviors in the workplace associated with new communication technologies, such as cyberbullying and cyber-surveillance.

Anne Linke's chapter on *Marketing and public relations* analyses the digital evolution from the Web 1.0 to the Web 2.0, which has created many challenges for enterprises and their corporate communication, and discusses how to find adequate communication management processes and structures in this new context.

Tomasz Janowski's chapter titled *From electronic governance to policy-driven electronic governance – Evolution of technology use in government* presents different ways in which ICTs have been integrated within government-related activities and how, in turn, the government itself has been shaped by them. It outlines the evolution of government use of technology, and introduces policy-driven eGov as the latest phase in this evolution.

Aziz Douai analyses the very hot topic of *Technology and terrorism: media symbiosis and the "dark side" of the web*. He discusses the complex relationship between technology and the deployment of terrorism as a political weapon in contemporary societies. While internet's own architecture has allowed terrorists to use it to evade detection, communicate, recruit and organize, the author claims that it is important to avoid new "moral panics" about the internet, and that intrusive surveillance of regular citizens' online activities should not be justified based on them.

Pauline Hope Cheong and Daniel Arasa's chapter on *Religion* provides a discussion of religion's dynamic developments alongside contemporary communication technology connections and appropriations. It spotlights key insights from prior reviews, and examines forces of interaction and tensions in the relationships between religion and the internet. Illustrations are drawn from various religions, particularly the Catholic Church, the largest religious institution in the world.

Thomas C. Reeves and Patricia M. Reeves approach the issue of *Learning*, and examine its nature and role in the information and communication age by addressing key questions related to both formal learning in schools and informal learning through experience. The theoretical perspectives on learning presented in this chapter include behaviorist, cognitivist, humanist, constructivist, constructionist, social, and connectivist orientations.

Gary L. Kreps presents a chapter on *Communication technology and health: the advent of eHealth applications*. Powerful new health information technologies are transforming the modern healthcare system by supplementing and extending traditional channels for health communication, and by enabling broad dissemination of relevant health information that can be personalized to the unique information needs of individuals.

Alessandro Inversini, Zheng Xiang, and Daniel R. Fesenmaier explore *New media in travel and tourism communication: toward a new paradigm*. They review and highlight the most important milestones in the last two decades that changed tourism communication, focusing on three main areas of development: the persuasive nature of tourism websites, social media conversations, and mobile computing.

John V. Pavlik presents a chapter titled *Journalism: from delivering information to engaging citizen dialogue*. This chapter provides an overview of the traditions and principles of journalism and places them in the evolving context of a digital, mobile and networked world, where citizen reporters operate alongside and sometimes collaboratively with professionally educated journalists.

Stephen M. Griffin's chapter discusses *Libraries in the Digital Age: Technologies, Innovation, Shared Resources and New Responsibilities*. In fact, ICTs have challenged and are challenging the role of libraries, in particular academic libraries: comprehensive reporting of digital scholarship requires new models of scholarly communication that cannot be based only/mainly on print media. Libraries as knowledge institutions are at a unique and opportune time for examining new services and resources that support digital scholarship across the disciplinary spectrum, while also developing innovative practices that serve patrons from the broader population.

Loet Leydesdorff presents a chapter titled *The sciences are discursive constructs: The communication perspective as an empirical philosophy of science*. New communication technologies have introduced a new dynamics in the sciences: in addition to the (local) context of discovery and the (global) context of justification, a third context of mediation enables scholars to reflect on the sciences as discursive constructs.

Acknowledgments

For sure, such an enterprise had not been possible without the contribution of many people. We would like to thank them individually.

First of all, Paul Cobley and Peter J. Schulz, editors of this series: without their invitation and impulse no single line would have been written.

Then all contributors, who have accepted to be part of this adventure, with its risks and demands, especially in terms of time to devote to writing and editing, as well as in terms of meeting deadlines ...

Two PhD candidates of Lorenzo Cantoni, from Università della Svizzera italiana (Lugano, Switzerland), have provided a major contribution: Asta Adukaite in the first part of the project, to ensure its coordination, and Marta Pucciarelli in the second part, to revise manuscripts and finalize the volume.

Last, but not least, we want to thank the great people at De Gruyter Mouton, and especially Barbara Karlson and Wolfgang Konwitschny, who have accompanied the project up to become the book you hold in your hands or you watch through a screen.

References

Bass, F. M. 1969. A new product growth for model consumer durables. *Management Science* 15(5). 215–227.

Bell, D. 1967. Notes on post-industrial society. 2. *Public Interest* 7. 102–118.

Danowski, J. A. 1993. Network analysis of message content. G. Barnett, & W. Richards (eds.). Progress in communication sciences XII (pp. 197–222). Norwood, NJ: Ablex.

Danowski, J. A. 2013. WORDij version 3.0: Semantic network analysis software [computer program]. Chicago: University of Illinois at Chicago. http://wordij.net

Danowski, J. A., & Ruchinskas, J. E. 1983. Period, cohort, and aging effects: a study of television exposure in presidential election campaigns, 1952–1980. *Communication Research* 10(1). 77–96.

Danowski, J. A., Riopelle, K., & Gluesing, J. 2011. The revolution in diffusion models caused by new media: The shift from s-shaped to convex curves. In G. A. Barnett & A.Vishwanath (Eds.) *The diffusion of innovations: A communication science perspective*, (pp. 123–144). NewYork: Peter Lang Publishing.

Fidler, R. 1997. *Mediamorphosis. Understanding new media*. Thousand Oaks: Pine Forge Press.

Oettinger, A. G. 1971. *Compunications in the national decision-making process. Computers, Communications, and the Public Interest*. Johns Hopkins Press, Baltimore (MD), USA.

Ong, W.J. 1982. *Orality and Literacy. The Technologizing of the Word*. London – New York, NY: Routledge.

Porat, M. 1977. *The Information Economy*. Washington, DC: US Department of Commerce.

Rogers, E. M., & Shoemaker, F. F. 1971. *Communication of innovations; a cross-cultural approach*.

Rosenkopf, L., & Abrahamson, E. 1999. Modeling reputational and informational influences in threshold models of bandwagon innovation diffusion. *Computational & Mathematical Organization Theory* 5(4). 361–384.

Smith, M., Milic-Frayling, N., Shneiderman, B., Mendes Rodrigues, E., Leskovec, J., Dunne, C. 2010. NodeXL: a free and open network overview, discovery and exploration add-in for Excel 2007/2010, http://nodexl.codeplex.com/ from the Social Media Research Foundation, http://www.smrfoundation.org

I. The history of communication technologies

Brett Oppegaard

1 From orality to newspaper wire services: Conceptualizing a medium

Abstract: Technologies have been aiding human communication for millennia, with the sign language of gesturing leading to layers upon layers of other communicative innovations, such as the alphabet, writing, and the telegraphic transmissions common today via digital media delivery systems. Connecting cave paintings to wearable computing is the concept of the medium, in which the media form creates a setting, or environment, in which communication takes place. That environment shapes, encourages, promotes, constricts, and restricts the messages in ways that affect cultural and social behaviors, meaning the medium – through which communication takes place – also is an important part of the message.

Keywords: medium theory, medium specificity, technological determinism, mass media history

Imagine entering a foreign land, disassociated from any familiar languages, customs, and cultures. To communicate, you might revert to infantile pointing and pantomime, as a way to somehow share your thoughts with others. In software development parlance, that was the beta 1.0 version of human communication in ancient times, as body language and gesturing gradually grew into the social-cognitive and social-motivational platforms upon which conventional linguistic systems could be cooperatively constructed (Tomasello 2008). While precise dates are in dispute, humans have existed as a distinct species for about 2 million years. Yet they only started speaking to each other about 100,000 years ago, possibly inspired by birdsong and other sounds in nature (Aitchison 1996; Stephens 2007; Changizi 2011). In context, the relatively recent surge of invention and innovation and expansion in human communication systems seems even more stunning. To put a point on it, mass communication technologies mostly have been created and developed within the past 200 years, or, in the broader sense of time, 0.1% of human history. From that perspective, the World Wide Web and social media platforms, such as Twitter and Facebook, are merely the latest of the slew of newcomers, including radio, television, and film. How did people generally communicate before the 1850s, when newspaper wire services – as a harbinger of things to come – started competing for audiences on an international scale? This chapter will focus upon such historical and theoretical foundations, and the progression of technologies that led us to the penny press and the emergence of a massive commercial marketplace for communication.

The evolution from gesturing and grunting to the expression and interpretation of complicated speech and symbol patterns demonstrates humankind's continual integration of technologies into the communication process, as a way to enhance our natural senses and abilities. While we might think of talking and writing as "natural", they both actually are complex learned behaviors, enhanced by the technological innovations and refinements of language, and, at an even finer level, a tool called an alphabet, which, in the English culture allows users to create any word at all, about any thought at all, from just 26 distinct symbols, used in various combinations (Sapir 2004; Stephens 2007). Language and all other communication technologies have been built specifically to expand or increase human faculties, either psychic or physical, Canadian media theorist Marshall McLuhan (1967) contended. He wrote that the book, for example, could be viewed as an extension of the eye and electric circuitry as an extension of the nervous system. The evolution from an entirely oral – and generally interpersonal – culture to a culture blending orality and mass media messages in many digital and analog forms has increased the complexity of theoretical understandings in the field, in part through recognition of the ever-expanding role of communication media.

1 Medium Theory

A broad paradigm to provide grounding in this mediated world is "medium theory", in which each medium is defined as a specific type of setting or environment that has relatively fixed characteristics, which influence communication in a particular manner, regardless of the choice of content elements and the particular manipulation of production variables. These unique features separate the medium from other media forms as well as from face-to-face interaction (Meyrowitz 1998). Think, for example, about how a written document differs from a face-to-face conversation. A written document has many attractive affordances. It can capture thoughts precisely in a systematic way and place those in a durable package that can be delivered over long distances, without the message changing. It requires literacy, though, to unlock those thoughts, and a written document remains silent, no matter how much the recipient might want to question it. This document also leaves behind on its journey a wealth of rich sensory information that typically gets conveyed when people talk to each other, including the body language of the participants, plus the tone, pitch, and intensity of their voices. When people began writing about ideas, instead of only talking about them, some hailed that emerging form of human expression while others raised questions about what also was being lost in the newfangled process.

Socrates, for example, was recorded as a critic of writing in Plato's dialogue titled "Phaedrus". The philosopher, called the "wisest man" in ancient Athens by the Oracle of Delphi, did not write down his thoughts at all and preferred to conduct with other people a dialectic process of inquiry, asking question after question until he felt the issue was fully understood, and any flaws of logic were unveiled, through the experience of speaking and hearing the series of answers. Plato, who was a student of Socrates, and who kept notes of what Socrates tended to say, ironically preserved his teacher's concerns on this subject in writing, which is how we know about them today. In this account, Socrates relayed the story of Thamus, an Egyptian king, who had an ingenious inventor in his realm named Theuth. Among his many marvelous ideas, Theuth brought to Thamus the concept of "letters". Theuth explained how valuable these could be for Egyptians, in terms of making them wiser, and enhancing their memories. Instead of celebrating the idea of writing, Thamus criticized it, stating that such a medium would create forgetfulness and a trust in the written word that supersedes trust in a person's own memories. Writing therefore would appear as a memory but function as reminiscence, and a semblance of truth, rather than truth itself. People who read, rather than gather knowledge through Socratic debate, Thamus argued, will be "hearers of many things" but will have learned, and generally know, nothing. Socrates furthermore compares writing to painting, adding that the creations of the painter have the attitude of life, and yet if you ask them a question, they preserve a solemn silence (Plato 1972: 67–70). His criticisms symbolized the schism created as an oral society transformed into a written one, and, gradually, as that written one expanded into a complex blend of innumerable multimedia channels pervading civilization.

History offers many more examples about how innovations in communication technologies have altered the course of human development. The speed at which communication traveled, for example, was limited to a fast runner's pace until the Chinese domesticated horses around 3,500 BC, allowing messengers to travel by horseback (Stephens 2007). Not until thousands of years later, in the 1860s, did the telegraph finally break the tether that communication always had to human transportation. The telegraph not only allowed messages to travel freely within wires at great speeds over vast distances, without a person along as a guide, it also created a new conceptualization of communication, as something that could be transmitted outside of traditional ideas of space and time. That distinction of media messages thought of as transmitted, rather than transported, and operating independently of both humanity and geography, has been exploited by virtually every communication technology since (Carey 1983). It also emphasizes the dilemma of technological determinism. Do we have the ultimate control over our destiny with technologies, mindfully deciding what we use or not, or do technological developments, ambitiously built and unleashed, instead ultimately drive and shape our direction as a species? (Smith and Marx 1994).

2 Technological Determinism

Therefore, technology – broadly conceived – could be considered the single most important factor in producing, integrating, and destroying cultural phenomena.

Bain (1937) added that the term encompasses all tools, machines, utensils, weapons, instruments, housing, clothing, communicating and transporting devices, and the skills by which we produce and use them. In turn, social institutions, and their so-called non-material concomitants, such as values, morals, manners, wishes, hopes, fears, and attitudes, are directly – and indirectly – dependent upon technology and also mediated by it. As an example, before the development of Johannes Gutenberg's printing press in Europe, in the 15th century, few people had direct access to the Bible, primarily just the Catholic clergymen who interpreted its meanings to their flocks. With mass production of the text, though, more people had access, and more questions were asked about the traditional interpretations, and behaviors of pastors, fueling the Protestant Reformation. In that case, and many others, particularly in the past century, the efficacy of technology as a driving force in history is apparent, as a specific innovation appears and causes tremendous change (Smith and Marx 1994).

Yet that determinist perspective also fails to account for the numerous technological flops and failures, some ahead of their time, some just ideas that don't connect with people. If technological determinism could be counted on every time, as a working grand theory, then superior technologies always would be successful the moment they left the inventor's workshop, because they were destined to change us. In practice, though, some great ideas take a long time to gestate in a society; some ideas are adopted by some cultures but not others; and some technological ideas, as great as they might appear to be, never make an impact on humanity. Leonardo da Vinci's trove of inventions and futuristic ideas in the late 1400s and early 1500s, such as flying machines, is a classic example of the envisioning of advanced technologies before a society was ready to adopt them, suggesting that social dynamics and other pragmatic issues, such as the development of production and distribution systems, the availability of raw materials, and the growth of a network of users, complicate matters of technological integration and impact. Lynn White described the situation this way: "A new device merely opens a door; it does not compel one to enter" (White 1964: 28). In relation, the interaction between a technology and the social ecology is unpredictable, even to the creators of it. Frequently, technologies have environmental, social, and human consequences that go far beyond the immediate purposes and practices initially imagined, and many of our technology-related problems arise because of the unforeseen consequences of seemingly benign technologies employed on a massive scale (Kranzberg 1986).

The Apple Newton, as an example, was released in the late 1980s and did not turn out to be a broadly successful mobile device. It might have been a harbinger of the mobile computing age. It might have been just another also-ran in the competition to create a viable mobile device for the masses. It might have been a technological amalgamation that simply was a major technology or two short of being widely useful. It might have suffered from the lack of a larger support network. Many possible reasons exist for its relatively tiny impact on society, despite the prodigious use of such small computers today clearly showing that a latent human need existed. Many years later, Apple also released the iPhone, to much initial acclaim, like with the earlier device, but something significant had changed since the days of the Newton. Maybe that difference was the better integration of improved technologies. Maybe it was the App Store, allowing third-party development. Maybe society broadly had become more accustomed to the mobile idea, after the breakthroughs of the Palm Pilot, the Blackberry, and the iPod. After nearly 20 years of trying, Apple ended up in the right place at the right time with the right technology for phenomenal commercial success. What customers originally envisioned as a handy convergence tool – combining the cellular phone, the personal digital assistant, and the music player – has been that, and more, including causing concerns about new societal issues, such as driving while texting, bullying via social media, and general complicity with the creation of a national surveillance state, problems that simply did not exist on a large scale before commercially viable mobile technologies.

By looking at each technology from a historical and a medium-focused perspective, important parallels begin to form around their emergences. Bain (1937), for example, noted that all human activities are conditioned by biological nature, the physical environment, and technological and other cultural limitations, meaning that there is constant reciprocal interaction between technology and the other aspects that affect culture, and meaning the implications of technological determinism often can be overstated. An example of such a determinist claim was made by Evans (1979: 13), when he argued of the transformation of world society, "at all levels", by the computer, in which "the computer revolution will have an overwhelming and comprehensive impact, affecting every being on earth, in every aspect of his or her life". Has the computer fulfilled that grand prophecy? Are the effects of a computer existing simply from the purposeful navigation of layers of technology (the alphabet, language, the operating system) by humans, in control of the situation, or, in a deterministic sense, did the invention of computers cause society to seismically and subconsciously shift in response?

To begin to untangle this issue in a technological context, consider that the word medium is derived from the Latin term "medius", which roughly translates to middle, as in something in the middle, like what lies between the sender and receiver of a message. Theorists initially assumed that the content of a message was much more important than the form in which it was delivered, a perspective known as "media transparency". Yet "media richness" theorists, inspired by the new media generated through computers, typically define media through its material hardware and software features, or its functional affordances and constraints, instead of focusing upon social functions and locations (Danowski, 1993).

3 Living a mediated life

Understanding how media work to create our environments, and how each medium works with others within those situations, is critical to comprehending and navigating contemporary society, because most of our experiences today are mediated. At the origins of humanity, most of our knowledge came directly from connections to natural sensations. But each new layer of mediation that we have created over the millennia – from language to writing to the printing press to the World Wide Web – has added swirling and overlapping social and cultural layers of information, contributing to our knowledge, too. Think for a moment about all of the information that you process in a day, from the time you wake up, until the time you go to sleep at night. How much of that are you directly accessing from observations of natural sources, through independent investigations? How much of that is being provided to you by someone else through a form of mediation? This mediated environment includes not only the traditional mass media, such as newspapers, radio, television, etc., but also the graphics on your cereal box, the road signs you pass by, the company memos, the water cooler conversations, the staff meetings, the executive summaries, the spreadsheets, the voice mail, the social media feeds, and so on. Bandura (2001: 2–3) stated that human self-development, adaptation, and change are embedded in such social systems, and people are "self-organizing, proactive, self-reflecting, and self-regulating, not just reactive organisms shaped and shepherded by environmental events or inner forces". Through interactions with our symbols, such as written language, people can process and transform transient experiences into cognitive models that serve as guides for judgment and action, giving meaning, form and continuity to our experiences. In short, we learn our cultures and our histories from our social interactions, and, in the heavily mediated environment of contemporary society, those interactions often happen through media, especially through mass media. The ability to learn socially, through symbols, is part of what makes humanity special, as Bandura (2001: 5) notes, "If knowledge and skills could be acquired only by response consequences, human development would be greatly retarded, not to mention exceedingly tedious and hazardous". By learning socially, through symbols, we quickly can learn how others have fared in similar or related situations and then use that information to shape our responses in the moment, or even adapt larger life philosophies.

How else does being immersed in all of this mediation affect us? That question leads to some of the other most fundamental communication concerns. To foreshadow for a moment, when this field was forming, in the early to mid-1900s, scholars initially were magnetized to the pull of a Mass Society Theory, which could conceptualize the idea of a single mediated message acting as a universal and unifying force in the world, affecting pretty much everyone in the same way. The metaphors for the Mass Society message included a "magic bullet" or "hypodermic needle", implying that the media message would enter our collective systems and course through the masses to a uniform effect (Webster and Phalen 2013). Scholars at Princeton and Columbia, though, among them Paul Lazarsfeld (1941), quickly began forming contrary theories, known as "limited effects" theories, that generally note how different people respond in different ways to any single media message and that to understand the masses, we need to first understand the individual as a segment of the masses. The mixture in just one person of many simultaneous messages and competing motivations to respond to them, as well as a lifelong history of managing such messages, complicates communication research on an individual level and on a societal level. To try to comprehend what really happens when people communicate, scholars apply various methodologies through theoretical frameworks based in the social sciences and humanities. They typically try to peel apart the layers of the situation, through reductionism, in order to find a focus for the study and to set the parameters of the specific inquiry. One of those approaches is to examine communication not by a close analysis of the content being transmitted but by isolating the medium, as part of the puzzle, and thinking about what can be attributed as its effects.

4 Medium specificity

Under the general label of "medium specificity" theories, these lines of thought further extend discussions about the importance of the medium, focusing on ideas that different media have "essential" and unique characteristics that form the core of how they can – and should be – used. A medium-specificity approach can be used to examine the unstable interface among ideology, technology, and desires, and it also can be used to acknowledge that there are identifiable differences between one medium and another while establishing the broader assemblage aspects of each medium, per Gilles Deleuze and Felix Guattari, as a cultural artifact, with a complex and intertwined lineage (Maras and Sutton 2000). When a new medium emerges, historical and cultural influences, in hybrid understandings, come along as well, as older media gets blended into new forms, in a process Bolter and Grusin (1998) describe as remediation. Richardson (2011) connects the concepts of remediation and convergence (Jenkins 2006) through Medium Specificity by noting that each interface, even when experiencing different kinds of services and content within a single apparatus, such as a mobile device, can be interpreted in terms of specific and differential effects, each demanding a particular mode of embodied interaction. In this process, what emerges is not a single all-purpose device but a seemingly endless iteration of handsets with varying capabilities and design features, each prioritizing a specific technosomatic arrangement. Studying communication technologies from such a medium-focused perspective therefore might make the research design challenges greater than those focused on particular media messages, but, potentially, the rewards are higher, too, in terms of transferability to other contexts and in terms of deeper understandings about evolving changes in human social interactions (Meyrowitz 1997).

Meyrowitz (2009) contended that a focus should be upon the new means of communication, which afforded new possibilities, and that those new forms, which could be creatively exploited for both old and new purposes, could help people to actively develop new content and new interactions that match the personalities and constraints of the new media. One of the primary considerations in Medium Theory therefore is the type of sensory information that the medium is able to transmit. While content is important, studying content alone does not generate sufficient understanding of the underlying changes in social structures encouraged or enabled by new forms of communication. From this viewpoint, media is a setting, or an environment, for social interaction, and it is examined, on the broadest level, by the ways in which particular characteristics of a medium make it physically, psychologically, and socially different from other media and from face-to-face interaction (Meyrowitz 1997).

5 Early "mediavolutions"

To ancient people, the concept of leaving artistic marks on cave walls to permanently show imagery from their lives probably was a radical proposition. The earliest known forms of this communicative art, from about 30,000 years ago, primarily involved people using their fingers to rub the three predominately available colors – charcoal (black), an iron-oxide mixture (red), and goethite (yellow) – on limestone walls (Chalmin, et al., 2003; Leroi-Gourhan, 1982). As a medium, in retrospect, cave painting seems relatively simplistic. But to those Paleolithic people, in that time period, the medium was the modern-day equivalent of a wearable computing headset, providing them with a dramatically new way to see and interpret their world, and to communicate with each other in an unprecedented manner. Someone, somewhere, must have ended up with charcoaled hands, maybe from tending a fire, and happened to touch a cave wall and left a mark. Maybe it was the same person, or an observant and curious companion, who started to play around with the making of marks, inventing different expressive techniques, such as using a charcoaled stick or a handprint instead of a finger. How did that medium open up new opportunities for expression and communication? Gibson (1979) created the term affordances to describe what an environment provided or furnished its inhabitants, for good or ill, relative to each individual. Terrestrial surfaces, for example, can be "climb-on-able, or fall-off-able, or get-underneath-able, or bump-able, relative to the animal", and a platform that is knee-high to an adult might be neck-high to a child, dramatically altering the dynamic for use" (Gibson 1979: 127–128). Cave painting allowed people to make permanent marks, in an otherwise oral culture, separating ideas from the minds that created them, and leaving those ideas in a public space, for others to encounter, even when the originator no longer was around. These paintings could have provided a visual image, to accompany a story, or had important symbolic and ritualistic connotations. They fundamentally allowed people to communicate in new ways.

A medium, such as cave painting, also has limits, or weaknesses. Norman (1988) called those "constraints", which can be both natural and cultural. Natural constraints are those physical properties that dictate certain behaviors, such as "physical features – projections, depressions, screwthreads, appendages – that limit relationships to other objects, operations that can be performed to it, what can be attached to it, and so on" (Norman 1988: 55). Cultural constraints are the learned and artificial conventions that govern acceptable behavior in the situation, such as turning screws clockwise to tighten. Both of these types of constraints, in turn, reduce the number of potential alternative actions for each situation. In the cave painting example, the available choices of colors could be considered a constraint. Another one would be the physical limits of the applicator tool, such as the widths of a line that a finger can make, and the lack of portability of the image, the difficulty of accessing the image in the cave, and any cultural restrictions that might have emerged around such access. Examining a medium in such a way reveals its essence, which otherwise can be difficult to describe.

From this medium perspective, each new communication platform emerges with distinct affordances and constraints. The medium also has a lineage, of sorts, with earlier technologies, as some of the characteristics of established forms are carried forward and integrated, at least temporarily, while others are left encased in the earlier iterations. The newspaper, for example, could not exist without human conversations, as a way of sharing information among people, layered underneath the theoretical framework of "news". Humans had to develop language before they could have conversations, and books and magazines could be considered the older cousins of newspapers, providing the medium with a path to follow until it developed its niche. Newspapers were not the first form to feature written accounts of local happenings, but they evolved, by focusing on their unique technological affordances, to claim a role in mass communication by being an informative and immediate medium while also being relatively cheap to produce and easy to distribute. To put the newspaper in a historical perspective, as a communication technology, consider the importance of these major chronological waypoints identified by Stephens (2007):

- 8,000 BC: Agriculture becomes widespread, and societies start to stabilize in particular places. How does news circulate when civilizations form and grow larger than a small group of family members and friends? Messengers, town criers, smoke signals, and drums, all emerge as forms of communication media to aid the flow and reach of information.
- 3,100 BC: the earliest known writing systems flourish, in Mesopotamia and Egypt; at this point, the symbols represent entire words, not sounds.

- 1,500 BC: the Canaanites develop the first alphabet.
- 443 BC: the first nonfictional account of Western history is written, by Herodotus, focusing upon the wars between Greece and Persia; could be considered an ancestor of news.
- 145 BC: Romans gather daily in the open-air public Forum to hear the latest news about the Republic.
- 59 BC: Julius Caesar creates what could be considered the first newspaper, the Acta Diurna, a written posting that describes the daily activities of the Senate.
- 1041 AD: Bi Sheng of China develops a system for printing with moveable type; because of the large number of Chinese characters, though, the system is impractical and unsuccessful.
- 1450 AD: with a much smaller set of characters to manage, through the use of a phonetic alphabet, Johann Gutenberg's letter press allows for mass production of texts; books at first, but, then, centuries later, newspapers.
- 1566 AD: handwritten newssheets circulate in Venice, as a direct ancestor to the modern newspaper.
- 1609 AD: a German creates a weekly printed newspaper, the oldest known publication of its type in Europe.
- 1620 AD: the first English printed newspaper appears, in Amsterdam.
- 1645 AD: a government-sponsored account of the wars with Native Americans is printed in Britain's American colonies.
- 1690 AD: America's first newspaper, "Publick Occurrences Both Forreign and Domestick", is published in Boston; it is shut down by public officials after the first issue.
- 1783 AD: America's first daily newspaper, The Pennsylvania Evening Post, starts publication; its owner, Benjamin Towne, is indicted for treason a few months later.
- 1789 AD: the United States Bill of Rights is approved, ensuring freedom of speech and the press.
- 1798 AD: the Sedition Act makes it a crime to "write, print, utter or publish" criticisms of the U.S. government. The acts are allowed to lapse two years later, when Thomas Jefferson becomes president.
- 1825 AD: the U.S. becomes the world leader in newspaper circulation.
- 1833 AD: the New York Sun starts selling for a penny a copy, attracting a large, working-class audience. Two more successful penny papers, The Philadelphia Public Ledger and the Baltimore Sun, are launched soon afterward.
- 1844 AD: Samuel Morse demonstrates the power of his new telegraph machine.
- 1851 AD: Yet one more penny press newspaper joins the crowded field; this one called The New York Times.

Those sketches of a historical timeline show how much had to happen, and how much time had to pass, in the course of humanity for newspapers, as we know them, to emerge and thrive. As other chapters in this book will document, newspapers now are fading in mainstream societal importance, and digital innovations in communication technology are arising at a much different pace. Yet our understanding of news, and information sharing, has been shaped in significant ways by the development of newspapers, and magazines, and radio and television broadcasting, so when digital news circulates online, traces of those traditions can be found throughout. In addition, many major issues related to communication technology can transcend the individual medium, meaning that comprehension of media issues in general can be improved, too, by applying global and historical perspectives.

A fundamental question every communicator should consider, in every medium, for example, is: Who is the audience for this message? The answer not only matters on an individual level, for completing a desired communication circuit, but it also has larger implications as well. With the newspaper medium, for example, the first publishers, or printers as they were called, in general, were politically inspired activists, not aspiring media tycoons. Just as people today might not know exactly what to do yet with social media as a news medium, the newspaper as a medium also was shrouded at first in mystery. To try to summarize the burgeoning communication industry of his era, Park (1923) described the process of the newspaper becoming a distinct communication medium as one, not wholly rational, guided by many individuals participating without foreseeing what the ultimate product of their labors would be. He added, "No one sought to make it just what it is. ... The type of newspaper that exists is the type that has survived under the conditions of modern life. The men who may be said to have made the modern newspaper – James Gordon Bennett, Charles A. Dana, Joseph Pulitzer, and William Randolph Hearst – are the men who discovered the kind of paper that men and women would read and had the courage to publish it" (Park 1923: 273–274). The struggle to exist as a communication medium, Park said, was the struggle for circulation, or, in essence, mass popularity.

Audiences become attracted to a medium for various reasons, and the rise of the American newspaper was an example of the amalgamation of many diverse technologies – such as the mechanized printing press, Morse Code, the telegraph, the Bill of Rights' expression of fundamental societal freedoms, and advertising-supported business models – generating and morphing into a dynamic new form of communication. When a new medium appears like that, regardless of what that medium is (with the term medium defined very broadly), McLuhan and McLuhan (1988) suggest we ask four critical questions about it. This tetrad of universal effects, functioning within the "laws" of media, can be determined by asking of the medium artifact: 1. What does it enhance? 2. What does it make obsolete? 3. What does it retrieve that had been obsolesced earlier? And, 4. What does it flip into when pushed to extremes?

From this perspective, a new medium amplifies or intensifies a particular aspect (or aspects) of communication, while simultaneously obsolescing others. In the newspaper example, in the mid-1800s, publishing in that form suddenly became a way to communicate written words immediately to the masses, amplifying both audience reach and timeliness. Public speaking, letter writing, one-on-one conversations, and the use of other communication media of the era, shrank in societal importance when compared to the possibilities of the upstart newspaper industry. In conjunction, newspapers also retrieved and returned to prominence civic discourse about broad public matters and the contestation of authority (Conboy and Steel 2008). At its reversal point, when pushed to its most extreme in the overheated digital era, the newspaper has flipped into a niche publication that operates at a relatively slow speed, and has become generally more attractive to the contemplative and highly educated elite than the masses.

The largest audiences (and the advertisers), in short, have shifted to the faster and more personal digital channels, with this trend indicated by the stock value of, say, the biggest social media companies versus The New York Times. Newspapers no longer can compete in speed and mass appeal. U.S. newspapers had risen to prominence through the commercialization of communication media more than 150 years ago, by gathering large audiences through the quick delivery of information. This business model served them well. It also formed the metanarrative for the future of the mass media, when other media forms surpassed their technological advantages.

Before the 1830s, American newspapers generally served political parties and businessmen, both trying to expand their realms of power. In the 1830s and 1840s, ambitious U.S. newspaper publishers, trying to find a commercial market, decided to lower the consumer cost of their daily publications to one penny, thereby making the product financially accessible to most of the country and opening the world of newspapers to the masses. In conjunction, distribution strategies shifted to emphasize populist, and arguably sensational content, creating unprecedented circulations for some adept publications, such as The New York Sun, which, in turn, attracted substantial advertising dollars (Emery and Emery 1984; Nerone 1987). As the newspaper craft shifted into a mass-media industry during this era, the growing audiences of the successful penny press publications attracted more and more advertisers, and the advertising dollars provided the resources for publications to attract even larger audiences, creating a business advantage. The evolution of journalism in the mid-1800s, with its emphasis on audience growth, led to – among other innovations – the pursuit of objectivity and the employment of the first professional reporters. A publication that included multiple sides of a story, from a business perspective, could sell it to all of those sides (instead of an audience limited in size by one particular viewpoint). News, in short, became a marketable commodity, and many publishers rushed to find ways to sell more of it to more people (Emery and Emery 1984; Nerone 1987).

Such an emphasis on audience size, in an age of new technologies, such as the railroad and the telegraph, also led to the expansion of circulation zones, or the primary area in which a publication is circulated. As newsgathering power increased, publishers discovered that the larger the circulation zone, the larger the potential audience. This understanding led to another significant source of newspaper expansion in scope and reach, the first wire services, called news agencies. Just like today, being first with the news was highly valuable, and news agencies at first used fast horses, and fast ships, and even carrier pigeons, before latching on to the high-speed transmission capabilities of the telegraph (Zelizer and Allan 2010; Silberstein-Loeb 2014). Exchanging news could be lucrative, and with the telegraph, publishers suddenly could gather and distribute dispatches from long distances away. News agencies, first in France, then in England and Germany, and eventually, as a coalition of New York newspapers (the Associated Press), served as the middle men, collecting interesting stories and selling them to their publisher subscribers, who then distributed the information to their mass audiences (Silberstein-Loeb 2014). All of which increased audience sizes, for those who were first; and increased profits, and increased the scope and scale of the successful news organizations. At this point in mass media history, newspapers had no significant challengers in the news-exchange business. As scale and profitability increased, though, and new communication technologies emerged, people began exploring what else had communication affordances and could amplify our information networks, extending the reach of the telegraph, into interpersonal technologies, such as the telephone, and broadcast media, such as the radio.

References

Aitchison, Jean. 1996. *The seeds of speech: Language origin and evolution*. Cambridge, UK: Cambridge University Press.

Bain, Read. 1937. Technology and state government. *American Sociological Review* 2(6). 860–874.

Bandura, Albert. 2001. Social cognitive theory of mass communication. In Jennings Bryant & Dolf Zillman (eds.), *Media effects: Advances in theory and research*, 2nd edn, vol. 2, 121–153. Hillsdale, NJ: Lawrence Erlbaum.

Bolter, Jay D. & Richard Grusin. 1998. *Remediation: Understanding new media*: MIT Press Cambridge, MA.

Carey, James W. 1983. Technology and ideology: The case of the telegraph. *Prospects*, 8(1). 303–325.

Chalmin, Emilie, Michel Menu & Colette Vignaud. 2003. Analysis of rock art painting and technology of Palaeolithic painters. *Measurement Science and Technology* 14(9). 1590.

Changizi, Mark A. 2011. *Harnessed: how language and music mimicked nature and transformed ape to man*. Dallas, TX: BenBella Books.

Conboy, Martin & John Steel. 2008. The future of newspapers: historical perspectives. *Journalism Studies* 9(5). 650–661.

Danowski, James. 1993. An emerging macrolevel theory of organizational communication: Organizations as virtual reality management systems. In Lee Thayer & George Barnett (eds.), *Emerging perspectives in organizational communication*, 141–174. Norwood NJ: Ablex.

Emery, Edwin & Michael C. Emery. 1984. The press and America: an interpretive history of the mass media. Englewood Cliffs, N.J.: Prentice-Hall.

Evans, Christopher R. 1979. *The mighty micro: The impact of the computer revolution*. London: Gollancz.

Gibson, James J. 1979. *The ecological approach to visual perception*. Hillsdale, NJ: Lawrence Erlbaum Associates, Inc.

Jenkins, Henry. 2006. *Convergence culture: Where old and new media collide*. New York: New York University Press.

Kranzberg, Melvin. 1986. Technology and History: Kranzberg's Laws. *Technology and Culture* 27(3). 544–560.

Lazarsfeld, Paul F. 1941. Remarks on administrative and critical communications research. *Studies in Philosophy and Social Science* 9(1). 2–16.

Leroi-Gourhan, A. 1982. *The dawn of European art: An introduction to Palaeolithic cave painting*. Milan, Italy: Cambridge University Press.

Maras, Steven & David Sutton. 2000. Medium specificity re-visited. *Convergence: The International Journal of Research into New Media Technologies* 6(2). 98–113.

McLuhan, Marshall & Quentin Fiore. 1967. *The medium is the massage*. New York: NY, Bantam Books.

McLuhan, Marshall & Eric McLuhan. 1988. *Laws of media: The new science*. Toronto, ON: University of Toronto Press.

Meyrowitz, Joshua. 1997. Shifting worlds of strangers: Medium theory and changes in 'them' versus 'us.' *Sociological Inquiry* 67(1). 59–71.

Meyrowitz, Joshua. 1998. Multiple media literacies. *Journal of Communication* 48(1). 96–108.

Meyrowitz, Joshua. 2009. Medium Theory: An Alternative to the Dominant Paradigm of Media Effects. In Robin L. Nabi & Mary Beth Oliver (eds.), *The SAGE Handbook of Media Processes and Effects*. Thousand Oaks, CA, 517–530.

Nerone, John. C. 1987. The mythology of the penny press. *Critical Studies in Mass Communication* 4. 376–404.

Norman, Don A. 1988. *The psychology of everyday things*. New York: NY: Basic books.

Park, Robert. E. 1923. The natural history of the newspaper. *American Journal of Sociology* 273–289.

Richardson, Ingrid. 2011. The hybrid ontology of mobile gaming. Convergence: *The International Journal of Research into New Media Technologies* 17(4). 419–430.

Sapir, Edward. 2004. *Language: An introduction to the study of speech*. Mineover, NY: Dover Publications.

Silberstein-Loeb, Jonathan. 2014. *The International Distribution of News: The Associated Press, Press Association, and Reuters, 1848–1947*. New York, NY: Cambridge University Press.

Smith, Merrit Roe & Leo Marx. 1994. *Does technology drive history?: The dilemma of technological determinism*. Cambridge, MA: MIT Press.

Stephens, Mitchell. 2007. *A history of news*. New York, NY: Oxford University Press.

Tomasello, Michael. 2008. *Origins of Human Communication*. Cambridge, MA: MIT Press.

Webster, James & Patricia F. Phalen. 2013. *The mass audience: Rediscovering the dominant model*. New York: NY: Routledge.

White, Lynn. 1964. *Medieval technology and social change*. New York, NY: Oxford University Press.

Zelizer, Barbie & Stuart Allan. 2010. *Keywords in news and journalism studies*. New York, NY: McGraw-Hill International.

Gabriele Balbi and Richard R. John*

2 Point-to-point: telecommunications networks from the optical telegraph to the mobile telephone

Abstract: This chapter surveys the history of telecommunications from a global perspective and highlights three influential interpretative traditions. It has two parts. The first part defines "telecommunications" and sketches the main dimensions of four telecommunications networks over a two-hundred-year period – the optical telegraph, the electric telegraph, the landline telephone, and the mobile telephone (and its predecessor, the wireless telegraph). The second part shows how historical scholarship on topics in the history of telecommunications has been shaped by three intellectual traditions: the Large Technical Systems (LTS) approach; political economy; and the Social Construction of Technology (SCOT).

Keywords: history of communications, telecommunications, optical telegraph, electric telegraph, wireless telegraph, radio, telephone, mobile telephone, Large Technical Systems (LTS), political economy

Every generation writes its own history. The history of telecommunications is no exception. Our goal in this essay is to survey a familiar topic from an innovative perspective. Our perspective is innovative in a dual sense: the temporal span is unusually long (1790s–present) and the spatial boundaries extend beyond Europe and North America. Telecommunications is an interpretative construct that historians use to group certain communications networks under a common rubric. The rationale for this essay is the subtle but fundamental shift in the character and significance of these networks that has been hastened by the recent emergence of the mobile telephone. In the past few decades, the mobile telephone has become ubiquitous not only in Europe and North America, the seedbed for all prior innovations in telecommunications, but also in many parts of Africa, Asia, and South America, regions in which, prior to the mobile telephone, access to telecommunications had been limited, if not altogether absent.

The word "telecommunications" is a French neologism that was invented long after three of the communications networks with which it has today come to be linked. The word itself, which means, literally "communications at a distance", was the brainchild of a French postal administrator, Édouard Estaunié, who coined

* In the preparation of this essay, we are grateful for the assistance of Colin Agur, Caroline Chen, Dwayne Winseck, and Nancy R. John.

it in 1904 as a convenient catchphrase to lump together the landline telephone and the electric telegraph (Huurdeman 2003; John 2010). Estaunié's construct would not gain wide acceptance in France until the 1920s, and would remain largely unknown in the rest of Europe until 1932, when it was included in the official title of a newly reorganized regulatory agency, the International Telecommunication Union (or ITU). In the United States, the word was infrequently used (for a rare exception, see Herring & Gross 1936) until after the Second World War (John 2010: 12–13). Today, of course, historians of communications use this construct as an umbrella term for a variety of communications networks that often include, in addition to the electric telegraph and the landline telephone, the optical telegraph, radio, television, the satellite, the mobile telephone, and the Internet (Noam 1992; Huurdeman 2003).

It may well be a fool's errand to lend coherence to such an amorphous construct. In this essay, we will try. It is our contention that the word telecommunications is best reserved for a limited, yet extremely important, constellation of long-distance communications networks that transmit messages from point to point. We use the term "network" – rather than possible alternatives, such as "media" or "system" – deliberately. The network form, as analyzed by scholars such as the Spanish-born historical sociologist Manuel Castells (Castells 1996–1998), the French political scientist Pierre Musso (Musso 1997), and the Hungarian-born physicist Lászlô Barabâsi (Barabâsi 2002), most accurately characterizes the defining features of this construct.

Telecommunications networks have three main features (Balbi 2013a). The first of these features concerns *users*. Telecommunications is a point-to-point (one-to-one) network that makes it possible to establish a unique link between a relatively small number of nodes (as few as two). A key dimension of this feature is privacy: the exchange of information within a small group (or between two individuals) is predicated on the assumption, which in our age of digitally mediated communication is often mistaken, that this information is *not* widely shared. Indeed, it is often assumed that the communication is secret. This feature excludes radio and television, which are broadcast (many-to-many) networks rather than point-to-point (one-to-one) networks. Radio and television are designed to reach mass audiences, rather than niche audiences. Publicity, and not secrecy, is the goal. (For an alternative, more capacious definition, see McChesney 1993). The second feature concerns *transmission*. Telecommunications networks do not transmit a physical message, but, rather, a coded signal that represents the message. This signal is encoded at a network node, transmitted through the network, and decoded at a network node. This feature excludes the mail, which transmits physical messages, rather than signals (John 1995), but includes the electric telegraph, even though it often combined the physical transportation of messages from a sender to a transmitting office with its electrical transmission from office to office (Downey 2002). The third feature concerns *directionality*. Telecommunications networks enable the recipient to respond to a message in a timely fashion, and in the case of the landline and mobile telephone, instantaneously. They are, in a word, interactive. This feature excludes individualized broadcast media, such as email blasts, live-streaming, and on-line video.

While this definition is provisional, we believe it can help us to shift the long-standing focus of media scholarship from radio and TV *broadcasting*, media that many twentieth-century historians of communications assumed to represent the wave of the future, toward *telecommunications*, a medium that, at the moment, is in the ascendency not only in Europe and North America, but also in much of the rest of the world.

The essay has two parts. First, we sketch from our innovative perspective certain features of the history of four telecommunications networks – the optical telegraph, the electric telegraph, the landline telephone, and the mobile telephone (and its predecessor, the wireless telegraph). Second, we show how our perspective has been informed by three academic traditions that have proved influential in promoting historical understanding of these networks in the past, and that we believe will continue to prove useful in charting their evolution in the future.

1 Historical overview

While mid-twentieth-century communications historians typically trivialized the optical telegraph (Carey 1989), a consensus has emerged since the 1990s that it deserves pride of place as the first telecommunications network (John 2013). The idea of an optical telegraph originated in the eighteenth century. Like "telecommunications", the word was a French neologism (from the Greek: *têle* = distance and *graphe* = writing). The first optical telegraphs were built almost simultaneously in the 1790s in Sweden and France (Wilson 1976; Flichy 1991; Holzmann & Pehrson 1995; Headrick 2000: Ch. 6; Matterlart 1992, 2000; Rosenfeld 2001: 199–203). The Swedish optical telegraph was the brainchild of the Finnish-born poet Abraham N. Edelcrantz; in France, its primary champion was Claude Chappe, a cleric who had been deprived of his benefice during the French Revolution. Chappe demonstrated his invention in 1792; it went into operation two years later, the event that historians of communications typically regard as the beginnings of telecommunications (Matterlart 2000). The French optical telegraph remained in operation for some decades following the commercialization of the electric telegraph in Great Britain (1839) and the United States (1845). In 1852, it linked 556 towers in a Paris-based hub-and-spoke network that extended over 2,900 miles (John 2010: 14).

The optical telegraph utilized only one technical contrivance (the telescope) that would have been unknown to the ancients. The novelty of this medium lay in its combination in a single ensemble of three distinct elements: a network of signal towers, a technical apparatus to relay coded signals from tower to tower, and a code book that translated short phrases into a numerical form. In so doing, it provided government administrators with a fast and reliable tool for coordinating administrative and military operations. The French optical telegraph was built before Napoleon came to power, yet it was Napoleon who most fully demonstrated its utility, using it to coordinate his armies in the field. To this day, it remains, along with the guillotine, interchangeable parts, and the metric system, one of the most consequential inventions to have been spawned by the French Revolution.

Though the optical telegraph relied for its motive power on human labor rather than electricity, it fits our criteria for a telecommunications network. That is, it transmitted coded signals rather than physical messages from one point to another in either direction. The French optical telegraph was built by and for the French government and was closed to merchants. In other countries, different arrangements prevailed. In Great Britain and the United States, for example, optical telegraphs were built to facilitate the point-to-point circulation of information on market trends (Wilson 1976; John 2010: 16–17). Unlike the French optical telegraph, these networks were open to the public and patronized largely by merchants. Yet it was the French optical telegraph that remains the best known and that has dominated historical scholarship on this topic.

The electric telegraph is often hailed as a major technical breakthrough. This is not surprising. Along with electroplating, it was the first major practical application of electricity. Yet as a communications network, it was, as its name suggests, merely an incremental advance over the optical telegraph.

The similarities between the electric telegraph and the optical telegraph were self-evident to Samuel F. B. Morse, the American portrait-painter turned-electric-telegraph promoter who in 1840 obtained the first U.S. electric telegraph patent. In fact, Morse modeled two components of his original prototype on its optical precursor. Like the French optical telegraph, Morse's original prototype compressed messages by employing an elaborate numerical codebook; and like the French optical telegraph, Morse's original prototype was designed to minimize the likelihood of sabotage. By burying wires underground instead of stringing them overhead, Morse hoped to build a network as secure as the French government's fortified signal towers (John 2010: Ch. 2). When each of these components proved unworkable, Morse improvised. The numerical codebook was slow and complicated and the underground burial of wires was expensive and unreliable. As an alternative, Morse strung wires overhead and devised a letter-based binary signaling scheme – the eponymous Morse code – which is often regarded as a precursor to the binary language of the digital computer.

The electric telegraph had by 1840 already been commercialized in Great Britain, as a result of the creative collaboration of inventor William Fothergill Cooke and scientist Charles Wheatstone (Kieve 1973). Yet Morse's name would forever after be linked with the new medium. This was partly due to the widespread global adoption of Morse code and partly due to the publicity his anxious financial backers lavished on his invention. The electric telegraph in Great Britain found a ready market in the railroad sector, where it was rapidly adopted to coordinate the scheduling of trains. In the United States, in contrast, railroad managers were much slower to recognize its potential. Had they been quicker to adopt the new medium, Morse's electric telegraph might well have been less aggressively hyped (John 2010: Ch. 2).

The electric telegraph in Great Britain, the United States, and many other countries eventually became an important adjunct to business, journalism, and the military (Matterlart 1992; Blondheim 1994; Mattelart 2000; Hermans & De Wit 2004; Hochfelder 2012). The network was widely used to consolidate imperial authority, and, in conjunction with the underwater cable – a related, yet distinct invention – would become an indispensable instrument of command and control for the leading colonial powers: Great Britain, France, and Japan (Headrick 1988, 1992; Noam 1999; Winseck & Pike 2007; Yang 2011). In countries such as China and the Ottoman Empire, which were never formally colonized, it became a major agent of economic development (Baark 1997; Bektas 2000). By 1870, most of the world's electric telegraph networks had become government monopolies. The principal exceptions were the international cable network, which for commercial and diplomatic reasons remained privately owned and operated (Headrick 1992), and the domestic electric telegraph network in the United States, which for political reasons remained privately owned (John 2010; Wolff 2013).

Outside of the United States, the electric telegraph would remain the most influential telecommunications network from the 1840s until the Second World War. In the United States, however, a third telecommunications network – the landline telephone – would eclipse the telegraph by the First World War.

The landline telephone was commercialized in the 1870s, following the near-simultaneous invention by several different people of a technical contrivance that was capable of transforming the human voice into a coded signal. Though the landline telephone was often hailed as a long-distance medium, for many decades following its commercialization it was used primarily for short-distance communications that were typically confined to a specific locality (Armstrong & Nelles, 1986; Weiman 2003; Calvo 2006; John 2010). This was true not only in small towns, but also in giant cities. The average distance of a telephone call originating in Chicago, Illinois, in 1900, for example, was a mere 3.4 miles (John 2010: 283).

The regulation of the landline telephone varied from place to place. In North America, the principal regulatory domain was local. Operating companies held municipal franchises that typically specified rates and limited entry. These franchises were modeled on the charters granted to gas works and water plants and had little to do with the regulations that had been devised for the electric telegraph (Armstrong & Nelles 1986; Horwitz 1989; Gabel 1995; Weiman 2003; John 2010).

In Europe, in contrast, the legacy of the electric telegraph was more direct. Most electric telegraphs were government owned and operated, and this circumstance cast a long shadow on the regulation of the landline telephone. In Great Britain, Spain, France, Switzerland and Italy, the government initially granted concessions to private ventures, before deciding to nationalize all or part of the network, a task that for technical, administrative, and financial reasons proved enormously challenging (Hazlewood 1953; Kobelt 1980; Bertho-Lavenir 1981, 1988; Calvo 2002; Millward 2005; Wallsten 2005; Balbi 2011). In the Scandinavian countries and the Netherlands, in contrast, governments permitted ownership and operation of the local network, but retained control over the long-distance network (De Wit 1998; Helgesson 1999; Millward 2005; Wallsten 2005). Finally, in Germany, Greece, and Romania, the government owned and operated the network from the start (Thomas 1988; Schneider 1991; Millward 2005; Wallsten 2005). The legacy of the electric telegraph often slowed network expansion. In Italy and Great Britain, for example, the government limited investment in the landline telephone to protect its investment in its telegraph network (Perry 1992; Fari 2008; Balbi 2011).

The landline telephone also differed from the electric telegraph in its relationship to its users. In the United States, the landline telephone would be technically and administratively reconfigured beginning around 1900 as a mass service for the entire population (John 2010: Ch. 8). In Europe, in contrast, with the partial exception of Sweden (Helgesson 1999), the medium would remain largely confined to an exclusive clientele until after the Second World War (Huurdeman 2003). The Soviet government proved to be particularly reluctant to permit the landline telephone to gain a foothold, fearful that it might prove subversive (Solnick 1991). Eventually, however, the landline telephone would find its way into homes as well as businesses in much of the industrialized world, making it the first telecommunications medium to foster new habits of sociability (Marvin 1988; Young 1991; Martin 1991; Fischer 1992; Kline 2000) that shifted the boundary between public and private (Bertho-Lavenir 1981; Flichy 1991). For Anglophone scholars, the sociological implications of this shift – a shift that until recently has been assumed to strengthen neither the family nor the group, but the individual – were clarified by the 1995 translation into English of Flichy's 1991 *Une histoire de la communication moderne: Espace public et vie privée* (Flichy 1995).

Two additional features of the early landline telephone are worth underscoring. First, the landline telephone was the first telecommunications network to presuppose real-time two-way communications. To minimize the call-connection delay, enterprising telephone managers invested heavily in manual telephone switchboards (John 2010: Ch. 7–8). If it took an excessive period of time to complete a connection, an interval that in the United States rapidly decreased in the 1880s and 1890s from minutes to seconds, users could be expected to seek out less time-consuming alternatives, such as the employment of a messenger, or even the scheduling of a face-to-face meeting. The heavy investment that American telephone companies made in manual switchboards slowed the transition to electromechanical (or automatic) switchboards. In several European countries, the transition from manual to electromechanical switchboards took even longer. In part for this reason, European telephone users came to associate the new medium with long wait times. Users waited not only for an open circuit, but also to be connected with a particular respondent once the circuit had been opened, and to be reconnected to their respondent should the open circuit fail. Criticism sometimes focused not on the limitations of the manual switchboards, but, rather, on the alleged laziness of the female telephone operators who operated them (Balbi 2013b).

A second notable feature of the landline telephone was its communicative bias. The landline telephone was the first telecommunications network that its users experienced primarily as an aural rather than a visual medium. Unlike the electric telegraph, it left no written trace. In 1877, Thomas A. Edison invented the phonograph to solve this problem. Edison envisioned the phonograph transforming a telephone call into a permanent record that would be analogous to a telegram (Gitelman 1999). In practice, however, the phonograph became a broadcast medium that was almost never used in conjunction with the telephone. (Telephone answering machines would not become commonplace until well after the Second World War).

The mobile telephone is often presumed to be a logical successor to the landline telephone, since, like the landline telephone, it facilitates real-time two-way communications. Yet if the mobile telephone is considered historically, then it can plausibly be contended that its most direct predecessor is not the landline telephone, but the wireless telegraph.

The wireless telegraph was the brainchild of Guglielmo Marconi, an Italian-born inventor of mixed Scottish-Italian ancestry. Marconi's fundamental innovation was to substitute for a wire network the air – that is, the electromagnetic spectrum, or what Marconi called the "ether." In so doing, he greatly decreased the cost of real-time point-to-point communication, since the greatest expense in the establishment of a wire network was the cost of its construction (Curien 2005).

The "killer app" for the early wireless telegraph was maritime logistics. Ships at sea were constantly in motion, making it impossible to link them into the already existing landline telegraph network. It was largely for this reason that the wireless telegraph received such lavish support from the British Admiralty, the German Navy, and the American Navy: no other communications network could match its flexibility (Friedewald 2000; Anduaga 2009; Evans 2010; Winkler 2008). By the 1910s, powerful nationally based corporations – Marconi in Great Britain, Telefunken in Germany, Société Général in France, and the Radio Corporation of America (RCA) in the United States – looked beyond maritime logistics to international communications (Griset 1996; Hugill 1999; Friedewald 2000). The wireless networks that the colonial powers established in this period were often the first telecommunications media to reach the more thinly settled and less economically developed parts of the world (Friedewald 2000; Winseck & Pike 2007; Anduaga

2009; Yang 2011). In so doing, these wireless networks linked localities that had yet to be reached by the hard-wired underwater cable corporations that were domiciled mostly in Great Britain (Headrick 1992; Finn & Yang 2009). Wireless was cheaper (it was expensive to lay cables underwater) and harder to monopolize. By the Second World War, wireless telegraph (or radio) corporations had broken Britain's cable monopoly, while an upstart rival, the United States, had, by virtue of its large and powerful radio network, emerged as a major power in global telecommunications (Headrick 1994, 1995; Hugill 1999).

The wireless telegraph was the first new telecommunications network to emerge during the twentieth century. By mid-century it would be joined by radar and by the 1960s by satellites. Yet wireless point-to-point telecommunications would remain a specialty service for an exclusive clientele until the 1980s, with the rise of the mobile telephone.

The first successful mobile telephone network dates from 1977, when the American telecommunications provider American Telephone and Telegraph (AT&T) demonstrated in Chicago an experimental wireless telephone. Two years later, the Japanese telecommunications provider Nippon Telephone and Telegraph (NTT) established in Tokyo the first mobile network that was open to users. Within five years, the NTT network would expand to cover all of Japan, making it the first nationwide first-generation (1G) network (Steinbock 2003). Similar ventures emerged at roughly the same time in Europe. In fact, Europe soon overtook the United States in mobile telephone penetration, a fact that, despite the emergence of an imaginative cohort of mobile telephone entrepreneurs (Galambos & Abrahamson 2002), has remained true up to the present.

The expansion of mobile telephone networks in Europe was particularly notable, since it reversed a historical pattern. For much of the twentieth century, the United States had been the world's leading telecommunication provider. Now, in the century's waning years, Europe took the lead. One key development was the establishment in the early 1980s of the Nordic Mobile Telephone Group (NMT), a consortium that linked Denmark, Norway, Sweden, and Finland. NMT was the first mobile telephone network to make possible international roaming. Within a few years, NMT-designed protocols would become commonplace in more than forty countries in Europe and Asia, including Eastern Europe, Russia and Ukraine (Goggin 2006).

Equally consequential for the future was the establishment in Europe in December 1992 of a continental common standard: the Global System for Mobile Communication (GSM). This innovation, aptly dubbed a "bureaucratic miracle" (Agar 2013), was initially confined to 8 countries. Within four years, 103 countries, many of them outside Europe, had opted in. The GSM standard was far from universal. Other standards, for example, were introduced in the United States and Latin America. Yet GSM quickly became the world leader, facilitating a raft of technical improvements in signal quality and frequency management. Further innovations followed the rollout in the 1990s of the subscriber identity module (SIM) card, which enabled subscribers to authenticate their identity if they switched telephone devices, and, eventually, to retain a unique telephone number if they switched network providers.

The mobile telephone is by any measure one of the most significant innovations in media history. By 2013, 6.7 billion mobile-cellular telephone subscriptions had been entered into worldwide. The rapid adoption of the mobile telephone is particularly notable in those parts of the world in which access to telecommunications had previously been highly limited or even unknown. Of the 15 countries with the largest number of mobile telephone subscribers today, nine are in Asia (China, India, Indonesia, Pakistan, Japan, Bangladesh, Philippines, Iran and Asian Russia), two are in Latin America (Brazil and Mexico), and one is in Africa (Nigeria). The two countries with the world's largest number of mobile telephone subscribers – that is, China and India – had been, in the pre-mobile telephone era, telecommunications backwaters in which access had been confined to a tiny elite (Harwit 2008; Jeffrey & Doron 2013). The present-day (2013) mobile-telephone penetration rate for both countries is impressive. In China, it hovers around 89 percent (1.2 billion telephones); in India, around 71 percent (886 million telephones). To be sure, these percentages overstate somewhat the telephone penetration rates in these countries, since well-to-do subscribers often have more than one mobile telephone. Even so, they testify to the rapidity with which this new medium has become a ubiquitous feature of everyday life in the developing world. Three other countries in which telecommunications had previously been little developed – namely, Russia, Indonesia, and Brazil – at present (2013) each boast collectively more than 200 million mobile telephone subscribers. From a global perspective, the mobile telephone has probably had its greatest impact in Africa, where it has enabled millions of people to talk at a distance for the first time in their lives. Its efficacy is probably enhanced by the oral nature of many African cultures (Hahn & Kibora 2008), though mobile telephone penetration rates are high throughout the continent (de Bruijn, Nyamnjoh & Brinkman 2009). The ubiquity of the mobile telephone raises challenging questions about the much-discussed digital divide: for millions of people around the world, the mobile telephone – rather than the computer monitor – is now the principal interface for connecting not only with other people, but also with the Internet.

2 Three interpretative traditions

The new perspective on the history of telecommunications that we sketched in the previous pages has been informed by three interpretative traditions: the history of technology; political economy; and social constructivism. To clarify our perspective, and to suggest avenues for future research, we believe it would be useful to detail some of the premises of these traditions, and to highlight some of the ways in which they have informed our analysis.

Historians of technology study communications networks in myriad ways. Some analyse the evolution of hardware; others the social context of invention. (Much of the English-language literature on these topics is cited in Sterling and Shiers 2000 and Sterling et al. 2006). Yet one approach – the large technical systems (or LTS) tradition – has proved for us to be particularly influential. While LTS ignores neither politics nor culture, it places particular emphasis on the internal logic of communications networks, and, in particular, their trajectory.

LTS comes in an American and a French variant (Balbi 2009). The American version was pioneered by historian Thomas P. Hughes; the French version – which is called "Macro-Systèmes Techniques" – by sociologist Alain Gras. For Hughes (1987), every large technical system has three components: technical contrivances; formal organizations; and rules. Gras reaches a parallel conclusion using somewhat different language. For him, the key components are industrial objects; complex organizations; and commercial intermediaries (Gras 1993, 1997).

Two of Hughes's key concepts (1987) are "momentum" and "load factor." The momentum of a large technical system is its propensity to become increasingly impervious to outside pressure, such as rivals or government regulation, as it expands. To adopt the language of the institutional economist, its evolution is path dependent: prior choices affect future outcomes. The load factor of a large technical system – as Hughes (1987) explains, using the electric power grid as his example – is the ratio of average output to maximum output during a specific time interval:

> Best defined by a graph, or curve, the load factor traces the output of a generator, power plant, or utility system over a twenty-four-hour period. The curve usually displays a valley in the early morning, before the waking hour, and a peak in the early evening, when business and industry use power, homeowners turn on lights, and commuters increase their use of electrified conveyance. Showing graphically the maximum capacity of the generator, plant, or utility (which must be greater than the highest peak) and tracing the load curve with its peaks and valleys starkly reveals the utilization of capacity (Hughes 1987: 72).

One example of institutional momentum is the continuing reliance of French government administrators on the optical telegraph following the commercialization of the electric telegraph in Great Britain and the United States. The French optical telegraph by 1840 had fallen into what the economic historian Mark Elvin (1972) calls a "high-level equilibrium trap". Since it had operated effectively for half a century, its administrators proved reluctant to abandon it in favour of the new medium. Interestingly, the French would confront an analogous dilemma in the 1990s, when government engineers retained allegiance to Minitel, an early on-line network, slowing the rollout of the Internet (Schafer & Thierry 2012). Once again, precocity proved to be an obstacle to change. A related example of institutional momentum has been the slow adoption of the mobile telephone in the United

States. The United States in the 1980s boasted the finest landline telephone network in the world; predictably, the switchover to the mobile telephone network proved more halting than it would in China or India, countries that had never made a comparable investment. Various factors slowed the adoption of the mobile telephone in the United States. Potential users had already become accustomed to using pagers and electronic beepers, which provided some of the functionality that would come to be associated with the mobile telephone. Potential users were also frustrated by the available calling plans, which charged even for incoming calls. Network expansion was further constrained by poorly designed handsets and the existence of multiple, incompatible technical standards (Agar 2013: Ch. 1).

The load factor concept has proved useful in explaining the early history of the landline telephone, which was decisively shaped by the high cost of manual switching. For example, the diseconomies associated with the scaling up of the landline telephone network obliged enterprising late-nineteenth-century operating company managers such as Angus Hibbard of the Chicago Telephone Company to devise elaborate techniques to calculate load factors in telephone usage, a concept that Hibbard called the "telephone day" (John 2010: Ch. 8). All telecommunications managers, of course, must strike a balance between costs, network expansion, and network usage. Yet the management of the big-city landline telephone network in the era of manual switching posed an unusual challenge: every doubling in network size quadrupled the costs of making connections, giving managers an obvious incentive to keep their network small. Enterprising managers like Hibbard overcame this obstacle by rationalizing work flow, purchasing new switching equipment, and rolling out innovative marketing strategies and calling plans (John 2010: Ch. 8).

LTS can also help explain the relationship between telecommunications networks and the so-called "info-structure" (Joerges 1988: 24; Gras 1997: 31–33). Consider, for example, the relationship between the electric telegraph and the railroad in Great Britain and the United States (Schivelbusch 1987; Schwantes 2008; John 2010: Ch. 3). In both countries, railroad managers initially coordinated the movement of trains through personal observation. Following the commercialization of the electric telegraph, managers supplemented personal observations with data transmitted by electric telegraph. This switchover occurred much more quickly in Great Britain than in the United States. This was partly because British railroads were more highly capitalized than American railroads, and partly because of the conservatism of American railroad managers. Following the switchover, it became possible to greatly increase the movement of rolling stock on the rights-of-way. In fact, railroad engineers sometimes contended that a single-tracked railroad with telegraphic dispatching was more effective than a double-tracked railroad that continued to rely on personal observation (Schwantes 2008).

A second tradition that we have found to be particularly useful in writing the history of telecommunications networks has been political economy. Historical writing in this tradition is highly variegated, sometimes obscure, and occasionally tendentious (Mosco 2009). Two useful concepts are the structuring presence of rate-and-entry regulation, a relationship that has been analyzed by political scientist Robert B. Horwitz (1989), and the influence on business strategy of political structure, a relationship that has been explored by the economic historian Richard H. K. Vietor (1994).

The political economy tradition can help explain why telecommunications networks have evolved differently in different times and places. National styles are often a determining factor. Indeed, when historians of telecommunications networks adopt an international comparative perspective, they often discover that governmental institutions and civic ideals play a larger role in shaping network architecture than technology and economics (Moyal 1984; Galambos 1988; Starr 2004). Neither technology nor economics can explain why the French optical telegraph was a government monopoly while the British and American optical telegraphs were not. This outcome, rather, was a by-product of what the historical sociologist Paul Starr calls the "constitutive choices" of lawmakers (Starr 2004). It was, similarly, neither technology nor economics, but political fiat, that explains why government administrators in Switzerland, Belgium, Great Britain, France, and Germany configured the electric telegraph as a mass service for the entire population while corporate managers in the United States designed the electric telegraph as a specialty service for an exclusive clientele. Or, for that matter, why the Argentinian telegraph network ended up as a curious public-private hybrid (Hodge 1984). The very different trajectories of the landline telephone in Italy, Canada, Great Britain, and the United States, similarly, cannot be understood without emphasizing the influence on operating company managers of municipal franchise politics (Armstrong and Nelles 1986; Horwitz 1989; John 2008, 2010; Balbi 2011; McDougall 2013).

Even at the international level, politics mattered. Consider, for example, the Telegraph Union, later called the International Telecommunications Union (ITU). The establishment of the ITU in 1855 was shaped not only by technical considerations, but also by geopolitics, and in particular by the determination of Belgian administrators to challenge the prerogatives of the French emperor, Napoleon III (Laborie 2010). Large nations, of course, remained disproportionately influential in setting international standards (Feldman 1974; Griset 1999). Yet small nations such as Switzerland have also helped to devise network protocols and sometimes even to align organizational priorities with cherished national ideals such as neutrality and internationalism – as was demonstrated, for example, in the establishment of the Telegraph Union (Balbi et al. 2014).

Among the topics that can be illuminated by the political economy tradition is the propensity of telecommunications networks to become increasingly entrenched as their user base expanded, a phenomena that is often attributed in the Anglophone literature to "network effects" and in the Francophone literature to "*effets de réseaux*" or "*effets de club.*" Telecommunications networks – unlike, say, gas or water networks – become (all things being equal) more valuable to individual users as they expand. This is because network expansion facilitates the linkage of a larger number of nodes. To be sure, one can point to historical instances in which telecommunications providers opposed network expansion as too costly, sometimes with the support of users unwilling to pay the higher rates that network expansion would bring (John 2010: Ch. 7). Even so, network expansion can develop a self-reinforcing logic that rewards early entrants, a phenomenon that has been analysed by many scholars, including the sociologist Manuel Castells (1996–1998) and the political theorist David Singh Grewal (2008).

The third and final tradition that we have found useful in our survey of the history of telecommunications networks is social constructivism (or SCOT, an acronym for the "social construction of technology"). Social constructivists highlight the extent to which technological artefacts, such as telecommunications networks, are built and maintained by social groups that include lawmakers, engineers, and managers. Among the social groups that the social constructivists devote special attention to are users, a group that other historians often neglect.

For the social constructivists, no analysis of a telecommunications network could pretend to be comprehensive if it failed to analyse the social matrix in which it was embedded, a matrix that included not only the social groups responsible for the construction, operation, and regulation of the network, but also the network's users. Each of these actors tries to shape the technology, with the outcome being a product of the resulting negotiation. The importance of social groups for the social constructivists was underscored in a celebrated essay by the sociologists Trevor Pinch and Wiebe Bijker (1984). Social scientists, they contend, should be mindful of all of the actors who contributed to the construction of a particular artefact, whether or not these social groups were insiders. What mattered, instead, were the mental maps that the actors shared: "The key requirement is that all members of a certain social group share the same set of meanings, attached to a specific artefact" (Pinch & Bijker 1984: 414).

The influence of users for telecommunications networks has been widely documented. Users are important for at least two reasons. First, users sometimes invent applications for a network that its designers had not anticipated. The rollout of new communications networks is often accompanied by a great deal of uncertainty regarding its potential utility. The early landline telephone, for example, was sometimes configured not only as a one-to-one medium for personal communication, but also as a broadcast medium for music, news, and entertainment (Marvin 1988; Balbi 2010). In the case of the wireless telegraph, hobbyists popularized the controversial idea that a point-to-point medium might be transformed into a broadcast medium (Douglas 1989). And in the case of the mobile telephone, network managers failed to anticipate that text-messaging might be a popular feature until subscribers began to demonstrate its utility for social communication (Taylor & Vincent 2005; Goggin 2011). Text-messaging is but one of the many unanticipated new social practices that the mobile telephone has spawned. Some of these social practices have attracted the attention of media sociologists, who contend that their ritual dimensions are fostering not the unbridled individualism predicted by media sociologists like Flichy (1991), but, instead, compelling new forms of social cohesion (Castells at al. 2007; Ling 2008).

Users are significant for a second reason. In certain instances, and contrary to a widespread assumption, they can block network expansion. In the case of the landline telephone, business users fearful that network expansion would increase rates and degrade the level of service to which they had become accustomed lobbied *against* the transformation of the medium from a specialty service for an exclusive clientele into a mass service for the entire population (John 2010: Ch. 7).

To understand the social consequences of the mobile telephone, it can be very useful to adopt a user-centric lens. To an extent that may well be unmatched by any previous telecommunications medium, with the possible exception of the early landline telephone (Kline 2000), mobile telephone users have proved ingenious in devising unanticipated uses for the medium. In India, fishermen, factory workers, and farmers are using the mobile telephone to improve their ability to compete in the marketplace (Jensen 2007; Jeffrey & Doron 2013). In Russia, nomadic reindeer hunters are using the medium to remain in touch with their families (Stammler 2009). And in Kenya, urbanites use the mobile telephone as an electronic wallet to send money to friends and family members in distant villages – a convenience called M-PESA ("M" stands for mobile; "PESA" is the word for money in Swahili) that is highly valued in a country in which reliable cash machines are few and far between (Agar 2013: Ch. 16). Even so, a single-minded focus on the user can be misleading. It is a mistake, for example, to conflate the proliferation of the mobile telephone with a concomitant increase in the power (or "agency") of users to increase their autonomy. In fact, as Italian scholar Michela Nacci (2000) has posited, this conflation owes much to what she terms the "transparency paradox." Mobile telephone users interact primarily with their own, often highly personalized, mobile devices, which can lead them to underestimate the vulnerability of their on-line behaviour to data mining by businesses, governments, and other individuals. As a consequence, they can easily exaggerate their autonomy and underestimate the extent to which they are subject to mental manipulation by marketers, scam artists, or even sexual predators. Network managers often conceal the full extent of this electronic surveillance from users for two reasons: first, to simplify their interaction with the network; and, second, to lull them into giving the network their unqualified trust.

Closely related to yet distinct from social constructivism is cultural studies, a convenient shorthand term for scholarship that focuses on the ideational dimensions of historical phenomena. While many social constructivists distance themselves from cultural studies, the two traditions share an interest in social relationships, ideology, and group identity. This tradition has been particularly fruitful for historians interested in gaining critical perspective on the rhetoric deployed by promoters, popularizers, and publicists. To a greater extent than historians of technology who focus on large technical systems or political economy, these scholars are attentive to issues of language and rhetoric. Among its most distinguished practitioners have been the British sociologist Raymond Williams (1974); the West Indian-born Anglophone social theorist Stuart Hall (1980); the American literary critic Leo Marx (1994, 2010); and the American communications scholar James W. Carey (1989).

The cultural studies tradition has provided historians of communications with the tools to gain a critical perspective on fundamental, yet sometimes implicit, values and norms. For example, the ideological currents that shaped network-building in France have been probed by Armand Matterlart (1992, 2000), Rosalind Williams (1993), and Pierre Musso (1997). In a similar vein, the mythologizing of the mobile telephone as a spiritual – and, indeed, transcendent medium – has been sketched by James E. Katz (2006) and Rich Ling (2008). The ideology of "universal service"– long an influential norm for telecommunications providers in the United States – is clarified by recognizing its embeddedness in a cultural discourse, as are such seemingly commonsensical ideas as the "invention" of blockbuster devices such as the electric telegraph or the telephone, the establishment of global telecommunications networks, and the transformation of the telephone from a specialty service for the few into a mass service for the entire population (John 2010; Hampf & Müller-Pohl 2013; McDougall 2013; Beauchamp 2014).

3 Conclusion

The rapid emergence of the mobile telephone in the past several decades has fundamentally altered not only the present and future of telecommunications but also its past. To render this past intelligible, we have surveyed the history of four point-to-point networks – the optical telegraph, the electric telegraph, the landline telephone, and the mobile telephone – using different interpretative traditions – the history of technology, political economy, and social constructivism. No longer can it plausibly be assumed that telecommunications is a detour on the road toward broadcasting or that the computer monitor (as distinct from the mobile telephone) will remain the primary digital interface for the world's netizens. The commercialization of the Internet, the convergence between broadcast and point-to-point media, and the proliferation of the mobile telephone all underscore the indispensability of point-to-point networks to public and private life. To understand today's media ecology and the digital culture it has spawned, we need to know how and why these networks emerged.

References

Agar, Jon. 2013. *Constant touch: A global history of the mobile phone*. Rev. ed. Cambridge: Icon Books.

Anduaga, Aitor. 2009. *Wireless and empire: Geopolitics, radio industry, and ionosphere in the British Empire, 1918–1939*. Oxford: Oxford University Press.

Armstrong, Christopher & H. V. Nelles. 1986. *Monopoly's moment: The organization and regulation of Canadian utilities, 1830–1930*. Philadelphia: Temple University Press.

Baark, Erik. 1997. *Lightning wires: The telegraph and China's technological modernization, 1860–1890*. Westport, CT: Greenwood Press.

Balbi, Gabriele. 2009. Studying the social history of telecommunications: Between Anglophone and continental traditions. *Media Histo* 15. 85–101.

Balbi, Gabriele. 2010. Radio before radio: Araldo telefonico and the invention of Italian broadcastsing. *Technology and Culture* 41. 786–808.

Balbi, Gabriele. 2011. The origins of the telephone in Italy, 1877–1915: Politics, economics, technology, and society. *International Journal of Communications* 5. 1058–1081.

Balbi, Gabriele. 2013a. Telecommunications. In Peter Simonson, Janis Peck, Robert Craig, & John Jackson (eds.), *Handbook of communication history*, 209–222. London: Routledge.

Balbi, Gabriele. 2013b. 'I will answer you, my friend, but I am afraid': Telephones and the fear of a new medium in nineteenth and early twentieth-century Italy. In Siân Nicholas & Tom O'Malley (eds.), *The media, social fears and moral panics: Historical perspectives*, 59–75. London and New York: Routledge.

Balbi, Gabriele, with Simone Fari, Giuseppe Richeri & Spartaco Calvo. 2014. *Network neutrality: Switzerland's role in the genesis of the Telegraph Union, 1855–1875*. Bern: Peter Lang.

Barabási, Albert-Laszlo. 2002. *Linked: The new science of networks*. Cambridge, MA: Perseus Publishing.

Beauchamp, Christopher. 2014. *Invented by law: Alexander Graham Bell and the patent that changed America*. Cambridge, MA: Harvard University Press.

Bektas, Yakup. 2000. The sultan's messenger: Cultural constructions of Ottoman telegraphy, 1847–1880. *Technology and Culture* 41. 669–696.

Bertho-Lavenir, Catherine. 1981. *Télégraphes et téléphones: de Valmy au microprocesseur*. Paris: Livre de Poche.

Bertho-Lavenir, Catherine. 1988. The telephone in France 1879–1979: National characteristics and international influences. In Renate Mayntz & Thomas P. Hughes (eds.), *The development of large technical systems*, 155–177. Boulder: Westview Press.

Blondheim, Menahem. 1994. *News over the wires: The telegraph and the flow of public information in America, 1844–1897*. Cambridge MA: Harvard University Press.

de Bruijn, Mirjam, Nyamnjoh, Francis & Inge Brinkman (eds.). 2009. *Mobile phones: the new talking drums of everyday Africa*. Langaa, Bamenda (Cameroon); African Studies Centre, Leiden (the Netherlands).

Calvo, Angel. 2002. The Spanish telephone sector (1876–1924): A case of technological backwardness. *History and Technology* 18. 77–102.

Calvo, Angel. 2006. The shaping of urban telephone networks in Europe, 1877–1926. *Urban History* 33. 411–433.

Carey, James. W. 1989. *Communication as culture: Essays on media and society*. Boston: Unwin Hyman.

Castells, Manuel. 1996–1998. *The Information age: Economy, society and culture*. 3 Vols. Oxford: Blackwell.

Castells, Manuel, Mireia Fernández-Ardèvol, Jack Linchuan Qiu & Araba Sey. 2007. *Mobile communication and society: A global perspective*. Cambridge, MA: MIT Press.

Curien, Nicolas. 2005. *Economie des réseaux*. Paris: Découverte.

De Wit, Onno. 1998. *Telefonie in Nederland, 1877–1940*. Den Haag: Cramwinckel.

Douglas, Susan. J. 1989. *Inventing American broadcasting, 1899–1922*. Baltimore: Johns Hopkins University Press.

Downey, Gregory John. 2002. *Telegraph messenger boys: Labor, technology, and geography, 1850–1950*. New York: Routledge.

Elvin, Marc. 1972. The high-level equilibrum trap: The causes of the decline of invention in traditional Chinese textile industries. In William E. Willmott (ed.), *Economic organization in Chinese society*, 137–172. Stanford: Stanford University Press.

Evans, Heidi J. S. 2010. 'The path to freedom?' Transocean and German wireless telegraphy, 1914–1922. *Historical Social Research*, 35, 209–233.

Fari, Simone. 2008. *Una penisola in comunicazione: Il servizio telegrafico italiano dall'Unità alla Grande Guerra*. Bari: Cacucci Editore.

Feldman, Mildred. L. B. 1974. *The United States in the International Telecommunication Union and in pre-ITU conferences: Submarine cables, overland telegraph, sea and land radio, telecommunications*. Baton Rouge: n. p.

Finn, Bernard & Daging Yang (eds.). 2009. *Communications under the seas: The evolving cable network and its implications*. Cambridge, MA: MIT Press.

Fischer, Claude S. 1992. *America calling: A social history of the telephone to 1940*. Berkeley: University of California Press.

Flichy, Patrice. 1991. *Une histoire de la communication moderne: espace public et vie privée*. Paris: Decouverte.

Flichy, Patrice. 1995. *Dynamics of modern communication: The shaping and impact of modern communication technologies*. Liz Libbrecht (trans.). New York: SAGE.

Friedewald, Michael. 2000. The beginnings of radio communication in Germany, 1897–1918. *Journal of Radio Studies* 7. 441–462.

Gabel, David. 1995. Federalism: An historical perspective. In Paul Teske (ed.), *American regulatory federalism and telecommunications infrastructure*, 19–31. Hillsdale, NJ: Lawrence Erlbaum Associates.

Galambos, Louis. 1988. Looking for the boundaries of technological determinism: A brief history of the U.S. telephone system. In Renate Mayntz & Thomas P. Hughes (eds.), *The development of large technical systems*, 135–153. Boulder: Westview Press.

Galambos, Louis & Eric J. Abrahamson. 2002. *Anytime, anywhere: Entrepreneurship and the creation of a wireless world*. Cambridge: Cambridge University Press.

Gitelman, Lisa. 1999. *Scripts, grooves, and writing machines: Representing technology in the Edison era*. Stanford: Stanford University Press.

Goggin, Gerard. 2006. *Cell phone culture: mobile technology in everyday life*. London: Routledge.

Goggin, Gerard. 2011. *Global mobile media*. London: Routledge.

Gras, Alain. 1993. *Grandeur et dépendance: sociologie des macro-systèmes techniques*. Paris: Presses universitaires de France.

Gras, Alain. 1997. *Les macro-systèmes techniques*. Paris: Puf.

Grewal, David Singh. 2008. *Network power: The social dynamics of globalization*. New Haven: Yale University Press.

Griset, Pascal. 1996. *Technologie, entreprise, et souveraineté: Les télécommunications transatlantiques de la France*. Paris: Rive-Droite.

Griset, Pascal. 1999. Technical system and strategy: Intercontinental telecommunications in the first quarter of the twentieth century. In Olivier Coutard (ed.), *The governance of large technical systems*, 58–72. London: Routledge.

Hall, Stuart. 1980. Encoding/decoding. In Stuart Hall, Dorothy Hobson, Andrew Lowe, & Paul Willis (eds.), *Culture, media, language: Working papers in cultural studies 1972–79*, 128–138. London: Hutchinson.

Hampf, M. Michaela & Simone Müller-Pohl (eds.). 2013. *Global communication electric: Business, news, and politics in the world of telegraphy*. Chicago: University of Chicago Press.
Hahn, Hans Peter & Ludovic Kibora. 2008. The domestication of the mobile phone: Oral society and new ICT in Burkina Faso. *Journal of Modern African Studies* 46. 87–109.
Harwit, Eric. 2008. *China's telecommunications revolution*. Oxford: Oxford University Press.
Hazlewood, Arthur. 1953. The origin of the state telephone service in Britain. *Oxford Economic Papers, new series* 5. 13–25.
Headrick, Daniel R. 1988. *The tentacles of progress: Technology transfer in the age of imperialism, 1850–1940*. New York: Oxford University Press.
Headrick, Daniel R. 1992. *The invisible weapon: Telecommunications and international politics, 1851–1945*. New York: Oxford University Press.
Headrick, Daniel R. 1994. Shortwave radio communication and its impact on international telecommunications between the wars. *History and Technolog* 11. 21–32.
Headrick, Daniel R. 1995. Public-private relations in international telecommunications before World War II. In Bella Mody, Johannes M. Bauer & Joseph D. Straubhaar (eds.), *Telecommunications politics: Ownership and control of the information highway in developing countries*, 31–49. Hillsdale, NJ: Lawrence Erlbaum Associates.
Headrick, Daniel R. 2000. *When information came of age: Technologies of knowledge in the age of reason and revolution, 1700–1850*. Oxford: Oxford University Press.
Helgesson, Claes-Fredrik. 1999. *Making a natural monopoly: The configuration of a techno-economic order in Swedish telecommunications*. Stockholm: Stockholm School of Economics.
Hermans, Janneke & Onno De Wit. 2004. Bourses and brokers: stock exchanges as ICT junctions. *History and Technology* 20. 227–247.
Herring, James Morton, & Gerald Connop Gross. 1936. *Telecommunications: Economics and regulation*. New York: McGraw-Hill.
Hochfelder, David. 2012. *The telegraph in America, 1832–1920*. Baltimore: Johns Hopkins University Press.
Hodge, John E. 1984. The role of the telegraph in the consolidation and expansion of the Argentine Republic. *The Americas* 41. 59–80.
Holzmann, Gerard. J. & Bjorn Pehrson 1995. *The early history of data networks*. Hoboken, NJ: John Wiley & Sons.
Horwitz, Robert Britt. 1989. *The irony of regulatory reform: The deregulation of American telecommunications*. New York: Oxford University Press.
Hughes, Thomas P. 1987. The evolution of large technical systems. In Thomas P. Hughes & Trevor Pinch (eds.), *The social construction of technological systems*, 51–82. Cambridge MA: MIT Press.
Hugill, Peter J. 1999. *Global communications since 1844: geopolitics and technology*. Baltimore: Johns Hopkins University Press.
Huurdeman, Anton. 2003. *The worldwide history of telecommunications*. Hoboken, NJ: John Wiley & Sons.
Jeffrey, Robin & Assa Doron. 2013. *The great Indian phone book: How the cheap cell phone changes business, politics, and daily life*. Cambridge, MA: Harvard University Press.
Jensen, Robert. 2007. The digital provide: Information (technology), market performance, and welfare in the south Indian fisheries sector. *Quarterly Journal of Economic*, 122. 879–924.
Joerges, Bernward. 1988. Large technical systems: Concepts and issues. In Renate Mayntz & Thomas P. Hughes (eds.), *The development of large technical systems*, 9–36. Boulder: Westview Press.
John, Richard R. 1995. *Spreading the news: The American postal system from Franklin to Morse*. Cambridge, MA: Harvard University Press.

John, Richard R. 2008. Telecommunications. *Enterprise and Society* 9. 507–520.
John, Richard R. 2010. *Network nation: Inventing American telecommunications*. Cambridge: Belknap Press of Harvard University Press.
John, Richard R. 2013. Communications networks in the United States from Chappe to Marconi. In Angharad N. Valdivia (ed.), *International Encyclopedia of Media Studies*, 310–332. Oxford: Blackwell.
Katz, James Everett. 2006. *Magic in the air: Mobile communication and the transformation of social life*. New Brunswick, NJ: Transaction.
Kieve, Jeffrey. 1973. *The electric telegraph: A social and economic history*. Devon: David and Charles Newton Abbot.
Kline, Ronald B. 2000. *Consumers in the country: Technology and social change in rural America*. Baltimore: Johns Hopkins University Press.
Kobelt, C. 1980. One-hundred years of telephone service in Switzerland. *Bulletin Technique PTT* 10. 344–363.
Laborie, Léonard. 2010. *L'Europe mise en réseaux: La France et la coopération internationale dans les postes et les télécommunications (années 1850–années 1950)*. Bern: Peter Lang.
Ling, Richard Seyler. 2008. *New tech, new ties: How mobile communication is reshaping social cohesion*. Cambridge, MA: MIT Press.
Martin, Michèle. 1991. *'Hello central?': Gender, technology, and culture in the formation of telephone systems*. Montréal: McGill-Queen's University Press.
Marvin, Carolyn. 1988. *When old technologies were new: Thinking about electric communication in the late nineteenth century*. New York: Oxford University Press.
Marx, Leo. 1994. The idea of 'technology' and postmodern pessimism. In Merrit Roe Smith & Leo Marx (eds.), *Does technology drive history? The dilemma of technological determinism*, 237–258. Cambridge, MA: MIT Press.
Marx, Leo. 2010. Technology: The emergence of a hazardous concept. *Technology and Culture* 51. 561–577.
Matterlart, Armand. 1992. *La communication-monde: Histoire des idées et des stratégies*. Paris: Découverte.
Matterlart, Armand. 2000. *Networking the world, 1794–2000*. J. A. Cohen and L. Carey-Libbrecht (trans.). Minneapolis: University of Minnesota Press.
Mayntz, Renate and Thomas P. Hughes (eds.). 1988. *The development of large technical systems*. Boulder: Westview Press.
McChesney, Robert W. 1993. *Telecommunications, mass media, & democracy: The battle for control of U.S. broadcasting, 1928–1935*. New York: Oxford University Press.
McDougall, Robert. 2013. *The people's network: The political economy of the telephone in the gilded age*. Philadelphia: University of Pennsylvania Press.
Millward, Robert. 2005. *Private and public enterprise in Europe: Energy, telecommunications, and transport, 1830–1990*. Cambridge: Cambridge University Press.
Mosco, Vincent. 2009. *The political economy of communication*. London: Sage.
Moyal, Ann. 1984. *Clear across Australia: A history of telecommunications*. Melbourne: Nelson.
Musso, Pierre. 1997. *Télécommunications et philosophie des réseaux: la postérité paradoxale de Saint-Simon*. Paris: Presses universitaires de France.
Nacci, Michela. 2000. *Pensare la tecnica: Un secolo di incomprensioni*. Roma-Bari: Laterza.
Noam, Eli M. 1992. *Telecommunications in Europe*. New York: Oxford University Press.
Noam, Eli M. (ed.). 1999. *Telecommunications in Africa*. New York: Oxford University Press.
Perry, C. R. 1992. *The Victorian post office: The growth of a bureaucracy*. Suffolk, England: Boydell Press.
Pinch, Trevor & Wiebe E. Bijker. 1984. The social construction of facts and artefacts: Or how the sociology of science and the sociology of technology might benefit each other. *Social Studies of Science* 14. 399–441.

Rosenfeld, Sophia. 2001. *A revolution in language: the problem of signs in late eighteenth-century France*. Stanford, CA: Stanford University Press.
Schafer, Valérie & Benjamin G. Thierry. 2012. *Le Minitel: l'enfance numérique de la France*. Paris: Nuvis.
Schivelbusch, W. 1987. *The railway journey: The industrialization of time and space in the nineteenth century*. Berkeley: University of California Press.
Schneider, Volker. 1991. The governance of large technical systems: The case of telecommunications. In Tedd R. L. Porte (ed.), *Social responses to large technical systems: Control or anticipation*, 19–41. Dordrecht: Kluwer Academic Publishers.
Schwantes, Benjamin. 2008. *Fallible guardian: The social construction of railroad telegraphy in nineteenth-century America*. Unpublished doctoral dissertation. University of Delaware, Newark, DE.
Solnick, Steven. L. 1991. Revolution, reform and the Soviet telephone system, 1917–1927. *Soviet Studies* 43. 157–175.
Stammler, Florian. M. 2009. Mobile phone revolution in the tundra? Technological change among Russian nomads. *Folklore* 41. 47–78.
Starr, Paul. 2004. *The creation of the media: Political origins of modern communications*. New York: Basic Books.
Steinbock, Dan. 2003. *Wireless horizon: Strategy and competition in the worldwide mobile marketplace*. New York: AMACON.
Sterling, Christopher H. & George Shiers. 2000. *History of telecommunications technology: An annotated bibliography*. Lanham, MD: Scarcecrow Press.
Sterling, Christopher, Phyllis Bernt & Martin B. H. Weiss. 2006. *Shaping American telecommunications: A history of technology, policy, and economics*. Hillsdale, NJ: Lawrence Erlbaum Associates.
Taylor, Alex S. & Jane Vincent. 2005. An sms history. In Lynne Hamil & Amparo Larsen (eds.), *Mobile world: Past, present and future*, 75–91. London: Springer.
Thomas, Frank. 1988. The politics of growth: The German telephone system. In Renate Mayntz and Thomas P. Hughes (eds.), *The development of large technical systems*, 179–213. Boulder: Westview Press.
Vietor, Richard. 1994. *Contrived competition: Regulation and deregulation in America*. Cambridge: Cambridge University Press.
Wallsten, Scott. 2005. Returning to Victorian competition, ownership, and regulation: An empirical study of European telecommunications at the turn of the twentieth century. *Journal of Economic History* 65. 693–722.
Weiman, David F. 2003. Building 'universal service' in the early Bell system: The co-evolution of regional urban systems and long distance telephone networks. In William Sundstrom, Timothy W. Guinnane & Warren C. Whatley (eds.), *History matters: Essays on economic growth, technology, and demographic change*, 328–33. Stanford, CA: Stanford University Press.
Williams, Raymond. 1974. *Television: Technology and cultural form*. London: Fontana.
Williams, Rosalind. 1993. Cultural origins and environmental implications of large technological systems, *Science in Context* 6. 377–403.
Wilson, Geoffrey. 1976. *The old telegraphs*. London: Phillimore.
Winkler, Jonathan Reed. 2008. *Nexus: Strategic communications and American security in World War I*. Cambridge, MA: Harvard University Press.
Winseck, Dwayne R. & Robert M. Pike. 2007. *Communication and empire: Media, markets, and globalization, 1860–1930*. Durham: Duke University Press.
Wolff, Joshua D. 2013. *Western Union and the creation of the American corporate order, 1845–1893*. Cambridge: Cambridge University Press.

Yang, Daqing. 2011. *Technology of empire: Telecommunications and Japanese expansion in Asia, 1883–1945*. Cambridge, MA: Harvard University Press.
Young, Peter. 1991. *Person to person: The international impact of the telephone*. Cambridge: Granta.

Alejandro Pardo

3 Cinema and technology: From painting to photography and cinema, up to digital motion pictures in theatres and on the net

Abstract: The relationship between cinema and technology has been present since the very inception of motion pictures, looking to offer a more immersive and believable experience. This chapter presents a comprehensive compilation of key scholarly literature – mainly in the English language – while identifying some of the theoretical issues, emerging concepts, current and further research, as well as lists of references with regard to this topic. In order to accomplish this task, it is divided into three main sections. The first one focuses on the relationship between the arts and technology, and specifically between cinema and the arts, and between cinema and technology. The middle section draws a brief historical summary on the technological development of the (audio)visual media, moving from the primitive canvas to the first photographic plates and from the birth of cinema to the digital image. Finally, the third part is a synthesis of some of the most relevant theoretical and critical issues regarding the imbrication of art, technology and cinema, all of it in the words of well-known experts and scholars. An epilogue with some final thoughts closes this chapter.

Keywords: cinema, technology, audiovisual media, art, film, movies, motion pictures, photography

1 Introduction

The relationship between cinema and technology has been present since the very inception of motion pictures. In fact, the cinema as a medium (moving image) was invented as a further technological step away from photography (still image), in the same way that photography (real image) was a revolutionary innovation, with respect to painting (artistic image), in the human being's secular search for means of representing reality (indexicality). In this sense, the evolution of the cinema has been closely linked to technological development (sound, colour, widescreen, 3-D, digital effects, etc.) seeking to offer a more immersive and believable experience. Therefore, as Steve Neale summarizes, "technology is a basic component of cinema, a condition of its existence and continuing factor in its development" (Neale 1985: 2).

This chapter attempts to accomplish a rather difficult goal: to present a comprehensive compilation of key scholarly literature while identifying some of the theoretical issues, emerging concepts, current and further research, as well as lists of references with regard to this topic of cinema and technology. Since this is almost an insurmountable task for such a limited number of pages, I will try to narrow the field of research according to the following criteria. In the first place, with a few exceptions, I will opt mainly for the literature published in English (or English translations of authors in other languages). Secondly, I will give pre-eminence to collective works such as film readers and other edited volumes, which are usually more comprehensive than monographic studies – although I will also mention some of the latter. Lastly, I will try to keep cinema (movies) as the core subject, especially in regard to its relation to other arts and techniques (painting, photography, computer sciences, etc.). Other audiovisual media (television, videogames) are widely covered by different chapters within this volume.

This chapter is divided into three main sections. The first one offers a comprehensive as well as concise literature review about the relationship between the arts and technology, and specifically between cinema and the arts, and between cinema and technology. The middle section draws a brief historical summary on the technological development of the (audio)visual media, moving from the primitive canvas to the first photographic plates and from the birth of cinema to the digital image. Finally, the third part is a synthesis of some of the most relevant theoretical and critical issues regarding the imbrication of art, technology and cinema, all of it in the words of well-known experts and scholars. An epilogue with some final thoughts closes the chapter.

2 Literature review

Cinema represents, by any means, the quintessential cross point between art and technology, between "life" image and reproducibility. For this very reason, it can be stressed that every book on film history is a book on audiovisual technology history as well.

Before going into detail with this account, it is necessary to pay attention – at least in a general way – to those art historians who have researched the evolution of the artistic and mechanical processes related to the creation and reproduction of images, from painting to photography. I am not going to mention here the books and encyclopaedias on art history, most of them quite well known; I will only give some references of works that are especially convenient for our approach – painting and photography as precedents of cinema. One crucial book in this regard is *The Painter and the Photograph: from Delacroix to Warhol*, written by American photographer and scholar, F. Van Deren Coke, a pictorial and verbal history of the various ways in which artists of many countries have used photographs. The introductory chapter offers a useful summary of the progressive symbiosis of both arts (Coke 1964: 1–15). In a close connection with this last book, *Cinema and Painting: How Art is Used in Film*, by American art historian Angela Dalle Vacche, is worthy of mention. In a very discerning way, this scholar discusses how filmmakers have used the imagery of paintings to shape or enrich the meaning of their films, as well as underlines the value of intertextuality to appreciate this close relationship between both arts (Dalle Vacche 1996: 1–12). A third recommendable book in this regard, also edited by Angela Dalle Vacche, is *The Visual Turn: Classical Film Theory and Art History*, which is set as a dialogue between art historians and film theorists throughout the twentieth century (Dalle Vacche 2003: 1). With a total number of 14 articles – most of them canonical texts of classical film theory – this collection attempts to dip into the links between art (and especially painting) and cinema. Most of the contributors are renowned film theorists and philosophers (Walter Benjamin, Edwin Panofsky, Gilles Deleuze, Andre Bazin, Béla Balázs, Rudolf Arnheim) and filmmakers (Sergei Eisenstein).

Moving forward, a number of authors have accurately narrated the history of photography, exhaustively recollected by Laurent Roosens and Luc Salu in their bibliographical compendium, *History of Photography: A Bibliography of Books* (Roosens and Salu 1989). In the following years, new books have appeared (Rosenblum 1997; Frizot 1998). For the purpose of this chapter, it will suffice to mention two key works by Brian Coe: *The Birth of Photography: The Story of the Formative Years 1800–1900* and *The History of Movie Photography* (Coe 1976, 1981). This last one is probably the most recommendable for the purpose of this chapter since it covers the natural steps forward from the still image to the moving image and beyond. In particular, this book follows the main lines of technical advances from the primitive optical devices aimed at enhancing the illusion of reality, through the birth of photography and the emergence of the cinematic machines, to the first moving pictures and the inclusion of sound, colour and the development of different wide screen formats as well as of home cinema.

Similarly, the list of film historians and scholars who have researched the origins and further technological innovations in the motion picture industry in detail is endless. For this reason, the bibliographical compendium compiled by Frank Manchel, *Film Study: An Analytical Bibliography*, is quite helpful. The third volume of this major multi-volume work includes a selection of books about the origins of cinema (Manchel 1990: 1,587–1,595). To mention just a few, I would underline the classic text by C. W. Ceram *Archaeology of the Cinema* and Steven Neale's *Cinema and Technology*, an account of the major developments in the technological history of cinema as well as of the interrelationships between technology, economics and aesthetics (Ceram 1965; Neale 1985). Other European and American scholars have also covered the birth of motion pictures (Thomas 1964; Musser 1990; Toulet 1995; Tosi 2006) or have focused their attention on particular innovations as sound (Cameron 1959), colour (Ryan 1977) and widescreen formats (Wysotsky 1971). In another particularly interesting study, *The Classical Hollywood Cinema*, David Bordwell, Janet Staiger and Kristin Thompson explain how technology conditioned the mode of production during the Hollywood studio-system era (Bordwell et al. 1985: 241–308; 339–364). Finally, particularly valuable is the anthology compiled by Raymond Field from the pages of the *Journal of the Society of Motion Picture and Television Engineers* and published under the title of *A Technological History of Motion Pictures and Television* (Fielding 1967).

Another set of books worthy of mention is that of film handbooks that cover a vast array of issues, from history to theory and from criticism to technical aspects, in a very didactical way. Three of the most well known are James Monaco's *How to Read A Film*, Joseph M. Boggs' *The Art of Watching Films*, and Louis Giannetti's *Understanding Movies*. The three of them are published regularly in updated editions (latest ones: Monaco 2009; Boggs and Petrie 2012; Giannetti 2014). All of them devote substantial space to summarizing the technological evolution of the cinematic art.

A last group of bibliographical references would be those that focus their attention on the very last stage of this technical evolution, in particular, the move from analogue to digital cinema. The number of works in this field, although more recent, are equally abundant. Some books focus their attention on the technical and professional aspects of moviemaking in the digital age, either in a general way – as in Brian McKernan's *Digital Cinema: The Revolution in Cinematography, Postproduction, and Distribution* (McKernan 2005) and Charles Swartz's *Understanding Digital Cinema: A Professional Handbook* (Swartz 2005) – or pay attention to particular aspects such as visual effects (Wright 2010), sound (Kerins 2011) or 3-D projection (Zone 2012). In addition, some scholars express their critical views on this transition from celluloid to pixels in a more theoretical way, thereby enriching the debate about the gain and looses of cinema as a medium. This is the case of David Rodowick's *The Virtual Life of Film*, where the author discusses the "philosophical consequences of the disappearance of a photographic ontology for the art of film and the future of cinema studies" in this current climate of technological change (Rodowick 2007: vii). Similarly, Nicholas Rombes, in his book *Cinema in the Digital Age*, examines the fate of cinema in this new era, paying special attention not only to the technologies that are reshaping film, but also to the cultural meaning of those technologies (Rombes 2009). A third example is Stephen Prince's *Digital Visual Effects in Cinema: The Seduction of Reality*, a critical view of computer-generated movies from an integrated perspective – aesthetic, historical, and theoretical analyses (Prince 2012). Lastly, on the specific case of the relationship between Hollywood and the Internet, we can also find a series of books published in the last decade, which represent a critical account of the different attitudes the majors studios have developed towards convergence and digital media (Geirland and Sonesh-Kedar 1999; Dekom and Sealey 2003; Lasica 2005; Tryon 2009; Brookey 2010; Tryon 2013).

Most of the previous paragraphs contain a representative sample of significant contributions, which is useful to understand the technological evolution of the visual arts in a linear way, from painting to photography, from moving images to digital cinema. Nevertheless, this literature review would be incomplete without mentioning some other books which offer a more transversal approach. All of them are edited collections of articles or essays specifically focused on the intertwinement of visual arts, technology and culture, and they constitute the basis for the summary of theoretical and critical issues addressed in the last section of this chapter.

The first volume is *Cinema Futures: Cain, Abel or Cable? The Screen Arts in the Digital Age*, co-edited by Thomas Elsaesser and Kay Hoffman. It is a compilation of 21 essays written mostly by European scholars (like Pierre Sorlin, John Ellis or both editors), some American contributors (Lev Manovich), and a group of avant-garde filmmakers and specialized journalists. Examining the complex dynamics of convergence and divergence among the different audiovisual media in this period of technological change, this collective work presents a careful and forceful argument about the future of cinema in its relation with new media, defending cinema's aesthetic identity, the opening up of new cultural spaces and the increment of opportunities for fresh creative input (Elsaesser and Hoffmann 1998: 8).

A second seminal book is *Technology and Culture: The Film Reader*, edited by Andrew Utterson, a historical collection of 12 articles plus one general introduction by the editor. These contributions are written by renowned scholars from different countries and historical periods (Walter Benjamin, Andre Bazin, Douglas Gomery, Vivian Sobchack, Lev Manovich), together with established filmmakers (Dziga Vertov, Lars Von Trier and Thomas Vinterberg) and even some film pioneers (Henry V. Hoopwood and Morton Heilig). In this way, this collective book brings together key theoretical texts from more than a century of writing on film and technology. It begins by exploring the intertwined technologies of cinematic representation, reproduction, distribution and reception, before locating the technological history of cinema as one component of an increasingly complex technological culture. The selected articles encompass a range of disciplines, perspectives and methodologies, reflecting the multiplicity of contemporary approaches to technology (Utterson 2005: 1–10). They are grouped into four thematic sections – Origins and evolution, Definitions and determinism, Projections and aesthetics, Contexts and consequences – each with an introduction by the editor.

A third necessary reference is *Cinema and Technology: Cultures, Theories, Practices*, co-edited by Bruce Bennett, Marc Furstenau and Adrian Mackenzie. This book is made up of 14 essays grouped into four sections – Format, Norms, Scanning and Movement – and is mostly focused on the intersection between cinema and new media. It is in the open debate between so-called 'old' and 'new' media camps that this collection of essays makes its most incisive contribution in setting out to claim that film theory must include new media. In this sense, the argument that cinema is primarily a technological phenomenon is constantly tested, alongside the proposition that "changes to the medium are the direct effects of technological forces" (Bennett at al. 2008: 1). Many of the essays are penetrated by debates around technological innovations – based on the work by Deleuze and Manovich in particular – propelled by the introduction of cinema and other contemporary interactive media. In addition, this book addresses an examination of the web as a new window for movies, both as a new cinematic experience and as an expanding market. Finally, there is a discussion of fan culture and new media interactivity, which also permeates various chapters. In summary, this collection of essays shows that technology has been used to great effect to materialize the use-value of film and as a commodified aesthetic experience. Therefore, film studies have certainly been greatly sustained by technological developments across the spectrum, as this volume effectively demonstrates.

Lastly, one more volume should be added to this ultimate group of collective works, *The Film Theory Reader: Debates and Arguments*, edited by Marc Furstenau. With a total number of 20 essays, also divided into four sections plus one introduction, this book combines classic debates on film theory (Münsterbeg, Carroll, Bazin, Metz, Deleuze, Wollen) with recent arguments (Rodowick, Manovich, Gunning, Friedberg, Belton), under the premise that "[t]he arguments about the present and future status of film have enlivened the discipline, and have generated significant debates about the present and future status of film theory" (Furstenau 2010: 15).

3 A brief historical approach: from canvas to plates, from celluloid to pixels

As seen in the previous section, the technological evolution of the different media and languages of expression such as photography, film or television have promoted a very filling and specialised literature, both theoretical and historiographical, some of which will be discussed later on.

Now it is time to draw a brief historical summary of the technological development of the audiovisual media. To accomplish this, I will mainly follow the synthesis made by Jaime Brihuega in his contribution to a general history of the arts, with a few contributions by some other previously quoted authors. According to Brihuega, the history of the technological evolution of the audiovisual media lies on three milestones: the reproducibility of images (painting and engraving), the mechanization of the production of images (printing) and the photogenesis of images (photography and cinema). This very last step runs in parallell with the Industrial Revolution, and its consequences, the development of communications, the impulse of international trade and the emergence of new forms of leisure (Brihuega 1997).

3.1 First forms of image reproducibility

Writing in 1936, Walter Benjamin chose the term *reproducibility* as one of the key concepts to understand the nature of the contemporary visual culture, in which audiovisual media – and film and television in particular – have played an undisputed central role (Benjamin 2005; quoted here as included in Utterson's collection of essays). The history of the reproducibility of images and sounds has developed in parallel to the means of communication and, therefore, of the human capacity for knowledge. Thus, a new phenomenon hitherto unknown was consolidated: the power of media as a means for socialization.

Although this chapter focuses on the transition from pictorial to the photographic image, and from this to the filmic one, it is worth listing some of the milestones in the history of the reproducibility of images. Actually, it is a phenomenon as old as urban civilization itself, which began from the Neolithic onwards. In Ancient Middle East, glyptic embodiments give us a proof of this, especially through those cylinder-seals produced in Mesopotamian cultures, from which is derived the whole of the Western world's sigillography (from the Roman-ring sealing to wax seal that, anachronistically, is still used today to certify some documents). Something similar happens with the objects studied by numismatics, as iconic models that, in turn, represent the abstract pattern of economic value artificially preset. Also, metal casting or clay materials developed during the Middle Ages are part of the area of these technical procedures involving the reproduction of objects (weapons, brooches, appliques, *sigillated terra*, small sculptures).

The diffusion of reproducible images restarted an upswing in the late Middle Ages. It happened thanks especially to the development of commercial routes, which allow trade and cultural exchange between cities. In this context, the widespread use of the techniques of engraving and printing of two-dimensional icons helped to increase the traffic of images. However, with regard to Western culture, the starting point must be placed in the woodcut process launched in Europe from the mid-fourteenth century onwards. Also, during the fifteenth century the printing of images started to involve copper plates engraved with sharp instruments. Johannes Gutenberg was the first to use movable type printing (1439), and he developed a technique for the mass production of printed works. These techniques gained ground, thanks to the expressive possibilities they deployed, and during the early sixteenth century caused the decline of woodcut engraving. Soon after, from the sixteenth century onwards, the technique of etching was generalized.

With the various techniques of copper engraving, the printing of reproduced images became an expressive form fully in the hands of artists, and it served as a nascent form of massive distribution. However, the runs allowed by the engraving were still small. By the last quarter of the eighteenth century, the woodcut was reborn in England from the hand of Thomas Bewick, who developed the technique of woodcut *a la testa*. In the late eighteenth century, the Prague-born German composer Alois Senelferden devised a procedure to play music that became known as lithography during the next century. Its freedom of execution as well as its great resistance during runs converted lithography into an ideal medium for artistic expression and a powerful tool for the reproduction of images. From 1816 onwards, it was possible to get lithographic prints in several colours. Only the unstoppable emergence of photography would gradually move lithography aside as a main technique for the reproduction of images.

3.2 From artisan paintings to printed images: the mechanization of image (re)production

This gradual development of the reproducibility of images was a first step in the dissemination of culture. However, the physical act of the reproduction of images was – until the early nineteenth century – an artisan work in which all traditional utensils and technical resources involved were manually operated. It would be as a result of the Industrial Revolution that the discovery of new machines and new forms of energy drove the reproduction of images to expand almost without limit, achieving a greater democratization of tools and a wider diffusion capacity.

The printing of engravings was usually done by wooden hand presses, even with the development of new forms of efficient energy. It was not until 1798 that Lord Stanhope devised the first iron hand press designed for print. However, it would be in 1811 that Friedrich Koenig, Thomas Bensley and Andreas Bauer would build the first steam-powered cylindrical press, giving a true quantum leap to the process, especially with regard to the circulation of newspapers. Progress in this field was successive (printing GP Minerva Gordon in 1850, Marinoni Rotary printing in 1863) and coexisted with the complex process of industrialization in which the mechanization of the various production processes permitted the massive production of an increasing number of objects; this would be the seed of the consumer society.

At this historic moment, photography made its breakthrough, and the techniques were those of the industrialized printing field in use since 1860, the year in which Thomas Bolton managed to photographically sensitize some wooden blocks used for woodblock printing.

3.3 The birth of photography: the image "photogenesis"

Photography revolutionized the world of visual communication as no other means of producing images had done before. According to Monaco, photography "marks an historical shift as important as the Gutenberg revolution" (Monaco 2009: 80). The photographic image had been surrounded by an aura that blended the real magic of an act qualified as quasi "natural" (the sunlight as causal factor) with the cognitive truth of an objective mimesis. However, every photograph is an intermediated act and therefore manipulable: factors such as framing, focus, lighting and, ultimately, the two-dimensional translation of space or the possibility of freezing time, converted this new art into a unique one. However, in its origins, the extraordinary resemblance between photography and photographed object was so dazzling that the first photographers tried to rhetorically emulate the paintings to overcome their poetic inferiority complex, giving origin to pictorial photography.

Like most scientific and technical achievements, the invention and implementation of photography was developed through a complex chain of events. The achieving of mirror images using the camera obscura had been known since the Renaissance, but it was not until 1816 that Niepce Nicéphone performed the first paper negative photography. Soon afterwards, in 1829, Louis J. Mande Daguerre invented the Daguerreotype, a relatively simple procedure that allowed one to impress very precise images on sensitized metal plates. In 1839, the French government bought the patent, and a book on the Daguerreotype was published and translated into eight languages, reaching as many as thirty editions. Thus, photography was born and spread throughout the world in a fulminant way.

From this moment on, several technical advances succeded in an uninterrupted way. In 1835, William H. Fox Talbot made small pictures on sensitized paper through tiny pinhole cameras. Four years later, coinciding with Daguerre, Talbot published his investigations, which culminated in 1841 with the Calotype, patented that very year. Although less sharp than the Daguerreotype, sensitized paper using the Calotype allowed one to obtain negative and positive images consecutively which, coupled to the lower weight of the device, cheapened the cost of the process. So, photography became remarkably popular during the subsequent decade. In 1849, continuing the work that emulated the bifocal vision of human ocular apparatus, David Brewster tuned equipment for capturing and viewing stereoscopic photography, which consumed the illusion of three-dimensional vision. His success would be so complete that, in 1863, the London Stereoscopic Company managed to sell over a million copies of stereoscopic plates. In 1850, Scott Archs unveiled the sensitization method using wet collodion, which combined Calotype's economy with Daguerreotype's sharpness, and allowed the possibility of a large number of copies. In 1871, Richard L. Maddox invented the gelatin emulsion, which in 1878 allowed Charles Harper Bennet to perform the first picture "snapshot", able to capture a moving image photographically. The discovery of gelatin on celluloid in 1888 opened the doors to the manufacture of portable cameras. George Eastman designed the famous Kodak that year and, in 1895, he presented the Kodak Pocket, so that photography came into the home. Thousands of citizens became amateur photographers and began to build the collective memory of portraits of everyday life.

Obviously, the idea of printing photographs in the periodical press was an immediate goal. Already in 1824, Nicéphore Niepce had made photogravures on metal plates that could be printed. We also mentioned Bolton procedures for recording woodcut blocks thanks to photography. In 1867, Alphonse Louis Poitevin got standardized procedures of photolithography. The major step was in 1880, when *The New York Daily Graphic* managed to publish photographs with halftone, made ready for printing by using grids to translate their tones into points, known as the basis for photogravure. Ten years later the basis for trichromatic was consolidated, and that allowed the printing of colour photography.

At this point, painting and photography were destined to collide. At the bottom of the debates was the hidden fear of the figurative dethronement propelled by the popularity of photography. But there was also the convergence between the demystifying power of the photographic "objectivity" and the poetic realism, between the new look that photography enshrined in the collective cultural consciousness and most of the concerns that guided the look of the Impressionist painters.

3.4 The parallel development of communications

The technological evolution of recording media (images and sounds) runs parallel with those of transportation. The nineteenth century is the time when the globe began to shrink thanks to the development of communication routes by land, sea and air: the first railway line (1825), the introduction of the steam engine to ships (1858), the invention of the automobile (1880) or the first long distance flight (1908). The ability to travel exponentially increased the traffic of images – people, places and events.

Neverthless, communication was not only flowing via physical presence, but also through the distance. To the primitive heliographic communication systems or mobile signals were summed the first long distance electric telegraph line (1839), the telephone (1876), the wireless telegraph (1894) and the radio (1896). Since 1863, images could be transmitted by telegraph, and in 1902, phototelegraphy was in place. To this we must add – in the field of urban development – gas (from 1830) and electric lighting (from 1879, the year Edison invented the light bulb), which ended up breaking the thick barrier separating day and night in the big cities.

All previous innovations provided the basis for the gradual eclosion of the true global village, in which any citizen could gain access, physically or by images or sounds, to any place in the world, as well as come into contact, similarly, with novelties or curiosities of all kinds. Thus the universal exhibitions arose from 1851 onwards and toured the capitals of Europe and the United States during the nineteenth century and first half of the twentieth. Culture, discoveries, machines, objects or pure extravagances originated anywhere, could be known immediately by the rest of the world.

3.5 The birth of cinema

This tangible designation of space and time, of real and imaginary movements, would have its paradigmatic counterpart in a new revolutionary contribution to the history of media and to the consolidation of popular culture: the birth of the kinetic image or cinematograph. As mentioned in the previous section, "[t]he existence of cinema is premised upon the existence of certain technologies, most of them developed during the course of industrialisation in Europe and America in the nineteenth century" (Neale 1985: 1).

In fact, the antecedents of image projection are even older. In the late seventeenth century Father Kircher invented the Magic Lantern, a device which projected still images by multiplying the size of transparent slides. This invention became widespread in the eighteenth century as a living curiosity and, in the nineteenth, as a toy. Moreover, the concern about how to analyse and reproduce the image kineticism captured by the human eye or by transient persistence of the image on the retina was old. Between 1831 and 1833, Joseph Plateau launched the Phenakistiscope and the Zoetrope, devices capable of producing the optical sensation of a certain movement, accumulating in the retina a series of images with small consecutive phases of a movement. The discovery of the photographic snapshot in 1872 allowed Eadweard Muybridge to perform photographs breaking down the phases of movement. The invention of the "photographic gun" allowed Étienne Jules Marey to incorporate the Zoetrope into the Muybridge studies. In 1888, Charles-Émile Reynaud succeeded in projecting the Zoetrope on a wall.

According to Manovich, "[t]hese earlier techniques shared a number of common characteristics. First, they all relied on hand-painted or hand-drawn images" and, on top of that "[n]ot only were the images created manually, they were also manually animated". In addition,

> Cinema's most immediate predecessors share something else. As the nineteenth-century obsession with movement intensified, devices which could animate more than just a few images became increasingly popular. All of them – the Zootrope, the Phonoscope, the Tachyscope, the Kinetoscope – were based on loops, sequences of images featuring complete actions which can be played repeatedly ... (Manovich 2010: 247).

All these factors converged in achieving different procedures and equipment which, almost simultaneously, lead to the invention of cinema. As this author keeps on explaining,

> It was not until the last decade of the nineteenth century that the automatic generation of images and their automatic projection were finally combined. A mechanical eye became coupled with a mechanical heart; photography met the motor. As a result, cinema – a very particular regime of the visible – was born (Manovich 2010: 247).

In 1894, Thomas A. Edison – who had already invented the celluloid film, although Eastman patented his own in 1889 – created the Kinetoscope, a device that permitted the watching of movies through a single viewfinder. Something similar was conceived by William Friese-Greenhe in England or Max Skladanovsky in Germany. However, the final consecration of cinema took place on December 28th, 1895, when the Lumiere brothers organized the first public screening of their cinematograph at the Grand Café in Paris. The lush and dense history of cinema as a mass spectacle and popular entertainment had only started. Its successive technical developments would be produced with overwhelming speed. As Brihuega explains,

> As a large blanket spread over the whole civilized world, cinema was revealed as the backbone of a new mass visual culture, most powerful and influent in its sociological scope than painting, literature, theater or opera. Simultaneously art, entertainment and industry, cinema was able to gather on the screen all the possibilities and levels of cultural behavior: from utter banality to the deeper poetic, from the most harmless amusement to the stronger ideological action (Brihuega 1997: 421; translated by author).

And James Monaco concludes:

> Recording technology offered the opportunity of capturing representation of sounds, images, and events and transmitting them directly to the observer without the necessary interposition of the artist's talents. A new channel of communication had been opened, equal in importance to written language (Monaco 2009: 81).

From 1895 onwards, the technical innovations accompanied cinema along the twentieh century. The advent of sound together with colour and the improvement in camera lenses, camera models, film stock and projection systems transformed cinema into the more plausible and espectacular recreation of life (see Monaco 2009: 75–167). To mention with more detail some of these milestones, in 1922 Lee DeForrest developed a method for recording sound on the edge of a film strip. In 1927, Warner Brothers presented the first spoken movie (*talkies*), *The Jazz Singer*, and patented the Vitaphone (recording sound on discs). In 1935, Technicolor introduced a three-colour process in the film *Becky Sharp*. In 1952, the widescreen formats Cinerama and 3-D were introduced. One year later, in 1953, Cinemascope and the first Dolby sound format arrived at theatres.

Only the advent of television, which began to spread from the Second World War, was able to compete with the power of the big screen. And not only that, but it also facilitated the birth and consolidation of home entertainment.

3.6 From analogue to digital

The development of television and other electronic media (videocassettes) runs in parallell way with the invention and development of computers and digital technology. In the 1950s, IBM built the first machines, programmed by feeding in decks of punched cards. In the 1960s, it was proved that cathode ray tube (CRT) screens provided a more efficient link betewen electronic images and computers. In 1961, filmmakers James and John Whitney used mainframes to produce abstract images for the first time. In the late 1960s, the first music synthesizer (Moog) appeared and ten years later (1971) Lexicon offered the first digital audiotape recorder. By the late 1970s CBS had developed its own machine for digital editing videotape.

Few years before, in 1968, Douglas Engelbart designed an effective user interface. In the early 1970s, Xerox combined the graphics in CRT screens with a remote pointing device called "mouse". This point was crucial since it represented "the invention of a coherent visual and physical metaphor for a complex and subtle interaction between humans and their first true intellectual tools" (Monaco 2009: 585). From this point onwards, interface design experienced a rapid development. As Monaco concludes,

> It isn't often that a new basic and universal language is invented. The twentieth century saw two: first film, then the graphical interface. As new systems of communication, it was only a matter of time before the languages of film and computers merged (Monaco 2009: 581–582).

In the 1970s, the first word processors were developed and computers were ready. Nevertheless, it was in the mid 1980s when the introduction of the Apple's Macintosh computer marked the birth of multimedia. Those first Macs included word-processing as well as painting programmes software, much more user-friendly than the dot-matrix character-based screens of that time.

From this point onwards, the development of computer generated images (CGI) was a question of time. CGI origins date as far back as 1968, when a group of Russian mathematicians and physicists led by Nikolay Konstantinov designed a programme, based on mathematical models, for moving a cat across a screen. This programme ran on a specialized computer named BESM-4, which printed hundreds of frames to be later transformed into usable film material. During the next decade, CGI experienced a fast growth. In 1971, animator Peter Foldes created the first CGI animated short film drawn on a data tablet, using the world's first key frame animation software, invented by Nestor Burtnyk and Marceli Wein. Two years later, the first CGI 2-D animated effect was realized by the company Triple-I in the movie *Westworld* (showing the point of view of one character). The sequel, *Futureworld* (1976) incorporated the first 3-D CGI effect.

The next milestone was the first *Star Wars* movie (*A New Hope* 1977), which included top edge CGI effects, never seen before. George Lucas's special effects company, Industrial Light and Magic (ILM) became one of the pioneers in the field of visual digital effects. Some other movies completed the CGI landmarks at the end of the 1970s and during the 1980s: *Superman* (1978) offered the first CGI title sequence; *Alien* (1979) and *Black Hole* (1979) pushed the boundaries of CGI even further by conceiving 3D wireframe rasters that created more detailed CGI effects; *Star Trek 2: The Wrath of Khan* (1982) showed the first CGI human character (with the first use of 3D shaded CGI) and the invention of the Genesis effect for creating alien-like landscapes; *Tron* (1982) made use of 15 minutes of fully rendered CGI footage including the famous light cycle sequence; *Young Sherlock Holmes* (1985) was the first movie to include a completely generated character interacting with a living one (the stained glass warrior); in *The Abyss* (1988), ILM moved forward the CGI technique creating a realistic worm-like subsea pseudopod, enhanced in *Terminator 2: Judgement Day* (1992), with the liquid T-1000; earlier, *Indiana Jones: The Last Crusade* (1989) contained the the first digital composite.

The 1990s were also exponents of the rapid evolution of CGI effects, with more companies competing to reign in this realm. *Jurassic Park* (1993) offered the first eye dropping photo realistic CGI creatures. Two years later Pixar's *Toy Story* (1995) became first fully CGI animated movie. Shortly after, worldwide audiences were shocked by the realism of *Titanic* (1997), whose visual effects were created by James Cameron's Digital Domain. Also *The Matrix* (1999) was the first to use the so-called 'bullet time' effect. It is also worthy of mention that as the movie industry matured in the use of CGI, the game industry started to flourish on its own. In the so-called fifth generation gaming consoles, fully 3-D playable games became more and more popular. With the release of the Playstation (1994) and the Nintendo 64 (1996), games got their first fully 3-D supported gaming platforms. Titles as *Super Mario 64, Doom, Final Fantasy* and *Crash Bandicoot* set the standard for many CGI games that followed.

As we entered into the twenty-first century, CGI techniques expanded almost endlessly, becoming more and more mixed with real/live film footage. *The Lord of the Rings* trilogy (2001–2003) through the visual effects company Weta Digital made use of artificial intelligence for its digitally created characters (a software called 'Massive') and created the first fully developed digital character (Gollum), able to offer a convincing performance interacting with live characters, through motion capture techniques. *The Matrix Reloaded* (2003) was the first to use Universal Capture to capture more frames in an image. One year later *The Polar Express* (2004) was the first animated film to use motion capture on all its movie characters. Peter Jackson's Weta Digital would became the top digital effects company creating CGI characters as shown in movies like *King Kong* (2005) and *Avatar* (2009).

This progressive transformation of cinema from celluloid to pixels was not a traumatic process but a natural and smooth one. In Manovich's words:

> Cinema not only plays a special role in the history of the computer. Since the late nineteenth century, cinema was also preparing us for digital media in a more direct way. It worked to make familiar such 'digital' concepts as sampling, random access, or a database – in order to allow us to swallow the digital revolution as painlessly as possible. Gradually, cinema taught us to accept the manipulation of time and space, the arbitrary coding of the visible, the mechanization of vision, and the reduction of reality to a moving image as a given. As a result, today the conceptual shock of the digital revolution is not experienced as a real shock – because we were ready for it for a long time (Manovich 2005: 28–29).

3.7 Towards the complete digitisation of cinema

In a parallell way, the two vital components of film – images and sounds – were going to experience the same profound transformation (or reinvention), from analogue technological standards to digital ones.

Digital sound was introduced at the beginning of the 1990s, with the release of *Dick Tracy* (1990), *The Doors* (1991) and *Terminator 2* (1992). These innovations obliged theatrical sound companies like Dolby and others to develop their own digital systems. In 1993, Dolby introduced the Digital Theater Systems (DTS) with the release of *Jurassic Park*. That very year Sony also presented its own system called Sony Dinamic Digital Sound (SDDS). Both formats became the standard for theatrical releases.

The next step was digital projection. In 1999, *Star Wars Episode I: The Phantom Menace* became the first movie digitally projected in theatres. Soon after, many other Hollywood blockbusters followed that path. As had happened with sound systems, many hardware manufacturers began the battle to get the right cinema projector. Texas Instruments Digital Light Processing (DLP) was the initial winner, followed by Christie, Barco, NEC and some other firms. During the last decade, digital projection has become the standard in most of the world, improving the image quality from 2k to 4k.

Even prior to projection, the process of recording images experienced an early transformation from celluloid to pixels. It was in 1998, when Sony introduced the HDCAM recorders and the first high definition (HD) video cameras (1920 × 1080 pixels) based on CCD technology, that the term "digital cinematography" was coined. Three years later, *Once Upon a Time in Mexico* (2001) became the first movie shot in 24 frames-per-second HD digital video – partially developed by George Lucas using a Sony HDW-F900 camera. In 2002, *Star Wars Episode II: Attack of the Clones* was also shot using a Sony HDW-F900 camera. Today, most of the film camera manufacturers (Sony, Vision Research, Arri, Panavision and Red) offer a variety of choices for shooting high-definition video at the highest standards.

As a last milestone, in 2012 *The Hobbit: An Unexpected Journey* became the first movie shot and projected in High Frame Rate (HFR) 3-D (48 frames-per-second), doubling the traditional 24 frames and getting an unprecedented image quality.

As an epilogue for this section, it is also worthwhile mentioning, as John Belton does, that "[p]erhaps the most important concern about the digitization of cinema is its implication for film preservation" (Belton 2010: 292). Apparently, polyester safety film is the ideal medium for long-term storage of motion pictures (lasting around one hundred years). Digital formats tend to last from five to ten years. Therefore, "[g]iven the rapid obsolescence of various past digital formats, it is not clear that digital information can be retrieved in the future" (Belton 2010: 292).

4 Cinema and technology: theoretical issues, critical views

As it can be deduced from the previous sections, the relationship between technology and the cinematic art is as inherent as it is manifold. In this sense, it is not surprising that it has provoked a consequent amount of scholarly literature. As Steve Neale has stated,

> [U]nderstanding the place of technology in the cinema requires not only a knowledge of science and of the evolution of machines. It is a question also of aesthetics, psychology, ideology and economics; of a set of conditions, effects and context which affect, and are in turn affected by, the technologies employed by the cinema (Neale 1985; 2).

According to this view, the present section tries to condense some key issues around the central question of this chapter. As a comprehensive and expositive text, this is not the place to distil a refined synthesis of the different contributions. I will just present them as headlines, followed by some excerpts taken from a selection of the bibliographical references previously mentioned, trying to offer contrasting views about these topics.

4.1 Art, film and technology: the reproducible artwork

According to James Monaco, in the lines quoted above, "(e)very art is shaped not only by the politics, philosophy, and economics of society, but also by its technology" (Monaco 2009: 76). Nevertheless, contrary to what could be expected,

> The relationship isn't always clear: sometimes technological development leads to a change in the aesthetic system of the art; sometimes aesthetic requirements call for a new technology; often the development of the technology itself is the result of a combination of ideological and economic factors. But until artistic impulses can be expressed throughout some kind of technology, there is no artefact (Monaco 2009: 76).

In this regard, the development of the so-called recording arts mark a singular milestone in the history of this relationship, as this author also points out:

> The great artistic contribution of the industrial age, the recording arts – film, sound recording, and photography – are inherently dependent on a complex, ingenious, and ever more sophisticated technology. No one can ever hope to comprehend fully the way their effects are accomplished without a basic understanding of the technology that makes them possible, as well as its underlying science (Monaco 2009: 77).

This statement leads us to the seminal essay written by Walter Benjamin in 1936 and entitled *The Work of Art in the Age of Mechanical Reproduction* (reprinted in Utterson 2005: 105–126). Although every work of art has been reproducible in some way or another since primitive times, it lacked the possibility of being present in various times and spaces. Only the invention of the first systems of mechanical reproduction allowed the artwork to be endlessly reproduced and become object of mass consumption, losing along the way some of its magical aura. In Benjamin's words,

> With the different methods of technical reproduction of a work of art, its fitness for exhibition increased to such an extent that the quantitative shift between its two poles turned into a qualitative transformation of its nature. This is comparable to the situation of the work of art in the prehistoric times when, by the absolute emphasis on its cult value, it was, first and foremost, an instrument of magic. Only later did it come to be recognized as work of art. In the same way today, by the absolute emphasis on its exhibition value, the work of art becomes a creation with entirely new functions, among which the one we are conscious of, the artistic function, later may be recognized as incidental. This much is certain: today photography and film are most serviceable exemplifications of this new function (Benjamin 2005: 110).

Despite this observation, cinema has placed itself in the realm of the arts, influencing them as no other newcomer before, as Monaco explains:

> The 'art' of film, then, bridges the older arts rather than fitting snugly into the preexisting spectrum ... But as this revolutionary mode of discourse was applied, in turn, to each of the older arts, it took a life on its own ... Indeed, for the past hundred years, the history of the arts is tightly bound up with the challenge of film. As the recording arts drew freely from their predecessors, so painting, music, the novel, stage drama – even architecture – had to redefine themselves in terms of the new artistic language of film (Monaco 2009: 44–45).

As a consequence, art theorists and other scholars were attracted by this new medium, and a new field of knowledge was born. In Marc Fusternau's words,

> Theories of film began as expressions of wonder. As soon as moving, photographic images were projected on to screens, critics, writers, poets, philosophers, artists, and even filmmakers themselves began describing the new medium, speculating about film's nature, debating its various effects, and arguing for its value and significance. Almost all of the early observers agreed that they were witnessing the advent of something new and unprecedented, and they sought immediately to provide some sort of account of it (Fusternau 2010: 1).

4.2 The technological nature of cinema: benefits and limits

Among the different arts, film presents a closer link to technology. It has, in fact, been defined as the art of reality, or the art of the plausible reality. In this sense, as Steve Neale explains, "[t]he history of technology in the cinema is seen as the history of an even greater approximation to reality, with sound, colour, widescreen and the rest adding to the basic visual ontology" (Neale 1985: 160).

Theoretically as well as critically, this close relationship between technology and film has been studied from different angles, giving birth to thought-provoking insights. Andrew Utterson summarizes them quite accurately. After assessing that "[t]hroughout the period the history of cinema spans, the world has witnessed a mass proliferation of technologies" (Utterson 2005: 1), this author underlines how "[d]uring this sustained period of expansion, technologies have taken on reconfigured meaning with regard to human experience, impacting on the fundamental processes by which we make sense of, and interact with, the world around us" (Utterson 2005: 1). Then, he explains with further detail:

> In an archetypal or traditional scenario, the machines of cinema connect the processes of representation (the role of the camera in capturing the world around us and the subsequent manipulation, processing and reshaping of this representation through editing and other practices), reproduction (the means of duplication and distribution) and exhibition (the dynamic within the space of the cinema itself, between the spectator and the parade of light emitted by the projector). Theorists have endeavoured to extrapolate and explain the precise nature of these technologies ... while accounting for the forces that shape and determine their form and function ... (Utterson 2005: 1).

Nevertheless, this author also gives the reminder that as "an equally significant aspect of cinema's technological history, and how this history might be theorized" the presence of "a range of social, cultural, political and other contexts [is] likewise steeped in technology" (Utterson 2005: 1). Therefore, film theorists have been obliged to expand their field of knowledge, as Utterson points out:

> In an era replete with ramifications and reverberations that surround the more familiar presence of physical devices, theorists have broadened their scope, shifting their attention from the machines of cinema per se to cinema's engagement with technologies other than ts own, and with technological contexts beyond the purely cinematic (Utterson 2005: 1).

And he goes further to assess:

> Like all arts or cultural practices, cinema does not exist in vacuum. From its representations, the images we see on screen ..., to the consequences, both foreseen and unforeseen, of the actual uses of cinema's machines ..., the moving image is elemental to a much broader evolution that has seen the influx and influence of technological process across many aspects of our lives. Crucially, it is only in the relation of cinema to this broader context that we can begin to comprehend the relevance of its own technological lineage (Utterson 2005: 2).

Finally, by way of conclusion, Utterson draws a challenging scenario, as every new technology produces a double-effect – positive and negative – on the previous one.

> In certain respects, the technological history of cinema appears to be coming full circle. Just as the early machines of cinema refined the existing technologies and aesthetics of photography and other forms, giving rise to a medium predicated on mechanized movement, cinema is now undergoing an equivalent period of transition, as its technologies and expressions are reconfigured in relation to those of the digital computer. Optimists point to ways in which cinema has appropriated these technologies in its own practices. Pessimists, by contrast, suggest the computer has absorbed cinema as one constituent element within a universal hyper-

media. Where theorists once celebrated the birth of cinema, they now point to its potential passing" (Utterson 2005: 10).

By way of proof, several authors mentioned in the above literature review address this issue (see Belton 2010; Elsaesser 1998; Friedberg 2010; Gunning 2010; Hoffmann 1998).

4.3 The future of cinema: from kino-eye to kino-brush

Utterson's previous words lead us towards this final theoretical and critical issue. Is cinema going to disappear as devoured by digitisation or will it just be transformed into a new medium? David Rodowick has explored "the philosophical consequences of the disappearance of a photographic ontology for the art of film and the future of cinema studies" (Rodowick 2007: vii). In particular, he states:

> In the current climate of rapid technological change, "film" as photographic medium is disappearing as every element of cinema production is replaced by digital technologies. Consequently, the young field of cinema studies is undergoing a period of self-examination concerning the persistence of its object, its relation to other time-based spatial media, and its relation to the study of contemporary visual culture. The film industry also roils with debate concerning the aesthetic and economic impact of digital technologies, and what the disappearance of film will mean for the art of movies in the twenty-first century (Rodowick 2007: vii).

Together with Rodowick, Lev Manovich offers very interesting thoughts about this change of pattern, from analogue to digital.

> In the twentieth century, cinema has played two roles at once. As a media technology, cinema's role was to capture and to store visible reality. The difficulty of modifying images once they were recorded was exactly what gave cinema its value as a document, assuring its authenticity... [Nevertheless] [t]he mutability of digital data impairs the value of cinema recordings as documents of reality. In retrospect, we can see that twentieth century cinema's regime of visual realism, the result of automatically recording visual reality, was only an exception, an isolated accident in the history of visual representation which has always involved, and now again involves the manual construction of images. Cinema becomes a particular branch of painting – painting in time. No longer a kino-eye, but a kino-brush (Manovich 2010: 252).

Therefore, as this author underlines, "digital media redefines the very identity of cinema" (Manovich 2010: 245). If cinema had been defined as "the art of the index" (Manovich 2010: 245) – in reference to the reality of the object represented – digital technologies break that indexicality, creating parallel worlds and characters (Gunning 2010). As Manovich (2010) continues to explain,

> The privileged role played by the manual construction of images in digital cinema is one example of a larger trend: the return of pre-cinematic moving images techniques. Marginalized by the twentieth century institution of live action narrative cinema which relegated them to the realms of animation and special effects, these techniques reemerge as the foundation of

digital filmmaking. What was supplemental to cinema becomes its norm; what was at its boundaries comes into the center. Digital media returns to us the repressed of the cinema.

[Therefore] the directions which were closed off at the turn of the century when cinema came to dominate the modern moving image culture are now again beginning to be explored. Moving image culture is being redefined once again; the cinematic realism is being displaced from being its dominant mode to become only one option among many (Manovich 2010: 253).

5 Epilogue

In the beginning, there was the sign. The sign was spoken and sung. Then it was written, first as picture, then as word. Eventually the sign was printed. The ingenious coding system that was writing allowed ideas and feelings, descriptions and observations to be captured and preserved. The technology of printing liberated these written records from isolated libraries, allowing them to be communicated to thousands, then millions ...

As the scientific revolution took hold in the nineteenth century, we discovered methods to capture images and sounds technologically. Photography, then records and films, reproduced reality without the intervention of words ...

But now we find ourselves on the verge of a new phase in the history of media. The languages we invented to represent reality are merging. Film is no longer separate from print. Books can include movies; movies, books. We called this synthesis 'multimedia' or 'new media' ... (Monaco 2009: 578).

These words by Monaco – written in an intended biblical tone – may very well summarize what this chapter has tried to accomplish: to offer a bibliographical, historical and theoretical synthesis of the deep connections between art and reproductibility, between cinema and technology. Nevertheless, our point of arrival is in fact a point of departure. Digitisation is changing our culture in a way never seen before, and in the time to come the marriage between cinema and technology will give birth to more innovations and technical advances.

Note: An interesting chronology of technical innovations in film can be found in Monaco, 2009: 640–676.

Acknowledgements

I would like to thank Jorge Latorre, a scholar expert in Visual Culture and History of Photography, for providing me with helpful comments and inputs when preparing this text. Equally, I am deeply grateful to Ike Obiaya for revising and correcting my written English.

References

Belton, John. 2010. Digital Cinema: A False Revolution. In Marc Furstenau (ed.), *The Film Theory Reader: Debates and Arguments*, 282–294. London; New York: Routledge.

Benjamin, Walter. 2005. The Work of Art in the Age of Mechanical Reproduction. In Andrew Utterson (ed.), *Technology and Culture: the Film Reader*, 105–126. London; New York Routledge.

Bennett, Bruce, Marc Furstenau & Adrian Mackenzie (eds.). 2008. *Cinema and Technology: Cultures, Theories, Practices*. Basingstoke [England]; New York Palgrave Macmillan.

Boggs, Joseph M. & Dennis W. Petrie. 2012. *The Art of Watching Films*. 8th Ed. New York: McGraw-Hill.

Bordwell, David, Janet Staiger & Janet Thompson. 1985. *The Classical Hollywood Cinema: Film Style and Mode of Production to 1960*. New York: Columbia University Press.

Brihuega, Jaime. 1997. Origen y desarrollo de la cultura icónica de masas. In Juan Antonio Ramírez & Adolfo Gómez (eds.), *Historia del Arte, vol. 4: El mundo contemporáneo*. 415–431. Madrid: Alianza Editorial.

Brookey, Robert Alan. 2010. *Hollywood Gamers: Digital Convergence in the Film and Video Game Industries*. Bloomington, IN: Indiana University Press.

Cameron, James R. (ed.). 1959. *Sound Motion Pictures*. New York: Cameron Publishing Company; Manhattan Beach.

Ceram, C. W. 1965. *Archaeology of the Cinema*. London: Thames & Hudson.

Coe, Brian. 1976. *The Birth of Photography: The Story of the Formative Years 1800–1900* London: Ash & Grant.

Coe, Brian. 1981. *The History of Movie Photography* Westfield (NJ): Eastview Editions.

Coke, Van Deren. 1964. *The Painter and the Photograph: from Delacroix to Warhol*. Albuquerque: University of New Mexico Press.

Dalle Vacche, Angela (ed.). 1996. *Cinema and Painting: How Art is Used in Film*. Austin, TX: University of Texas Press.

Dalle Vacche, Angela (ed.). 2003. *The Visual Turn: Classical Film Theory and Art History*. New Brunswick, NJ; London: Rutgers University Press.

Dekom, Peter J. & Peter Sealey. 2003. *Not on My Watch ... Hollywood vs. the Future*. Beverly Hills: New Millennium Press.

Elsaesser, Thomas. 1998. Digital Cinema: Delivery, Event, Time. In Thomas Elsaesser & Kay Hoffmann, (eds.), *Cinema Futures: Cain, Abel or Cable? The Screen Arts in the Digital Age*, 201–222. Amsterdam: Amsterdam University Press.

Elsaesser, Thomas & Kay Hoffmann (eds.). 1998. *Cinema Futures: Cain, Able or Cable? The Screen Arts in the Digital Age*. Amsterdam: Amsterdam University Press.

Fielding, Ramond (ed.). 1967. *A Technological History of Motion Pictures and Television: An Anthology from the Pages of the Journal of the Society of Motion Picture and Television Engineers*. Berkeley; Los Angeles: University of California Press.

Friedberg, Anne. 2010. The End of Cinema: Multimedia and Technological Change. In Marc Furstenau (ed.), *The Film Theory Reader: Debates and Arguments*, 270–281. London; New York: Routledge.

Frizot, Michel (ed.). 1998. *A New History of Photography*. Köln: Könemann.

Furstenau, Marc (ed.). 2010. *The Film Theory Reader: Debates and Arguments*. London: Routledge.

Geirland, John & Eva Sonesh-Kedar. 1999. *Digital Babylon: How the Geeks, the Suits and the Ponytails fought to Bring Hollywood to the Internet*. New York: Arcade Publishing.

Giannetti, Louis. 2014. *Understanding Movies*. 13th Ed. New York; London: Pearson.

Gunning, Tom. 2010. Moving Away from the Index: Cinema and the Impression of Reality. In Marc Furstenau (ed.), *The Film Theory Reader: Debates and Arguments*, 255–269. London; New York: Routledge.

Hoffmann, Kay. 1998. Electronic Cinema: On the Way to the Digital. In Thomas Elsaesser & Kay Hoffmann (eds.), *Cinema Futures: Cain, Abel or Cable? The Screen Arts in the Digital Age*, 241–250. Amsterdam: Amsterdam University Press.

Kerins, Mark. 2011. *Beyond Dolby (Stereo): Cinema in the Digital Sound Age*. Bloomington, IN: Indiana University Press.

Lasica, Joseph Daniel. 2005. *Darknet: Hollywood's War Against the Digital Generation*. Hoboken, New Jersey: Wiley & Sons.

Manchel, Frank (ed.). 1990. *Film Study: An Analytical Bibliography* (Vol. 3). London; Toronto: Associated University Presses.

Manovich, Lev. 2005. Cinema and Digital Media. In Andrew Utterson (ed.), *Technology and Culture: The Film Reader*, 27–30. Oxon; New York: Routledge; Taylor & Francis.

Manovich, Lev. 2010. Digital Cinema and the History of a Moving Image. In Marc Furstenau (ed.), *The Film Theory Reader: Debates and Arguments*, 245–254. London; New York: Routledge.

McKernan, Brian. 2005. *Digital Cinema: The Revolution in Cinematography, Postproduction, and Distribution*. New York: McGraw-Hill.

Monaco, James. 2009. *How to Read a Film: Movies, Media and Beyond*. 4th Ed. Oxford, New York: Oxford University Press.

Musser, Charles. 1990. *The Emergence of Cinema: The American Screen to 1907* Berkeley; Los Angeles: University of California Press.

Neale, Steven. 1985. *Cinema and Technology*. Bloomington: Indiana University Press.

Prince, Stephen. 2012. *Digital Visual Effects in Cinema: The Seduction of Reality*. New Brunswick, NJ; London: Rutgers University Press.

Rodowick, David N. 2007. *The Virtual Life of Film*. Cambridge, MA; London: Harvard University Press.

Rombes, Nicholas. 2009. *Cinema in the Digital Age*. London: Wallflower.

Roosens, Laurent & Luc Salu. 1989. *History of Photography: A Bibliography of Books*. London: Mansell.

Rosenblum, Naomi. 1997. *A World History of Photography*. New York; London: Abbeville Presst.

Ryan, Roderick T. (ed.). 1977. *A History of Motion Picture Colour*. London; New York: Focal Press.

Swartz, Charles S. 2005. *Digital Cinema: The Revolution in Cinematography, Postproduction, and Distribution*. Amsterdam: Focal Press.

Thomas, David B. (ed.). 1964. *The Origins of the Motion Picture*. London: HMSO.

Tosi, Virgilio. 2006. *Cinema Before Cinema: The Origins of Scientific Cinematography*. London: Wallflower Press.

Toulet, Emmanuelle. 1995. *Discoveries: Birth of the Motion Picture*. New York: Harry N. Abrams.

Tryon, Chuck. 2009. *Reinventing Cinema: Movies in the Age of Media Convergence*. New Brunswick (New Jersey): Rutgers University Press.

Tryon, Chuck. 2013. *On-Demand Culture: Digital Delivery and the Future of Movies*. New Brunswick (New Jersey): Rutgers University Press.

Utterson, Andrew (ed.). 2005. *Technology and Culture: the Film Reader* London; New York Routledge.

Wright, Steve. 2010. *Digital Compositing for Film and Video*. Amsterdam: Focal Press.

Wysotsky, Michael Z. (ed.). 1971. *Wide Screen and Stereophonic Sound*. London; New York: Focal Press.

Zone, Ray. 2012. *3-D Revolution: The History of Modern Stereoscopic Cinema*. 3rd Ed. Lexington, KY: University Press of Kentucky.

Tom McCourt

4 Recorded music

Abstract: Recordings fix performances in space and time, enabling sound to be bought and sold as a commodity. The history of recording can be separated into acoustic, electric and digital eras; however, each of these periods have common characteristics. First, a shifting oligopoly of record companies has controlled this process. Second, each era claimed to more accurately capture sound through greater technological intervention. Third, changes in recording and distribution have repurposed and decentralized music, affecting its creation and reception.

Keywords: acoustic, analog, copyright, digital, file sharing, music, realism, recording, sampling

> *"Record listening is a séance in which we get to choose our ghosts."*
> Evan Eisenberg, *The Recording Angel* (2005: 46)

A recording captures a unique event, a performance, and separates the sound from its source, disseminating it as a fixed, unchanging form across space and time. The history of recorded music is complex and contradictory, filled with unanticipated and unintended consequences; however, we may usefully follow a few common threads. First, recording made it possible for sound to be bought and sold. A shifting oligopoly of record companies, with stakes in both hardware (technology) and software (the rights to the recordings themselves), has controlled this process. Lately it has developed new methods of recording, storage and playback that may, paradoxically, compromise rather than consolidate that control.

Second, we may trace the consequences of what Sterne (2003: 4) calls "the dream of verisimilitude", in which each technological advance will capture "accurate" sound more effectively (and thereby sell new hardware and software). As Morton (2000: 177) notes, "[T]he drive to achieve 'fidelity' in recording involved a clash of cultures, and the combination of science and aesthetics pulled recording technology in different ways." While the goal may be transparency between source and receiver, the means is often ever-greater technological intervention. Our standards of realism change and contradict one another; as Eisenberg (2005: 92) states, while live recording "sometimes conveys a real sense of occasion ... Aggressive mixing and overdubbing, especially in rock, can give a sense of conscious intelligence and so of life."

Third, we may usefully examine how the repurposing and decentralization of music made possible by recording have affected both its creation and reception. Frith (1986: 272) claims that successful innovations have decentralized music production and consumption. Morton (2000: 178) elaborates: "users and consumers of sound recording consistently broke the rules set for them by inventors, engineers, managers, policy makers and corporations." When a user can program music at will, its use may have nothing to do with its creator's aesthetic intentions; it is "less about an artist's self-expression than a customer's desire for self-reflection" (Goldberg 2000: 6). Today, more recorded music is released each year than there is time to listen to it. Music is the ambience of modern life; its ubiquity renders it nearly invisible. Recordings recently have shed their materiality altogether, yet our sense of ownership has grown as what had been public (the live performance that recording makes unnecessary) has become increasingly private. Recordings are endlessly recycled, evoking nostalgia and influencing new forms. As Chanan (1995: 22) notes, "the old hierarchies of aesthetic taste and judgment may have broken down but music continues to breathe and to live according to its own immanent criteria."

1 The acoustic era

1.1 Edison, Bell and Berliner

Thomas Edison, often credited as the "inventor" of audio recording, envisioned a device to record Morse Code from the telegraph and realized that with modification it could record the voice as well. The "phonautograph", or sound writer, created by Leon Scott in 1855, used a stylus attached to a diaphragm to trace sound waves onto a paper-covered cylinder. Edison developed this principle to inscribe the sound itself onto a cylinder. Unlike the telephone, his device used no electricity: a cylinder covered in tin foil was turned by a hand crank. The user would shout into a funnel connected to a diaphragm that would vibrate in response. The diaphragm was connected to a steel stylus that would indent the rotating cylinder's surface. For playback, the process was reversed; the stylus reproduced the sound wave inscribed on the cylinder's surface, and a faint recording of the voice could be heard through the funnel. Edison unveiled the "phonograph" at the offices of the Scientific American on December 6, 1877. Although it attracted widespread attention, the device had poor sound quality and no clear commercial use, and Edison shifted his attention to the incandescent electric light, which offered a greater potential market.

In 1880, Alexander Graham Bell took up the phonograph and made several improvements on Edison's design. He substituted wax for tinfoil to add durability and used a more accurate cutter to engrave rather than indent the signal onto the surface of the cylinder, increasing the quality of the recording. Bell's "graphophone" was patented in 1886. Like the telephone, the "graphophone" was initially intended for business use, recording telephone messages and taking dictation.

However, transcribing was difficult as the device lacked an efficient start/stop mechanism and the recordings lacked fidelity (Sterne 2003: 201). In 1887, Edison returned to his device, incorporating Bell's improvements and substituting an electric motor for the hand crank.

By 1889, with improvements in fidelity, the phonograph was being adapted for use in entertainment rather than business. A San Francisco entrepreneur, Louis Glass, patented a phonograph equipped with coin-operated listening tubes, in which listeners could hear a song for a fee – the progenitor of the jukebox. Like Edison's kinetoscopes, these machines were located in hotels, railway stations, arcades, cafes and saloons; they "began to push the phonograph in the direction of music" (Chanan 1995: 26). Bell's organization established the Columbia Phonograph Company, and by 1891, Columbia had issued a ten-page catalogue of recordings, featuring Sousa marches, comic monologues, and "artistic whistling" (Chanan 1995: 26). Edison also formed a record company and began supplying cylinders for these machines. Replicating cylinders was highly problematic, however; initially, every recording was an original, since only one cylinder could be recorded on a machine at a time.

A different approach to recording was developed by Emile Berliner, a German immigrant who patented his "gramophone" in 1887. Instead of cylinders, Berliner's device used a flat disc rotating on a horizontal plane, in which the stylus moved from side to side in a consistently deep groove, rather than in Edison's "hill and dale" vertical manner. In 1897, Berliner opened a recording studio in Philadelphia and began selling players and discs. As Morton (2000: 19) notes, Berliner's player was "simple, relatively inexpensive, and marketed to consumers only as a form of home entertainment – it did not appear in the form of a business machine." Berliner also devised a means of mass-duplication by chemically etching the recording onto a metal disc, then producing a reverse metal master that could be used to stamp copies onto shellac, a much harder material than the wax used by Edison's cylinders. Yet the etching and replicating process was quite complicated, making records virtually impossible for people to produce in their homes.

1.2 Corporate strategies

With Eldridge Johnson, Berliner formed the Victor Talking Machine Company in 1901 as a rival to Edison, and the first "war of the formats" was on (the other member of the Big Three, Columbia, released recordings in both cylinder and disc format). Johnson further refined the phonograph and masterminded the Victrola, in which the sound horn and all movable parts were concealed within a massive mahogany cabinet. Introduced in 1906, its cost of $ 200 initially limited the Victrola to the luxury market. Yet the high-end Victrola legitimated the phonograph as an emblem of culture and refinement, as did Victor's aggressive marketing of classical recordings under its Red Seal line: a 1904 recording by Caruso was the first to sell a million copies. The "prestige" of classical music helped make the phonograph a "reputable" form of leisure, unlike early film – and, in the process, created demand for phonographs. However, as Kenney (1999: 50) notes, "a significant portion of what passed for 'opera' records actually presented folk, semipopular, and popular songs interpreted in operatic style by famous opera singers who lent their cultural prestige to nonoperatic music."For example Caruso released "O Sole Mio" in 1916 and "Over There" in 1918 for Victor.

Edison resisted his competitor's strategy of releasing "star" recordings, going so far as to not list the names of his performers; as Millard (2005: 62) notes, he "thought the emphasis should be on the quality of the recording rather than the reputation of the singer, maintaining that the public would prefer faithful reproduction to 'a rotten scratchy record by a great singer.'" Frith (1986: 269–270) notes the repercussions of focusing on performers: "One immediate consequence was that star performers began to take over from composers as popular music 'authors' ... but, more importantly, recording gave a public means of emotionally complex communication to otherwise socially inarticulate people – performers and listeners". Ironically, while Edison touted the accuracy of his recordings, beginning in 1900 advertisements for archrival Victor featured a painting titled "His Master's Voice", in which a fox terrier named Nipper listened intently to a recording of his owner. Kenney (1999: 54) notes that "the picture actually contains two revealing inconsistencies: first, and most significantly, the dog could only have been listening to his master's voice if his master were a 'recording star' ... Second, and perhaps less important, the painting shows the turntable braking mechanism in the On position."

Despite Edison's misgivings, the popularity of Victor machines led him to abandon the cylinder format and introduce the Edison Diamond Disc in 1913. The new machine used a diamond-tipped stylus, and the discs themselves were formed on a new, hard plastic called Condensite, which produced far less surface noise on playback: "The reproduction was so good, it was claimed, that it was superior to listening to live music in the imperfect acoustic environment of the opera house. The Diamond Discs were marketed not as mere recordings but 'recreations' of the original sounds" (Millard 2005: 78). Edison staked his device's reputation on its fidelity, its exact replication of the original performance, a poignant claim given Edison's hearing loss: he was known to sink his teeth into the base of the phonograph as a record played in order to better hear certain frequencies (Milner 2009: 39). Between 1915 and 1925, Edison's company held over four thousand public "tone tests." A performer would begin singing or playing a solo before an audience, the phonograph would then accompany him or her, and then the performer would stop playing while the phonograph continued. The curtain would then rise to reveal the phonograph. However, the Diamond Discs, which used the hill and dale method rather than Berliner's, were incompatible with Victrolas. Sales gradually declined, and Edison abandoned the phonograph industry in 1929. He did leave one important legacy: he popularized the concept of fidelity, or faithfulness to the original performance (Morton 2000: 45).

1.3 The recording process

Yet the recording process made performances anything but "natural." As Sterne (2003: 26) notes, "Performers had to develop whole new performance techniques in order to produce 'originals' suitable for reproduction." They sang or played into a single large horn, which offered no means of controlling volume. As Katz (2004: 82) explains, "Depending on the instrument, some performers had to play right into the horn, some were put up on risers, and others had to face away from the machine or even play in an adjoining room." When hitting high notes, a soprano would have to step away from the horn in order to avoid distorting the recording. According to Morton (2000: 21), "During the session, the director motioned to vocalists to indicate when to lean in close and when to duck or step away from the horn during instrumental solos, allowing the musicians to come forward." Musicians consciously minimized the dynamics of their playing, and classical musicians in particular often felt constrained by the rigor and artificiality of the recording process. Stage fright was common. Musicians could hear themselves detached from the act of performance, and many alluded to the alienation we often feel when confronted with the sound of our own voices.

These difficulties were compounded by the complexities of the acoustic recording process. Loud volume or low notes could force the stylus outside the groove, ruining the recording, and since playback required the stylus to be run back through the soft wax groove, eradicating the original, a "test" recording would have to be made and sacrificed. Pianos were difficult to record, particularly in ensemble pieces, and banjos often substituted for keyboards – their sharp attack and rapid decay was well suited to acoustic recording. Tubas supplanted string basses; other instruments, such as the Stroh violin, which replaced the body with a diaphragm and horn, were designed to address the deficiencies of early recording. Katz (2004: 93) argues that violinists employed a much more pronounced vibrato in response to the limitations of acoustic recording: "It could obscure imperfect intonation, which is more noticeable on record than in live performance. And ... it could offer a greater sense of the performer's presence on record, conveying to unseen listeners what body language and facial expressions would have communicated in concert." Both classical and popular composers tailored their work to the three-minute limit of the ten-inch, 78-rpm record: Katz (2004: 3) notes that Igor Stravinsky wrote each of the four movements of his 1925 "Serenade for Piano" so that it would fit onto a 78.

While symphonies and operas were released in bulky "albums" of 78-rpm discs, the limits of the medium favored arias, marches and brief popular songs. Katz (2004: 34) finds that performers were inclined to edit pieces rather than rush their tempos to suit the limitations of the recording medium, and the same held true of composers. The limited reach of the recording horn precluded recording large orchestras, and the limited dynamics and frequency response of early recording favored vocals, particularly those by trained operatic singers, over instruments. As Chanan (1995: 30) notes, “Caruso’s strong tenor voice (with its baritone quality) helped to drown out the surface noise, so that even on the inadequate apparatus of the time, his records sounded rich and vibrant.”

1.4 Culture and copyright

Classical music has been used to legitimate new formats throughout the history of recording: long-playing records in the ’40s, stereo in the ’50s and compact discs in the ’80s. However, Chanan (1995: 40) claims that despite the success of opera singers, “sales of the popular repertoire far outstripped, *in toto*, those of the classical.” Kenney (1999: 61) adds, “In marketing three times as many popular records as Red Seal discs, and in creating its own recorded mixture of genres on many of its ‘opera’ records, Victor actually promoted the dissemination of American popular music far more than it did European concert hall music.”

This dissemination had enormous impact on musical culture, as new forms were assimilated and then hybridized. Ragtime and instrumental novelty numbers were recorded in the 1890s, and a 1917 recording by Original Dixieland ‘Jass’ Band (a group of five white musicians who journeyed from New Orleans to New York) was the first million-selling jazz record. These forms were thus diffused far from the urban centers in which they arose (Bix Beiderbecke learned jazz from records while growing up in Davenport, Iowa). Recording allowed technique and tonality, the subjective measures that cannot be noted in scores, to be directly communicated for the first time. As Katz (2004: 78) notes, “in jazz the values of the classical world are inverted: the performance is the primary text, while the score is merely an interpretation.” Yet pieces that in live performance were extended for improvisation and dancing had to be abbreviated to suit the three-minute capacity of 78 rpm. Paradoxically, the solo improvisations that were a hallmark of jazz were cut short: many early jazz records end abruptly, with a cymbal crash.

Although the Big Three (Victor, Edison and Columbia) dominated the early recording industry, low entry costs and potentially high profits led entrepreneurs to establish labels catering to small or niche markets. By 1920, the U.S. had nearly two hundred labels (Chanan 1995: 54). The discovery of a record market among the African American population led to the development of “race records”, or classic blues; the first such release, Mamie Smith’s “Crazy Blues”, was recorded by Okeh Records on August 10, 1920. Record companies found these recordings particularly attractive because the songs were based on uncopyrighted material. The 1886 Berne Convention, intended to standardize international copyrights, treated the gramophone as akin to musical boxes and held that mechanical reproduction of music would not infringe on copyrights: "If anything, publishers generally thought that there was good advertising in records, which would result in increased sales of sheet music" (Chanan 1995: 34). But as the record industry gathered steam this attitude changed. The 1909 Copyright Act gave music publishers a royalty of two cents per copy, and gave record companies the right to record any song after an initial recording release, without permission from the publisher. In 1914, the American Society of Composers, Authors and Publishers formed to license songs.

The result was that many labels, particularly small "race" labels, pressured songwriters to sign over copyrights, reasoning that if a record were successfully covered by another label, the label that owned the copyright would at least get some royalty payments. Early jazz recordings were, therefore, as Eisenberg notes, largely of ad hoc ensembles and spur-of-the-moment compositions rather than published songs. By circumscribing these improvised popular songs through formulas of verse/chorus/verse, record companies were able to copyright them as compositions. According to Kenney (1999: 150–151), producer Ralph S. Peer pushed country and blues music "in the direction of melodic and lyric innovation within a generally familiar-sounding style" so that he could copyright the results (traditional "folk" material would be preserved later, in field recordings undertaken in the '30s by the father-son team of John and Alan Lomax and others under the aegis of the Library of Congress). Kenney (1999: 119) notes also that Bessie Smith, whose "sales of around 6.5 million discs kept the perpetually floundering Columbia label afloat during the Twenties ... had signed away her copyright royalties when signing her recording contracts", although she did receive royalties from sales. Not all performers were given even this compensation. Since recordings were classified as "works for hire", performers had no absolute right to royalties for record sales. This placed them at an economic disadvantage that their "employers" the record companies have exploited ever since.

While recorded sound changed the culture and economics of music making, it led some to worry that it posed a threat to the very nature of the art. Once, any performance of music (aside from a musician playing for him/herself) was a social event; now recordings separated listening from performance, and the audience was dispersed and atomized. In a 1906 essay (cited in Katz 2004: 68), John Philip Sousa inveighed against what he termed "The Menace of Mechanical Music", predicting that "when music can be heard in the homes without the labor of study ... it will simply be a question of time when the amateur disappears entirely." Kenney (1999: 57) finds that "Sousa really had two interrelated criticisms: first, the phonograph encouraged a passive relationship to the world of music; second, it transformed what he believed to be the intensely human and interpersonal world of music into a soulless machine." These concerns were echoed by Claude Debussy, who wrote in 1913, "Should we not fear this domestication of sound, this magic preserved in a disc that anyone can awaken at will? Will it not mean a diminution of the secret forces of art, which until now have been considered indestructible?" (cited in Eisenberg 2005: 45).

2 The electric era

2.1 Radio, electricity and realism

These concerns were intensified with the transition from acoustic to electrical recording and the development of radio in the 1920's and '30s. Radio's first incarnation, wireless telegraphy, was developed in the late 19th century by Guglielmo Marconi, who envisioned radio as a point-to-point, coded medium and had no conception of "broadcasting" (Douglas 1989). Reginald Fessenden developed technology that allowed for sound to be modulated upon continuous waves in 1906. By the 1920s, when radio developed into a popular broadcast medium, its sound technology had advanced beyond that of gramophone recordings. Instead of horns, radio performers used diaphragm-and-moving-coil microphones, and radio listeners heard the results through dynamic speakers made of thick cones of paper and moving coils driven by vacuum tubes. Fidelity was thus greatly improved. Radios and speakers were housed in free-standing cabinets and retailed through dealers and department stores. In consequence of this "radio boom", sales of recordings and phonographs, which had risen steadily between 1914 and 1921, suddenly collapsed in the early '20s. Victor's sales dropped by 50 percent, and Columbia went bankrupt, although it continued to issue records.

The recording industry had no choice but to catch up in both recording and playback technology. A team at Western Electric headed by Joseph Maxfield and H. Harrison began making experimental electrical recordings in 1920. In 1925 Victor introduced the Orthophonic "folded horn" system. It provided a "warm" sound to listeners accustomed to radio. They noticed "the dramatic increase in volume, the clear sibilants, and most of all, the amazing reproduction of the bass notes" (Millard 2005: 143). Record sales picked in consequence; as would prove to be the case throughout the history of recorded music, the introduction of new hardware was crucial to economic recovery. Cabinets featuring radio and phonograph sets that shared an amplifier were marketed by 1928, and Morton (2000: 27) claims that these probably contributed more to a revival of record sales than the introduction of electrically recorded disks. Nevertheless, the new "electrical" sound was not universally welcomed. Presaging the critiques of digital compact discs in the 1980s, Compton McKenzie, the editor of *The Gramophone*, claimed that "the exaggeration of sibilants by the new method is abominable, and there is a harshness which recalls some of the worst excesses of the past" (Millard 2005: 307). Edison, contemplating the decline of his beloved Diamond Disc in 1926, groused that "people hear what you tell them to hear and not what they really hear" (Millard 2005: 306).

Electrical recording allowed for the design of the modern recording studio, in which the recorder and engineer were distanced from the performance and housed in a separate control room. Rather than grouped for volume, performers could be positioned naturally around the studio, and microphone levels could be adjusted, or mixed, in the control room to address imbalances. While engineers of acoustic recordings pursued faithfulness to sources (despite the contortions required by the acoustic recording process), engineers of electrical recordings sought to create soundscapes, or aural images. A division of labor developed between engineers, who placed and balanced the multiple microphones involved in recording; the recording operator, who supervised the recorder itself; and musical directors or "producers" who hired performers and arranged music. The role of the recording engineers changed as well, from subjectively directing performers to objectively measuring and manipulating audio signals. Film engineers in the 1930s began to attenuate, or "equalize", the strength of certain frequencies during the recording process to minimize hiss from high frequencies and to reduce low-frequency rumble. In addition they began to compress signals to limit peak volumes that otherwise would distort the recording. Both compression and equalization were soon a standard part of the recording process. However, the "balance" their expertise sought to achieve remained subjective nevertheless, as their efforts served "to 'enhance' the sound rather than be satisfied with preserving the original" (Morton 2000: 32).

The desire for greater realism ironically led to greater mediation between source and listener in other ways. Traditionally, microphones were placed at a distance to include studio ambience. In a new approach, adapted from radio, small groups of performers were miked closely with minimal room ambience. "Naturalness" was supplanted by sounds that could only be heard through loudspeakers. Frith (1986: 270) claims that the microphone served the same function as a close-up in film: it moved the focus from the song to the singer. "Crooners" such as Rudy Vallee, Bing Crosby and Perry Como, "provided a sense of intimacy between artist and audience, collapsing the technologically imposed distance that would seem to preclude such a relationship" (Katz 2004: 40–41). Frith (1986: 264) adds, "Microphones enabled intimate sounds to take on a pseudo-public presence, and, for crooners' critics, technical dishonesty meant emotional dishonesty – hence terms like 'slushy.'" He cites the efforts of the controller of programs at the BBC, Cecil Graves, to keep "crooners" off the airwaves on grounds that they "rouse more evil passions in certain breasts than anything else" (cited in Frith 1986: 263). The intimacy of crooning implied "knowability", lending itself readily to the burgeoning star system that characterized the recording industry in the wake of the Great Depression.

Meanwhile, the Depression decimated the industry. Sales of phonographs dropped from nearly one million in 1927 to 40,000 in 1932, while record sales dropped from 128 million in 1926 to only six million in 1932 (Chanan 1995: 65). The industry leader, the Victor Talking Machine Company, was taken over by the Radio Corporation of America in 1929, and the Columbia Phonograph Company was absorbed by the Columbia Broadcasting Company in 1938. Small, independent labels catering to minority audiences with jazz and blues records were particularly hard hit by the Depression, as was classical and other niches. Warner Brothers purchased newcomer Brunswick Records in 1930, underscoring the growing liaison between the music and film industries. As Millard (2005: 7) notes, "The development of recorded-sound technology was often the result of the diffusion of ideas and techniques between film [studios] and record companies ... In many cases the recording engineers in film studios and their counterparts involved in producing popular songs were all working for the same business organization."

2.2 The hit record

The "hit" record phenomenon was created in part by the jukebox, a machine that combined electric amplification and multi-record changers. It was introduced by the Automatic Music Instrument Company in 1927; 225,000 jukeboxes were operating in the United States by 1930, and by 1936, over half of all U.S. record production was destined for jukeboxes. Their large dynamic speakers gave listeners "the highest level of sound reproduction outside the movie theater" (Millard 2005: 169). The English Decca record company established an American subsidiary in 1934, and "Decca quickly established itself as a major producer of popular records, especially those destined for jukeboxes" (Millard 2005: 168). The head of Decca, Jack Kapp, introduced differential market-by-market targeting to reduce promotional uncertainty. He also emphasized jukebox airplay, which featured rapid turnover of a limited stock of records, and aggressively promoted "stars" like Bing Crosby and Paul Whiteman. Kapp also cut the price of his records to 35 cents, half what competitors charged (although discounting was common, particularly on slower-selling titles). According to Kenney (1999: 165), Kapp turned the industry "away from the relatively long-term preservation of 'immortal' and 'timeless' recorded concert hall music ... and toward short-run profits from the quick sale of the latest recordings of popular music." Seeking to reduce risk, the major record companies adopted increasingly hierarchical business structures, and increasingly relied on musical formulas.

In the 1930s the major music labels were integrated into companies that provided music for jukeboxes, radio and films. As Kenney (1999: 158) notes, "These multimedia consolidations led to the simultaneous playing of a limited number of popular songs on movie sound tracks, radio broadcasts, and jukeboxes, saturating the media with hit songs, overwhelming ethnic and race music traditions into popular music formulas." Writing in the 1940s, critic Theodor Adorno claimed the standardization promoted by the recording industry worked by incorporating a crucial element of supposed novelty, in a star's personality or a song's musical "hooks." To Adorno, this conditioning of the listener served the fundamental goal of advertising, to encourage the standardized consumption of other "new" products. Adorno's views have been influential, but many even of his later admirers (like Chanan 1995: 152) take issue with his apparent supposition that all popular music was incapable of "emancipating itself from exchange value."

2.3 Radio and records

Broadcasts of records began with independent radio stations in the early 1930s; on February 13, 1935, Martin Block began broadcasting "Make Believe Ballroom", an adaptation of a West Coast show that aired records, on WNEW in New York City. Despite radio airplay's potential role in promoting records, the major record companies (and many bandleaders) at first resisted it, since radio stations did not share their revenue with labels and performers. Decca, for one, stamped "Not to be used for Radio Broadcasting" on their records. By 1942, however, the broadcasting of recorded music had become so widespread that the American Federation of Musicians (AFM) instigated a two-year ban on recording against the major record companies so as to prevent it and thereby ensure the survival of live music. This benefited "small independent labels, without backlists or vested interests", which were willing to meet the AFM's terms (Chanan 1995: 86).

Other developments spurred the growth of "indie" music. The American Society of Composers, Authors and Publishers (ASCAP), created in 1914 to monitor mechanical and performance rights, was closely tied to Broadway and Hollywood and as Kenney (1999: 140) notes, refused membership to blues and hillbilly artists because their works "were not really compositions in the formal, written and printed sense." Radio stations, chafing under ASCAP control and noting a popular shift away from the standards licensed by ASCAP, set up Broadcast Music Incorporated (BMI) in 1940 and boycotted ASCAP music. "Race" and "hillbilly" recordings were thus given increasing airplay in the years following World War II. While Columbia, Decca, Victor and Capital (founded in 1942 by songwriters Johnny Mercer and Buddy DeSylva) dominated the recording industry, independent record companies began to serve market segments ignored by the majors. "Rhythm and blues" replaced "race" music on the *Billboard* charts beginning in 1949, just as "country" replaced "hillbilly." Exposure created sales, and sales conferred legitimacy. However, these "rhythm and blues" and "country" songs were often covered by artists signed to the majors, and the original artists were often forced by their labels to sign over the copyright on their compositions.

Meanwhile, radio was broadcasting not just commercial records, but longer recordings of its own. "Transcription" discs, invented by the Vitaphone Company for synchronized film sound but quickly superseded by optical soundtracks, were first broadcast by New York's WOR in 1929 (Kenney 1999: 188). These were 16 inch,

33 1/3 rpm electrical recordings pressed on shellac discs, with about 15 minutes per side. The transcription discs "gave advertisers greater efficiency in targeting specific areas of the country with carefully recorded messages. Electrical transcriptions allowed local radio stations to broadcast independently of the networks and, for that reason, the major broadcasting companies did not get into the business until 1934" (Kenney 1999: 188). In fact, though the technology was adapted internationally by the BBC and used increasingly in local American broadcasts, NBC and CBS discouraged airplay of records on their affiliate stations, until, after 1945, they "began to admit that transcription recording technology had 'progressed' to the point of commercial acceptability" (Morton 2000: 67).

Transcription discs wore out quickly, developing surface noise that made them unlistenable after about 100 plays. Wartime scarcities of shellac led to a search for alternatives, until CBS researchers found that vinyl (first produced by Union Carbide in 1930s) could accommodate narrower grooves and more recorded material, while producing better frequency response and sustaining more use. Long-playing "microgroove" records (in which the grooves were cut nearly three times smaller than a shellac disc) were developed by Columbia under the direction of Dr. Peter Goldmark and released in 1948. The LP format was particularly well-suited for classical music; listeners no longer had to change records every three to four minutes, since a single vinyl LP side could hold up to 25 minutes. Columbia intended the LP to supplant the 78-rpm single; in 1949, a year after the LP's introduction, RCA countered with its own micro-groove technology, the seven-inch 45-rpm single. This lent itself more readily than the LP to the short length of popular music recordings, for which it soon became the standard format. These records were smaller and more durable than 78s, and so better suited to jukeboxes and radio play. At first, few popular artists took advantage of the LP format; Frank Sinatra was a pioneer in using "the LP to build up moods and atmospheres in ways impossible on three-minute singles" (Frith 1986: 271).

2.4 The transition to tape

In 1898 Valdermar Poulsen, an engineer at the Copenhagen Telephone Company, devised a way to save telephone messages using thin magnetized wire, in which magnetic pulses would match the variations of the telephone's electrical current. Poulsen termed his device the "telegraphone", a combination of telephone, telegraph and phonograph (Millard 2005: 35). However, wire was prone to knotting and tangling. The first "modern" recording tape, created by the German firm BASF, used a plastic base coated with magnetic oxide. AEG, a German manufacturer of electrical equipment, unveiled a revised version of the "magnetophon" recorder in 1935 that was quickly adopted by Third Reich radio stations to broadcast speeches and classical music programs. After World War II, this German technology was launched in the U.S. with recorders made by Ampex (whose founder, A. M. Poniatoff, was one of three American servicemen who discovered the German magnetic tape recorders when Radio Luxembourg was captured in 1944) and tape made by 3M. Ampex was financed in part by Bing Crosby, who sought a way to record several shows in the course of a week for delayed broadcast rather than going live every week. Crosby had moved from NBC to ABC in 1944 in part because the struggling network allowed him to record several shows at one time for later broadcast. The network switched to tape to record his show in 1947.

Tape offered radio producers many advantages. Tape recorders were more rugged than disk recorders. Tape itself was less susceptible to dust, heat and humidity. Most notably tape could easily be spliced and edited. It was no longer necessary to choose between several entire takes of a performance to choose the best one; now, an ideal version could be assembled from sections of several takes. While some regarded this as trickery, others welcomed its empowerment of musicians. Pianist Glenn Gould spliced two takes of a fugue from Bach's *Well-Tempered Clavier*, one legato and the other staccato, and proclaimed the results to be "far superior to anything we could at the time have done in the studio" (cited in Chanan 1995: 132–133). Broadcasters and record companies also began to switch from disc to tape for mastering; Ampex introduced video tape recorders in 1956.

The relatively low costs of tape recording spurred entrepreneurs to start up small recording studios in regional cities (Sam Philips' Memphis Recording Service was one of the most notable). These independent studios were seldom unionized, so engineers had greater role flexibility in production. The tape recorder could be housed in the control room so that the engineer could operate the recorder as well as the mixing console. In addition, the recording engineer, no longer required to master discs onsite, could focus on signal processing. Delays, reverberation units, filters, equalizers, limiters and compressors became commonplace. Engineers began to develop characteristic sounds at particular labels: for example, Philips' engineers perfected the "slapback" style of artificial reverb that provided the audio signature of Sun Records, Atlantic Record's Tom Dowd worked to clarify the voice of each instrument, and major labels such as RCA used a small number of microphones to record orchestral backings that buoyed close-miked vocalists.

Although independent record companies continued to release 78-rpm shellac record discs, they took advantage of the streamlined operations and lowered costs of tape recording, and proliferated in consequence. Millard (2005: 230) notes that by the mid-'50s, a recording venture, from recording to pressing and distribution, could be financed for less than $ 1000. Recordings were often made in radio stations, and the independent labels leaned heavily on radio for promotion as well as production. The old system of record stores allied with major record companies was undercut by the development of new distribution systems that included rack jobbers (entrepreneurs who maintained racks of popular records in supermarkets and drug stores). The growth of independent companies skyrocketed. Millard (2005: 229) adds that "by 1960 there were around 3,000 labels in the United States, of which only about 500 were operated by established companies."

2.5 Multi-tracking and hi fi

By the end of the 1940s multi-microphone setups had become common in recording studios. Portable sound baffles were used to minimize leakage between microphones, isolating musicians. While ostensibly offering greater fidelity, recordings became more and more artificially constructed, until they presented music that could not be performed in a live setting. Multi-track recording enabled synchronic editing; instruments could now be added individually, or "overdubbed", on adjacent tracks, and "punch-ins" allowed for parts to be re-recorded onto the work in progress. "Dubbing" had been pioneered in movie soundtracks to add music, sound effects and dialogue; in music recording, the key figure was guitarist Les Paul, who was given an Ampex recorder by Bing Crosby. As Millard (2005: 289) states, "Inspired by the dubbing techniques he had witnessed on Hollywood sound stages, Paul modified his tape machine to make recordings of the recordings", which he "overdubbed" onto one another, allowing him to add body to his playing and his wife Mary Ford to sing in harmony with herself.

Musicians could now compose in the studio, rather than working from pre-arranged parts. Multi-track recording also resulted in heightened attention to detail. Effects such as equalization and reverb could be added and discarded after recording, rather than carefully shaped beforehand. This led quickly to a reaction. Presaging Autotune, a journalist wrote in 1959, "Recording techniques have become so ingenious that almost everyone can seem to be a singer ... The gadgetry dam really burst after Elvis Presley's recorded voice was so doctored up with echoes that he sounded as though he were going to shake apart" (Chanan 1995: 107). According to Morton (2000: 37), "John Hammond of Vanguard Records sought a 'natural sound,' using a single microphone, and he denounced popular record producer Mitch Miller's artificial reverberation as 'horrible' and 'phony.'" Edison had chosen the term "record" carefully; the phonograph recording was literally a "record" of a performance, in which the medium was to be as transparent as possible. However, Eisenberg (2005: 89) notes, "The word 'record' is misleading. Only live recordings record an event; studio recordings, which are the great majority, record nothing. Pieced together from bits of actual events, they construct an ideal event." Cultural theorists continue to debate the importance of the resulting loss of spontaneity and "authenticity."

The "mixing" process further entrenched the role of the producer in the recording studio. At the major labels, the recording producer began assuming overall supervision of the recording process, adding an additional layer of input and challenging the technical control exercised by engineers. As Morton (2000: 37) notes, "Postwar recording directors often emerged from the ranks of artists and repertoire men, the agents of record companies who put together new talent and songs." This development underscored the recording industry's growing bureaucratization. The producer now "was not only responsible for paying the bills and organizing the session efficiently but also for supervising the post-production stage, which now became more important" (Chanan 1995: 104–105). Given the possibilities afforded by tape editing and multi-track recording, the studio itself became an instrument, and producers became a new kind of musical creator. Although dismissing most of Phil Spector's recordings as "perfect trash", Eisenberg (2005: 103) hails Spector as the first *auteur* among producers: "In its urgent solipsism, its perfectionism, its mad *bricolage*, Spector's work was perhaps the first fully self-conscious phonography in the popular field". These developments further erased the line of authority in music recording; as Frith (1986: 265) notes, "One effect of technological change is to make problematic the usual distinction between 'musician' and 'sound engineer,' with its implication that musicians are creative artists in a way that engineers are not."

These developments also problematized the traditional recording goal of "realism." The "audiophile" of the 1950s sought transparency through "high fidelity", a term used since the 1930s that "referred to the faithfulness of the machine's reproduction of the original music: wide frequency response, flat frequency response (in that all sounds are reproduced at equal levels), wide dynamic levels, and low distortion" (Millard 2005: 208). Yet "accuracy" remained problematic for listeners, who grew accustomed to listening at higher volumes than in live venues and could modify bass or treble in ways unintended by conductors and composers. Indeed, Eisenberg (2005: 90) argues that:

> Fidelity itself is a vexatious concept. A producer might attempt to make a record in Carnegie Hall that, when played back in Carnegie Hall, would fool a blindfolded audience. To do this he would have to remove most of the hall's natural resonance from the recording, lest it multiply itself and muffle the music. The resulting record would sound dismal in a living room.

The introduction of transistors in the mid-1950s further lowered the costs of recording and playback equipment. Cheaper and easier to mass-produce than vacuum tubes, transistors played a major role in transforming radio into a portable, privatized medium and profoundly influenced music listening in the process, resulting in small, inexpensive record players targeting the youth market. At the same time, audio "realism" was enhanced with the introduction of stereo. The technology had been showcased in a few theaters wired for multi-channel playback and featured in Walt Disney's *Fantasia* (which, again, used classical music to introduce audio novelty). Stereo recordings were marketed as pre-recorded tapes in 1954 and LPs in 1958, the first ones showing off the new "hi fi" effect with the sounds of passing trains or of ping pong balls bouncing between speakers (Milner 2009: 140). By the late 1960s pop performers were making increasing use of the studio as a compositional tool and the LP as an extended medium: the Beatles' *Sergeant Pepper's Lonely Hearts Club Band*, released in June 1967, is often cited as a milestone of the studio-as-instrument. The first four-track recorders were introduced in 1958; by the late '60s, eight-tracks were commonplace and 16-track recorders available in leading studios.

2.6 Eight-tracks and cassettes

As studios became increasingly sophisticated, their recordings became increasingly accessible through technologies that promoted decentralization. Two tape cartridge formats, in which the tape was encased in a plastic housing that eliminated the handling problems that plagued the medium for home use, were in development by the mid-1960s: the eight-track tape, designed for playback, and the Philips Compact Cassette, intended primarily for portable low-fidelity recording. The eight-track tape player was developed in 1966 by Lear, the manufacturer of executive jet airplanes, together with Motorola, for use in luxury Ford automobiles, while RCA released the recorded tapes (Millard 2005: 316). This led to a significant shift in the use of recorded music. The eight track "introduced Americans to the idea of using multiple audio systems at home and on the move and of programming their own music to suit their tastes and activities" (Morton 2000: 169).

The Philips Compact Cassette was introduced to the U.S. in 1965 and marketed as a rugged and inexpensive portable sound recorder. It used half the width of a standard ¼-inch tape and recorded at half the three and 3/4 inches-per-second speed used by eight tracks and home tape recorders. While at first sales were small compared to eight tracks, Morton (2000: 164) finds several reasons why cassettes eventually succeeded despite their "low-fi" status. Philips licensed the technology to other manufacturers, who quickly offered their own models; noise reduction systems upgraded the quality of cassette recordings; and revised formats were "backwards compatible" – older, mono tapes could be played on newer stereo machines. Millard (2005: 320) describes the attraction of cassettes: "Smaller and more durable than a long-playing disc, a cassette could exceed it in playing time and almost match it in sound quality. But most importantly it had a recording capability, which gave it the commercial edge over records and eight-track cartridges."

The cassette allowed listeners to create their own compilations. Morton (2000: 176) argues that "its popularization bucks the trend of modern capitalism toward centralized, automated, factory production of goods previously made by hand or at home." Tape formats obviously lent themselves to unauthorized duplication: Ampex Corporation, the largest tape duplicator in the United States in the late 1960s, led a campaign in Congress to end tape piracy, estimating that illegal sales of cassettes and eight-tracks amounted to $100 million in lost sales annually. The resulting 1972 copyright law "created stiff penalties for unauthorized duplication and led to threats of sanctions against countries where pirates were known to operate" (Morton 2000: 162). By the late '70s, the cassette had become a global technology. Much of the world's music produced by indigenous groups and subcultures was distributed on cassettes. In addition to home units, large battery-powered cassette players, or "boom boxes", were popularized in the 1970s, as were "Walkman" cassette players. The Walkman, introduced by Sony in 1979, dramatically transformed the listening environment through portability and mobile privatization, representing another step away from music as public ritual, in a process that dates from the commodification of sound in the late 19th century.

3 The digital era

3.1 The introduction of compact discs

In the 1960s the major record companies were absorbed into integrated conglomerates with diverse portfolios; these in turn absorbed small independent labels (for example, the Dutch-based PolyGram label included Polydor, Mercury, Smash, MGM and Verve). By the end of the '60s, the "Big Six" labels were CBS, Warner Brothers, RCA, Capitol-EMI, PolyGram and MCA (the latter formed from a merger of Decca, MCA and Universal earlier in the decade). Profits grew steadily from 1955 through 1978, but flattened toward the end of the decade as catalogue sales waned. The market was ripe for a new format.

Digital technology, like so many recording technologies, was developed by the telephone industry for signal processing. Pulse Code Modulation, in which a continuous signal is sampled thousands of times per second and each sample converted to a binary code, was first mentioned in a Western Electric patent in 1926 (Millard 2005: 347). A signal is broken into samples, and each sample is assigned a series of binary numbers to represent its amplitude. On playback, the digital series thus generated is converted back to analog. The standard sampling rate is 44,100 per second, which in theory can capture frequencies up to 20 kilohertz; each of these samples is assigned a sixteen-digit binary number, which can measure up to 65,000 levels of voltage.

Philips developed an experimental digital laser disc in 1964; 15 years later the company had produced a working technology in which a laser optical decoder read sound through a series of pits and lands on the surface of the disc. Working with Sony, it developed a digital playback system termed the "compact disc", and began demonstrating it to audio executives in 1981. The compact disc was commercially released in 1982. As Millard (2005: 353) notes, "the compact disc has a signal-to-noise ratio of 96 db, which in effect makes it noiseless recording". The CD was chosen to have a size of 12 centimeters, slightly larger than that of a cassette, so that a CD player could fit into the center console of a car in the slot designed for a cassette player. According to an oft-told (and perhaps apocryphal) tale, the 74-minute length of a CD was selected because Sony President Norio Ohga wanted the entirety of Beethoven's Ninth Symphony to fit on one CD (Sterne 2012: 12).

True or false, the Beethoven story legitimated the CD as a vehicle for high culture. Classical archives again led reissues, as they had when 78s and LPs were introduced, and digital was touted as promoting greater transparency and realism.

However, Eisenberg (2005: 210) notes, "Sonically speaking, the jump from analogue to digital is narrower than that from mono to stereo, to say nothing of the leap from acoustic to electric", and the widely-touted "realism" of CDs was somewhat suspect. Early digital classical recordings were often close-miked to provide the listener with greater intimacy; to Eisenberg (2005: 212), "some are so closely miked that an attentive rhinologist could make a map of each player's sinuses." Since CDs retailed at significantly higher prices, they resulted in much greater profits for record companies, which quickly stopped shipping vinyl records to retail stores. Sterne (2012: 141–142) argues, "Consumer demand for digital content had to be created – it did not simply exist out there waiting to be tapped. In the case of compact discs, sales lagged in the United States until record labels made it very financially difficult for stores to continue stocking LP records."

3.2 Digital recording

All musicians "borrow" from their predecessors. Yet digital "sampling" allows for the exact duplication of any prerecorded sound. This possibility further extends the process of assembling recordings from fragments that began with the advent of multi-tracking in the late 1950s. To Katz (2004: 157), sampling "transformed the very art of composition" from a score requiring interpretive realization to "a document of binary numbers requiring electronic conversion". The attention to detail that characterized music recording advanced to a nearly unimaginable granularity: "With rhythm quantization, for example, a performance with an unsteady tempo becomes metronomically precise as all notes are forced to fall on the closest beat. Pitch correction follows a similar principle, pushing pitches up or down to the nearest specified level. Moreover, both can be applied in real time" (Katz 2004: 43). Yet attention to detail is intrinsic to recordings, which remove the visual element of performance. Itzhak Perlman noted that "people only half listen to you when you play – the other half is watching." (cited in Katz 2004: 20). Recording enables the unique features of a performance to become regarded as integral to the music as well as the performance: "In other words, listeners may come to think of an interpretation as the work itself" (Katz 2004: 25). Concerts must live up to the expectations of recordings, rather than vice-versa; hence the contested growth of sampling and lip-synching in concert. At the same time, when an improvisation is reproduced, it becomes a composition itself.

Drums were among the first instruments to be synthesized; they are among the most difficult instruments to record, and no human can keep perfect time. Samplers and sequencers address this by allowing the transfer of digital information via Musical Instrument Digital Interface (MIDI). The developer, Yamaha, promoted the MIDI standard by making the patent freely available to all interested parties, enabling users to employ "a wide range of different equipment to produce digital sounds because they all communicated in the same way ... [T]he MIDI sequencer became a word processor for music" (Millard 2005: 357). Digital synthesizers able to mimic analog instruments were available by the early '80s (a benchmark was the Yamaha DX-7, introduced in 1983 for $ 2000). Abetted by the development and lowered costs of microprocessors, by the end of the 1980s synthesizers were playing a larger and larger role in the recording process.

The same critiques that were leveled at acoustic recording were leveled again at digital technology: it was "inauthentic", antithetical to art and somehow immoral. In fact sampling led to much artistic innovation. Even before the proliferation of digital recording, Jamaican "dub" music in the early 1970s had experimented with "remixing." Dub remixers like King Tubby would strip vocal tracks from recordings, then create wholly new versions by adding effects like echo and tape delay to the instrumentals and dropping in snatches of vocals. Dub's "use of the record as a musical instrument" (Frith 1986: 275) was taken up, and, through digital technology, greatly extended, by hip hop, which "treats records – typically finished musical products – as raw material" (Katz 2004: 132).

In its free-ranging borrowing, sampling quickly proved legally problematic. Ideas cannot be copyrighted, but expressions, which embody ideas, can be. Yet sampling arguably transforms the expressions it samples, recombining preexisting sources as musicians have always done. Chanan (1995: 163) notes the irony: "Conflict with authority is almost inevitable. The corporations have placed on the market devices that invite people to transgress the laws on which the market operates". In 1990, record companies began to require that recording artists identify and clear all samples before release. In a landmark 1992 case, a U.S. Federal judge ruled against Warner's WEA for using without permission a sample from a Gilbert O'Sullivan track on a Biz Markie album called "I Need a Haircut" (Chanan 1995: 164).

3.3 Studios in the home

Digital home recording systems further decentralized the process of music recording: they allowed a single amateur to accomplish what previously had required several professionals, hundreds of hours and thousands of dollars in state-of-the-art recording studios. In 1992, Alessis introduced the pioneering digital home recording system, ADAT, using inexpensive videocassettes that held approximately 60 minutes of recorded material. These modular eight-track recorders sold for $ 3995, and their impact was such that the October 1992 edition of *Electronic Musician* claimed that "ADAT is more than a technological innovation, it's a social force" (cited by Millard 2005: 381). Throughout the decade personal computers increased in processing ability and storage capacity, and software engineers began producing programs that emulated the capabilities of multi-track recording.

A particularly noteworthy event was the introduction of Digidesign's Pro Tools, which "provided a new and easier methods of managing multi-track recordings. It presents two modes of operation: the 'mix' interface reproduces the mixing board console on the computer's screen, and the 'edit' interface shows the sound as a waveform that swirls horizontally across the screen" (Millard 2005: 382). Tracks could now be cut and pasted much more expeditiously. These developments further democratized production, increased capabilities and erased old categories. As Chanan (1995: 165) notes, producers might "alter radically the musical material even in late stages of production," but musicians might as easily take over "control of the apparatus whether in self-interest or the interests of expanding the frontiers of musical creation". And the malleable tracks thus produced might be adapted to myriad listening contexts, "one for each format: a short mix for AM radio, a longer more elaborate mix for FM radio, a long mix with many effects and edits added for dance clubs, and a version of the FM radio mix with effects and 'sweetening' added specifically for combining the song with a video" (Jones, cited by Chanan 1995: 148). No final version of a recording existed as producers and record companies "windowed" releases for their listening contexts.

3.4 MP3s and file sharing

MP3 digital file technology offered great promise and peril to the recording industry. Motion Picture Experts Group 1, Layer 3, developed from efforts to standardize digitization of video and audio. The Motion Picture Experts Group convened in 1988 and engaged the Fraunhofer Institute for Integrated Circuits in Germany to devise a scheme for compressing digital audio files without perceptible effect. A series of listening tests held for experts in 1990 and 1991 to decide among competing standards harkened back to the "Tone Tests" of Edison's era: "Listening tests show the degree to which a professionally-defined aesthetic of "good sound" shaped the format as much as more scientific or technical determinations did" (Sterne 2012: 26).

In 1992, Fraunhofer released an audiovisual standard for digitization, MPEG-1, and a free "demo" program, in which the third "layer" of MPEG-1 was used to compress music files to about one-twelfth the size they would occupy on a compact disc. One hundred twenty eight kilobits per second was chosen for MP3s, since ISDN protocols, which used telephone lines for transmission, had a capacity of 128K. The release of MP3 went largely unnoticed, but the format exploded with the development of Internet peer-to-peer file sharing networks in the late '90s, of which Napster was the most notable. Napster relied on a central server to distribute files over the Internet; its centrality made it an easy target for prosecution on the grounds that copyrighted files were being "stolen." Napster was shut down in February 2001, but was quickly replaced by decentralized file sharing networks which hampered copyright enforcement.

The history of the recording industry is characterized by a string of economic and technological crises. The record industry blamed declining profits in the late

'70s on home taping, although industry studies cited by Frith (1986: 264) suggest that it was carried on only by "people spending as much money on music as they can." In the 2000s, sales of compact discs collapsed, and more recent new formats (HDCD and DVD-A) have failed in the marketplace. After all, how many times can people be expected to buy the same non-essential goods? Meanwhile, industry consolidation continued apace. In December 1998 Seagrams bought Polygram Records from Philips and folded the label into its Universal operation; in 2004 BMG merged with Sony, and 2013 Universal acquired EMI (whose subsidiary labels were dispersed to Warner and publishing assets sold to Sony). Today the recording industry is dominated by three labels that collectively control over 87 percent of the market for recorded music: Universal, Sony, and Warner.

In the last decade, the music industry has turned for revenue from recordings to live performances (with record companies increasingly seeking a percentage of revenues from artist tours and merchandise through so-called "360 deals"), third-party sponsorship (beginning with the Rolling Stones/Jovan tour deal in 1981), merchandise, and song publishing (for songs used in ads, video games and other contexts). It also has begun to charge fees to Internet webcasters for playing recordings, providing a new revenue stream. However, the music industry's initial failure to come up with a model for digital distribution allowed an outside firm, Apple's iTunes, to become the world's largest music retailer. The industry has arguably been hampered by concentrating its energies not on the development of new business models but on largely futile attempts at preserving the old by preventing unauthorized access to copyrighted recordings. Given millions of web sites, and their constantly changing nature, it's impossible to ferret out all unlawful activity. And, as Sterne (2012: 187) notes, the public has not accepted the industry's concept of "piracy", which "collapses people who make mix CDs for their friends with kidnappers who operate off the coast of Somalia."

The industry has, however, sought turn the Internet to its own advantage through customer relationship management (CRM) technologies which monitor music downloads and streams in real time, creating user profiles for marketing purposes. CRM technologies include collaborative filtering, which recommends recordings based on user behavior; and genre/mood matching, which recommends recordings based on "experts" categorization. Pandora and Spotify, two recent music streaming services, seek to direct the "right" recordings to the "right" listeners, one through automatic algorithms, the other through recommendations. Yet predicting consumer behavior is just as difficult on-line as it is through traditional channels. CRM cannot tell why a customer decided to choose to listen to a song; it can only make correlations that result in tautologies. Comprehensive definitions and maps for musical genres cannot be created, as they are continuously proliferating and evolving and fusing with other genres ((Burkart and McCourt 2006: 99).

4 Conclusion

Throughout its short history, musical recording has offered successive formats each of which has occupied less physical space while allowing for more storage and flexibility of use. The 78 required listeners to change records every three to four minutes. The LP and cassette allowed for two contiguous halves of up to 24 and 45 minutes per side respectively, CDs allowed 74–80 minutes per disc. An iPod can store up to ten thousand songs in a gleaming white box smaller than a pack of cigarettes and, unlike LPs and CDs, allows the user to determine their flow. Some argue that through digital formats, music may return to an intangible essence altogether, in which it "would stop being something to collect and revert to its age-old transience: something that transforms a moment and then disappears like a troubadour leaving town" (Pareles 1998: 22). "Bodiless" music predates recording; Katz (2004: 21) notes the "age-old practice in Christian churches of placing the organist and sometimes the choir out of sight of the congregation. The removal of visual cues, certainly no accident, separates body from sound, heightening the sense that the music comes not from humans but from heaven."

Paradoxically, the lack of materiality in digital files heightens our desire to sample, collect, and trade music in new ways through playlists and other means. Technology is always in flux; the outcomes are never pre-determined. The ways in which technologies are implemented, not the technologies themselves, determine their effects. Recording democratized access to "high culture" and also disseminated vernacular culture; in the process, it created communities as well as commodities. However, and by whomever, they are packaged and traded, recordings will have the power both to summon our collective memory and to transport us beyond ourselves. No technological innovation, and no process of commodification, will ever rob the ghosts of their power.

References

Burkart, Patrick & Tom McCourt. 2006. *Digital Music Wars: Ownership and Control of the Celestial Jukebox*. Lanham, MD: Rowman and Littlefield.

Chanan, Michael. 1995. *Repeated Takes: A Short History of Recording and Its Effects On* Music. London: Verso: 1995.

Douglas, Susan. 1989. *Inventing American Broadcasting, 1899–1922*. Baltimore: The Johns Hopkins University Press.

Eisenberg, Evan. 2005. *The Recording Angel: Music, Records and Culture from Aristotle to Zappa*. Second Edition. New Haven, CT: Yale University Press.

Frith, Simon. 1986. Art Versus Technology: The Strange Case of Popular Music. *Media, Culture and Society* 3(8). 263–279.

Goldberg, Michelle. 2000. Mood Radio: Do On-line Make-Your-Own Radio Stations Turn Music Into Muzak?. *San Francisco Bay Guardian*. November 6, 2000.

Jones, Steve. 1992. *Rock Formation: Music, Technology, and Mass Communication*. Beverly Hills, CA: Sage.

Katz, Mark. 2004. *Capturing Sound: How Technology Has Changed Music*. Berkeley, CA: University of California Press.

Kenney, William Howard. 1999. *Recorded Music in American Life: The Phonograph and Popular Memory, 1890–1945*. New York: Oxford University Press.

Millard, Andre. 2005. *America on Record: A History of Recorded Sound*, Second Edition. Cambridge, UK: Cambridge University Press.

Milner, Greg. 2009. *Perfecting Sound Forever: An Aural History of Recorded Music*. New York: Faber and Faber.

Morton, David. 2000. *Off the Record: The Technology and Culture of Sound Recording In America*. New Brunswick, NJ: Rutgers University Press.

Pareles, Jon. 1998. With a Click, a New Era of Music Dawns. *The New York Times*, November 15, 1998. http://www.nytimes.com/1998/11/15/arts/music-with-a-click-a-new-era-of-music-dawns.html?src=pm & pagewanted=3 (Accessed 19 March 2014)

Sterne, Jonathan. 2003. *The Audible Past: Cultural Origins of Sound Reproduction*. Durham, NC: Duke University Press.

Sterne, Jonathan. 2012. *MP3: The Meaning of a Format*. Durham NC: Duke University Press.

Marko Siitonen

5 Communication in video games: From players to player communities

Abstract: Digital games research and communication studies intertwine at several points. Gaming, and play in general, is a social activity. For those motivated to participate, digital gaming and online game worlds offer near endless ways for self-expression and socializing. This chapter looks at questions of social interaction within the realm of online multiplayer games. The topics introduced proceed from motivations of individual players to the social dynamics of player groups and communities to exploring games as communication systems and platforms. The way players utilize the affordances provided to them in online games and game-like virtual worlds vary. What is typical is that players often self-organize into what can be called collaborative groups or communities. At the heart of these groups are emergent negotiations of shared norms and rules, a shared purpose. In combination with game mechanics, these negotiations create a rich window of opportunity for creative human interaction. Throughout the chapter examples related to theoretical issues, methodology, and future research directions are provided.

Keywords: digital games, game studies, computer-mediated communication, multiplayer communities, video games

Alongside the popularization and domestication of information technologies from the 1960s onward there is a clear trend in the increasing popularity of gaming. Developing computers and computer networks opened the door to digital gaming, which has evolved from its original niche to a global business and rich soil for contemporary culture. For example, virtual (game) worlds developed from the early Multiple User Domain (MUD) and its contemporaries during the 1970s and 1980s, through *Ultima Online* and *Everquest* in the 1990s into the different versions of *World of Warcraft* and *Eve Online* of the 2000s, captured the imagination of tens of millions of players worldwide (Van Geel 2012; see also Bartle 2004; Koster 2002).

As the impact of games and the cultures that surround them gained momentum, scholars from various fields of academia started to show interest in 'game studies'. In the 1990s, there emerged a strong body of academic literature on MUDs, MOOs (Multi[user domain] object oriented), and various early versions of virtual worlds and communities. But it was in the 2000s that the field truly gained maturity. For example, special interest groups focused on digital games research emerged within larger associations, such as European Communication Research and Education Association (ECREA) and International Communication Association (ICA). The decade also saw the birth of an international scientific association concentrating specifically on game studies, the Digital Games Research Association (DiGRA), as well as a plethora of publications dedicated to the topic.

Game studies, also known as digital games research, is a decidedly multi-disciplinary field of inquiry. A survey in 2012 of 544 academics identifying themselves as being connected to the field shows that communication sciences (labeled in the survey as communication studies, media studies, and information studies) comprise approximately a quarter of the body of researchers (Mäyrä et al. 2013). This is an area of interest, in which communication sciences pair up with disciplines like psychology and educational sciences in search for answers to interesting questions, for example: how player communities can operate as communities of practice in developing expertise (Chen 2012); whether and how playing games could have a short or long term effect on users (Ferguson et al. 2013); or how journalism and games can benefit from each other in the form of newsgames (Bogost et al. 2010). Research related to gamification, i.e. the use of game design elements in non-game contexts (Deterding et al. 2011), is a good example of a research topic where a multidisciplinary approach is useful. Combining theories of motivation to those of learning and instruction (for a review, see Kapp 2012), gamification provides an interesting focal point for communication scholars as well.

The field of game studies in general is much too broad to be covered fully in this chapter. Järvinen (2003) presented a basic division of three distinct, but often overlapping approaches into the field. First, games can be approached from the point-of-view of the games themselves, where the focus is on the rules and mechanics within the game. Secondly, there is a significant body of research concerning the interaction between a game and its player, for example focusing on user experience or questions of user interface design. Thirdly, an approach favored by communication sciences' studies has focused on the rich cultures surrounding games and play, including player-to-player interaction. This chapter concentrates of this third approach.

This chapter will revolve around three main topics. First, we will take a look at individual players, their motivations and viewpoints. What do individual users seek from games, how do they utilize the affordances offered to them? Second, we will concentrate on the social dynamics of player communities and groups. What functions do these groups fulfill, what kind of communication takes place within? Third, games as communication systems and platforms are explored. What affordances are there for players to use, and what kind of new directions and possibilities for studying communication might games offer? Throughout the chapter provides examples related to theoretical issues, methodology and future research directions.

1 Players of games

Basic questions such as who plays video games, what motivates them, and what do they do within games have intrigued scholars over the years. Over the next few paragraphs, we look at each of these questions in turn.

Individuals inhabiting contemporary virtual worlds and game spaces are often referred to as *players* or *gamers*. Some have argued that since contemporary online game worlds may not truly fit into classic definitions of games, because of their infinite existence, the term *user* might fit better (Filiciak 2006). This chapter uses the two terms interchangeably.

Looking at the broad trends, while video gaming may have been traditionally seen as a realm for adolescent males, the picture of a typical player in the 2010s is much harder to portray, for as Bryce et al. (2006) indicate most adolescent girls play digital games. At the same time the age profile of players has continued to become more diverse, up to a point that you can say that people of all ages play digital games. Currently, the typical average age of players of digital games is somewhere between 30–35 years (Entertainment Software Association 2014). Of course the profile of the average gamer is dependent on the genre or game in question.

Within the widely studied genre of MMOs (Massively Multiplayer Online Games), which includes games such as *World of Warcraft* and *Eve Online*, a study combining survey data with unobtrusively collected game-based behavioral data showed that players of the MMO *EverQuest* 2 were on average just over 30 years old (Williams et al. 2008). The same study showed that approximately one in five players was female, and came from relatively wealthy backgrounds and were more educated than the general population in the US (Williams et al. 2008). While the results certainly cannot be generalized to global audiences or across game genres, *EverQuest 2* represents a typical example of the MMO market, i.e. fantasy role-playing. Unfortunately, studies that rely on more than just self-reported data are rare. Indeed, there is a general lack of reliable data on player demographics across genres and beyond the largest markets.

With a wide variety of players, it is only to be expected that the motivations driving gaming are as varied. In a seminal paper on player types in online multi-player games, Richard Bartle (1996) outlined, and later refined (Bartle, 2004), a typology consisting of four different orientations towards gaming. The typology categorized players as belonging to one of four groups: Killers, Achievers, Socializers, and Explorers. The motivations of these player types could be further explained by their relationship towards two behavioral dimensions. The first of these concentrated on the interaction between the player and the game (elements), and examined whether the player acted on them or interacted with them. The second behavioral dimension concerns whether the player focused more on other players or on the virtual world. In this framework, the Killers' emphasis is on acting on players (i.e. killing their characters), the Achievers' goal is to act on the game world (i.e. to 'win' over the course of the game), Socializers aim to interact with other players, and Explorers want to interact with the game world (i.e. by exploring its boundaries and trying out new things).

Since Bartle's typology, there have been several attempts at solving the question of player motivation or orientation, specifically in the context of online multiplayer games. Using factor analysis, Yee (2007) proposed three non-exclusive motivational factors, each with their own set of subcomponents. These three broad factors were labeled achievement, being social, and experiencing immersion. Players who value achievement are interested in competition, and seek mastery of the game. Players who value sociability look for opportunities to interact with other players, and are interested in building relationships with other players. Finally, players who seek immersion want to become a part of the story, for example through role-playing, or use the game as a means of escapism. This classification has been supported by later research (Williams et al. 2008).

Studies exploring player motivation have experienced a gradual move towards more widely validated and generalizeable explanations. For example, self-reported data have been compared to actual in-game behavior data in the form of behavioral validation, strengthening the assumption that there exist associations between players' motives and their in-game actions (Billieux et al. 2013). Also, validating scales across (national) cultures and game genres has been attempted (Kahn et al. 2013). While there exists a great deal of variation across models, certain elements come up regularly enough to increase their general validity. For example, the archetypes of socializer and competitor appear in several forms across studies. Also factors such as escapism, immersion, and interest in exploring as much as possible of the game's content or game world often occur. Underlying motivations may also change through time, for as Kahn et al. (2013) note an increasing number of players in the 2010s are approaching games from a utilitarian viewpoint, as they appreciate the possibilities for cultivating transferable skills or aim to develop their intelligence in general.

When trying to understand player behavior and experience, it is worthwhile to remember that there are often no simple answers to be found even when one is looking at a single player. For example, studies on player motivation typically simplify matters by focusing on the so-called main character a player has, or on a single dominating mode of play, instead of trying to capture all the different configurations. However, as Billieux et al. (2013) observe, players might have several characters that they consider 'main'. Players can also approach the game in many ways depending, for example, on whether they are currently engaged in PvP (player-versus-player) or PvE (player-versus-environment). Similarly, the orientation to playing can change even within one gaming session depending on a number of factors.

For those motivated to participate, digital gaming and online game worlds offer near endless ways for self-expression. The lure of these environments and the communities within has been well documented. There exists a large body of players who dedicate as much, or indeed more, time on their chosen games than they do on studying or working (Castronova 2001; Kolo and Baur 2004; Yee 2006). While it is certainly possible for an individual to spend scores of hours a week playing single player games, it is the membership in player groups and communities that is often connected to such devotion. From the viewpoint of communication sciences, player-to-player interaction has been a fruitful area of interest, spawning many insights into the dynamics of online social interaction.

2 Player groups and communities

Gaming, and play in general, is often social activity, and as far back as 1938 the play element in culture was recognized as a factor promoting the formation of groups and communities (Huizinga 1938). From negotiating rules and boundaries to cheering a teammate for their efforts, it is the presence of other players that draws people into contemporary digital realms. As video games have evolved in tandem with communication networks, it is only natural that players have used the affordances of information and communication technologies to form and maintain both interpersonal ties and larger social aggregates. The study of social dynamics within player communities, often called guilds or clans, has been a major thread in game studies.

One can approach player-to-player interaction in groups and communities from a variety of viewpoints. As Warmelink and Siitonen (2013) observe in their systematic review of research into player communities in the 2000s, there are three partly overlapping levels of interest that scholars have chosen to concentrate on – the micro, meso and macro levels. Closest to issues of interpersonal communication, the micro level focuses on groups and teams, such as individual groups doing 'raids' or coordinated co-operative tasks within the game world. Some studies focus on the meso level of larger social aggregates such as whole clans or player organizations. The macro level perspective is interested in larger networks of players, possibly whole populations that inhabit certain games or game genres. On this level, we can actually be talking of societies and sub-cultures instead of groups and communities.

Generally, there has been a heavy bias towards qualitative studies, especially ethnography and participant observation. On the other hand, large-scale surveys have also been utilized, and approaching the 2010s more and more studies have used various forms of data mining with large data sets, or so-called *big data* (Warmelink and Siitonen 2013).

As with individual player motivations, there are significant variations in how player communities organize themselves, and the kind of orientation they have. The most basic division occurs between militaristic communities where competiveness, rules and hierarchical power structures are emphasized, and casual communities, where the emphasis is on equality, close interpersonal ties and a relaxed sense of fun (Williams et al. 2006). These basic orientations are connected to communication practices within the communities, such as the preferred kind of leadership communication, or the ways conflicts arise and are dealt with. For example, guilds formed of competition-oriented players and with short-term objectives in mind can be rapidly dissolved or deserted by their members if those objectives are not met (Chen, Sun & Hsieh 2008).

Already early on in the development of digital game research it was noted, that as player experience and history with team mates develops, some community members may begin to value social interaction with other players over playing the game or exploring a virtual world (Schiano and White 1998). In these cases, the game environment can become a mere setting for establishing and maintaining interpersonal relationships. On the other hand, not every one wants to connect playing a game with socializing – at least all the time – highlighted by the continuing popularity of single player games and players choosing to stay out of established groups and communities.

While formulating abstract, generalizable constructs about player motivations and behavior can be useful, they do not represent the be-all and end-all of insight into the life of player communities. Referring to his experiences during an ethnographic study of *World of Warcaft* (WoW), Chen (2012) reminds us that, "... real social situations – like the ones I experienced in WoW – are messy and complex and problematize the very notion of constructs as convenient ways of modeling player behavior" (Chen 2012: 58). In order to understand the life worlds of players of online games, research needs to focus on actual communicative practices.

Online games offer many possibilities for examining the dynamics of social interaction. As a part of the current technology-mediated communication environment, online games are examples of 'third places', sites where people are engaged in a variety of informal social processes with familiar others, and where they spend a significant amount of their free time (Steinkuehler and Williams 2006). In these 'third places', players typically self-organize into what can be called *collaborative groups* (Stohl and Walker 2002). These groups are characterized by having formed naturally, having a shared goal that no group member on their own can reach, and displaying a distinct need for communication in order to reach that goal. There are other characteristics that fit these groups as well, such as having permeable boundaries, and a freedom to negotiate the structure (i.e. the need and distribution of leadership and power) as the group members see fit.

Another way of understanding player communities is to see them as communities of practice (Lave and Wenger 1991), in which core issues concern expertise and learning – i.e. how do new members learn to act and behave legitimately, according to the requirements of the group? In his ethnography of a *World of Warcraft* raiding group, Chen (2012) describes the long and arduous learning process through which a group of people continuously negotiate who they are, and the purpose of the group. Looking into player communities can teach us about how shared practices and coordination emerge in interaction. An example of this can be observed in the emergence of Dragon Kill Points (DKP), which basically is a system whereby players earn points or shares for their group by participating in its efforts (Chen 2012; Malone 2009). These points can then be used to bid for goods (spoils of war) that the group has collected. Alternatively, if a player has neglected to participate in shared tasks in a given time frame, they will not have such good access to the group's spoils of war. As a system, DKP answers a common problem of distributing goods within a collective. This means that it is not really specific to any one game in particular, but rather an abstract solution to a social dilemma that can be encountered in a variety of settings. Furthermore, DKP is also not a feature that would have been originally programmed into the game, but rather represents a social contract that has emerged through player interaction.

Whatever the approach of the study, player groups and communities in online games can offer an interesting window into social life online. Looking at player interaction on both the interpersonal and the group level has provided insight into how players' interaction operates as a basis from which norms and rules (culture) emerge (Taylor 2006a). This idea of emergence comes up time and again in game studies, highlighting that it is through interaction – whether it is between the player and the game, or between players in general – that games reach their potential. For example, roles in a WoW raiding group can be understood as coming into being as a "... combination of game mechanics and emerged social practice" (Chen 2012: 63).

Online games can be seen as spaces where relationships and trust are developed, and where players are able to provide each other with social support, and a feeling of belonging. What makes games such as MMOGs especially intriguing is that they

> ... offer greater interdependence, persistence of identity, and strength of reputation systems than general online environments (Ratan et al. 2010: 10).

> The point about persistence of identity is especially pertinent here, as the question of reputation is so central to the operation of groups and communities in general. Having persistent identities means that there is no true anonymity in the strictest sense of the word, and that it is possible to tarnish a player/character's reputation should they break social norms. Naturally, on the other – more positive – side of this coin is the possibility of gradually building up one's reputation by helping others out and being a reliable member of the play community (Jakobsson and Taylor 2003).

Expanding the view from player groups to larger social aggregates reveals an interesting panorama of life online. Similar to many other online environments, game systems offer possibilities for data collection that are hard to compare to or match in face-to-face reality. On the macro level of community, for example, it is possible to gather very large sets of data that can be used in data mining. Indeed, many companies do this automatically, even though it may not be all that easy for researchers in academia to negotiate access to such data. Often one can utilize programs and add-ons that interact with the game system, effectively making an automated log of players' in-game activities.

Macro-level analyses have suggested that player behavior in virtual environments follows, at least in some aspects, the patterns found in the "real" world. Castronova et al. (2009) found out that real-world categories and metrics could explain economic behavior in *EverQuest* 2. On the other hand, there were differences as well, such as more dramatic fluctuations of the gross domestic product than one would expect to see in real economies.

Games and virtual worlds have also been suggested as possible sites for studying human behavior in real-world pandemics. In what came to be known as the Corrupted Blood outbreak in *World of Warcraft*, a glitch in the game made it possible for a dangerous virus to spread outside its intended area of effect, causing unforeseeable panic and destruction. This led some scholars to look into the possible similarities in how people behave in an epidemic in a virtual environment and the physical one, and whether the spread of infectious diseases could for example be somehow modeled with the help of massively multiplayer online games (Balicer 2007; Lofgren and Fefferman 2007).

Of course what exactly is recorded on the server side of online games and virtual worlds is not always immediately useful for research purposes, for example to be used in behavioral validation of self-report data (see Kahn et al. 2013). Still, the idea of using online games as a sort of laboratory for studying human behavior is worth further exploration. Games can offer affordances similar to controlled experiments in laboratories, where the number of variables is – or at least is thought to be – controllable, making it possible to tease out cause and effect.

3 Games as communication systems and platforms

As an art form native to the digital environment, digital games employ the whole scope of communication possibilities made available by today's computer networks. Often, these affordances are utilized in imaginative and playful ways that give users great freedom of expression, such as avatars in virtual worlds, and more recently, the use of ubiquitous and geo-locating technologies.

In addition to everyday communication modalities such as text chat and voice over IP (VoIP), games can include mechanics that form a crucial part of the flow of interaction. For example, a simple function like making an avatar jump in a game world can be used in a variety of ways, communicating everything from excitement to frustration to camaraderie. An example of such a 'creative player action' (Wright et al. 2002) that can be achieved with a simple jump-command is when players pile on other players, trying to form 'towers' that can consist of dozens of players, achieving heights otherwise impossible to reach.

As the example about jumping demonstrates, communication in online games does not need to bear immediate connection to or similarity with face-to-face situations. Sometimes the way communication pans out in virtual environments could even be downright ludicrous if translated into an exact face-to-face copy. Two paradigm examples are 'gagging' and 'idling', both of which can be traced back to early virtual environments such as MUDs (Curtis 1997). Gagging refers to the possibility of silencing or ignoring another player altogether, often without any notification to the gagged player. When gagging a player, it is as if that player's communication immediately ceases to exist. Idling, on the other hand, refers to those times when a player's character is in-game but does not do anything. It is practically impossible to know whether the player is actually there, observing what is happening around the character, or whether they have left the game for a while in order to do something else.

In many ways, the whole act of playing a game can be viewed through a communication lens. Every move a player makes, and every action they implement, can be seen as a communicative act. This is true not only for video games but for games in general. In football, for example, *the way* one passes the ball or shoots it toward the goal can carry meaning. Similarly, in online video games interaction between players can be seen as mostly taking place through nonverbal behavior (Manninen 2003).

While it is important to avoid resorting to (technological) determinism, it has to be acknowledged that changes in game features, be they rules, mechanics, or technologies, can have a significant effect on communication and the social dynamics of players. For example, the original version of the MMOG *World of Warcraft* supported up to 40 players joining in on a joint venture (e.g. exploring a dungeon and fighting the monsters within). Later on, an expansion saw the maximum amount of players reduced to 10 and 25. This change affected areas of social interaction such as interpersonal relationships and social alienation and directly contributed to smaller communities becoming more viable. Players who found themselves displaced or left on the sidelines formed guilds of their own, resulting in a larger number of smaller guilds (Chen, Duh and Renyi 2008).

The creation of virtual worlds pushes us to appraise and articulate anew the many physical and interaction rules that we are used to in the physical world (Yee 2009). Rules such as how far a voice carries, how many players can participate in joint ventures, or indeed even be in the same place at the same time, have to be thought of and explicated. Some of these rules might encourage certain kinds of behavior, while others might make them difficult to carry out. Yee (2009) posits that, "In the same way that code is law in cyberspace, the rules of social interaction in EQ (*EverQuest*) – its social architecture – define the ways in which players can communicate and interact with each other. And these rules can be designed to shape social interactions and encourage cooperation, altruism or distrust" (Yee 2009). As simple as this sounds in theory, in practice even well designed and long-lasting MMOGs can fail in these goals. Based on data from direct player behavior in *EverQuest*, Shen (2014) demonstrates how it may come to be that, "[T]he very game mechanisms designed to encourage social play create a new set of constraints on social interactions" (Shen 2014: 689).

Thinking about the social architectures of virtual worlds is interesting from a design perspective, as well as an analytical one. The theory of Transformed Social Interaction (TSI) posits that collaborative virtual environments have the power to change the nature of social interaction in new ways (Bailenson 2006). The key idea here is that technologically mediated communication allows us an unprecedented level of control over the dynamics of interaction, as long as we think outside of the box of i.e. the laws of physics. For example, technology allows us to systematically alter our appearances, or to amplify or suppress nonverbal signals. Bailenson (2006) uses the example of how we could use virtual worlds to solve the inability to orient eye contact in traditional video conferences. Calling the idea a non-zero-sum-gaze, Bailenson describes how it is possible to make it appear as if an avatar is directing its gaze at more than a single interactant at a time. In an instructional situation with one teacher and twenty students, it is possible to make all twenty students have the illusion that the teacher is looking directly at them. It is also possible to make the situation appear differently for each participant, or to automate certain behaviors such as nonverbal mimicry in order to facilitate certain kinds of responses.

An interesting application of this line of thinking comes in the form of The Proteus effect (Yee et al. 2009). In brief, the Proteus effect explains what happens when people infer their expected behaviors and attitudes from observing their avatar's appearance in a virtual environment. For example, users who are given taller avatars will tend to negotiate more aggressively than those given shorter avatars. What is especially interesting in the experiment by Yee, Bailenson and Ducheneaut (2009) is that they noticed that the behavioral changes could transfer over, affecting subsequent face-to-face interactions. While it is certainly true that in some respect we can always try to "transform" interaction, for example when applying makeup or learning to be overtly aware of our nonverbal behavior, the idea of exploring outside the boundaries of our established notions concerning social interaction is a refreshing one. At the very least, these viewpoints remind us that we may be as little aware of nonverbal communication in virtual environment as we are in the "real life" of face-to-face interaction. Combined with the affordances that virtual environments have for altering our self-representations, we can start to make out whole new lines of research that not only describe communication behavior in technologically mediated settings, but may help us understand our face-to-face reality better as well.

As the previous examples have illustrated, game spaces and players' behavior in them can help us understand and appreciate the complexities of our contemporary communication environment, while opening up new avenues of thinking. As ever with human communication, deciding how it's best to try to understand the multiple forces at work in any given context is difficult. To what extent should one try to incorporate the technological dimension, and to what extent should one concentrate on the human actors. Returning to *World of Warcraft*, Chen (2012) offers an account on how a new technological add-on to a game, created by players and not the game company, had a tremendous influence on the raid groups' processes and capabilities, allowing them to be more coordinated in their efforts. "Not only did the add-on help us with our cognition, its use also changed who communicated with whom and about what (...)" (Chen 2012: 105). There have been other similar accounts, showing how changes in game design, or the tools that comprise the communication environment, can result in changes in player expectations, their collaboration and social interaction in general (Taylor 2006b; Chen, Sun and Hsieh 2008).

Naturally, it is often hard to tell how players use the affordances a game system presents them, especially since many contemporary online games are increasingly complex and procedural, meaning that the players are relatively free to choose how the game proceeds and what kind of an experience they create within it. In addition, games do not exist in a vacuum, but players are adept at creating go-arounds and replacements for features they wish to be there. This is similar to the concept of emergence introduced in the earlier section. One possible way of finding a balance in the issue of human *versus* nonhuman actors is to adopt a viewpoint that embraces the roles of both. For this purpose we will briefly turn to sociology and the Actor-Network-Theory (ANT) of Latour (2005).

While the focus of studies into social interaction is often on human actors, ANT proposes that nonhuman actors should be seen as just as important. This is a different way of thinking, asking us to appreciate the way in which nonhuman objects can act on us, for example by enabling or disabling certain choices, and how roles and responsibilities are distributed across multiple actors. These 'actants' can even be semiotic, that is, ideas, values, etc. anything that might make a difference in a situation. The ANT approach in examining what happens in games and player communities emphasizes that everything is in constant negotiation. For example, every time there is a new inclusion to a network, there must be a process of translation, of reassembling, that may just as well end up changing the whole network. The point of the analysis, then, is to make an assemblage – meaning all of its actors and their relationships – visible to the reader.

An approach like this naturally has its limits. For example, it is practically impossible to describe and analyze every actant, present in and absent from a game, meaning that the researcher has to make choices throughout the process. However, holistic analyses that try to take into consideration as many factors as possible are necessary in order to avoid determinism of any kind, or letting preassumptions dictate what we see when we look at communication behavior in the realm of digital games.

4 Conclusions

Looking into the immediate future, it is unlikely that the impact of games and playfulness will vanish. This is due to many reasons, such as the continued advances in communication technologies, as well as the macro-economic realities of industrialism and globalization continuing the drive towards the minimal employment of staff needed for basic production work, while more and more people provide and seek entertainment. Similarly, technology-mediated communication of the digital kind has become a prevalent part of our everyday communication landscapes. Digital games, or video games, are an integral part of that reality, and subsequently offer many opportunities for understanding contemporary communication practices.

As Lehdonvirta (2010) argues, much of what we know of online video games and their players is based on a dichotomous 'real world vs. virtual world' model, if only implicitly. What this means is that "cyberspace", or indeed smaller instances of it such as individual games, are seen as inherently separate and different from other aspects of reality. Instead of adopting this approach for future studies, Lehdonvirta (2010) suggests acknowledging the 'messy' reality that players live in, where boundaries are not set by distinct games or communication technologies, and where research that sees games or virtual worlds as independent mini-societies is flawed. Instead, Lehdonvirta reminds us of the interconnected nature of communication and relationships, taking a view not unlike systems theory, in which player behavior in video games cannot be understood in isolation from their "... other social worlds, such as families and workplaces, [that] penetrate the site of the MMO and are permanently tangled with the players' world" (Lehdonvirta 2010). Four years earlier, Taylor (2006a) presented a similar critique, reminding scholars that a virtual world is "... not a tidy, self-contained environment but one with deep ties to value systems, forms of identity and social networks, and always informed by the technological structures in which it was embedded" (Taylor 2006a: 18). A similar position can be found from contemporary scholars in similar interdisciplinary fields of inquiry such as intercultural communication, where the long-standing essentialist underpinnings of research based on easy and intuitively appealing categories, such as nationality, has been extensively criticized (see Holliday 2011). This is not to say that studies should never concentrate on a single game world or community, but rather it serves as a reminder of the dangers of oversimplification and losing sight of the larger picture in all its diversity.

There are many possible new directions where communication sciences and game studies can head. The debate between formalist or structuralist approaches into game systems and more player-centered approaches highlighting resistance and emergence will most probably continue to be vibrant. New data collection and analysis techniques allow for new types of questions to be raised regarding tensions and discrepancies between reported and observed behavior (Williams et al.

2008). Parallels will continue to be drawn between the 'real world' and the virtual one in a variety of ways ranging from economics (Castronova et al. 2009) to cross-cultural adaptation (Ward 2010).

Interaction is at the heart of games. Often, the interaction may be between a player and the game, but it is equally possible that the game's rules and mechanics encourage or force player-to-player interaction to occur. In addition to designed interactivity, it is possible that players utilize the affordances of all communication systems and channels at their disposal, creating complementary, parallel, or even contradictory dimensions of communication that entwine with game play. Because of these emergent qualities, interactivity remains fluid, difficult to predict, and immensely intriguing as a topic of scholarly attention.

References

Bailenson, Jeremy N. 2006. Transformed social interaction in collaborative virtual environments. In Paul Messaris & Lee Humphreys (eds.), *Digital media: Transformations in human communication*, 255–264. New York: Peter Lang.

Balicer, Ran. 2007. Modeling infectious diseases dissemination through online role-playing games. *Epidemiology* 18(2). 260–261.

Bartle, Richard. 1996. Hearts, clubs, diamonds, spades: Players who suit MUDs. *Journal of MUD Research* 1(1). Retrieved October 3, 2006, from http://www.mud.co.uk/richard/hcds.htm

Bartle, Richard A. 2004. *Designing virtual worlds*. Berkeley, CA: New Riders.

Billieux, Joël, Martial Van der Linden, Sophia Achab, Yasser Khazaal, Laura Paraskevopoulos, Daniele Zullino & Gabriel Thorens. 2013. Why do you play World of Warcraft? An in-depth exploration of self-reported motivations to play online and in-game behaviours in the virtual world of Azeroth. *Computers in Human Behavior* 29(1). 103–109.

Bogost, Ian, Simon Ferrari & Bobby Schweiz. 2010. *Newsgames: Journalism at play*. Cambridge: MIT Press.

Bryce, Jo, Jason Rutter & Cath Sullivan. 2006. Digital games and gender. In Jason Rutter & Joe Bryce (eds.), *Understanding Digital Games*, 185–204. London: Sage.

Castronova, Edward. 2001. Virtual worlds: A first-hand account of market and society on the cyberian frontier. *CESifo Working Paper Series No. 618.*

Castronova, Edward, Dmitri Williams, Cuihua Shen, Rabindra Ratan, Li Xiong, Yun Huang, Brian Keegan & Noshir Contractor. 2009. As real as real? Macroeconomic behavior in a large-scale virtual world. *New Media & Society.* 11(5) 685–707.

Chen, Mark. 2012. *Leet noobs: The life and death of an expert player group in World of Warcraft.* New York: Peter Lang.

Chen, Vivian Hsueh-hua, Henry Been-Lirn Duh & Hong Renyi. 2008. The changing dynamic of social interaction in World of Warcraft: The impacts of game feature change. *Advances in Computer Entertainment Technology 2008*, Yokohama, Japan. 356–359.

Chen, Chien-Hsun, Chuen-Tsai Sun & Jilung Hsieh. 2008. Player guild dynamics and evolution in massively multiplayer online games. *CyberPsychology & Behavior* 11(3). 293–301.

Curtis, Pavel. 1997. Mudding: Social phenomena in text-based virtual realities. In Sara Kiesler (ed.), *Culture of the Internet*, 121–142. Mahwah: Lawrence Erlbaum Associates.

Deterding, Sebastian, Dan Dixon, Rilla Khaled & Lennart Nacke. 2011. From game design elements to gamefulness: Defining "gamification". *Proceedings of MindTrek'11,* September 28–30, 2011, Tampere, Finland.

Entertainment Software Association. 2014. *Essential facts about the computer and video game industry*. Retrieved May 29, 2014, from http://www.theesa.com/facts/pdfs/esa_ef_2014.pdf
Ferguson, Christopher J., Adolfo Garza, Jessica Jerabeck, Raul Ramos & Mariza Galindo. 2013. Not worth the fuss after all? Cross-sectional and prospective data on violent video game influences on aggression, visuospatial cognition and mathematics ability in a sample of youth. *Journal of Youth and Adolescence* 42(1). 109–122.
Filiciak, Miroslaw. 2006. Hyperidentities: Postmodern identity patterns in massively multiplayer online role-playing games. In Mark Wolf & Bernard Perron (eds.), *The Video Game Theory Reader*, 85–102. New York: Routledge.
Holliday, Adrian. 2011. *Intercultural communication and ideology*. Los Angeles, CA: Sage.
Huizinga, Johan. 1938. *Homo Ludens: Proeve eener bepaling van het spel-element der cultuur* [Homo ludens: A study of the play-element in culture]. Groningen: Wolters-Noordhoff cop. Original Dutch edition.
Jakobsson, Mikael & T. L. Taylor. 2003. The Sopranos meets Everquest: Socialization processes in massively multiuser games. *fineArt forum* 17(8). 81–91.
Järvinen, Aki. 2003. Verkkopelien ABC. [The ABC of online games] *Mediumi* 2(2).
Kahn, Adam S., Cuihua Shen, Li Lu, Rabindra Ratan, Sean Coary, Jinghui Hou, Jingbo Meng, Joseph Osborn & Dmitri Williams. 2013. *The trojan player typology: A cross-genre, cross-cultural, behaviorally validated scale of video game play motivations*. Paper presented at the 63rd Annual Conference of the International Communication Association (ICA), London, UK.
Kapp, Karl M. 2012. *The gamification of learning and instruction: Game-based methods and strategies for training and education*. San Francisco: Pfeiffer.
Kolo, Castulus & Timo Baur. 2004. Living a virtual life: Social dynamics of online gaming. *Game Studies* 4(1).
Koster, Raph. 2002. Online world timeline. Retrieved 15. 9. 2013 from the URL: http://www.raphkoster.com/gaming/mudtimeline.shtml
Latour, Bruno. 2005. *Reassembling the social: An introduction to Actor-Network-Theory*. Oxford, NY: Oxford University Press.
Lave, Jean & Etienne Wenger. 1991. *Situated Learning: Legitimate peripheral participation*. Cambridge: Cambridge University Press.
Lehdonvirta, Vili. 2010. Virtual worlds don't exist: Questioning the dichotomous approach in MMO studies. *Game Studies* 10(1).
Lofgren, Eric T. & Nina H. Fefferman. 2007. The untapped potential of virtual game worlds to shed light on real world epidemics. *The Lancet Infectious Diseases* 7. 625–629.
Malone, Krista-Lee M. 2009. Dragon kill points: The economics of power gamers. *Games and Culture* 4(3). 296–316.
Manninen, Tony. 2003. Interaction forms and communicative actions in multiplayer games. *Game Studies* 3(1).
Mäyrä, F., Van Looy, J., & Quandt, T. 2013. Disciplinary identity of game scholars: An outline. *Proceedings of DiGRA 2013: DeFragging Game Studies*.
Ratan, Rabindra, Jae Chung, Cuihua Shen, Marshall Scott Poole & Dmitri Williams. 2010. Schmoozing and smiting: Trust, social institutions and communication patterns in an MMOG. *Journal of Computer-Mediated Communication* 16(1). 93–114.
Schiano, Diane & Sean White. 1998. The first noble truth of cyberspace: People are people (even when they MOO). *CHI '98, Proceeding of the CHI 98 Conference on Human Factors in Computing Systems*. Los Angeles, CA, 352–359.
Shen, Cuihua. 2014. Network patterns and social architecture in massively multiplayer online games: Mapping the social world of EverQuest II. *New Media & Society* 16(4). 672–691.
Steinkuehler, Constance & Dmitri Williams. 2006. Where everybody knows your (screen) name: Onlinevgames as "third places". *Journal of Computer-Mediated Communication* 11(4). 885–909.

Stohl, Cynthia & Kasey Walker. 2002. A bona fide perspective for the future of groups: Understanding collaborating groups. In Lawrence R. Frey (ed.), *New directions in group communication*. Thousand Oaks, CA: Sage.
Taylor, T. L. 2006a. *Play between worlds: Exploring online game culture*. Cambridge, MA: The MIT Press.
Taylor, T. L. 2006b. Does WoW change everything?: How a PvP server, multinational player base, and surveillance mod scene caused me pause. *Games and Culture* 1(4). 318–337.
Van Geel, Ibe. 2013, August 3. *MMO Data*. Retrieved August 15th, 2013 from http://mmodata.net/
Ward, Mark. 2010. Avatars and sojourners: Explaining the acculturation of newcomers to multiplayer online games as cross-cultural adaptations. *Journal of Intercultural Communication* 23.
Warmelink, Harald & Marko Siitonen. 2013. A decade of research into player communities in online games. *Journal of Gaming & Virtual Worlds* 5(3). 271–293.
Williams, Dmitri, Nicolas Ducheneaut, Li Xiong, Yuanyuan Zhang, Nick Yee & Eric Nickell. 2006. From tree house to barracks: The social life of guilds in World of Warcraft. *Games and Culture* 1(4). 338–61.
Williams, Dmitri, Nick Yee & Scott Caplan. 2008. Who plays, how much, and why? A behavioral player census of virtual world. *Journal of Computer Mediated Communication* 13. 993–1018.
Wright, Talmadge, Eric Boria & Paul Breidenbach. 2002. Creative player actions in FPS online video game: Playing Counter-Strike. *Game Studies* 2(2).
Yee, Nick. 2006. The demographics, motivations and derived experiences of users of massively-multiuser online graphical environments. *PRESENCE: Teleoperators and Virtual Environments* 15. 309–329.
Yee, Nick. 2007. Motivations of play in online games. *Journal of CyberPsychology and Behavior* 9. 772–775.
Yee, Nick. 2009. Befriending ogres and wood-elves: Relationship formation and the social architecture of Norrath. *Game Studies* 9(1). http://gamestudies.org/0901/articles/yee (Accessed 14 April 2014).
Yee, Nick, Jeremy N. Bailenson & Nicolas Ducheneaut. 2009. The Proteus Effect: Implications of transformed digital self- representation on online and offline behavior. *Communication Research* 36(2). 285–312.

Stefano Tardini and Lorenzo Cantoni

6 Hypermedia, internet and the web

Abstract: In this chapter, two of the most important instances of ICTs (Information and Communication Technologies) are introduced: internet and the web, together with the concept of hypermedia/hypertext, which played a pivotal role in the theoretical discussions about ICT-mediated communication as well as in the widespread diffusion of the internet and the web. In the first paragraph the concept and the history of hypertext are presented, and some relevant interpretations of it are provided, borrowed from the field of communication sciences: a linguistic and semiotic approach, a rhetorical one, and a literary one. The Internet, its history, diffusion and different layers are then presented, to introduce the result of the application of hypertext to the internet: the World Wide Web, which decreed the success of the internet as the most widespread and powerful communication technology in the start of the third millennium. A model to design and interpret websites and – more generally – applications of online communication (OCM – Online Communication Model) is then explained. Finally, in the last two paragraphs some most recent developments in the field are presented: the so-called Web 2.0 and the 'pragmatic' turn of internet search engines.

Keywords: hypertext, hypermedia, world wide web, websites, internet, web 2.0, search

In the ecological system of the media market, Information and Communication Technologies (ICTs) have gained in the last decades a dominant position, bringing along very deep changes in the way we interact and communicate, as it happened in the past centuries with other "technologies of the word" (Ong 2002) like handwriting and the letterpress print. ICTs, for instance, allow for synchronous written communications among persons who are spatially very far from one another, or for the (relatively) easy publication of rich multimedia contents by people who do not need to have specific technical skills. In this chapter, two of the most important instances of ICTs will be introduced: internet and the web, together with the concept of hypermedia/hypertext, which played a pivotal role in the theoretical discussions about ICT-mediated communication as well as in the widespread diffusion of the internet and the web.

1 Hyper-text and hyper-media

The concept of hypertext is crucial to understand the World Wide Web (WWW) and the internet, which started to spread worldwide thanks to the invention of the

WWW. As we will see in the next paragraph, at the beginning of the 90s' Tim Berners-Lee, an English researcher at CERN in Geneva (Switzerland), could invent the WWW by applying the concept of hypertext to computer networks.

Hypertexts can be defined as textual structures that allow and encourage different fruition paths. They are basically composed by two elements: nodes, i.e. content units, and links between them. Landow and Delany, who are among the first hypertext theorists, defined it as "a variable structure, composed of blocks of text (or what Roland Barthes terms *lexia*) and the electronic links that join them" (Landow and Delany 1994: 3). Two remarks on Landow and Delany's definition of hypertext are necessary here: first, nodes can be composed not only of text, but also of other kinds of contents, such as images, videos, audio, graphics and so on; for this reason, we prefer to speak of 'content units' rather than 'blocks of text' and we will use also the term 'hypermedia' as interchangeable with 'hypertext'. Second, not only electronic texts can have a hypertextual structure: in printed texts, for instance, page numbers in table of contents or numbers in footnotes or endnotes are sort of hypertextual links that refer to specific content units; again, some gamebooks allow readers to choose how to continue the story by letting them select the next content unit among different options. However, the debate on hypertext started with the advent of electronic texts, as in those hypertextuality is often one of their peculiar features. Thus, in this chapter we will speak of 'hypertexts' always referring to digital hypertexts.

1.1 The history of hypertext

The word 'hypertext' was coined by Theodor Holm Nelson in 1965. However, its concept is usually dated back to 1945, when Vannevar Bush, an important American scientist, conceived the *memex*, a device that should act as the user's memory expander. The memex is a "device for individual use, which is a sort of mechanized private file and library. [...] A memex is a device in which an individual stores all his books, records and communications, and which is mechanized so that it may be consulted with exceeding speed and flexibility. It is an enlarged intimate supplement to his memory" (Bush 1945: 45). The essential feature of the memex is the 'associative indexing', i.e. "a provision whereby any item may be caused at will to select immediately and automatically another" (Bush 1945: 45). The possibility to create associations among the elements in the memex, thus being able to access one element starting from another one, is clearly a forerunner of the hypertext. However, the memex remained only an idea in the mind of Vannevar Bush, as it was never realized.

In 1965 the words 'hypertext' and 'hypermedia' were coined by Ted Nelson: "Let me introduce the word 'hypertext' to mean a body of written or pictorial material interconnected in such a complex way that it could not conveniently be presented or represented on paper" (Nelson 1965: 144). It is worth noting here that in Nelson's definition, hypertext is characterized mainly by having elements with complex interconnections and by the difficulty to represent them on paper. As a matter of fact, Nelson was working at a project to create a hypertext software, which he named Xanadu: his goal was to develop "a huge library available to a computer network, which takes into account, by means of links, the different relationships existing among its texts/nodes" (Cantoni & Tardini 2006: 91). However, also Xanadu, like Vannevar Bush's memex, was never realized: in 1987 a prototype was released, but no final release was published afterwards.

Nonetheless, in 1968 the first prototype of a hypertextual system was realized by Douglas Engelbart, an American researcher who invented, among other things, also the mouse, screen windowing and the word processor: NLS (oNLine System). NLS was the result of the researches of Engelbart and his team at the Augmentation Research Center, whose goal was to augment human intellect. More specifically, NLS was part of the Augmentation System, a system where information was hierarchically organized and made accessible through indexes and directories, and where users could create links among documents. NLS was demonstrated in December 1968 during a conference in San Francisco, in an event that is remembered still now as 'the mother of all demos' (see Engelbart and English 1968).

The first real hypertext systems were released only in the middle of the 80s': the more important ones were Storyspace and Hypercard. The former was invented by Jay David Bolter and Michael Joyce, and was presented in the first international meeting on hypertext in 1987; it is specifically targeted to the creation of fiction hypertexts, and is currently distributed by Eastgate (www.eastgate.com/storyspace/index.html). The latter was created by Bill Atkinson in 1985, and was given a few years later to Apple, under the condition that Apple would release it for free on all Macs; it was withdrawn from sale in 2004. Also Hypercard was targeted to the creation of fiction and educational hypertexts, although it was very flexible and it could be used for many purposes (presentations, databases, etc.).

However, the most important and successful application of hypertext is no doubt HTML language (HyperText Markup Language), which is at the basis of the whole World Wide Web. The basic features of HTML consist in "its being able to represent hypertextual connections, and in its use of marking elements (tagging)" (Cantoni and Tardini 2006: 65). HTML was invented by Tim Berners-Lee at the beginning of the 90s' as an application of SGML (Standard Generalized Markup Language), and was released first in 1993. Together with HTML, also HTTP protocol (HyperText Transfer Protocol) – the protocol the World Wide Web is based upon – has to be mentioned here as one of the most important applications of hypertext. The most relevant features of both HTML and HTTP will be described in the next paragraphs.

1.2 The structure of hypertext

As we have quickly introduced above, hypertexts are textual structures that are composed by content units (nodes) and links between them. *Content units* can be defined as the minimal fruition units of a hypertext, i.e., they are units that readers have necessarily to receive at once, without the possibility to choose only one part of them; they are composed of (combinations of) texts, images, videos and audios. *Links* are connections between a content unit (or a part of it) and another content unit or another part in the same content unit. Where a link is made available, usually this is signaled by some visual markers, for instance by underlining or boldfacing a piece of text, by having the cursor change its appearance when rolling over it (e.g., the arrow becomes a hand), and in other ways. Links always have an activation point and a destination point; both of them can be constituted by any kind of contents (text, image, audio, video), although activation points are most frequently constituted by texts or images.

Hypertexts can be roughly divided into *static* and *dynamic* hypertexts: in the former the appearance of each single node is completely pre-determined by the author/designer, while in dynamic hypertexts designers do not "define the content of each and every node, but rather they 'just' set the rules for their arrangements. (…) [H]ypertext designers have control over the rules for presenting contents in the hypermedia's nodes, but not over the nodes as they will be accessed and seen by users" (Cantoni and Tardini 2006: 73–74; see also Cantoni and Vittadini 2003, Cantoni and Paolini 2001). In the World Wide Web, most of the available websites are now dynamic websites, in particular all websites that have a certain complexity, as, for instance, news websites, eCommerce websites, educational websites, and so on. In all these websites, designers set the rules for the appearance of contents to be displayed on the page, establishing – for instance – that in the item's page the name of the item to be sold appears on top, then one image of the item and its description are displayed immediately below the title, then a detailed description follows, then the comments of users who have bought that item are presented, and so on. In other words, designers cannot determine exactly which contents will appear, as these are stored in a database and will be displayed in the web page according to the rules defined by the designer. In some cases, different contents are presented depending not only on the designers' instructions, but also on some external parameters, such as the device, the operating system or the browser used by a visitor, the country from where s/he is accessing the website, the language of the visitor, the items s/he has viewed or bought before, and so on. Hypertexts of this kind are called *adaptive* hypertexts, because they try to adapt the contents that are presented to the visitor's needs (Brusilovsky and Nejdl 2005). Adaptive hypertexts are more and more diffused in the World Wide Web, particularly in contexts where it is important to keep track of users' behaviors in order to offer them as much tailored and customized communications as possible, such as in educational and in eCommerce websites.

Dynamic and adaptive hypertexts clearly show that the structure of hypertexts is not a simple one. Jakob Nielsen (1995: 132–133) has singled out three main levels in the architecture of hypertextual structures: 1. the Database level; 2. the Hypertext Abstract Machine (HAM) level; and 3. the Presentation level. The Database level is the level where all the information (contents and links) are stored; this level has nothing to do specifically with hypertext. At the HAM level, the structure of the content units and of the links among them is defined: this level is where "the hypertext system determines the basic nature of its nodes and links and where it maintains the relation among them" (Nielsen 1995: 132). Finally, the Presentation level sets the rules for the presentation of the contents to the users, i.e., it defines how information has to be displayed in the user interface. These levels of a hypertextual structure can be compared to the different levels of linguistic production, through which speakers produce linguistic texts; in the next section of this paragraph some possible interpretations of and approaches to hypertexts borrowed from different disciplines will be presented, namely interpretations taken from linguistics and semiotics, from literary studies, and from rhetoric.

1.3 Interpretations of hypertext

Although the concept of hypertext has been conceived in the field of computer science, it plays an important role in different disciplines, for both its theoretical implications and practical applications. As for the latter, it is worth mentioning here only the World Wide Web, which will be presented in more detail in the last part of this chapter. As regards the theoretical implications of the concept of hypertext, they range across very different disciplines: from Human Computer Interaction, where the concept of hypertext contributed to the discussion about interactivity, to cognitive and education sciences, where new learning models have been conceived based on hypertextual structures. In this section, we will focus on three disciplines, all belonging to the field of communication studies: 1. linguistics and semiotics, where the concepts of text itself and dialogue have been affected by the diffusion of hypertexts; 2. literary studies, where the roles of authors/writers and readers have been seriously challenged and re-defined; and 3. rhetoric, which can provide a useful approach to a better understanding of the main dynamics of hypertext.

1.3.1 Linguistic and semiotic approach

The diffusion of hypertexts, especially through websites, has re-proposed the discussion about the nature itself of texts, in its relations with other texts (intertextuality and hypertextuality). If a text is a sequence of linguistic signs that performs a unique and relatively autonomous communication task (Rocci 2003) or that can be seen as a global unit of meaning (Andorno 2003; Conte 1980), how can these autonomous communication tasks or global units of meaning be recognized when the main feature of this kind of texts is the reference to other texts?

As already anticipated, discussions on hypertextuality and inter-textuality have started long time before the advent of digital hypertexts: the linguistic tradition has always acknowledged the presence of linguistic elements whose function is to create links between (parts of) texts or to refer to the external reality, i.e., linguistic elements whose function cannot be entirely performed within the borders of a text. Let us think, for instance, of deixis, of anaphoric and cataphoric elements, and so on. However, the diffusion of hypertexts in the WWW has posed the question whether a new form of textuality has emerged, which is based on the continuous reference to other texts. This is the idea of 'unlimited semiosis' central to Peirce's conception of the sign; in a sense hypertexts seem to have materialized these kinds of semiotic models, like Quillian's model Q:

> "Quillian's model (...) is based on a mass of nodes interconnected by various types of associative links. (...) The model, in all its complexity, is based on a process of unlimited semiosis. (...) We can imagine all the cultural units as an enormous number of marbles contained in a box; by shaking the box we can form different connections and affinities among the marbles. This box would constitute an informational source provided with high entropy, and it would constitute the abstract model of semantic association in a free state. (...) A system is a rule which magnetizes the marbles according to a combination of mutual attractions and repulsions on the same plane" (Eco 1979: 122–6).

In hypertexts "elements can be arbitrarily connected to each other; by means of links every object can be made the sign of any other object. In this way, the hypertextual structure can proceed endlessly" (Cantoni and Tardini 2006: 95). However, these approaches do not consider that hypertexts are first of all communications, and these processes of infinite and self-referential semiosis are not able to create real communications.

In a communicative perspective, hypertexts can be seen as dialogues. In a sense, any text can be seen as a dialogic structure: as a matter of fact, every text is generated by a high-level question, posed by the reader/receiver. This question, which corresponds to the general goal/desire of the reader, defines the task of the (hyper)text, and can be more or less generic: "Can you give me some information about St. Peter's Cathedral in Rome?", "How much does a ticket for the next match of Chelsea FC cost?", "I want to have some fun. Which games can you offer?", and so on. So, if in a broad sense in hypertexts, as in any kind of text, the reader poses the first question, then, in the dialogue that follows, it is the system that poses questions to the reader/user: "readers are continuously requested to answer hypertext's questions: "what do you want afterwards?" From this point of view, a hypertext can be seen as a (partially foreseen) dialogue, being actualized by (partially foreseen) dialogical exchanges" (Cantoni and Paolini 2001: 43). In this perspective, all the links that are offered to the reader in a hypertext page can be seen as a question that asks the reader to choose among different options: "do you want to go back to the Home page or to the Projects page or to the People page or to the Services page or do you want more info on this project or do you want to download the slideshow (and so on)?". Every interaction allowed by the hypertextual system through links, forms, and buttons is a request to the reader/user to choose an option or to perform a specific action.

1.3.2 Literary studies

As observed by Landow, in the field of literary studies, hypertext "has much in common with recent literary and critical theory. For example, like much recent work by poststructuralists, such as Roland Barthes and Jacques Derrida, hypertext reconceives conventional, long-held assumptions about authors and readers and the texts they write and read. Electronic linking (...) also embodies Julia Kristeva's notions of intertextuality, Mikhail Bakhtin's emphasis upon multivocality, Michel Foucault's conceptions of network of power, and Gilles Deleuze and Félix Guattari's ideas of rhizomatic, 'nomad thought'" (Landow 1994: 1). As a support for the diffusion of knowledge, hypertext has been approached in this field after the models of a book and of literature itself. In the first approach, as a new space of writing, hypertext can be easily interpreted after the model of printed encyclopedias or handbooks: as a system of references and links among its entries, electronic hypertexts are a re-mediation of printed encyclopedias and handbooks; furthermore, the organization of the World Wide Web itself is encyclopedic, as it provides, usually through a limited number of portals, access to a huge amount of information (Bolter 2001). In the second approach, hypertext is seen as a whole literary system, thanks to its capability to link potentially every text into a single metatext. In this perspective, two basic assumptions of "traditional" literary theories are challenged: the idea of text as an authority, and the relation between the author and the reader of a text. As for the former, as we have seen in the linguistic and semiotic approach presented above, hypertext has challenged "the idea of the text as an authority (*auctoritas*), i.e. as a complete work, with a beginning and an end, which can be easily kept separate from other texts and cannot be modified through time" (Cantoni and Tardini 2006: 94). As for the latter, earlier hypertext theories have stressed the fact that the author of a hypertext is no longer the ultimate responsible for the creation of a text, because this is co-created, in a sense, by the author and the reader, who participates in the co-creation of the text by selecting a link and choosing in this way his/her own path through it. In this perspective, it can be said that "in hypertext the function of reader merges with that of author and the division between the two is blurred" (Landow 1994: 14). The following rhetoric approach can help us further elaborate this issue.

1.3.3 Rhetoric

Ancient rhetoric developed a five-step process for the production of a speech. The five traditional steps were called: *inventio*, *dispositio*, *elocutio*, *memoria*, *actio*. In this section, we will just introduce them in their relevant relations with hypertexts (Liestøl 1994), focusing only on the *dispositio*.

1. *Inventio* is the activity through which all the concepts, ideas, arguments and topics that are to be presented in the speech are discovered and gathered together; it can be compared to the activities of brainstorming and brain mapping. In hypertexts, *inventio* can be seen as the phase where the designer finds out ideas about the contents of the hypertext, how they are structured and connected to one another, how navigation is structured, and so on.
2. *Dispositio* is the activity of ordering the concepts and arguments in a linear sequence; in this phase, the orator has to decide which arguments will be presented at the beginning of his speech, which arguments in the middle, which arguments at the end. The issue of *dispositio* is crucial in the debate about hypertext, as one of the main features that is always attributed to hypertexts is non-linearity or multi-linearity (see, for instance, Aarseth 1994, 1997; Liestøl 1994; Slatin 1994; Landow 1997; Bolter 2001; Fagerjord 2003), i.e., the possibility to offer different fruition paths to readers. In hypertexts, the *dispositio* is not defined by authors/designers, but it is performed by readers. In this sense, it must be remarked that, on the one hand, from the point of view of the reader, the fruition of a hypertext is always linear, because through his/her choices the reader necessarily puts one content unit after the other. On the other hand, however, readers perform the *dispositio* by selecting the content unit to access among different possible options, i.e. they perform one linear sequence selecting it among many other possible linear sequences. This fact has often been interpreted as a re-balancing of the roles of author and reader in hypertexts, where authors would lose part of their power, while readers would rise to the level of authors because of their ability to co-create the text by performing different *dispositiones*: "no longer an intimidating figure, an electronic author assumes the role of a craftsperson, working with prescribed materials and goals. She works within the limitations of a computer system, and she imposes further limitations upon her readers. Within those limits, however, her reader is free to move. (...) [T]he reader participates in the making of the text as a sequence of words. (...) The author writes a set of potential texts, from which the reader chooses, and there is no single univocal text apart from the reader. The role of the reader in electronic fiction therefore lies halfway between the customary roles of author and reader in the medium of print" (Bolter 2001: 168–173). However, also in this case it must be remarked that in some cases in digital hypertexts authors might have much more power over their readers than in printed texts: we can think, for instance, of some educational applications where the availability of some resources is conditioned to getting a minimum score in a previous activity (e.g.: a quiz): in a printed handbook it would be impossible for an author to prevent the reader

to read one page if s/he has not obtained a certain score in a quiz! Again, in some web applications it happens sometimes to be forced to watch an advertising before having the possibility to access the requested content: "You can skip this ad in 5 seconds"; this would not be possible with any printed media. These examples clearly show that if in hypertexts the relationship between author and reader has to be rethought, this does not necessarily go in the direction of an empowerment of the reader, because hypertext designers have more control over linearity than authors of printed texts, and can decide to allow the reader to go through more or less possible paths, i.e., they can decide to give the reader more or less freedom in his/her choices. This makes the task of hypermedia designers "very important and difficult, since they have to guarantee that all good *dispositiones* are generated, and at the same time that all bad ones are avoided" (Cantoni and Tardini 2006: 83).

3. *Elocutio* is the activity of providing all the concepts and arguments that have been arranged in a sequence with a linguistic form. In hypermedia, *elocutio* refers to the decision on how to present all the contents to the user, i.e., how many elements must be put in each content unit, which kind of elements (text, audio, images, movies, animations, etc.), how to dispose them reciprocally in the display space of the content unit, and so on.
4. *Memoria:* in ancient rhetoric, this was the activity of memorizing the speech, for which many techniques and strategies were developed (mnemotechniques). In hypermedia design, *memoria* plays an important role in different aspects, such as the possibility of the hypermedia application to 'remember' the content units a reader has already visited, to 'remember' the objects a visitor has put in his/her basket or in his/her list of favorites in an eCommerce web application, to 'remember' the user navigation and 'learn' from it in adaptive hypermedia, and so on.
5. *Actio*, finally, is the performance of the speech. In hypertexts, "*actio* coincides with the *dispositio*: since the readers actualize by means of their choices one of the potential *dispositiones* offered by the designer in the same time they perform the *actio*. In other words, the interaction of the reader with the system creates the actual (or 'acted') *dispositio*, i.e. the *actio*: this convergence is an important novelty introduced by hypermedia" (Cantoni and Tardini 2006: 90; see also Cantoni and Paolini 2001: 50).

This approach borrowed from ancient rhetoric leads us to propose a further interpretation of hypertexts: hypertexts as languages (Cantoni and Paolini 2001). In this perspective, hypertexts' designers "do not design actual texts or dialogues, but only sets of syntactic rules and basic elements, i.e. they produce a sort of grammar. In this sense, hypertexts can be considered as new languages. In other words, hypertexts' authors design potential dialogues, which will become an actual communication, i.e. a complete exchange of meaning, only when one reader navigates through the hypertext's structure" (Cantoni and Tardini 2006: 78).

In the next paragraphs, the most popular and important application of hypertext will be presented: the World Wide Web. Before addressing it, it is worth introducing the infrastructure to which the concept of hypertext has been applied to create the WWW: the internet.

2 Internet

2.1 History and diffusion of the internet

The internet is a networking infrastructure that connects millions of computers all over the world. Its origins are usually traced back to 1957, when the Soviet Union launched *Sputnik I,* the first artificial satellite, into space; as a reaction, in 1958 the United States founded the Advanced Research Projects Agency (ARPA) and, within it, in 1962 the Information Processing Technology Office (IPTO), to contend the supposed technological supremacy of the Soviets. One goal of ARPA and IPTO was to develop a telecommunication network that could survive in the case of an attack to the US or of a nuclear war.

The first core of a computer network (ARPANET) was active in 1969 and connected four US universities: the University of California Los Angeles, the University of California Santa Barbara, the University of Utah, and the Stanford Research Institute in Palo Alto. This was made possible by the development of different concepts and processes that occurred in the field of computer science in the early 1960s: 1. the invention of *time sharing*, which allowed more users to work in the same time on the same computer through different terminals, thus making the use of computers much more effective; 2. the invention of *packet switching*, which allowed information to be sent over the network (e.g., files) to be divided into small pieces (packets) that were put together again at the receiver's end, thus allowing the transmission of larger and larger pieces of information avoiding the congestion of lines; 3. the development of the concept of *distributed network*, i.e. information systems that do not have any hierarchy among their nodes, thus being much more secure than centralized ones, because if one node is destroyed information pieces can follow another route among the remaining nodes (Baran 1964).

The application of time sharing and packet switching to the concept of distributed network led to the birth of the first computer network. It is still debated if the origins of the internet are to be found in the military field, in the scientific/academic one or in the commercial one. According to some authors, the military origin of the internet is "a myth that has gone unchallenged long enough to become widely accepted as a fact" (Hafner and Lyon 1998: 10); nonetheless,

Baran's researches on distributed networks had strict military aims and exerted big influence on the researches conducted at ARPA.

The origins of the internet are likely to be found in the convergent contributions given by the military, scientific and commercial fields, if not in their very cooperation: ARPANET could see the light in 1969 thanks to the strict collaboration among the four universities, the ARPA agency, a private organization (BBN – Bolt Beranek and Newman) and a computer industry (Honeywell) (Cantoni and Tardini 2006: 29, Blasi 1999: 28–29). Furthermore, in addition to ARPANET and to the concept of a military network developed at RAND (Research and Development) Corporation by Paul Baran (Baran 1964), also the commercial network developed at the National Physical Laboratory (NPL) in England and the scientific network Cyclades, in France, can be considered as the foundations for the modern internet. Specifically, NPL contributed to the development of packet switching, while Cyclades focused on the connection among networks rather than computers, contributing to the further development of the communication protocols in use (NCP – Network Control Protocol at the beginning, TCP – Transmission Control Protocol afterwards), which ended in the definition of the protocol TCP/IP (Transmission Control Protocol / Internet Protocol). TCP/IP is the protocol still in use in the internet today, and was adopted by ARPANET in 1983: "the transition to TCP/IP was perhaps the most important event that would take place in the development of the internet for years to come. After TCP/IP was installed, the network could branch everywhere" (Hafner and Lyon 1998: 249).

After its foundation, ARPANET started to grow: in 1971 it had 23 hosts, in 1982 235, in 1990 313,000. In 1990 ARPANET was officially decommissioned; at that time the backbone of the internet was an academic network: NSFNET – National Science Foundation Network. However, still in the first half of the 1990s, the internet was only one among different available networks, such as Usenet, AmericaOnLine, Prodigy, Fidonet, Videotel, Minitel, and others. What decreed the success of the internet over the other networks was the invention of the World Wide Web (WWW), made by Tim Berners-Lee in 1993 at the European Organization for Nuclear Research (CERN, in Geneva, Switzerland). Basically, Berners-Lee put together hypertext and computer networks, thus making the internet shift from the 'messenger' model to the 'hypermedia' one: "the internet was no longer intended primarily as a tool for information exchanges (via textual chat systems) between human beings, but as a place for searching, retrieving and consulting documents of all kinds" (Cantoni and Tardini 2010: 223; see also Berners-Lee 2000). Although the Web was "a relatively primitive hypertext system (...) [it was] overwhelmingly successful in linking and making accessible a world-wide wealth of information, more than has ever been contained in any physical library" (Montfort and Wardrip-Fruin 2003: 791). The invention of the WWW was crucial to fill the gap that existed until the end of the 1980s between the world of computers and that of networks,

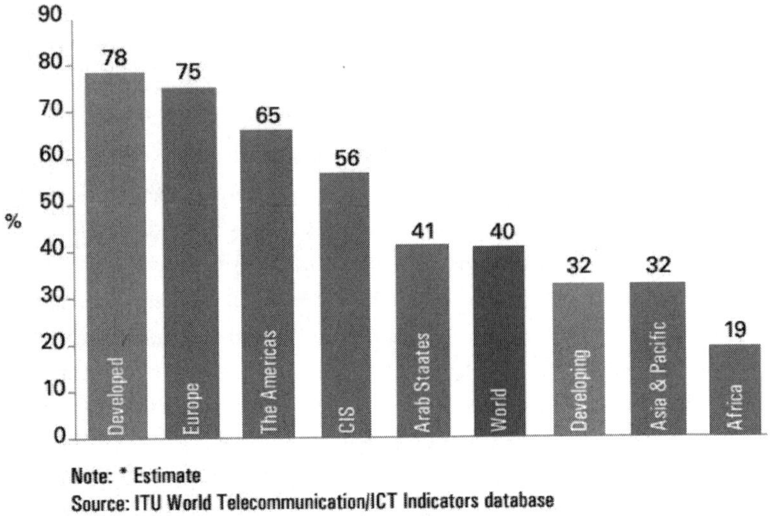

Fig. 1: Percentage of individuals using the internet, by region, 2014 (estimate). Source: The World in 2014. ICT facts and figures, ITU, available online at: http://www.itu.int/en/ITU-D/Statistics/Documents/facts/ICTFactsFigures2014-e.pdf.

due to the fact that the graphic level and multimedia richness of offline computers could not be reached by network applications. After the invention of the Web, the internet became the predominant networking system, counting about 50 million of users in 1995.

As a further evolution of the WWW, in 2004 the term web 2.0 appeared for the first time (O'Reilly 2005), referring to "a second generation of the World Wide Web that is focused on the ability for people to collaborate and share information online. Web 2.0 basically refers to the transition from static HTML Web pages to a more dynamic Web" (http://www.webopedia.com/TERM/W/Web_2_point_0.html). Web 2.0 refers mainly to a new use of the web itself, which has three main features: 1. a huge enlargement of the possibilities for people to publish content online (so-called User Generated Contents – UGC); 2. a shift from the 'library' model of the Web to the 'square' model, i.e., "a public space where people go to meet, to share and discuss knowledge" (Cantoni and Tardini 2010: 223); 3. the fulfillment of the multimedia promises of the Web, made possible also by an always increasing availability of large bandwidth connections.

It is not easy to establish how many internet users are there in the world today; according to ITU – International Telecommunication Union, the United Nations specialized agency for ICTs, by end 2014 the number of internet users globally has reached almost three billion, which is around 40 % of the world's population. However, the penetration of the internet is still unbalanced between developed and developing countries: internet user penetration has reached 78 % in developed countries and 32 % in developing countries. The region with the highest penetration of the internet is Europe (75 % of European people using the internet), followed by the Americas (65 %), CIS (Commonwealth of Independent States – 56 %), the Arab States (41 %), with Asia & Pacific and Africa being below the world average (32 % and 19 %, respectively).

2.2 Internet and its different layers

Before approaching the web and its endless websites, a brief introduction to the internet structure is needed, not to discuss in detail all its technical aspects and issues, but to get the needed understanding of the involved complexity of the mother of all networks.

While several architectural views of computer networks have been proposed (e.g. the OSI Reference Model – The ISO Model of Architecture for Open Systems Interconnection), we will approach this task by presenting the structure of the Internet as it has been described and defined by the Internet Engineering Task Force (IETF), which defines itself as "a large open international community of network designers, operators, vendors, and researchers concerned with the evolution of the Internet architecture and the smooth operation of the Internet" (www.ietf.org/about/). IETF publishes documents called *Requests for Comments* (RFC), which contain technical and organizational notes about the Internet. Two such RFC are to be approached here: 1122 (October 1989) "Requirements for Internet Hosts – Communication Layers" as well as 1123 (October 1989) "Requirements for Internet Hosts – Application and Support", which collectively address the four layers internet is based upon: (i) link, (ii) IP, (iii) transport, and (iv) application.

(i) The *link layer* manages the direct connection between two host computers/apparatuses, so that they can exchange data between them. At this level, for instance, you might have heard of the MAC: Media Access Control, a communication protocol used to reference individual physical devices. To ensure that data can move across a complex network, not only through direct point-to-point connections, the (ii) *IP (Internet Protocol) layer* is needed, which allows to address all individual nodes in the network. In order to move data over the internet, all hosts have to be identified through a unique address, the IP number; IANA – Internet Assigned Numbers Authority (www.iana.org) "is in charge of all 'unique parameters' on the Internet, including IP (Internet Protocol) addresses. Each domain name is associated with a unique IP address, a numerical name consisting of four blocks of up to three digits each, e.g. 204.146.46.8, which systems use to direct information through the network" (www.ietf.org/glossary.html#IANA). (iii) The *transport layer* provides end-to-end communication services for applications. The most known transport layer protocol is TCP: Transmission Control Protocol: "a reliable connection-oriented transport service that provides end-to-end reliability, resequencing, and flow control" (RFC 1122: 1.1.3). (iv) The highest level, the application one, manages the actual use of data on the network and is responsible for creating and transmitting user data between applications. "We distinguish two categories of application layer protocols: user protocols that provide service directly to users, and support protocols that provide common system" (ibid.). Even if not foreseen in 1989, nowadays one of the most important protocols within the application layer is the HTTP: HyperText Transfer Protocol, which forms the basis for communication on the web (HTTP has been addressed by RFC 2616, dated June 1999).

2.3 The web and its websites

"The World Wide Web (WWW, or simply Web) is an information space in which the items of interest, referred to as resources, are identified by global identifiers called Uniform Resource Identifiers (URI)." ("Architecture of the World Wide Web, Volume One", W3C Recommendation, 15 December 2004, 1, www.w3.org/TR/webarch/). Web-related standards are discussed and developed by the World Wide Web Consortium (W3C), "[l]ed by Web inventor Tim Berners-Lee and CEO Jeffrey Jaffe, W3C's mission is to lead the Web to its full potential" (www.w3.org/Consortium/). It's worth reading a citation from the "Architecture of the World Wide Web, Volume One" (W3C Recommendation, 15 December 2004, www.w3.org/TR/webarch/), which well explains the different levels of the web:

Story

While planning a trip to Mexico, Nadia reads "Oaxaca weather information: 'http://weather.example.com/oaxaca'" in a glossy travel magazine. Nadia has enough experience with the Web to recognize that "http://weather.example.com/Oaxaca" is a URI and that she is likely to be able to retrieve associated information with her Web browser. When Nadia enters the URI into her browser:

1. The browser recognizes that what Nadia typed is a URI.
2. The browser performs an information retrieval action in accordance with its configured behavior for resources identified via the "http" URI scheme.
3. The authority responsible for "weather.example.com" provides information in a response to the retrieval request.
4. The browser interprets the response, identified as XHTML by the server, and performs additional retrieval actions for inline graphics and other content as necessary.
5. The browser displays the retrieved information, which includes hypertext links to other information. Nadia can follow these hypertext links to retrieve additional information.

This scenario illustrates the three architectural bases of the Web [...]:

1. Identification [...]. URIs are used to identify resources. In this travel scenario, the resource is a periodically updated report on the weather in Oaxaca, and the URI is "http://weather.example.com/oaxaca".

2. Interaction [...]. Web agents communicate using standardized protocols that enable interaction through the exchange of messages which adhere to a defined syntax and semantics. By entering a URI into a retrieval dialog or selecting a hypertext link, Nadia tells her browser to perform a retrieval action for the resource identified by the URI. In this example, the browser sends an HTTP GET request (part of the HTTP protocol) to the server at "weather.example.com", via TCP/IP port 80, and the server sends back a message containing what it determines to be a representation of the resource as of the time that representation was generated [...].
3. Formats [...]. Most protocols used for representation retrieval and/or submission make use of a sequence of one or more messages, which taken together contain a payload of representation data and metadata, to transfer the representation between agents. [...]

The following illustration shows the relationship between identifier, resource, and representation.

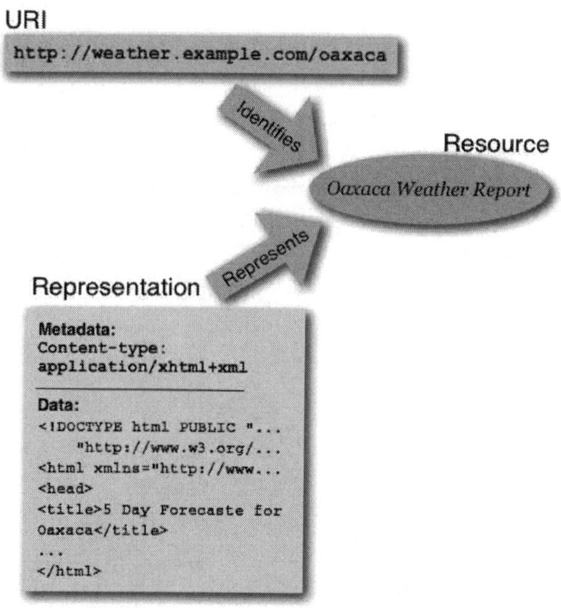

Fig. 2: (Architecture of the World Wide Web, Volume One, W3C Recommendation 15 December 2004, Introduction, http://www.w3.org/TR/webarch/).

The structure created by the web has become by far the largest multi-media hypertext, with endless information available in it, a structure to which theoretical approaches developed for hypertexts (as presented above) can be fully applied.

Hereafter, we will approach the web with a communicative perspective: after having offered an overall framework of understanding – the Online Communication Model – it will be followed by in-depth discussion of two communication-related issues.

3 The Online Communication Model (OCM)

While websites and other online internet services are for sure a matter of technologies, along the various layers outlined above, they should also be framed as communication media. To that goal, the OCM offers a quite comprehensive framework (Cantoni and Tardini 2006; 2010).

It suggests to distinguish four main pillars and a fifth element: (i) contents and services; (ii) accessibility tools and publication outlets; (iii) people managing the online resources; (iv) people accessing them. The fifth element (v) are the info-competitors: the information market surrounding a given resource.

Like in every symbolic system, also online every element takes its own meaning and role not only by its peculiar nature, but also by the universe of meanings it is part of. Nevertheless, *when online* such universe of concurring/competing meanings is much more easily accessible and present: let's think, for instance, of a search on a search engine (e.g.: Google): thousand if not million resources will be presented, each of them concurring/competing to answer your information need. The web has provided a technical embodiment of the 'semiosphere' (Lotman 2001), which has speeded up all its dynamics.

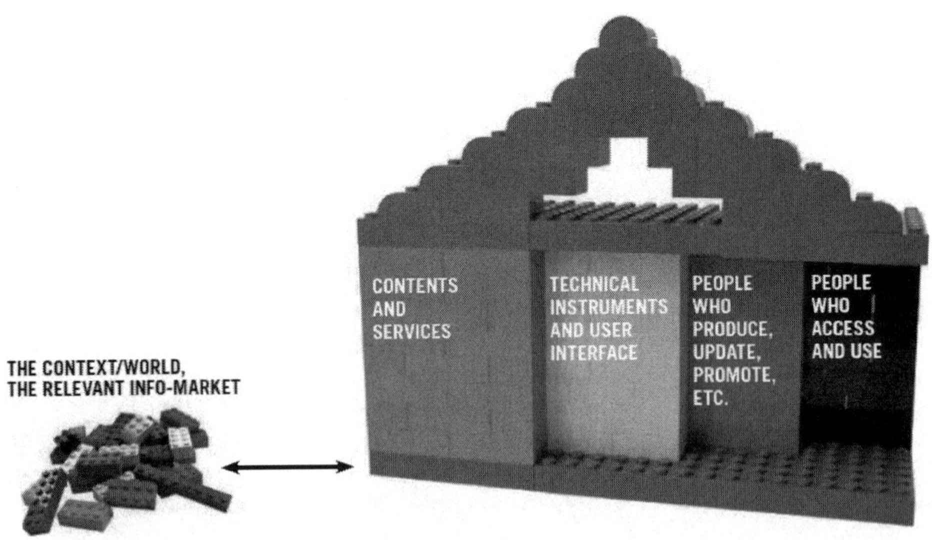

Fig. 3: The Online Communication Model (OCM).

Main communication-related issues connected with the four pillars can be summarized as follows: (i) information quality, information architecture; (ii) human-computer interface, media and publication outlet choice; (iii) organizational maturity when it comes to online communication, user requirements elicitation; (iv) user experience / usability, analysis of usages (webanalytics), online promotion, reputation analysis.

The following table presents an overview of OCM and related communication issues.

online communication model

Cluster	Pillar/ Element	Outline	Description	Relevant communication issues	Main issues in web2.0
Things	I pillar	Contents and functionalities	It includes all contents and services provided, their structure and features	– Information quality – Information architecture – Semiotic code choice – Formats/styles of web writing (new 'literary genres') – Online service design – Intercultural communication/Localization – Communication benchmarking – Arguments' design – Ethics of/in communication	User Generated Contents (UGC)
Things	II pillar	Accessibility tools and publication outlet	It includes everything that is needed to make available the elements belonging to the first pillar: hardware, software and human computer interface. It also includes the choice of where to publish such elements, be it an owned media (e.g.: a corporate website), an earned one (e.g.: a Facebook page), or a paid one (e.g.: a paid campaign hosted by a third party).	– Search Engine Optimization (SEO) – Online communication strategy (incl. evaluation of adequate publication outlet) – Human Computer Interface/Interaction – Usability and User Experience	– Earned media – Multimedia in its fullness

online communication model

Cluster	Pillar/ Element	Outline	Description	Relevant communication issues	Main issues in web2.0
Persons	III pillar	People who manage online contents and services	It includes all involved people having a stake in the publication of contents/services: from those who have the idea, up to those who design, develop, implement, test, maintain, promote, evaluate it, not to forget those who interact with users.	– User requirements' engineering/ elicitation – Organizational Maturity when it comes to online communication – Training in relevant communication issues (*see all other pillars*)	– Always on (mobile) – Accessing info + meeting people (library + public square)
Persons	IV pillar	People who access online contents and services	It includes all people who access published messages. In fact, in the so-called web2.0, users do become quite easily publishers – while producing UGC: User Generated Contents – so to blur the border in-between pillar III and IV, and to co-create the online representation of companies, institutions etc.	– Human Computer Interface/Interaction – Usability and User Experience – Online promotion – Usages' analysis (webanalytics) – Search Engine Marketing	
Context	V element	Information competitors	All global communication players that are likely to be accessed by users interested in a given topic/issue.	– Reputation in online media – Online image – Communication benchmarking	

A single introductory chapter is not enough to present them all in detail; in the following pages two dynamics will be singled out and discussed: the so-called web2.0, and the "pragmatic" shift of search engines.

4 The web2.0

Usually attributed to Tim O'Reilly (2005), the term – which has enjoyed a great success – suggests that we are now experiencing a new, better, more advanced web, as it happens in the standardized naming of software releases, which use a subsequent figure to suggest a more stable piece of software, which offers extended, more advanced, and new functionalities. For the sake of correctness, it's important to mention that years earlier George Landow used the same metaphor applied to hypertext in his book, dated 1997 and titled ...

4.1 Hypertext 2.0

In fact, in recent years there haven't been substantial changes in the technologies that back the web: what has dramatically changed is the way people live the online communication, the way it has been socialized. Three main elements can be stressed here.

1. The emergence of UGC: User Generated Contents (also named "user created contents"). While recent technologies – especially desktop publishing – had socialized self-publishing, internet has offered a new platform where to share contents with a (theoretically) endless number of people. At first, such publishing opportunities did require still a certain level of technological competence, necessary to produce html pages and to upload them to a web-server. Recently, the publication threshold has been dramatically lowered: new applications have been developed, which do not require (almost) any technical competence: to publish on a blog, to share pictures and videos on one's own Facebook profile, on YouTube or Flickr, one has just to fill-in forms. Even the computer itself is not any longer necessary to enable such publication process: you can take a picture with your smartphone, and share it directly online via a single button click.
2. A second qualifying aspect of web2.0 is the fact that internet is accessed not only to get/find information, but also – and sometimes mostly – to meet people, to stay in touch with relevant persons. At first, the web has been interpreted as the best approximation to a universal library (something like the Library of Babel, in the story by Jorge Luis Borges) where to find almost every piece of information. Nowadays, with the emergence of social media and other similar tools, it has to be framed also as the largest public square, where people go to meet other people, and to have conversations with them. In ancient and modern libraries, visitors are requested to stay silent, not to disturb the reading process, on the contrary, in public squares people chat. In fact, UGC are closely connected with this second aspect of web2.0: while having conversations with their friends (whatever large this term can be), people publish documents, which in many cases are publicly available and persistent, and which become in their turn information findable on search engines ...
3. Two technological affordances should be presented as the last characterizing aspect of web2.0: mobile connection and available bandwidth. The first one has made possible being always connected, merging web activities within all other existential activities: people access social media, e-mails, and other internet/web-services in every moment, from every place, with the mediation

of a single, small – really personal – device. The second technical condition has made possible the fulfilment of the multimedia promise of the web: from its very beginning the web has been a multi-media hyper-text, but in fact – due to technical constraints as well as to communicative practices/patterns – it has been made for long mostly by texts and a few images. Nowadays, people access the internet not only to read, but also to see high quality images and videos, and to listen to music.

5 The 'pragmatic' turn of search engines

As mentioned above, to better understand the web as it is now, we need to add the metaphor of a public square to that of an endless library, but still this last image is of the utmost importance to interpret it. As in every library, a librarian is needed to find relevant information. Search engines do play such role for the web, providing relevant matches between information needs and available resources. Their operations are based on indexing activities, which try to interpret the most relevant keywords for each document. Such operations are based on natural language processing (NLP) algorithms, which operate at the syntactic and the semantic levels. While such algorithms do operate on 'internal' elements (i.e.: a page's source code, and its URI), also 'external' elements, belonging to the information market, have started to be considered, and have granted search engines a quality quantum leap. One of the better known is for sure the so-called link-popularity (which is called by Google Page Rank): it considers every link pointing at a given page as a recommendation to access it, thus: the more backlink a page enjoys, the higher its position in the rank (if, naturally, all syntactic/semantic conditions are fulfilled). As it can be seen here, pragmatic is entering the playfield: search engines are not selecting pages only depending on their fitness to certain queries, but also considering how other people – those managing other pages linking them – are evaluating such pages. While, of course, link-popularity has been revised and refined to avoid possible search engine spam (e.g.: people creating websites just to link other websites), several other *pragmatic* elements have entered the algorithms of search engines. Freshness of content is being used as an indicator of the activity of the publisher, geo-localization of both publisher and user doing the query are used to provide better matches (e.g.: if you look for a restaurant, you are likely to be interested in restaurants near you), actual clicks by users are used as a feedback, to improve results – click-popularity, your past search behavior and the preferences of your social media friends. All those elements, and many more, are bringing not only contents (pillar I in OCM) in the practices of search engines, but also providers (pillar III), users (IV), and the whole information system (V), so to reconstruct a live information/communication landscape, full of documents as well as of people producing, accessing, evaluating them.

References

Aarseth, Espen J. 1994. Nonlinearity and Literary Theory. In George P. Landow (ed.) *Hyper/Text/Theory*, 51–86. Baltimore, MD–London: The Johns Hopkins University Press.

Aarseth, Espen J. 1997. *Cybertext. Perspectives on Ergodic Literature*. Baltimore, MD–London: The Johns Hopkins University Press.

Andorno, Cecilia. 2003. *Linguistica testuale. Un'introduzione*. Roma: Carocci.

Baran, Paul. 1964. On Distributed Communications. *Memorandum RM-3420*, RAND Corporation. Available online: http://www.rand.org/publications/RM/RM3420/.

Berners-Lee, Tim. 2000. *Weaving the Web. The Original Design and Ultimate Destiny of the World Wide Web by Its Inventor*. New York, NY: Harper Collins.

Blasi, Giulio. 1999. *Internet. Storia e futuro di un nuovo medium*. Milano: Guerini Studio.

Bolter, Jay David. 2001. *Writing Space. Computers, Hypertext, and the Remediation of Print*, Mahwah, NJ: Lawrence Erlbaum Associates.

Brusilovsky, Peter and Wolfgang Nejdl. 2005. Adaptive Hypermedia and Adaptive Web. In P. Singh Munindar (ed.), *Practical Handbook of Internet Computing*, 1.1–1.14. Baton Rouge: Chapman Hall & CRC Press.

Bush, Vannevar. 2003. As we may think. In Noah Wardrip-Fruin and Nick Montfort (eds.), *The New Media Reader*, 37–47. Cambridge, MA–London: The MIT Press. [Originally published in *Atlantic Monthly* 176(1): 101–8; and in *Life*, 19 (11), September 1945)]. Available online: http://www.ps.uni-sb.de/~duchier/pub/vbush/vbush-all.shtml.

Cantoni, Lorenzo and Paolo Paolini. 2001. Hypermedia Analysis. Some Insights from Semiotics and Ancient Rhetoric. *Studies in Communication Sciences* 1(1). 33–53.

Cantoni, Lorenzo and Stefano Tardini. 2006. *Internet (Routledge Introductions to Media and Communications)*. London–New York: Routledge.

Cantoni, Lorenzo and Stefano Tardini. 2010. The Internet and the Web. In Daniele Albertazzi and Paul Cobley (eds.), *The media. An introduction*. 3rd ed., 220–232 Harlow et al.: Longman.

Cantoni, Lorenzo and Nicoletta Vittadini. 2003. L'ipertesto digitale. In Gianfranco Bettetini, Sergio Cigada, Savina Raynaud and Eddo Rigotti (eds.), *Semiotica II. Configurazione disciplinare e questioni contemporanee*, 321–51. Brescia: La Scuola.

Conte, Maria-Elisabeth. 1980. Coerenza testuale. *Lingua e Stile* 15. 135–154.

Eco, Umberto. 1979. *A Theory of Semiotics*. Bloomington: Indiana University Press.

Engelbart, Douglas and William English. 1968. A Research Center for Augmenting Human Intellect. In Noah Wardrip-Fruin and Nick Montfort (eds.), *The New Media Reader*, 233–46. Cambridge, MA–London: The MIT Press, 2003. Originally published in *AFIPS [American Federation of Information Processing Societies] Conference Proceedings*, 33, part 1, Fall Joint Computer Conference, 395–410.

Fagerjord, Anders. 2003. Rhetorical Convergence. Studying Web Media. In Gunnar Liestøl, Andrew Morrison and Terje Rasmussen (eds.), *Digital Media Revisited. Theoretical and Conceptual Innovation in Digital Domains*, 295–325. Cambridge, MA – London: The MIT Press.

Hafner, Katie and Matthew Lyon. 1998. *Where Wizards Stay Up Late. The Origins of the Internet*, New York, NY: Touchstone.

ITU – International Telecommunication Union. 2014. *The World in 2014. ICT facts and figures*, ITU, available online at: http://www.itu.int/en/ITU-D/Statistics/Documents/facts/ICTFactsFigures2014-e.pdf

Landow, George P. 1994. What's a Critic to Do?: Critical Theory in the Age of Hypertext. In George P. Landow (ed.) *Hyper/Text/Theory*, 1–48. Baltimore, MD – London: The Johns Hopkins University Press.

Landow, George P. 1997. *Hypertext 2.0*, Baltimore, MD–London: The Johns Hopkins University Press.

Landow, George P. and Paul Delany. 1994. Hypertext, Hypermedia and Literary Studies: the State of the Art. In Paul Delany and George P. Landow (eds.), *Hypermedia and Literary Studies*, 3–50. Cambridge, MA – London: The MIT Press.

Liestøl, Gunnar. 1994. Wittgenstein, Genette and the Reader's Narrative in Hypertext. In George P. Landow (ed.) *Hyper/Text/Theory*, 87–120. Baltimore, MD – London: The Johns Hopkins University Press.

Lotman, Yuri M. 2001. *Universe of the mind. A semiotic theory of culture*, London – New York: I. B. Tauris.

Montfort, Nick and Noah Wardrip-Fruin. 2003. The World Wide Web (Introduction). In Noah Wardrip-Fruin and Nick Montfort (eds.) *The New Media Reader*, 791. Cambridge, MA–London: The MIT Press.

Nelson, Theodor Holm. 2003. A File Structure for the Complex, the Changing, and the Indeterminate. In Noah Wardrip-Fruin and Nick Montfort (eds.), *The New Media Reader*, 134–45. Cambridge, MA – London: The MIT Press. Originally published in Lewis Winner (ed.). 1965. *ACM [Association for Computing Machinery]: Proceedings of the 20th National Conference*, 84–100. New York, NY: ACM Press.

Nielsen, Jakob. 1995. *Multimedia and Hypertext. The Internet and Beyond*, Cambridge, MA: AP Professional.

O'Reilly, Tim. 2005. What Is Web 2.0. Design Patterns and Business Models for the Next Generation of Software, 30 September 2005, available at: http://oreilly.com/web2/archive/what-is-web-20.html?page=1

Ong, Walter J. 2002. *Orality and literacy. The technologizing of the word.* 3rd Ed. London–New York: Routledge.

Rocci, Andrea. 2003. La testualità. In Gianfranco Bettetini, Sergio Cigada, Savina Raynaud and Eddo Rigotti (eds.) *Semiotica II. Configurazione disciplinare e questioni contemporanee*, 257–319. Brescia: La Scuola.

Slatin, John. 1994. Reading Hypertext: Order and Coherence in a New Medium. In Paul Delany and George P. Landow (eds.), *Hypermedia and Literary Studies*, 153–69. Cambridge, MA–London: The MIT Press.

Rita M. Lauria and Jacquelyn Ford Morie

7 Virtuality: VR as metamedia and herald of our future realities

Abstract: This chapter examines the concept of virtual reality (VR) as an advanced telecommunications medium that transcends all that has gone before, forming, in essence, a new and advanced metamedium we term *virtuality*. Virtuality, due to its tightly coupled interactions with our perceptual and cognitive systems and the inclusion of an embodied self, blurs any distinctions between simulation and reality, creating in essence new levels of reality. By acknowledging the porous boundaries between the simulated and the "real", virtuality constitutes a phenomenological structure of "seeming", where the computer-constructed reality delivered through advanced, interactive telecommunications systems feels experientially authentic. However, our direct awareness of the merging of the physical and the virtual precipitates a radical shift in our understanding over previous forms of media. As background, we present a brief history of developments in VR, from both technological and more philosophical viewpoints. We explore the complementary concepts of the computer system as an active participant, and the embodiment of the human actor within the simulated reality and discuss how the concept of virtuality serves to fuse these potential dichotomies. Finally, we examine a future where such systems become so tightly coupled to our selves that they form an indivisible whole, contributing to the evolution of the human condition of being in the world.

Keywords: virtual reality, virtuality, virtual environments, virtual worlds, metamedia, sensory immersion, embodiment, avatars, telepresence

1 Introduction

While the Internet may be considered the latest and in many respects the most powerful driver in the continuum of evolving media from the telegraph to television, in the last three decades virtual reality (VR) has evolved as a powerful and unique communication media in its own right. VR has precipitated a powerful evolution towards the emergence of a metamedium – one that we call *virtuality* – which transcends the ability to merely statistically represent and present data, but increasingly alters our very perception of reality. As a constituent of virtuality, VR allows one to enter into a constructed world and be present there, with all the attendant meaning accorded to our living self. This "presence" within a VR is a concept of intense discussion as scholars attempt to tease out its phenomenological nature. Virtuality media dynamically simulate a range of human experiential possibilities, such as presence, that can serve to enhance and augment our understanding of reality. According to VR scholar Frank Biocca, VR explicitly embodies "a destination for the evolution of this metamedium" (Biocca and Levy 1995: 16).

In this chapter, we are focused on the notion of VR within the larger concept of virtuality. We recognize virtuality media as metamedia that are capable of immersing the user in an interactive computational communications environment along a wide continuum of perceived immersion, which we introduce to define both the realities and the potential of VR. We then briefly discuss the historical development of virtual reality and note this history suggests that, among other goals, one hope for VR included a desire for intelligence augmentation (IA).

Following this, we define how different forms of presence tie into these interactions and how VR can provide experiences that are "real" – emotionally, physically, and socially resonant. We then present the evolution of interface design for VR, which has involved advances in displays, data-gloves, motion-tracking and other forms of embodied interfaces enabling users to enter within the virtuality. Next we discuss VR in relation to the various forms of embodiment experienced by a VR user and how they contribute to the ongoing creation of a dynamic self. We touch on the concept of avatars and their increasing importance to the user who inhabits a virtual space. Lastly we touch on the future nature of human reality as it may evolve with and be impacted by virtuality.

2 Virtuality

Virtuality is a more encompassing term we will use to indicate the larger connectivity and phenomenological reach of VR as communication media based on computer constructed realities. Virtuality is the property of a virtual system (operating inside the computer) to become extended into the non-virtual world, which then behaves according to the template dictated by the virtual system. In philosophical terms, the property of virtuality speaks to a system's potential evolution from being merely a novel and separate descriptive technology to its advancement as an integrated and normal aspect of our living.

Within VR the computer plays a key role in how we perceive reality. In some sense the computer itself serves as an active partner in creating the experienced realities in ways not possible in other forms of media.

Brian Cantwell Smith has worked in computer science, cognitive science, and artificial intelligence for over twenty-five years. He argues that computers participate in their subject matter and that for "general participatory systems" the boundary between sign and signified, and the corresponding theoretical boundary between syntax and semantics is about as "far from sharp as it is possible to be" (Smith 1994: 8).

Computers, Cantwell Smith says (1994), are not at all separate from the worlds they represent. Nor is it possible to delineate their interaction with those worlds into the traditional distinct activities of reason, action, and perception, or even to generalize to a broader notion of experience. Computers, he says, "muck around in, create and destroy, change, and constitute, to say nothing of represent and reason and store information about, a hundred realms – new realms, some of them, that owe their existence to the very computers that interact with them" (Smith 1994: 8). Ironically computers, while often hailed for their objectivity, are concurrently "candidates for a theory of what it is to be a subject" because of their "manifest intentional character" (Smith 1994: 9).

Smith believes the connection between virtual reality and reality is much stronger than most people think it is and that how we design virtuality media is of key importance.

> There is no doubt that the ability to design artistic works and to do things, like in virtual reality, where in fact you have an experience that transcends what can actually happen in the ordinary physical world, is tremendously potentially creative and powerful ... I think ... that both the ethics and the aesthetics of those experiences are much more continuous with our aesthetics and ethics of ordinary life.
>
> It isn't like a black and white distinction that there is sort of virtual reality, which is unreal and you can do anything and then there is real reality ... It's not a sort of false way of being in touch ... It's perfectly real and can be weighty (Brian Cantwell Smith quoted in interview in Lauria 2000: 34).

Brenda Laurel, digital interface designer, researcher, and writer focusing on human-computer interaction, culture, and technology also recognizes the extent that computational media can deliver experiences that correspond to reality. Laurel believes there is potential for these experiences to usurp our ordinary existence. She states: "The ability to synthesize images that are not representations of the world is heavy duty and digital media lets us create them more easily and transmit them more easily. So the sort of danger is that it is easier and easier to live in virtuality, if you will" (Brenda Laurel quoted in interview in Lauria 2000: 35).

William Bricken, an expert in VR software architectures, also believes virtualities are capable of expanding reality, and that computers can and do generate entire multi-sensory environments that include us as interactive participants. "VR is the body of techniques that apply computation to the generation of experientially valid realities" (Bricken 1990: 2). With VR, Bricken argues that computers are no longer just symbol processors. They are reality generators. Computers as computational media become communications media that generate new forms of realities. Marshall McLuhan foreshadowed this possibility in 1964 when he wrote that the content of any medium is always another medium (McLuhan 1964).

Computer prophet Ted Nelson, whose work ultimately served as the basis for the World Wide Web, envisioned early on the potential of virtuality media (Nelson 1980). Enlarging upon Vannevar Bush's 1945 design proposal for a computer-based system that would serve as a tool to augment human intelligence (Bush 1945), Nelson invented hypertext as a "way of linking up all the world's knowledge into a kind of automated network ... accessible to everyone everywhere" (Rheingold 1991: 180). In elaborating how one should approach designing for such systems, Nelson defined the meaning of virtuality as a *structure of seeming* and the "central concern of interactive system design" (Nelson 1980: 57). The important things are not the data structures, or the hardware on which the system runs. What is important and meaningful is what does the experience feel like – what do we think it is, as we live it? Therefore Nelson declares that the virtuality of a thing refers to ...

> ... the *seeming* of it, as distinct from its more concrete "reality", which may not be important. An interactive computer system is a series of presentations intended to affect the mind in a certain way, just like a movie. This is not a casual analogy; this is the central issue ... A "virtuality", then, is a *structure of seeming* – the conceptual structure and feel of what is created (Nelson 1980: 57. Emphasis original).

Virtuality is therefore more a phenomenological structure constituted by what the interactive computational communication system provides, which we then assimilate within our lived selves.

According to Murray Turoff (1997), considered one of the founders of the computer-mediated communications field, there are three stages in the evolution of virtuality. The first stage in this evolution, *Technological Progress*, concerned the impacts such systems would impose on all aspects of our social milieu. The second stage, *Social System Design*, maintained that in creating such systems in effect, they could become part of the user's reality becoming, in essence, what he termed "prescriptive in nature." We have now entered the third stage, *Control System Design*, wherein the system itself has some controlling influence on the reality of the world (Turoff 1997: 41).

The process of being human as it existed before the emergence of computers has given way to our current process of virtuality in which computer created virtual environments become part of daily life. Thus virtuality becomes reality and guides our understanding of the models and representations we now require to describe the world. Our daily living now must seek agreement between our mental models and the variety of virtual environments that are now part of our active existence. This integration, Turoff explains, means that "virtuality is a process of 'negotiated reality'" (Turoff 1997: 40). He also states that the design of these third stage evolutionary systems must take this into account.

> The essence of the power of virtuality is in allowing the design of systems with high resiliency in a social and behavioral sense – systems allowing their users to adapt to a range of changing environmental conditions so their organizations can survive and even perform well in an uncertain future (Turoff 1997: 42).

He firmly believes that "what is possible with computers is not a representation of reality as we know it but a new essence or a new reality that may be very different from anything we have known before" (Turoff 1999).

Communication scholar Katherine Hayles defines virtuality as a modern, pervasive condition – a "mind-set that finds instantiation in an array of powerful technologies" (Hayles 1999: 69). She, too, emphasizes that it goes beyond being merely virtual, or in her terms, psychological, and no longer can one make a distinction between what the computer generates and our perception of reality. Our modern understanding of the world is predicated on this tight coupling, or "interpenetration" of the material and the informational contributions from our technologies. Rita Lauria maintains that this perception coupled with continual advances in technology drive more changes along every societal vector – social, political, legal, and commercial, and that these changes in turn reinforce the needs and expectations we demand (Lauria-White and White 1988). Portending the end state of this evolutionary cycle and the nature of our future needs and expectations perhaps would best be approached by examining the nature of some of the technologies that generate virtualities.

3 VR: A brief history

Virtual reality was first defined through a curious blend of ideas emerging from creative science fiction literature and then later manifested through a heyday of technological innovation in a variety of disciplines. As is often the case, science fiction projected astounding ideas into the future, unencumbered by material or engineering realities. Popular notions for conceptualizing what would become known as virtual reality includes Ray Bradbury's (1951) short story *The Veldt*, that tells the story of parents who provide their children with the latest and greatest media room, where they can walk into simulated realities such as an African veldt, which turns out to be all too real. In the late 1950s through the mid-1960s, Hollywood cinematographer Mort Heilig created an immersive multi-sensory experience system he called *Sensorama*, which he hoped would constitute a new form of cinema (Heilig 1977). The United Sates Department of Defense funded various VR-related research in the early 1960s, including research on CAD world-building tools (Sutherland 1965), wearable visual displays (Sutherland 1968), and flight simulators (Watkins and Marenka 1994). Artist-scientist Myron Krueger pioneered the development of early interactive computer artworks in the mid-1970s with his *Videoplace*, where one's body became the unencumbered interface to a computer-generated environment via video cameras. Krueger described these computer-created telecommunication experiences arising from full-body interaction with digital projections as *Artificial Reality* and *Responsive Environments* (Krueger 1983). In the 1980s many engineering advances began to support a different kind of immersive system, which unlike Kreuger's, required a number of devices to be worn on one's person. Head-mounted displays, or HMDs, had become smaller and lightweight enough to be worn somewhat comfortably on the head (compared to Sutherland's original version) and new tracking systems could keep realtime updates on a person's movements. In the mid-1980s Jaron Lanier, CEO of VPL Research, introduced a full suite of VR gear, including an HMD called the Eyephone, a flexible Data Glove that could be worn on the hand for navigation and gestures in a computer-constructed space, and a full body-tracking suit. He also coined the popular phrase *Virtual Reality* (Sherman and Craig 2003).

During this heady period, science fiction took the idea of living within a virtual reality to a broader public with stories such as Vernor Vinge's 1981 *True Names* (Vinge 1981), William Gibson's *Neuromancer* (Gibson 1984) and *Burning Chrome* (Gibson 1986), and Neal Stephenson's *Snow Crash* (Stephenson 1992). Gibson invented the term *cyberspace* to describe an artificial reality that could simultaneously be experienced by many people worldwide (Krueger 1991). But by the 1990s, hype usurped what could actually be done in VR with dystopian films and wild promises about what the technology could actually do. The public became disenchanted with the gap between promise and actuality. Researchers, however, increased their focus and systematic work to find applications for which VR was ideally suited. Many of these scientists also worked on uncovering the reasons *why* VR applications could affect people so profoundly in psychological, emotional, and sensory manners.

A driving force for some members of the early VR communities was the concept of intelligence augmentation (IA). This was widely seen as culminating in the creation of a machine that would surpass our current human capabilities and bring us to a new period in human evolution. As described by I. J. Good (1965), who was a colleague of Alan Turing's during the Second World War:

> Let an ultraintelligent machine be defined as a machine that can far surpass all the intellectual activities of any man however clever. Since the design of machines is one of these intellectual activities, an ultraintelligent machine could design even better machines; there would then unquestionably be an 'intelligence explosion', and the intelligence of man would be left far behind. Thus the first ultraintelligent machine is the last invention that man need ever make, provided that the machine is docile enough to tell us how to keep it under control (Good 1965: 34).

Many of the pioneers in this field had these futuristic projections. VR was described in a variety of phrases that attempted to capture its revolutionary and evolutionary impact. Within Howard Rheingold's 1991 book *Virtual Reality* we find these phrases: "Tools for thought", "Mind amplifiers", "Media for the transmission of intention", "Reality-sculpting tools", "Electronic experience factory", "Cyberspace systems" (Rheingold 1991: 19, 19, 254, 245, 229, 234, 181 respectively). Brenda Laurel equated computers with the type of theater that could render an epiphany, matching even the rapturous state induced by the ceremony and ritual (Laurel

1991). Other descriptions include: The "ultimate display" leading to the interdimensional window of "Wonderland into which Alice walked" (Sutherland 1965: 508). A "powerful idea with possible implications for every human transaction", "An incarnation of imagination", "A projection hallucinogen that can be shared by any number of people", "A laboratory for philosophy", "Responsive environments" (Krueger 1991: xv, xvii, 86). "An opportunity to sensitize people to the subtlety of the real world" (Lanier and Biocca 1992: 162). "A potent tool for understanding the physical universe" (Brooks 1988: 1). "A microscope for the mind", and an "unreal, alternate reality in which anything could happen" (Robinett 1991: 72, 16).

While no one knew precisely what VR was, or where it would ultimately end up, the early pioneers knew it was something powerful, deep, and world changing.

In spite of the forecasted moment of *singularity* popularized by Ray Kurzweil and other transhumanists (Kurzweil 2005; More and Vita-More 2013), wherein we become one with our machines and enter into the next stage of human evolution, we argue that virtual reality technologies, or virtuality media, will continue to co-develop and co-evolve *with* us, and we with it. It will provide the experience of multiple layers of reality where a continuum of virtuality experience overlaps, or augments, reality to seamlessly become part of our experiential lives, functioning to augment our intelligence, change our perceptions and move us along the human evolutionary spiral. In other words, the continuing pace of virtuality development, coupled with our changing perception of our lived reality, may preclude any definitive moment of abrupt change.

4 Being within virtuality

What is it that makes VR so special and elicits such poetic language as we saw in the previous section? Researchers have tried to zero in on the uniqueness of VR – what makes possible this new reality, or what Turoff (1997: 38) called this new social order of the real world? The overarching concept they have latched onto is that of *presence*, a more succinct form of the original term for the concept, which was *telepresence*. Telepresence meant, simply, that one could be in a particular location, but be able to affect things at a location far removed from one's center, perhaps a location that was even virtual or computer constructed. The notion of presence has been considered central to the understanding of virtual realities over the past three decades. Presence itself is more like a psychological state in which an individual using a VR medium fails to be aware of its existence (Slater 2003). Essentially, the medium becomes transparent, and the individual responds to objects in the environment as if they were actual and not computer-generated. There is a perception – some say an illusion – of non-mediation, and the experience is accepted as part of ordinary reality (Lombard and Ditton 1997). Despite all the work on teasing out the determinants of presence over the years, there continues to be much animated discussion about its subtleties (ISPR 2000).

Lombard and Ditton (1997) identified and defined various kinds of presence that can ensue from the use of virtuality media. Broadly enumerated these are: 1) Telepresence – the ability to be present at a distance; 2. Co-Presence – the ability to be present with other people while not physically collocated, and 3. Social-Presence – the ability to share social cues with others in a virtual space while being collocated only within the virtual space (Biocca and Nowak 2001; Lombard and Ditton 1997). The first type of presence is often connected to an individual's personal experience in a VR. Social presence and co-presence become important factors in achieving a high telepresence experience for collaborative applications, where humans work together in virtual reality environments.

People familiar with VR have often heard the term *immersion* used to describe being within a virtuality medium and sometimes see this word as being the equivalent to presence, often using the two words interchangeably (Bowman and McMahan 2007). However, immersion and presence are distinct concepts. Mel Slater, who has contributed much to the discourse on the nature of VR systems, defines presence as a user's subjective psychological response to a VR system, whereas immersion refers rather to the objective level of sensory fidelity (Slater 2003).

Immersion is not all or nothing as the terms immersive and nonimmersive might suggest. In fact, a better way to conceive of the immersive levels of virtual reality is not as a single construct, but instead as a multidimensional continuum along which combinations of input and output components create more, or less, immersion (Bowman and Mahan 2007). The position along the continuum depends upon the user's response to and the fidelity of the synthetic stimuli provided by the complex of technologies that replace real-world sensory information. In fact virtuality media can have low immersiveness, such as provided by many computer games and online worlds such as *Second Life*. These can be experienced via conventional computing capabilities such as a computer, screen and game controller or mouse and keyboard (Bolin et al. 2009). With these, the normal world is still obvious to the user at some level of awareness, even though they may be fully engaged in the virtuality.

VR that replaces one's standard sensory inputs, and shuts out the normal ones, can be considered to have high immersion. Sensory-immersive VR requires advanced technologies to replace real-world sensory information with synthetic stimuli like 3D imagery, spatialized sound, and force and tactile feedback. The various VR interface devices, HMDs, high end headphones or room-sized speakers, special gloves coupled with position tracking permit a user to traverse the virtual space and to manipulate objects within it. Because the normal world sensory inputs have been replaced, the person is isolated, seemingly totally within the virtual reality, that is within the virtuality, to the exclusion of what is happening in the physical space around them. This isolation and replacement is a key tenant of immersion and may be one of the factors that elicits that illusive feeling of presence.

Communications researcher and theorist Jonathan Steuer (1992) emphasizes that the locus of a virtual reality experience is on the perceptions of the individual that serve to create an experience for that person, rather than on what the machine does. For Steuer, the essence of presence is the extent to which one feels present in the mediated environment, rather than in the immediate physical environment (Steuer 1992: 36). However, at this stage of technology development, it is rare that one can truly feel a full sense of presence in a mediated environment for at least two reasons. First, the systems are far from full fidelity and our senses know the difference, and secondly because of the mechanisms by which we experience virtualities (gear, scarcity of systems, limited choices, etc.) we tend to remain aware, in some sense, of the outside world (Morie 2007: 114).

5 How to be virtually embodied

For many practitioners, assessing the degree of potential presence that can be experienced is an ongoing and essential quest. Suffice it to say we cannot go into the myriad ways that have been proposed to determine presence, but as a general example, we present two vectors, sensory and motor engagement, proposed by Biocca (1997) and Biocca and Nowak (2001). Sensory engagement can be measured along three dimensions: 1. the number of sensory channels engaged by the virtual environment, 2. the level of fidelity of the sensory cues, and 3. the saturation of engaged senses coupled with the suppression of non-engaged senses (Biocca 1997). Motor engagement also involves three dimensions: 1. the number of motor channels engaged by the virtual environment, 2. resolution of body sensors, and 3. sensorimotor coordination.

What both sensory and motor vectors share is a respect for the physical body of the participant in a virtuality medium, which, as Biocca articulates, is our primordial communication medium, which rests as the "fleshy gateway to the mind" (Biocca 1997: 13). This concept of the body as a display device for mind, integrates it to create a whole body/mind ensemble. Neuropsychologist Antonio Damasio (1994) asserts that the body constitutes the indispensable form of reference for the neural processes of the brain, which he says we experience as "mind." Our very organism is that which is used as the ground of reference for the myriad of constructions we make of the world. Indeed, our sense of subjectivity uses the body as a "yardstick", making it fairly clear the body and mind, reason and emotion, are not separate systems, but rather a whole (Damasio 1994).

VR media, more than any other computer media, is highly cognizant of and acknowledges our body as an integral part of our interactions. The movement of our bodies (thanks to the trackers) can be situated firmly in the simulated space.

Signals input to the sensing parts of our bodies are the connection between our mind and the virtual representations. In VR we are embodied: we not only need our body to experience the simulated environment, we project our physicality into the virtual space.

In the early days of the technology, there was simply not sufficient computing power to display anything that closely resembled a person's own body in the display device. Viewing was done "through the eyes" or in what was called "first person." This meant that, even if the view the computer calculated and provided to us was accurate, when we looked down, we would not see our own feet represented in the computer-generated space. Some early systems that used a device like the VPL DataGlove did allow for a graphic image of a hand, whose movements corresponded with those made by our physical one within the instrumented glove. Today more systems do have the power to render a person's body, in the form of an *avatar*, or 3D representation of self, and this increases the sensations of embodiment we experience.

Thus the emphasis in VR, until very recently, has been on providing a believable *spatial* simulation, focusing on the environment – its navigability, its physics, its fidelity to the actual world. For this reason, immersive virtual environments are often considered spatial virtual reality (Qvortrup 2002). However, the increasing capabilities of the virtuality systems mean that our virtual bodies now become more important as a determinate of the virtual experience. In some sense, we can, more than ever, be said to *inhabit* the virtual body, making it a more direct form of lived experience.

Avatars are powerful constructs that help us experience a fuller sense of virtuality – of that tight coupling between the actual and the virtual. However, both in representation and in that coupling, today's systems based on the sensory and motor equipment described above still tend to provide an impoverished version of our bodily self in the virtual space. Nevertheless, while specific interfaces come and go, the process of what Biocca (1997) calls "progressive embodiment" – the gradual immersion of the body into computational environments – continues to advance. New sensors and techniques are key to the continued evolution of embodied self in the virtual space.

6 Once and future interfaces

6.1 Enhancing current technologies

For this embodied coupling to move into the future, many new technologies will need to be developed. Current hallmarks of VR technology, in use since the birth of VR as a communications medium, tend to encompass limited, purpose-built types of equipment. Besides the computer chips and graphics processors that actu-

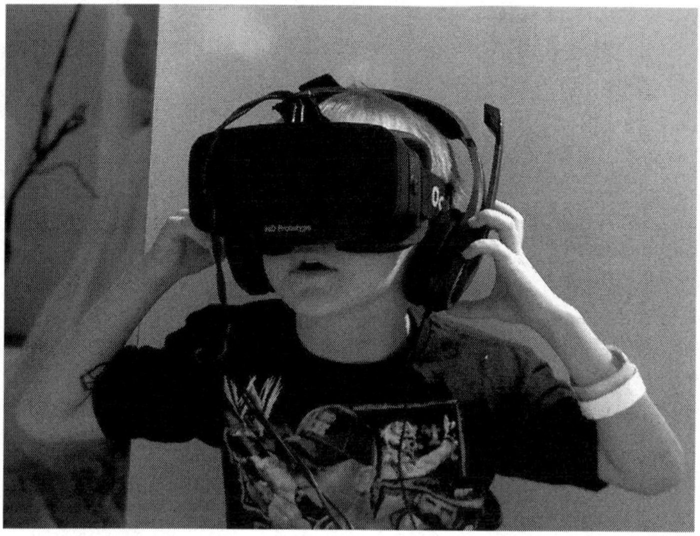

Fig. 1: One of the early Oculus Rift devices being enjoyed by a young user. Image taken by the author Morie, and used with permission.

ally compute the sensory elements that are presented to a VR participant, there are unique devices designed to replace the various human sensory inputs. Head mounted displays replace the photons that might come from the physical world to one's eyes with light patterns produced by a computer – often replicating the binocular vision and depth perception most people enjoy. Sometimes the visuals are projected on huge curved screens or domes that surround the person. Sounds are also produced via the computer, typically in multiple channels so that audio elements seem to emanate from the proper virtual location surrounding the participant, forming a consistent and coherent virtual space. Navigation devices run the gamut from game controllers, gloves and 3D mice (for the hands, gesturing and sometimes for walking or flying) to treadmills, human sized spheres, and clever techniques for redirected walking where a person traverses a small physical space but feels they are going longer distances (Suma at al. 2012). There are even devices to present smells and odors to the user (Washburn and Jones 2004), though for our final sense, that of taste, very few viable replica have been created.

While the types of available devices have changed little since the early days of VR, the trend has been to make these devices 1. faster, lighter, or more ergonomic, 2. have increased fidelity, and 3. be more affordable for the average user. A good case in point is the newly launched Oculus Rift HMD from the California-based company Oculus VR. The Rift is a basic HMD with a field of view more than double that of previous headsets, a promised resolution of 1080p (a high definition video mode), and built-in positional trackers. The most impressive aspect of this HMD, however, is that its entry-level price point is $ 300, making it one of the most affordable HMDs available, while offering better technology than its nearest competitors. Its Kickstarter campaign (where startup funding is raised through the contributions of ordinary people) sold over 15,000 of the Oculus Rift startup kits. Never in the history of VR have there been that many HMDs in existence, and these facts make this device a game changer in every sense. Other types of VR interface devices described above will most likely take a similar trajectory.

6.2 Going beyond

Beyond making current forms of VR technology better, more affordable and increasingly available, new devices and techniques on the near horizon promise to more tightly connect the human body and computer simulations. As per Arthur C. Clarke's famous statement that advanced technologies may be indistinguishable from magic (Clarke 1962: 32), we can anticipate that what will come may be beyond our current comprehension. But we can see what is on the near horizon with the following developments.

For example, depth-sensing cameras emerged as a new technology around 2007 (Lowensohn 2011) and are now incorporated into many gaming systems. They track the motions of a user's body, differentiating arms, torso, legs etc., and mapping those movements onto a computer graphic representation of the human in the game. This has proved useful not only for making physical action a viable input for games, but for applications designed for fitness, and rehabilitation.

Physiological sensors that collect signals such as skin conductance, heart rate and muscles signals have been part of VR research for decades, primarily as a means to measure and quantify a person's response to the VR environment (Morie et al. 2008; Usoh et al. 1999; Wiederhold et al. 2002). Some applications have used such measurements to enhance therapeutic applications, such as helping a psychologist ascertain stress levels on someone undergoing desensitization or exposure therapy (Parsons and Rizzo 2008). Others have taken this beyond mere measurement and actually feed the results of the collected signals back into the computer-generated scenarios to influence what is presented back to the participant so it is more relevant or responsive (Bersak et al. 2001). Though this use is still fairly new it can be used to bring about a reactive feedback loop that can tightly couple the human and the computer. In the last few years, the Quantified Self movement has accelerated the development of such physiological sensors so they can be utilized at a consumer level, leading to a wide variety of available fitness bands, heart rate monitors, and even some purported brain signal measuring devices (Swan 2012).

In fact, consumer level brain-computer interfaces, or BCIs, are being widely promoted today, especially in gaming applications. However, in their consumer-grade versions, furnished with few (typically 2–6) connectors, they come nowhere near measuring brain activity with any useful fidelity. They rely mostly on a person trying to project their intent onto some parameters of the computer program, such as having their character walk in the virtual world, or use a weapon. The most reliable devices (with 32–256 connectors) are still very much the domains of research or medicine (Vausanen 2008). While devices at this level do capture a huge amount of data from the brain, extracting meaning from that data requires extensive analysis involving sophisticated algorithms to make useful sense of those signals.

Interesting advances that use the computer's ability to analyze information about a person through their facial expressions, eye movement, voice prosody, and body activity are very promising for the future of human-computer coupling. A very recent example of this from the University of Southern California's Institute for Creative Technologies (ICT) is a Defense Advanced Research Agency (DARPA) funded research program called SimSensei (ICT 2011). In this demonstration program, a camera and microphone collect these human actions and convert them into responses that an Artificially Intelligent (AI) agent, in the form of graphically-presented virtual human, can direct back to the user in an interactive conversation.

Such agents can ascertain a person's intent and emotional state without physical sensors that must be placed on a user's body. Because of this, the fact that the data is being gathered and acted upon is transparent to the human participant in the interaction. It could eventually become something we take for granted. It permits the machine to respond to human emotions, needs and desires, whether through a virtual human connection, or an output to an emotional trigger like music, or changes to the environment itself. These are just a few tantalizing glimpses into the emerging future of virtuality media.

7 The key to embodiment and our virtual selves

Neuroscientist Damasio says that our notion of self is not static, but rather a constantly constructed entity. "It is an evanescent reference state, so continuously and consistently reconstructed that the owner never knows it is being remade ..." (Damasio 1994: 240). Because mind, and the self that may reside there, arises, according to Damasio's work, from the interactions between our physical body and the inputs it perceives in the world, it stands to reason that the inputs received from any virtuality medium also provide the raw materials that help form that dynamic entity of self.

In the past we had only the inputs from the actual world space in which we lived. Virtuality gives us parallel spaces and new forms of input that add to or replace our normal sensory mechanisms. However the tight coupling we have now between body and world throughout time immemorial, even if we are not overtly conscious of it, is not as firmly established within VR media. We have the rudimentary means of replacing senses, as previously described, to be sure. But much more is within the realm of possibility.

The ideas in the previous section point towards ways that can enhance connections and provide feedback loops that more firmly bring our physical selves into the virtual and conversely, the virtual into our physical and ultimately into the constructed self within our mind.

To explore this more deeply we need to return to the concept of an avatar – our self's embodiment in the space of the virtual. Having an avatar provides a means to, as Damasio (1994) says, "continuously reconstruct" the self that inhabits the medium. An avatar gives us both a glimpse into our internal self and a means to reconstruct that self in ways previously unimaginable. As previously mentioned, in the beginning days of VR, most inhabiting of the VR was done without a virtual body. One really only needed to have such a graphical representation if there were to be others in the virtual space – to facilitate social or co-presence. Even into the second decade of the 21st Century, fully immersive VRs rarely provided avatars for their users. This was due, in large part, to the fact that most applications were designed for a single participant. However, a more recent and less immersive form of virtuality – desktop virtual worlds (VWs) – were designed for social interactions between users. This requires seeing the other (and them seeing you), making the use of avatars a key affordance of a VW system. In VWs, not only can one select a starting avatar from a range of types, looks and styles, they are allowed to continually customize this avatar as they see fit, from dressing it in various outfits to changing its height, skin color of hair style any time. This meant that avatars could grow, change and evolve as they were used (Morie 2014). This also meant that an avatar representation could take advantage of better graphics as the VW system made them available, as can be seen in Figure 2.

Being able to modify and adapt one's avatar has led to an interesting phenomenon: the avatar ceases being merely an interface to the virtual world and instead becomes a viable projection of the individual inhabiting it. The precise reasons underlying this shift in perception – from the avatar as an "other" to the avatars as "self" is not known. More research is needed to propose any conclusive ideas. What is known is that the coupling between a human and his or her avatar, even if in place for a short amount of time, can profoundly affect the human's behavior.

Fig. 2: An avatar that has changed over time. Image created by the author Morie, and used with permission.

Experiments done by Stanford University Professor Jeremy Baileson and his colleagues have repeatedly shown this to be true (Yee and Bailenson 2007; Fox and Bailenson 2009; Blascovich and Bailenson 2011).

Some people become so firmly connected to their avatar that it is perceived as a "true" or "truer" self than their actual physical entity (Taylor 2002: 54). Researchers have conducted a variety of studies attempting to get to the heart of how people feel about their avatars. Some of these are enumerated in Morie (2014). Their findings show that many people create avatars that are idealized versions of their actual worlds selves – perhaps younger, taller, or better looking, though some make avatars that allow them to role play personas they could not act out in real life, such as animals, fantasy characters or even members of the opposite sex. Artist Kristen Shomaker's 2011 project, *1000 Avatars* shows snapshots she took of 1000 different avatars in the virtual world of Second Life, which allows for extreme customization of one's avatar. A second phase of the project raised the total of photographed avatars to 2000 (Shomaker 2011a, 2011b). No two avatars are alike. Most fascinating and revealing are the anonymous quotes she collected from participants in the project. We share a few of these here to show how important the avatar has become for its maker/owner. "My avatar is what I would be without financial, physical, social, cultural, emotional and mental constraints of the real world. She is the 'real me.' She is my soul." "Through my avatar, I have rediscovered portions of me long lost, presumed dead." "My avatar was my true self, even before I knew it." For these people the status of their lived self has evolved. They now exist in the world as an integration of their physical and virtual selves as equally dynamic aspects of their perceived total self.

8 Conclusion

The brief introduction we have provided for virtuality media within this Chapter predicts that we will experience VR and virtuality media (yet to be invented) very differently from what has been available to date.

No prior generation has witnessed such an extraordinary acceleration of technological power over what we have traditionally referred to as reality. The capability of new communication technologies points to a revolution in social interactions that involves a fundamental reshaping of the way that people understand themselves and the world about them. Corresponding social changes and ethical responsibilities accompany this acceleration (Floridi 2002). To be sure, computing and new telecommunication technologies are well along in creating a global network of social communications with near instantaneous transmission of information, ideas, and value judgments in science, commerce, education, politics, religion, entertainment, and every other facet of human activity. The very fabric of human understanding is shifting as physical reality gives way to virtual reality, or more appropriately, to virtuality, concurrently transforming the way philosophers understand foundational concepts like mind, consciousness, experience, knowledge, truth, ethics, and creativity.

The average person experiencing a virtuality medium may state that, like fairy tales, they understand this is not real. But when our sensory inputs are gathering data from a simulation, rather than from worldly inputs, and that data affects us in ways that set up neural traces of the experience within our minds, where does that line get drawn? Maya – that eastern philosophical concept that can be interpreted to mean the world we experience is an illusion or a deception, has resonance here. Can we really know what is true and real, or is it what we experience it to be? Is there a deeper, more complex or more nuanced truth or reality that virtuality allows us to experience?

Richard Thompson (2003), in his book *Maya: The World as Virtual Reality*, presents the idea of full and total immersion (i.e. indistinguishable from ordinary reality) as having the property of *closure*. Closure simply means that all inputs to the human are coming from signals generated by the computer simulation, and that no others can get in to compromise our mental awareness. Likewise, while within fully immersive VR, we have no way to experience anything outside of the simulated world (Thompson 2003: 23). And yet, we still exist in both the physical and the simulated world. It becomes us and we become it. These aspects are in union, a totality.

We may yet come to a day when technology, especially within the nascent science of brain interfaces, matures to the point where a full and complete *closure* is possible. When it is, we may fulfill that old oft-raised idea that we actually are living in a simulation of which we have no awareness. But until that time, if it ever comes, it is the wonderfully ambiguous blurring of ordinary reality with the possibilities of the virtual which allow us to enjoy virtuality media as unique and meaningful experiences.

As we noted previously, our modern understanding of the world is predicated on this tight coupling, or "interpenetration" of the material and the informational contributions from our technologies. We believe that evolving virtuality communication media will converge in a variety of ways and provide new "service environments" as McLuhan and McLuhan (1988) presaged. These will affect not only psychic and social processes, but will ultimately serve to obliterate any distinctions between organism, environment, and artifact, and alter our understanding of reality forever.

Of course, it is impossible for anyone to be certain what the future holds. In this chapter we do not presume to predict the end state the evolution of virtuality portends, if there can even be one. Will there be Intelligence Amplification? New forms of Self? Unimagined social interactions? All this and more? The once and future technologies of virtuality are disruptive, undetermined, unpredictable and ultimately fascinating, and through them we now can see our own future as through a (computer) glass darkly.

Acknowledgements

The authors would like to thank the following people for their valuable input to this article: Mabel Tsui, Gustav Verhulsdonck, and Frank Biocca.

References

Bersak, Daniel, Gary McDarby, Ned Augenblick, Phil McDarby, Daragh McDonnell, Brian McDonald & Rahul Karkun. 2001. Intelligent biofeedback using an immersive competitive environment. Paper at the 2001 Designing Ubiquitous Computing Games Workshop at UbiComp.

Biocca, Frank & Mark R. Levy. 1995. Virtual reality as a communication system. In Frank Biocca and Mark R. Levy (eds.), *Communication in the Age of Virtual Reality*, 15–31. Hillsdale, NJ: Lawrence Erlbaum Associates.

Biocca, Frank. 1997. The cyborg's dilemma: Progressive embodiment in virtual environments. *Journal of Computer-Mediated Communication* 3(2). Available: http://web.cs.wpi.edu/~gogo/hive/papers/Biocca_1997.pdf [Retrieved October 20, 2013].

Biocca, Frank & Kristine Nowak. 2001. Plugging your body into the telecommunication system: Mediated embodiment, media interfaces, and social virtual environments. In Carolyn Lin and David Atkin (eds.), *Communication technology and society*, 407–447. Waverly Hill, VI: Hampton Press.

Blascovich, Jim & Jeremy Bailenson. 2011. *Infinite reality: Avatars, eternal life, new worlds, and the dawn of the virtual revolution*. New York: HarperCollins.

Bolin, Corey J., Charles B. Owen, Eui Jun Jeong, Bradly Alicea & Frank Biocca. 2009. In William F. Eadie (ed.), *21st Century Communication: A Reference Handbook*, 534–542. Thousand Oaks, CA: SAGE Publications, Inc.

Bradbury, Ray. 1951. The Veldt. In *The Illustrated Man*, 7–25. Garden City, New York: Doubleday.

Bricken, William. 1990. Virtual Reality: Directions of Growth: Notes from the SIGGRAPH "90 Panel". Available: http://www.hitl.washington.edu/publications/papers/m-90-1.html. [Retrieved October 21, 2013].

Brooks, Jr., Frederick P. 1988. Grasping reality through illusion – Interactive graphics serving science. In Elliot Soloway, Douglas Frye, and Sylvia B. Sheppard (eds.), *Proceedings of the ACM CHI 88 Human Factors in Computing Systems Conference June 15–19, 1988*, 1–11. Washington, DC, US: ACM Press.

Bowman, Doug A. & Ryan P. McMahan. 2007. Virtual reality: How much immersion is enough? *IEEE Computer* 40(7), 36–43.

Bush, Vannevar. July 1945. As we may think. In: *Atlantic Monthly* 176(1). 101–108. Available: http://www.theatlantic.com/ideastour/technology/bush-excerpt.html [Retrieved October 21, 2013].

Clarke, Arthur C. 1962 "Hazards of Prophecy: The Failure of Imagination" in Profiles of the Future; an Inquiry into the Limits of the Possible. London: Gollancz: 32.

Damasio, Antonio. 1994. *Decartes' Error: Emotion, Reason, and the Brain*. New York, NY: G. P. Putnam's Sons.

Floridi, Luciano. 2002. What is the philosophy of information? *Metaphilosophy* 33(1/2), 123–145.

Fox, Jesse & Jeremy Bailenson. 2009. Virtual self-modeling: The effects of vicarious reinforcement and identification on exercise behaviors. *Media Psychology* 12(1), 1–25.

Gibson, William. 1984. *Neuromancer*. New York, NY: Ace.

Gibson, William. 1986. *Burning Chrome*. New York, NY: Ace.
Good, Irving John. 1965. Speculations concerning the first ultraintelligent machine. In Franz L. Alt and Morris Rubinoff (eds.), *Advances in Computers* 6(31), 31–88. Waltham, MA: Academic Press.
Hayles, N. Katherine. 1999. The condition of virtuality. In Peter Lunenfeld (ed.), *Digital Dialectic*, 68–94. Cambridge, MA: MIT Press.
Heilig, Mort. 1977. Beginnings: Sensorama and the telesphere mask. In Clark Dodsworth Jr. (ed.), *Digital Illusion, Entertaining the Future with High Technology*, 343–351. New York, NY: ACM Press/Addison-Wesley Publishing Co.
Institute for Creative Technologies (ICT) of the University of Southern California. 2011. Prototype: SimSensei. Available: http://ict.usc.edu/prototypes/simsensei/. [Retrieved January 15, 2014].
International Society for Presence Research (ISPR). 2000. *The Concept of Presence: Explication Statement.* Available: http://ispr.info/about-presence-2/about-presence/ Retrieved January 4, 2014]
Krueger, Myron W. 1983. *Artificial Reality.* Reading MA: Addison-Wesley Publishing Company.
Krueger, Myron. 1991. *Artificial Reality II.* Menlo Park, CA: Addison-Wesley Publishing Company.
Kurzweil, Raymond. 2005. *The Singularity Is Near.* New York: Viking.
Lanier, Jaron & Frank Biocca. 1992. An inside view of the future of virtual reality. *Journal of Communication* 42(4), 150–172.
Lauria, Rita Marie. 2000. *Virtuality.* Ph.D. dissertation. UMI Dissertation Abstracts International, 61-07A, Bell & Howell Information and Learning No. 9979464. Ann Arbor, MI: UMI Dissertation Services. Available: http://disexpress.umi.com/dxweb. [Retrieved November 1, 2013]
Lauria-White, Rita & Harold M. White, Jr. 1988. *The Law and Regulation of International Space Communication*. Boston, MA: Artech House.
Laurel, Brenda. 1991. *Computers as Theater*. Menlo Park, CA: Addison-Wesley Publishing Company.
Lombard, Matthew & Theresa Ditton. 1997. At the heart of it all: The concept of presence. *Journal of Computed Mediated Communication* 3(2). Available: http://onlinelibrary.wiley.com/doi/10.1111/j.1083-6101.1997.tb00072.x/full [Retrieved October 21, 2013].
Lowensohn, Josh. 2011. Timeline: A look back at Kinect's history. CNET Online. Available: http://news.cnet.com/8301-10805_3-20035039-75.html. [Retrieved January 15, 2014]
McLuhan, Marshall. 1964. *Understanding Media: The Extensions of Man.* New York, NY: McGraw-Hill.
McLuhan, Marshall & Eric McLuhan. 1988. *Laws of Media: The New Science*. Toronto, Canada: University of Toronto Press.
Morie, Jacquelyn Ford. 2007. *Meaning and Emplacement in Expressive Immersive Virtual Environments*. Dissertation. University of East London, 2007.
Morie, Jacquelyn Ford, Rebecca Tortell & Josh Williams. 2008. Would You Like to Play a Game? Experience and Expectation in Game-Based Learning Environments. In Harold F. O'Neil, Ray S. Perez (eds.), *Computer Games and Team and Individual Learning*, 269–286. Amsterdam: Elsevier
Morie, Jacquelyn Ford. 2014. (in press) Avatar Appearance as Prima Facie Non-Verbal Communication. In *Nonverbal Communication in Virtual Worlds,* Joshua Tanenbaum, Magy Seif el-Nasr, and Michael Nixon (eds.), Pittsburg, PA: ETC Publishing.
More, Max & Natasha Vita-More (eds.) 2013. *The Transhumanist Reader: Classical and Contemporary Essays on the Science, Technology, and Philosophy of the Human Future.* Hoboken, NR: John Wiley and Sons.
Nelson, Ted. 1980. "Interactive Systems and the Design of Virtuality." *Creative Computing* (November 1980), 56–62.

Parsons, Thomas D. & Albert A. Rizzo. 2008. Affective outcomes of virtual reality exposure therapy for anxiety and specific phobias: a meta-analysis. *Journal of behavior therapy and experimental psychiatry* 39(3), 250–261.

Qvortrup, Lars. 2002. *Virtual Space: Spatiality in Virtual Inhabited 3D Worlds*. London: Springer-Verlag.

Rheingold, Howard. 1991. *Virtual Reality*. New York, NY: Simon and Schuster.

Robinett, Warren. 1991, Fall. Electronic Expansion of Human Perception. *Whole Earth Review* (72), 16–21.

Sherman, William R. & Alan B. Craig. 2003. *Understanding virtual reality: Interface, application, and design*. Amsterdam: Elsevier.

Shomaker, Kristine. 2011a. *1000 Avatars, Vol 1*. Self published. Available: http://www.blurb.com/b/2289173-1000-avatars-vol-1. [Retrieved January 13, 2014].

Shomaker, Kristine. 2011b. *1000 Avatars, Vol 2*. Self published. Available: http://www.blurb.com/b/2289173-1000-avatars-vol-2. [Retrieved January 13, 2014].

Slater, Mel. 2003. A note on presence terminology. *Presence Connect* 3, 3.

Smith, Brian Cantwell. 1994. Coming Apart at the Seams: The Role of Computation in a Successor Metaphysics. Position paper for workshop *Biology, Computers, and Society: At the Intersection of the 'Real' and the 'Virtual' – Cultural Perspectives on Coding Life and Vitalizing Code*, June 2–4, 1994, 8–9. Stanford, CA: Stanford University.

Stephenson, Neal. 1992. *Snow Crash*. New York, NY: Bantam Books.

Steuer, Jonathan. 1992. Defining virtual reality: Dimensions determining telepresence. *Journal of Communication* 4(2), 73–93.

Suma, Evan A., Gerd Bruder, Frank Steinicke, David M. Krum & Mark Bolas. 2012. A taxonomy for deploying redirection techniques in immersive virtual environments. In *Virtual Reality Workshops (VR), 2012 IEEE*: 43–46.

Sutherland, Ivan. 1965. The Ultimate Display. *Proceedings of IFIP CONGRESS* 65(2), 506–508.

Sutherland, Ivan. 1968. A head-mounted three dimensional display. *Proceedings of the 1968 Fall Joint Computer Conference, AFIPS Conference* Proceedings 33, part 1, 757–764.

Swan, Melanie. 2012. Sensor mania! The Internet of Things, wearable computing, objective metrics, and the Quantified Self 2.0. *Journal of Sensor and Actuator Networks* 1(3), 217–253.

Taylor, T. L. 2002. Living digitally: Embodiment in virtual worlds. In Ralph Schroeder (ed.), *The social life of avatars: Presence and interaction in shared virtual environments*, 40–62. London: Springer-Verlag.

Thompson, Richard L. 1997. *Maya: The World as Virtual Reality*. Alachua, FL: Govardhan Hill Publishing.

Turoff, Murray. 1997. Virtuality. *Communications of the ACM* 40(9), 38–43.

Turoff, Murray. 1999. Murray Turoff email communication to Rita Lauria 25 July 1999. Transcript in the hand of Rita Lauria.

Usoh, Martin, Kevin Arthur, Mary C. Whitton, Rui Bastos, Anthony Steed, Mel Slater & Frederick P. Brooks Jr. 1999. Walking> walking-in-place> flying, in virtual environments. *Siggraph 99*. 359–364.

Vausanen, Outi. 2008. Multichannel EEG methods to improve the spatial resolution of cortical potential distribution and the signal quality of deep brain sources. Tampere Univeristy of Technology Publication 741. Available: *http://www.bem.fi/edu/doctor/vaisanen.pdf* [Retrieved January 17, 2014].

Vinge, Vernor. 1981. True Names, Binary Star #5. New York: Dell.

Washburn, Donald A. & Lauriann M. Jones. 2004. Could olfactory displays improve data visualization? *Computing in Science and Engineering* 6(6), 80–83.

Watkins, Christopher D. & Stephen R. Marenka. 1994. Taking flight: history, fundamentals, and applications of flight simulation. New York: M & T Books.

Wiederhold, Brenda K., Dong P. Jang, Sun I. Kim & Mark D. Wiederhold. 2002. Physiological monitoring as an objective tool in virtual reality therapy. *CyberPsychology & Behavior* 5(1), 77–82.

Yee, Nick & Jeremy Bailenson. 2007. The Proteus effect: The effect of transformed self-representation on behavior. In *Human communication research* 33(3), 271–290.

Constance Elise Porter

8 Virtual communities and social networks

Abstract: Virtual communities and social networking sites (SNSs) are becoming ubiquitous among those who communicate via the Internet. Yet, usage is declining amongst users of SNSs and engagement of virtual community members is difficult to sustain. In this chapter, the author reviews findings from previous research and identifies key scholarly issues from historical, contemporary and forward-looking perspectives. Ultimately, the author calls for scholars to go beyond descriptive accounts of human behavior in virtual communities and SNSs by developing theoretical explanations for such behavior. In doing so, the author lays a foundation upon which marketing and communications scholars might build programmatic research.

Keywords: virtual communities, online communities, social networking, social network theory, engagement

Virtual communities and social networking sites (SNSs) are a "top research priority" for scholars of marketing and communications (Trusov et al. 2010: 643). While many refer to both as the same, the two phenomena are viewed differently by scholars because each has a different focal point around which individuals connect and communicate. SNSs are ego-centric in that a particular individual serves as the focal point around which individuals connect. However, virtual communities are passion-centric in that a preferred interest (e.g. brand, company, hobby) serves as the focal point around which individuals communicate (boyd and Ellison 2008; Sindhav 2011).

Regardless of their differences, both virtual communities and SNSs (e.g. Facebook, LinkedIn) are becoming ubiquitous among Internet users. Early on, more people went online to participate in virtual communities than to make purchases (Horrigan 2001) and currently, approximately two-thirds of Internet users participate in SNSs (Duggan and Brenner 2013). As is true for virtual communities, the social networking site is popular not only in the US but also around the world (boyd and Ellison 2008).

Despite the upsurge of virtual communities and SNSs in the first decade of the twenty-first century, recent evidence suggests that usage is declining among users of SNSs and engagement of virtual community members is difficult to sustain (Porter et al. 2011; Rainie et al. 2013). Yet, a core set of individuals and organizations remain attracted to such online spaces. For these reasons, scholars need a better understanding of how to foster and sustain engagement among members of virtual communities and SNSs to help these dominant, online spaces remain viable and valuable to members and the organizations that sponsor such sites.

The content of this chapter will help answer two fundamental questions about human behavior and interaction in virtual communities and SNSs: 1. What are the key findings from previous research? 2. What would be the most constructive areas of research to pursue in the future? Indeed, defining and demarcating the domains of virtual communities and SNSs as well as fostering and sustaining engagement in these online spaces have received considerable research attention in the past. In this chapter, key scholarly issues that have been resolved, as well as those that remain unresolved, are presented from historical, contemporary and forward-looking perspectives. Doing so will help marketing and communications scholars develop programmatic research efforts in the future.

1 Defining and demarcating the domains of virtual communities and social networking sites

1.1 Defining the virtual community

While virtual communities abound, there is no single definition of the term "virtual community" that is embraced universally. A virtual community has been conceptualized as "an *aggregation* of individuals or business partners who *interact* around a *shared interest*, where the interaction is at least partially supported and/or *mediated by technology* and guided by some *protocols or norms*. (emphasis added)" (Porter 2004: Para. 2). This definition is consistent with Cantoni & Tardini's (2006) conclusion that virtual communities require members to share common interests (the defining characteristic of so-called paradigmatic communities) as well as interact amongst each other as they actively refine the domain of their shared interests (the defining characteristic of so-called syntagmatic communities). Since Porter's definition identifies the core attributes most associated with a virtual community, including interaction around shared interests, it remains one of the most cited definitions among scholars of marketing and communications.

However, Porter's (2004) definition was developed prior to the proliferation of other types of social media platforms that emerged over the subsequent decade. Thus, the relevance and applicability of the definition in the current socio-technical environment is worthy of scholarly examination. Achieving greater clarity around the definition, as well as the boundaries of the research domain, would facilitate the pursuit of valid, programmatic research of this important phenomenon. For practitioners, a clear definition would provide a useful foundation upon which they could successfully launch and manage virtual communities to support their organizations and stakeholders.

Given the complex ecosystem of social media platforms today, the precise definition of virtual community remains elusive. Indeed, different types of virtual community are defined by a variety of contextual factors including the type/strength of member-relations, content of communication among members, circumstances under which an individual joins the community or the conditions under which the community was founded (Greenacre et al. 2013; Lee et al. 2003; Preece 2000). These factors, singularly or in combination, have given rise to various types of virtual communities that are the subject of scholarly examination, including communities of practice (Schwen and Noriko 2003), brand communities (Muniz and O'Guinn 2001) and peer-to-peer communities (Mathwick et al. 2008).

Further complicating the issue of gaining consensus about a definition is that many new social media platforms share some of the core attributes often associated with the traditional virtual community. For example, SNSs, microblogging websites (e.g. Twitter, Tumblr), virtual worlds (e.g. Second Life, Entropia Universe), videosharing sites (e.g. YouTube) and photo-sharing sites (e.g. Pinterest, Houzz.com) each have been referred to as a virtual community by scholars and practitioners. Yet, it is unclear whether each meets the definition put forth by Porter (2004) and/or whether a broader conceptualization of the virtual community is warranted in today's social media environment.

Since Porter (2004)'s definition suggests that a) community members can be individuals or organizations who could interact in both virtual and real space and b) member-interaction could be mediated by any technology – not only computer technology, the core of the definition remains relevant. For example, as newer devices (e.g. smart phones) and newer social media platforms (e.g. Facebook, Twitter, Pinterest) fast-become ubiquitous, the boundaries of Porter's original definition seem broadly inclusive and sustainable. Yet, persistent equivocality of the definition, in the face of social media propagation, gives rise to important questions for both researchers and practitioners: Should *every* aggregation of individuals whom interact via *any* social media platform, be conceptualized, studied and/or managed as a virtual community? Alternatively, would it be more useful to identify key similarities and differences between virtual communities and other types of online aggregations, in a quest to establish a robust platform for future, programmatic research? Next, the answer to this question will be explored by comparing and contrasting the traditional virtual community with one of the most popular social media platforms that is often referred to as a virtual community: the Social Networking Site.

1.2 Defining the social networking site: from theory to practical application

While the SNS is a modern phenomenon, the fundamentals of social network theory, upon which such sites are based, are well-established. Social network theory was developed first by sociologists who focused on analyzing social relations among multiple individuals (Wasserman and Faust 1994). Over the years, however, it has been advanced and applied widely in other disciplines including marketing and communications (Garton et al. 1997; Iacobucci 1996; Reingen et al. 1984; Rindfleisch and Moorman 2001). Before defining the SNS, it is necessary to understand the foundations of social network theory which underlie it.

Fundamentals of social network theory. Social network theorists believe that social structure, rather than the attributes of individuals themselves, drives individual cognitions and/or behaviors (e.g. preferences, choices) among socially-connected individuals. These individuals comprise a so-called network. Specifically, a network is a finite set of more than two entities (e.g. individuals, corporations) amongst which important information-based resources (e.g. ideas, opinions) are exchanged. A large body of literature produced by social network theorists tends to describe a) individual members and their patterns of communication within a given network, b) the flow of information resources among members of a network and c) the influence that particular members exert over other members of a network.

The relational approach to social network theory is vital to understanding how scholars might conceptualize the modern-day social networking site. In general, scholars who use this approach focus on describing the level of cohesion among network members and the extent to which relationships among various networked individuals are embedded relationally. Through their research, they attempt to uncover which individuals are central to the flow of information within a network, and they have developed terms to describe such phenomena. For example, the number of people with whom an individual has a direct connection is often referred to as "degree" and the extent to which an individual controls the flow of information within a network of often referred to as "between-ness." See Wasserman and Faust (1994) and Knoke and Yang (2008) for addition information about social network theory and analysis.

Strong and weak tie relationships in social networks. The concept of cohesion is an essential element of the relational approach to social network theory. Cohesion is defined as the extent to which every network member is interconnected or, alternatively, the extent to which separate cliques (i.e. unconnected groups) exist (Reingen and Kernan 1986). As described below, modern-day SNSs tend to encourage and facilitate cohesion across a given individual's various relationships. Such cohesion is generally controlled and welcomed by members of SNSs. However, some have noted that complete cohesion, across an individual's various social spheres, can be a source of tension for some, due to the broadcast nature of communication and the persistence of content in online spaces (Binder et al. 2012).

Finally, "tie-strength" is an important and well-studied dimension of cohesion (Granovetter 1973). Strong ties are defined as relations that emerge among network members who interact frequently and who have developed relational norms such as trust, reciprocity and cooperation. Strong ties are transitive such that if X is related to Y and Z in a strong tie network, then it is likely that Y and Z will also be linked strongly. The transitive nature of strong ties often leads to the formation of subgroups within a network.

Alternatively, a weak tie exists when at least one individual has a membership in more than one subgroup and such weak ties are characterized as more casual or distal relations where few social norms have emerged. Weak ties can be fragile in that individuals will often consider withholding valuable information that could be used opportunistically by weak tie relations (Frenzen and Nakamoto 1993). Yet, because it is possible for novel information to be shared across weak tie relations, since weakly tied individuals are exposed to different information, scholars broadly have recognized the "strength of weak ties" in facilitating the transfer of information across disparate subgroups within a network (Granovetter 1973, 1982).

What is a Social Networking Site? As is the case with the definition of a virtual community, there is no singular definition of a SNS about which scholars agree. For example, boyd & Ellison (2008: 211) suggest that a social media platform that "allows individuals to 1. construct a public or semi-public profile within a bounded system, 2. articulate a list of other users with whom they share a connection, and 3. view and traverse their list of connections and those made by others within the system" should be defined as a social network site rather than a social network*ing* site (emphasis added). According to the authors, the latter term suggests that relationship formation amongst strangers is the primary practice on such sites when, in their view, users are less interested in meeting new people and more interested in a) making their networks widely visible and b) communicating with people whom already are part of their existing, offline social networks. In sum, boyd and Ellison prefer to use the term social network site since self-presentation of one's "articulated network" (boyd and Ellison 2008: 211) is, in their view, the primary activity of such sites (see boyd and Ellison for a full history, definition and description of "user profile" and the "articulated network" attribute of SNSs).

Notwithstanding boyd and Ellison's (2008) eloquent argument, others believe that members of SNSs go beyond merely articulating and viewing network member-profiles. They argue that members also focus on initiating new relationships and sustaining existing relationships, since mutual information sharing is central to all relational connections. Accordingly, a social network*ing* site (emphasis added) has been defined as a place "where registered members can place information that they want to share with others" (Trusov et al. 2010: 643). Such interactive, information sharing is consistent with the paradigm of researchers who take the relational approach to social network theory in that the focus of this definition is on the information flow among members of a relationally-embedded network.

Consistent with a relationship-orientation, most SNSs provide socio-technical features that allow members to articulate, establish and maintain a network of relationships via the site (see boyd and Ellison 2008 for a description of various features). Such features provide users with the following capabilities:

- *Connection Capability:* The ability for members to invite others to become part of their network and to accept invitations to join another's network (e.g. Facebook Friends, LinkedIn Contacts)
- *Communication Capability:*
  - *Content Management:* The ability for members to provide and view updates about their interests and activities to others with whom they are connected.
  - *Messaging:* The ability for members to respond to updates provided by others with whom they are connected.
- *Privacy Management Capability:* The ability for members to manage access to information contained in self-created profiles or personal homepages, based on member-controlled input.

The notion that a SNS enables members to connect, communicate and manage relationships effectively is reflected in the mission statement posted by Facebook on its homepage: "Facebook's mission is to *give people the power* to share and make the world more open and connected." Clearly, empowering members to connect, communicate and manage relationships successfully is at the core of Facebook's purpose. Interestingly, empowerment is a core element of successful management of a virtual community (Porter et al. 2011).

1.3 Demarcating the domains: virtual community and social networking site

The relationship-orientation of members who are supported with socio-technical features has led many to refer to SNSs as virtual communities. Indeed, scholars have examined two types of virtual communities that align strongly with the concept of a SNS. First, computer-supported social networks (CSSNs) were thought to emerge among individuals with strong, weak or moderate ties who are connected via computers (Wellman et al. 1996). Second, scholars have examined so-called "networked-based" virtual communities that are conceptualized as being geographically and socially dispersed, where members seek functional or utilitarian benefits such as information acquisition and problem solving. These networked-based communities are contrasted with small-group-based communities where members seek social benefits more so than utilitarian benefits (Dholakia et al. 2004). Importantly, strong ties are likely but not necessary among members of a virtual community, particularly a network-based virtual community. For this reason, some SNSs that are comprised of large and disparate individual networks might be cast appropriately a virtual communities, especially those SNSs that target specific individuals whom have shared interests.

Yet, an important trend suggests that even if the SNS is a form of virtual community, unique characteristics might substantively distinguish a SNS within the broader category of virtual communities: Some SNSs are suffering membership declines and/or disengagment. For example, members of Facebook, one of the most popular SNSs, are decreasing their frequency of use in part because their interest has waned and/or they are finding the content irrelevant (Duggan and Brenner 2013; Rainie et al. 2013).

Interestingly, concurrent with the declining rates of Facebook use is a noticeable increase in the number and popularity of interest-based, photo-sharing sites such as Pinterest and Houzz, which boyd and Ellison (2008) might refer to as passion-centric social networks. Pinterest, for example, is a website that allows users to create and manage theme-based photos of subjects that are of interest to them, while *Houzz* is a website in which members share photos, opinions and advice about their interest home design, improvement and decoration.

The surge in popularity of such passion-centric SNSs suggests that one of the most defining characteristics of a sustainable social network is the existence of a focal shared interest among members. Notably, this characteristic is consistent with Porter's (2004) definition of virtual community. In sum, evidence suggests the following proposition: The more community-oriented a SNS, in terms of a focal shared interest amongst uers, the more sustainable the SNS. Indeed, it appears that without a sustained focus on shared interests, SNSs would fail to engage members long-term (Porter et al. 2011).

Yet, boyd and Ellison (2008) acknowledge a key difference between a traditional virtual community and many SNSs: the former are organized around interests while the latter are organized around people who form "ego-centric networks" where they are the center of their own community. Thus, *if* a SNS is to fully flourish, then the sponsors of such sites need to ensure that they understand how to foster and sustain engagement among members who are at least as passionate about interacting around a shared interest with other like-minded individuals as they are about self-presentation and consumption of the profiles of members of their own and others' networks. In sum, wherever people come together, passions emerge and engagement about a shared interest will seed and sustain a virtual community.

2 Fostering and sustaining engagement in virtual communities and social networking sites

For a virtual community that is sponsored by an organization, the sponsor determines or significantly influences how the community is organized, managed and supported with value-added features (Porter and Donthu 2008). In this way, the community sponsor provides extrinsic motivation for members to participate. However, a certain level of intrinsic motivation must also be recognized and addressed, if members are to be engaged. In this way, intrinsic motivations sit at the core of member engagement in a virtual community (Porter et al. 2011).

2.1 Defining engagement

"Engagement is like love – everyone agrees it's a good thing, but everyone has a different definition of what it is." (Epps 2009). Cognitive psychologists describe engagement as a state of mind where an individual senses a high level of positivity, energy, commitment and loyalty to and about those who foster it. Behavioral psychologists suggest that engagement extends beyond cognition and includes emotions that stimulate actions that are thought to reflect cognitive engagement (Halbesleben and Wheeler 2008; van Doorn et al. 2010). Engagement reflects not only *whether* community members participate but also *why* they participate. In order to foster engagement, therefore, a sponsor must address a member's intrinsic needs that could be gratified through participation in the community. Below we address the unique drivers of member engagement in both virtual communities and SNSs.

2.2 Motivating and engaging members of a virtual community

According to the uses and gratification paradigm, individuals choose to use media in a way that yields specific forms of gratification (Katz et al. 1974; Larose et al. 2001; Severin and Tankard 1997). This paradigm has been applied to how individual's choose to use the Internet and virtual communities (Dimmick et al. 2004; Ko et al. 2005; Porter et al. 2012). Although gratifications obtained from using media in particular ways are unique to every individual, the following summarizes functional, social and hedonic needs (or derived values) that generally motivate individuals to use a virtual community (Porter et al. 2011: 81):

- *Information:* Virtual community members find value in a community that provides access to information that helps them solve problems and make decisions.
- *Relationship building:* Virtual community members seek to build productive relationships through interaction with others within a community.
- *Helping others:* Virtual community members are gratified by helping others within a community, especially those with whom they have a personal connection.
- *Enjoyment:* Virtual community members are gratified by achieving flow states when interacting with others and having control over their experience within a community.
- *Status/influence:* Virtual community members seek status and influence among others within a community.
- *Self-Identity/Self-Expression:* Virtual community members want to achieve self-awareness that they are a member of the community and are gratified by the emotional and cognitive connection with the community, as a whole, as well as their ability to express such connection.

- *Belongingness:* Virtual community members desire a sense of attachment to a community, as a whole, and are gratified by having their contributions to the community respected by others.

While the abovementioned needs could apply to any virtual community, a community sponsor can foster member engagement only after understanding and responding to the intrinsic needs of their specific community members. In other words, it is important for sponsors to recognize that certain needs might be magnified or muted in a particular community such that fostering engagement requires a deep understanding of the needs of one's own particular community members.

After identifying member needs and motivations for participation in the virtual community, a sponsor can foster member engagement by acting in ways that promote participation and motivate cooperation among community members (Porter et al. 2011). For example, encouraging members to contribute high-quality content, encouraging interaction among members, and producing enjoyable experiences promote participation by offering functional, social and hedonic value, respectively, for members. Also, by tapping even more deeply into social needs, such as the need for embeddedness and empowerment, a sponsor can motivate cooperation. An embedded member of a community feels such an attachment and fit with the group that the thought of leaving the group activates negative emotions (Crossley et al. 2007; Grewal and Slotegraaf 2007; Halbesleben and Wheeler 2008; Mallol et al. 2007). Indeed, one of the higher-order motivational values listed above is the need for belongingness.

Community sponsors can take specific actions that foster member embeddedness and, thereby, motivate cooperation. For example, giving members exclusive access to particular information and privileges that are not accessible to nonmembers of the community makes members feel like an "insider" to the sponsoring organization (Porter et al. 2011). Efforts to embed members not only fulfill a need for belongingness and status but also lower the risk that the member will leave the community. Certainly, keeping members from defecting will become increasingly challenging given the plethora of choices available in an ever-expanding ecosystem of social media spaces.

Finally, to foster the deepest forms of engagement, members must not only feel embedded but also empowered by a community sponsor. While embedded members feel obligated to support the community and its sponsor, empowered members sense that their supportive efforts have a real impact in the community. In terms of gratifications, empowered members gain sense of freedom and access to a sponsor's resources that enable them to make a difference in the community. What types of sponsor-efforts empower members? Giving members the ability to participate in key decision-making processes, encouraging members to collaborate with other members in value-creating activities and convincing members to make contributions that matter to them personally. (Porter et al. 2011).

2.3 Motivating and engaging members of a social networking site

Scholars have determined a variety of needs that motivate member participation in SNS, including the following (Greenacre et al. 2013; boyd and Ellison 2008; Brandtzæg 2012; Greenhow and Robelia 2009):

- *Information:* Members find value in accessing/sharing information among networked others.
  - *Consume Content:* Members find value in regularly reviewing the content that others have contributed rather than contributing their own content.
  - *Contribute Content:* Members find value in shaping online-identity and self-presentation through self-profiling as well as creating and uploading content to share with others.
- *Identity:* Members find value in creating or appropriating private and public markers of identity that enable self-presentation and impression management among networked others.
  - *Consume Content:* Members find value in regularly reviewing the content that others have contributed rather than contributing their own content.
  - *Contribute Content* Members find value in shaping online-identity and self-presentation through self-profiling as well as creating and uploading content to share with others.
  - *Debate/Discuss Issues:* Members find value in interacting with others in ways that shape online identity and reputation.
- *Relationship-building:* Members find value in forming new relationships or growing existing strong-, weak- or latent-tie relationships.
  - *Consume Content:* Members find value in regularly reviewing the content that others have contributed rather than contributing their own content.
  - *Contribute Content:* Members find value in writing and uploading content to share with others.
  - *Debate/Discuss Issues:* Members find value in discussing and debating issues with others.
  - *Create Emotional Bonds:* Members find value in giving and receiving emotional support, approval and/or validation from others.

Overall, SNSs are becoming more community-oriented. Brandtzæg (2012) found that a) members are gratified by maintaining relationships, particularly socializing with friends and family and b) the number of sporadic users as well as the percentage of members who only consume, rather than contribute, content is decreasing. These findings suggest at least two propositions. First, member engagement in a SNS reflects a community orientation, and the most sustainable SNSs are not only places where members post and review profiles but also places where members enjoy sharing information with others, often those with whom they share a social bond. Second, it is likely that those who seek high levels of functional value (e.g. only consuming content contributed by others) decrease usage, over time, as content becomes less relevant because it is contextualized in socially-bonded relationships.

3 Directions for future research

Currently, the research of virtual communities and SNSs remains at an early stage of maturity, so there is ample territory for researchers to explore in the future. First, researchers should leverage different theoretical perspectives that might be effective in demarcating the virtual community from the SNS, namely by exploring the relevancy of tribal theory versus social network theory. Second, researchers should investigate the influence of race, gender and national culture on the behavior of individuals in virtual communities and SNSs. Finally, researchers should extend beyond their current focus on the importance of strong or weak ties to the potential importance of other types of relational bonds in SNSs – namely influential and latent ties.

3.1 Leveraging different theoretical perspectives: Tribal theory vs. social network theory

Given the lack of consensus on the definitions of and lines of demarcation between the virtual community and the SNS, it would be useful for researchers to explore the relevancy of different theories that might explain how individuals and groups function within these social spaces. Different theoretical perspectives could yield unique insights about phenomena, especially those that are as emerging and dynamic as virtual communities and SNSs. Recently, for example, Greenacre et al. (2013) offer a unique perspective on the role of tribal theory, versus social network theory, to explain the different domains of virtual communities and SNSs.

While both tribal and social network theories focus on the behavior of socially-connected individuals, the two theories differ in their emphasis on the key determinant of such behavior. As stated earlier, social network theorists focus on describing how the social structure of networks, rather than the attributes of individuals that comprise the network, determines whether and how information flows amongst network members. Indeed, a vast range of social bonds (i.e. spanning from weak ties to strong ties) exist within a social network and information about various topics is exchanged amongst members.

Tribal theory uniquely places the shared interest of a group at the core of understanding individual behavior (Cove and Cova 2002). Consistent with the concepts of syntagmatic and paradigmatic communities (Cantoni and Tardini 2006), a close-knit tribe is bound strongly by a particular shared interest about which members of the tribe interact. In contrast with social network theorists, tribal theorists emphasize an individual's need to interact with like-minded others about a shared passion, rather than social structure, as a key determinant of behavior.

Social network theory and tribal theory also differ in how groups of individuals – networks or tribes – are thought to be sustained, over time. In fact, social network theorists pay scant attention to the issue of how networks are sustained, over time, because their focus is on describing the existing social structure and flow of information. Alternatively, tribal theorists focus on how interaction about a focal shared interest sustains the collective, over time (Greenacre et al. 2013). Unlike members of social networks, tribe members do not wait for an information need to emerge before interacting with others. Rather, they are motivated to interact with each other proactively, in order to fulfill their intrinsic needs for self-identification and belongingness amongst like-minded others. Consistent with member behavior in syntagmatic communities (Cantoni and Tardini 2006), tribal theorist suggest that members actively interact and sustain their community by negotiating and refining their common ground.

In the future, researchers should test tribal theory against social network theory as "competing theories" to explain member behavior in virtual communities and/or SNSs. Given the central role of a focal shared interest within tribes, tribal theory could be proposed as a more appropriate theoretical lens for scholars to use when exploring how and why virtual communities, and some SNSs, thrive while others fail to engage members. Alternatively, social network theory might be an appropriate theoretical lens for those interested in understanding the behavior of members of SNSs, especially when those sites have less community-orientation.

Another question would be essential for future researchers to address: Are tribal theory and social network theory complementary, rather than competing, such that both could be used to explain the development of an online collective, over time? For example, a particular online collective might initially form as a social network based on social proximity (e.g. location, occupation, education) rather than shared interest. However, over time, certain members might begin to interact around a particular shared interest and the collective might evolve into an interest-based tribe. This type of evolution is evident on LinkedIn, a popular SNS. The sponsor considers itself the host of a social network but they also offer the capability for individuals to identify with interest-based subgroups. Longitudinal research would be appropriate for testing such an evolutionary model of online collectives.

3.2 Understanding the influence of race, gender and culture on motivations for participation

In the last decade of the twentieth century, scholars focused on the so-called Digital Divide, a phenomenon often characterized as the unequal access or use of the

Internet, across a given population. For example, in the United States such a divide persists, with lower internet usage cited among older, less educated and less wealthy individuals. However, while race and gender differences in Internet use were once prevalent, in 2013, "neither race nor gender are themselves part of the story of digital differences in its current form" (Zickuhr and Smith, 2012: 6) as the number of women and minorities using the Internet, including virtual communities and SNSs, has increased dramatically. In the future, researchers should focus on understanding how and why women, minorities and members of different national cultures participate differentially in virtual communities and SNSs, as scholars shift their focus from the digital divide to "the emerging world of 'digital differentiation'" (Modarres 2011: 5).

The influence of gender in online environments is "still relatively nascent" (Awad and Ragowsky 2008: 102), but early research findings suggest that there is ample ground for scholars to explore in the future. For example, prior research shows not only that SNSs are especially appealing to women (Duggan and Brenner 2013) but also that those women:

- Use SNSs more intensely than men (Hargittai and Hsieh 2010a)
- Engage in more strong-tie activities (e.g. maintaining existing social relationships, looking at friends' photo albums) and fewer weak-tie activities (e.g. forming new social relationships, looking at strangers' photo albums), than men (Hargittai and Hsieh 2010b)
- Use SNSs for entertainment and passing time more frequently than men (Barker 2009)
- Have significantly greater influence on others' decision to begin using a SNS which they themselves already use (Katona et al. 2010)

These findings primarily are descriptive but, in the future, researchers should conduct studies that explain why such gender differences exist and the potential impact that such differences have on individuals and social welfare.

Gender affects not only how individuals use SNSs but also how they think and behave in virtual communities. For example, Porter et al. (2012) integrated social role theory (Eagly and Wood 1987, 1991; Meyers-Levy 1988), common identity theory, common bond theory (Baumeister and Sommer 1997; Ren et al. 2007) and the uses and gratifications approach to media use (Katz et al. 1974; Larose et al. 2001) to show how gender influences an individual's process for developing trust in the sponsor of a virtual community. They found that while the direct determinants of trust are universal, gender significantly moderates the trust-building process via key variables that influence the direct determinants of trust. For example, they found that a community sponsor's effort to provide quality content to the community facilitates trust with men, but that its effort to encourage interaction facilitates trust with women.

Porter et al. (2012) explained their findings by identifying the different needs or gratifications that women seek via virtual communities (e.g. interpersonal connectivity, self-discovery) as compared to the needs or gratifications sought by men (e.g. information). However, the question is open as to whether their findings reveal the full depth of gender-based need-gratification in virtual communities. Going forward, researchers should examine the influence of other gender-based, psychological needs that could influence cognitions or behaviors in virtual communities. For example, research suggests that while men seek entertainment gratification from online experiences (Fallows 2005) women seek hedonic experiences (Wang and Benbasat 2007). Understanding a full range of gender-based gratifications could make significant contributions to both theory and practice.

Scholars also have begun to explore how race might influence behavior in virtual communities and SNSs, as different racial/ethnic groups begun to have a significant online presence in the United States. Indeed, there is strong link between the relevancy of community-content and community participation (Byrne 2007), so understanding the influence of race/ethnicity on cognitions and behaviors in online spaces in essential. Thus, scholars are now extending beyond the notion of "race-less-ness" that is possible in anonymous online interactions to explore the role of race in nonymous (i.e. where users are self-identified) online spaces. For example, Grasmuck et al. (2009) found that different racial groups devote significantly different levels of effort toward constructing and displaying identity in SNSs. Specifically, they found that more significantly than Whites, various racial/ethnic minority groups displayed the following behaviors in SNSs:

- African Americans, Latinos and Indians invest in projecting visual self-representations via self-posted photos and wall-posts/comments posted by others
- African Americans and Latinos invest in projecting self-representations by revealing favorite consumption preferences in categories such as music, television shows, movies and quotes.
- African Americans and Latinos invest in projecting self-representations via first person, "About Me" narratives

Scholars have offered some theoretical explanations for such differences but much of the prior research findings are descriptive rather than explanatory, in nature. Following the migration toward understanding digital differentiation, future researchers should conduct studies that explain why such racial/ethnic differences exist and the potential impact that such differences have on individuals and social welfare.

Finally, cross- cultural theorist suggest that two different dimensions of national culture could play a role in explaining how individuals behave in virtual communities and SNSs (see Hofstede (1980) and Triandis (1995) for additional details on various cultural dimensions). First, the power-distance dimension of national culture represents the extent to which people of a certain culture expect and accept hierarchical order or the unequal distribution of power amongst members of the population. Second, the individualism-collectivism dimension of national culture represents the extent to which people of a certain culture tend rely on self-support (i.e. individualism) or on the expectation of support from others within their close circle of friends or relatives (i.e collectivism).

In general, America's national culture is viewed as low power distance/individualistic while the China's culture is viewed as high power distance/collectivist, and recent studies reveal how these cultural dimensions could influence behavior in virtual communities and SNSs. For example, Siau et al. (2010) found that less knowledge sharing takes place in Chinese virtual communities than in American virtual communities. Indeed, while members of American communities often provided long, detailed and opinionated messages, members of Chinese communities were less forthcoming in sharing their knowledge. Siau et al. argue that the high-power distance/collectivist nature of Chinese culture motivates members of Chinese communities to avoid relinquishing any position of expertise through excessive knowledge sharing. In fact, they observed that members of Chinese communities often met face-to-face in order to establish relationships that might foster greater knowledge-sharing within the virtual environment, reinforcing the influence of collectivism in Chinese culture.

Cultural factors also should be examined more closely in future research of SNSs. Santos et al. (2014, p.3) suggest that many social networks "are constrained and affected by, or perhaps even derived from culture", but cultural factors often overlooked when social networks are modeled. Although the research setting for their studies were offline social networks, the same cultural factors could affect behavior in SNSs in unique and substantive ways.

3.3 Understanding the significance of influential ties and latent ties

Over the years, much attention has been paid to whether relations are strong or weak, but other types of relational ties could be important to explore, especially in the context of virtual communities and SNSs. For example, an *influential tie* exists when a particular member's increased participation in a virtual community or social network prompts others with whom that member is connected to participate, based on the expectation that the influential member has made a new, valuable contribution (Trusov et al. 2010). Also, to the extent that interaction is possible but non-existent (i.e. not yet activated) amongst networked individuals, a *latent tie* exists (Haythornthwaite 2011), which could become activated as a weak-tie upon an initial interaction.

Understanding the importance of influential and latent ties in SNSs could prove useful for researchers and practitioners alike. For example, targeting influential ties could be essential for organizations and policymakers that want an important message to spread rapidly throughout a network. Going forward, researchers should explore the individual attributes or conditions under which ties become influential, given that Trusov et al. (2010) found that 1. very few networked individuals actually are influential and 2. merely having a very large network does not make an individual influential.

Finally, researchers should explore which actions of an organizational sponsor of a virtual community or SNS could help convert latent ties into weak or strong ties that are sustainable. Identifying such actions could be relevant to organizational leaders who desire a more cohesive workforce that is enabled by social media platforms such as a SNSs or a virtual community. In sum, while previous research focuses on the role of strong ties in sustaining virtual communities and the strength of weak ties in providing novel information across social networks, in the future, researchers should explore other types of social ties, since they could play a vital role in sustaining thriving online social spaces.

4 Conclusion

Virtual communities and SNSs are fast becoming ubiquitous among Internet users who have an increasingly complex set of social media platforms at their disposal. For this reason, these phenomena remain a research priority for scholars of marketing and communication. Defining and demarcating the domains of the phenomena, as well as fostering and sustaining engagement within those domains, not only has been the focus of prior research, but also remains the inspiration for future research. This chapter lays the foundation for such programmatic study. Researchers should respond to this call for them to go beyond descriptive accounts of behavior in virtual communities and SNSs and to focus on developing theoretical explanations for such behavior. Yet, the challenge is for researchers to do so within the context of an ever-evolving social media ecosystem.

References

Awad, Neven F. & Arik Ragowsky. 2008. Establishing trust in electronic commerce through online word of mouth: An examination across genders. *Journal of Management Information Systems* 24(4). 101–121.

Barker, Valerie. 2009. Older adolescents' motivations for social network site Use: The influence of gender, group identity, and collective self-esteem. *CyberPsychology and Behavior* 12(2). 209–213.

Baumeister, Roy F. & Kristin Sommer. 1997. What do men want? Gender differences and two spheres of belongingness: Comment on Cross and Madson. *Psychological Bulletin* 122(1). 38–44.

Binder, Jens F., Andrew Howes & Daniel Smart. 2012. Harmony and tension on social network sites. *Information, Communication & Society* 15(9). 1281–1299.

boyd, danah m. & Nicole B. Ellison. 2008. Social network sites: Definition, history, and scholarship. *Journal of Computer-Mediated Communication* 13(1). 210–230.

Brandtzæg, Petter Bae. 2012, Social Networking Sites: Their users and social implications – a longitudinal study. *Journal of Computer-Mediated Communication* 17(4). 467–488.

Byrne, Dara N. 2007. Public Discourse, Community Concerns, and Civic Engagement: Exploring Black Social Networking Traditions on BlackPlanet.com, *Journal of Computer-Mediated Communication* 13(1). 319–340.

Cantoni, Lorenzo & Stefano Tardini, 2006. *Internet (Routledge Introductions to Media and Communications)*. London; New York, NY: Routlege.

Cova, Bernard & Véronique Cova. 2002. Tribal marketing: the tribalisation of society and its impact on the conduct of marketing. *European Journal of Marketing* 36(5/6). 595–620.

Crossley, Craig D., Rebecca J. Bennett, Steve M. Jex & Jennifer L. Burnfield. 2007. Development of a global measure of job embeddedness and integration into a traditional model of voluntary turnover. *Journal of Applied Psychology* 92(4). 1031–1042.

Dholakia, Utpal M., Richard P. Bagozzi & Lisa Klein Pearo. 2004. A social influence model of consumer participation in network- and small-group-based virtual communities. *International Journal of Research in Marketing* 21(3). 241–263.

Dimmick, John, Yan Chen & Zhan Li. 2004. Competition between the internet and traditional news media: The gratification-opportunities niche dimension. *Journal of Media Economics* 17(1). 19–33.

Duggan, Maeve & Joanna Brenner. 2013. The demographics of social media users. 2012. Retrieved from http://pewinternet.org/Reports/2013/Social-media-users.aspx

Eagly, Alice H. & Wendy Wood. 1987. *Sex Differences in Social Behavior: A Social Role Interpretation*. Hillsdale, NJ: Erlbaum.

Eagly, Alice H. & Wendy Wood. 1991. Explaining sex differences in social behavior: A meta-analytic perspective. *Personality and Social Psychology Bulletin* 17(3). 306–15.

Epps, Sarah Rotman. 2009. What engagement means for media companies. Citing Jeffrey Graham. Retrieved from http://www.forrester.com/What+Engagement+Means+For+Media+Companies/fulltext/-/E-RES53814

Fallows, Deborah. 2005. How Women and Men Use the Internet. Retrieved from http://www.pewinternet.org/2005/12/28/how-women-and-men-use-the-internet/

Frenzen, Jonathan & Kent Nakamoto. 1993. Structure, cooperation, and the flow of market information. *Journal of Consumer Research* 20(3). 360–375.

Garton, Laura, Caroline Haythornthwaite & Barry Wellman. 1997. Studying online social networks. *Journal of Computer Mediated Communication* 3(1). Retrieved from http://onlinelibrary.wiley.com/doi/10.1111/j.1083–6101.1997.tb00062.x/full

Granovetter, Mark S. 1973. The strength of weak ties. *American Journal of Sociology* 78(6). 1360–1380.

Granovetter, Mark S. 1982. The strength of weak ties: A network theory revisited. In Peter V. Marsden & Nan Lin (ed.), *Social Structure and Network Analysis,* 105–130. Beverly Hills, CA: Sage.

Grasmuck, Sherri, Jason Martin & Shanyang Zhao. 2009. Ethno-racial identity displays on Facebook. *Journal of Computer-Mediated Communication* 15(1). 158–188.

Greenacre, Luke, Lynne Freeman & Melissa Donald. 2013. Contrasting social network and tribal theories: an applied perspective. *Journal of Business Research* 66(7). 948–954.

Greenhow, Christine & Beth Robelia. 2009. Old communication, new literacies: Social network sites as social learning resources. *Journal of Computer-Mediated Communication* 14(4). 1130–1161.

Grewal, Rajdeep & Rebecca J. Slotegraaf. 2007. Embeddedness of organizational capabilities. *Decision Sciences* 38(3). 451–488.

Halbesleben, Jonathon R.B. & Anthony R. Wheeler. 2008. The relative roles of engagement and embeddedness in predicting job performance and intention to leave. *Work & Stress* 22(3). 242–256.

Hargittai, Eszter & Yu-li Patrick Hseih. 2010a. From Dabblers to Omnivores: a typology of social network site usage. In Zizi Papacharissi (ed.), *The Networked Self*, 146–168. London: Routledge.

Hargittai, Eszter, & Yu-li Patrick Hsieh. 2010b. Predictors and consequences of differentiated practices on social network sites. *Information, Communication & Society* 13(4). 515–536.

Haythornthwaite, Caroline. 2005. Social networks and internet connectivity effects. *Information, Communication & Society* 8(2). 125–147.

Hofstede, Geert. 1980. *Culture's Consequences: International Differences in Work-Related Values.* Beverly Hills, CA: Sage Publications.

Horrigan, John B. 2001. Online communities: Networks that nurture long-distance relationships and local ties. Retrieved on May 28, 2014 from http://www.pewinternet.org/2001/10/31/online-communities/

Iacobucci, Dawn. (ed.). 1996. *Networks in Marketing*. Newbury Park, CA: Sage.

Katona, Zsot, Peter Pal Zubcsek & Miklos Sarvary. 2010. Network effects and personal influences: The diffusion of an online social network. *Journal of Marketing Research* 48(3). 425–443.

Katz, Elihi, Jay Blumler & Michael Gurevitch. 1974. Utilization of mass communication by the individual. In Jay Blumler & Elihu Katz (eds.), *The Uses of Mass Communications: Current Perspectives on Gratifications Research*, 19–32. Beverly Hills, CA: Sage.

Ko Hanjun, Chang-Hoan Cho and Marilyn S. Roberts. 2005. Internet uses and gratifications. *Journal of Advertising* 34(2). 57–70.

Knoke, David & Song Yang. 2008. *Social Network Analysis*. 2nd edn. Thousand Oaks, CA: Sage.

Larose, Robert, Dana Mastro & Matthew S. Eastin. 2001. Understanding internet usage: A social-cognitive approach to uses and gratification. *Social Science Computer Review* 19(4). 395–413

Lee, Fion. S., Douglas Vogel & Limayem Moez. 2003. Virtual community informatics: A review and research agenda. *Journal of Information Technology Theory and Application* 5(1). 47–61.

Mallol, Carlos M., Brook C. Holtom, & Thomas W. Lee. 2007. Job Embeddedness in a Culturally diverse environment. *Journal of Business & Psychology* 22(1). 35–44.

Mathwick Carla, Caroline Wiertz & Ko De Ruyter. 2008. Social capital production in a virtual P3 community. *Journal of Consumer Research* 34(6). 832–849.

Meyers-Levy, Joan. 1988. The influence of social roles on judgment. *Journal of Consumer Research* 14(4). 522–30.

Modarres, Ali. 2011. Beyond the digital divide. *National Civic Review* 100(3). 4–7.

Muniz, Alber M. Jr., & Thomas C. O'Guinn. 2001. Brand community. *Journal of Consumer Research* 27(4). 412–432.

Porter, Constance Elise. 2004. A typology of virtual communities: A multidisciplinary foundation for future research. *Journal of Computer-Mediated Communication* 10(1). Retrieved from http://onlinelibrary.wiley.com/doi/10.1111/j.1083–6101.2004.tb00228.x/full

Porter, Constance Elise, & Naveen Donthu. 2008. Cultivating trust and harvesting value in virtual communities. *Management Science* 54(1). 113–128.

Porter, Constance Elise, Naveen Donthu Andrew & Baker. 2012. Gender differences in trust formation in virtual communities. *Journal of Marketing Theory and Practice* 20(1). 39–58.

Porter, Constance Elise, Naveen Donthu, William H. MacElroy & Donna Wydra. 2011. How to foster and sustain engagement in virtual communities. *California Management Review* 53(4). 80–110.

Preece, Jenny. 2000. *Online communities: Designing usability, supporting sociability*. Chichester, UK: John Wiley & Sons, Ltd.

Rainie, Lee, Aaron Smith & Maeve Duggan. 2013. Coming and going on Facebook. Retrieved from http://pewinternet.org/Reports/2013/Coming-and-going-on-facebook.aspx

Reingen, Peter H., Brian L. Foster, Jacqueline Johnson Brown and Stephen B. Seidman. 1984. Brand congruence in interpersonal relations: A social network analysis. *Journal of Consumer Research* 11(3). 771–783.

Reingen, Peter H. & Jerome B. Kernan. 1986. Analysis of referral networks in marketing: Methods and illustration. *Journal of Marketing Research* 23(4). 370–378.

Ren, Yuqing, Robert Kraut & Sara Kiesler. 2007. Applying common identity and bond theory to the design of online communities. *Organizational Studies*, 28(3). 377–408.

Rindfleisch, Aric & Christine Moorman. 2001. The acquisition and utilization of information in new product alliances: A strength-of-ties perspective. *Journal of Marketing* 65(2). 1–18.

Santos, Eunice E., Eugene Santos Jr., Long Pan, John T. Wilkinson, Jeremy E. Thompson, John Korah. 2014. Infusing Social Networks with Culture. *IEEE Transactions on Systems, Man and Cybernetics: Systems* 44(1). 1–17.

Schwen, Thomas M., & Hara Noriko. 2003. Community of practice: A metaphor for online design?. *Information Society* 19(3). 257–270.

Severin, Werner. J. & James W. Tankard. 1997. *Communication Theories: Origins, Methods, and Uses in the Mass Media.* 4th ed. White Plains, NY: Longman.

Siau, Keng, John Erickson & Fiona Fui-Hoon Nah. 2010. Effects of national culture on types of knowledge sharing in virtual communities. *IEEE Transactions on Professional Communication* 53(3). 278–292.

Sindhav, Birud. 2011. The strategic implications of consumer-centric virtual communities. *Journal of Marketing Development and Competitiveness* 5(3). 11–23.

Triandis, Harry C. 1995. *Individualism and Collectivism.* Boulder, CO: Westview Press.

Trusov, Michael, Arnanf V. Bodapati & Randolph E. Bucklin. 2010. Determining influential users in internet social networks. *Journal of Marketing Research* 47(4). 643–658.

van Doorn, Jenny, Katherine L. Lemon, Vikas Mittal, Stephan Nass Doréen Pick, Peter Pirner & Peter C. Verhoef. 2010. Customer engagement behavior: Theoretical foundations and research directions. *Journal of Service Research* 13/3. 253–266.

Wang, Weiquan & Izak Benbasat. 2007. Recommendation agents for electronic commerce: Effects of explanation facilities on trusting beliefs. *Journal of Management Information Systems* 23 (4). 217–246.

Wasserman, Stanley & Katherine Faust. 1994. *Social network analysis: Methods and applications.* Cambridge: Cambridge University Press.

Wellman, Barry, Janet Salaff, Dimitrina Dimitrova, Laura Garton, Milena Gulia & Caroline Haythornthwaite. 1996. Computer networks as social networks: Collaborative work, telework, and virtual community. *Annual Review of Sociology* 22(1). 213–238.

Zickuhr, Kathryn & Aaron Smith. 2012. Digital differences. Retrieved from http://www.pewinternet.org/Reports/2012/Digital-differences.aspx

Ulrike Gretzel

9 Web 2.0 and 3.0

Abstract: Web 2.0 and 3.0 are summary terms widely used to describe emerging trends in the development of Internet technologies. This chapter describes the technological foundations of Web 2.0 and 3.0 and discusses the economic factors and cultural/social practices that are typically associated with the two phenomena. Adopting a Marxist view, it assumes that the interplay of the technological and economic bases facilitates the development of distinct communication and exchange models. It describes changes already visible using examples addressing various aspects of life and tries to anticipate future developments based on current discussions of technological advances and visions of the Internet of the future.

Keywords: web 2.0, web 3.0, social web, semantic web, social media, cloud computing, sharing economy, service dominant logic, social machines

Web 2.0 and 3.0 are terms that have been adopted widely by businesses, academia and the general public to describe phenomena that relate to the development of the World Wide Web from its early days of static HTML websites, graphic interface browsers, search engines and basic eCommerce functions to a complex assemblage of technologies that have changed and continue to change the way information is created, stored, presented and consumed. The terminology itself implies a technological point of view, hinting at new versions of a software program. It also signifies progress – something inherently better than the older version. It further assumes a clear developmental path, meaning a uniform direction in which the technological advances are heading.

From a Marxist point of view, describing the technological and economic base underlying cultural phenomena is critical for understanding why certain things emerged and not others. From a social constructionist point of view (Bijker et al. 2012), however, such a deterministic view only provides a glimpse of a phenomenon. It cannot grasp the complexity of what is essentially a socio-technical organism – an interplay of technological, cultural, economic, social and political forces. A thorough understanding of the emergence and development of communication related to Web 2.0 and 3.0 therefore needs a multi-perspective lens and an appreciation for its intricate technological as well as social nature. It also calls for a historic analysis in order to define the context(s) that enabled it. In many ways, the crisis of the Web in the aftermath of the burst of the dot-com bubble facilitated a reconceptualization or "back-to-the-roots" with respect to the culture of the Web with business interests lying low for a while.

However, despite its name, Web 2.0 was not a development in a linear sense and not a stage clearly delineated from Web 1.0. What we now generally view as technically and culturally belonging to Web 2.0 was very much embedded in the original idea of the Internet and its early uses. The Internet was designed as a decentralized network to facilitate content creation, sharing and storage across a large number of nodes. Its early users almost immediately recognized its social potential and created tools and virtual spaces that enabled such socialization. Chatrooms and virtual communities made it technologically possible for users to connect and commune. Multi-user dungeon (MUD) games took advantage of early Internet technology in order to facilitate playful interactions online. However, these technologies and uses remained a sort of subculture in the background, reserved for the technophile. Further, it is important to not confuse the Internet with the Web. Although related and the latter being dependent on the former, they are distinctive from a technology point of view. While the Internet might have been social from its early stages on, Web 1.0 was clearly not. The one-to-many communication model of websites dominated and commercial interests shaped its overall culture. E-commerce and copyright issues, hypertext and hyperlinks, information search and consumption, traditional advertising models (especially banner ads) and spam, content management and a database rather than narrative logic characterize what we now refer to as Web 1.0 (Manovich 2001). Similarly, Web 3.0 does not start where Web 2.0 stops but has developed somewhat in parallel while clearly feeding off the possibilities revealed by Web 1.0 and the boundaries pushed by Web 2.0. However, the phenomena and their technological boundaries remain messy. Allan (2013) proposes that the discourse around versions of the Web suggests both continuity as well as change and serves important purposes of historicization. Rather than aiming at establishing clear definitions and classifying specific technologies as either 1.0, 2.0, or 3.0, this chapter aims at describing the technological core of each term, discussing the communication models primarily associated with them, and outlining some of the economic, social and cultural changes they brought/are bringing about.

1 The technological base

Although used to describe larger phenomena, both Web 2.0 and 3.0 have a strong technological connotation and clearly emerged from specific technological advances rather than being driven by strong, defined societal needs. Yes, modern society is yearning for new ways to connect, but who knew we had a need for constant updates on the bowel movements of babies, the quirky behaviour of pets or the whereabouts and emotional states of distant acquaintances? While the focus here is on core Internet and Web-based technology, it should be mentioned that of course there was technological progress on many fronts which ultimately supported the widespread adoption and ubiquitous use of Web 2.0 technology (according to a Nielsen report published in 2012, 32% of people aged 18–24 indicated using social media in the bathroom) as well as the emergence of Web 3.0. Examples that should be kept in mind are not only the general increase in computing power but also the drop in data storage cost, the availability of global positioning system (GPS) technology and the proliferation of mobile devices. Oinas-Kukkonen and Oinas-Kukkonen (2013) further argue that the seamless integration between relational databases and the Web was a significant driver.

Web 2.0 refers to Internet/Web technology and applications that allow users without technical skills to be actively engaged in creating and distributing Web content (Gillin 2007). Flew (2008) also defines it as essentially a move from Web publishing by experts and content consumption by mainstream users to participation on a large scale, all supported through a technological architecture of participation. Gehl (2010) proposes that it is more of a discursive concept than a particular technology. Tim O'Reilly (2005) claims that the term was first put to use during a brainstorming session at a conference in 2004 in recognition of new technologies emerging after the burst of the dot-com bubble. What he stated a year later, in 2005, about Web 2.0 in an effort to document and summarize the initial ideas essentially still holds true:

> "Like many important concepts, Web 2.0 doesn't have a hard boundary, but rather, a gravitational core. You can visualize Web 2.0 as a set of principles and practices that tie together a veritable solar system of sites that demonstrate some or all of those principles, at a varying distance from that core" (O'Reilly 2005: 3).

He describes these technological principles and practices as:

1. The Web as a platform supporting services/applications;
2. Harnessing collective intelligence;
3. Data is the next Intel Inside;
4. End of the software release cycle;
5. Lightweight programming models;
6. Software above the level of a single device;
7. Rich user experiences.

In accordance with these principles, Webopedia (2014) defines Web 2.0 as a second generation World Wide Web focused on collaboration and sharing and technological infrastructure that emphasizes open communication and serving Web applications to users. Similarly, O'Reilly and Battelle (as cited in Cabage and Zhang 2013) stressed that the emphasis of Web 2.0 is on providing technological infrastructure that supports building software upon the Web rather than a computer. Thus, cloud computing is a term closely connected with Web 2.0. While the Internet and the Web have always supported content creation and sharing, Web 2.0 technologies (e.g. XML, Ajax, API, RSS, mash-ups, etc.) make it a lot easier for data to be created and exchanged. Further, dynamic programming languages such as Perl, Python and PHP and new programming models that see users as co-developers are also typically used to characterize Web 2.0 technologies.

Following the Web 2.0 model, contents become much more moveable and interactions more visible, giving rise to what is often referred to as the Read-Write Web (Hendler 2009) as opposed to the Read-Only paradigm that dominated much of Web 1.0. Recognizing that Web 2.0 applications also supported new ways for users to connect, it can be referred to as the Social Web (Gillin 2007). The trend started with the simultaneous emergence of early versions of social networking, instant messaging, online gaming, blogging and online encyclopaedias, with GeoCities and Classmates.com, for instance, being launched in the mid-1990s (Oinas-Kukkonen & Oinas-Kukkonen, 2013). Typical Web 2.0 applications today are blogs/microblogs, wikis, social networking sites, media sharing sites such as Flickr and YouTube, folksonomies, hosted services such as Dropbox, and mashups based on Google Maps. Efforts to map Web 2.0 applications exist but mostly focus on one specific category, namely social media. Social media are websites and applications that use Web 2.0 technology to facilitate content creation, sharing and social networking. Overdrive Interactive (2014), for instance, publishes a social media map that classifies social media into 24 different categories, including social networks, professional social networks, social recruiting (e.g. elance), blogging, microblogging, location-based apps, wikis, url shorteners, video sharing, photo sharing, social gaming, social commerce, social Q & A, lifecasting, social bookmarking, review sites, etc. Across all applications, connecting, collaborating, creating, conversing and commenting are the drivers of online behaviours supported through Web 2.0 technologies.

Web 3.0 is also referred to as the Semantic Web or the Web of Data (Herman 2009) or the Intelligent Web (Spivack 2007). Web 3.0 is essentially about new ways of integrating and combining data (Allan 2013). It is aimed at establishing technological frameworks and standards that allow data to be shared and reused across application, database and community boundaries. Berners-Lee et al. (2001) argue that Web 1.0 and 2.0 focused on producing information and documents for human consumption while Web 3.0 or the Semantic Web will be focused on producing documents for machines to allow for automatic processing. According to the W3C (w3.org), the Semantic Web is about two things. First, it is about common formats for integration and combination of data drawn from diverse sources, while the original Web mainly focused on the interchange of documents and Web 2.0 fuelled an explosion of data by enabling content creation by a broad mass of users. Second, it is also about language for recording how the data relates to real world objects and concepts. Main languages for realizing Semantic Web efforts are the Resource Description Framework (RDF), Web Ontology Language (OWL), Gleaning Resource Descriptions from Dialects of Languages (GRDDL) and Extensible Markup Language (XML). Therefore, Web 3.0 is inherently about uncovering and establishing meaningful associations. It is about making data more open and more useful. Berners-Lee et al. (2001) proclaim that such efforts will allow machines to better process and actually understand the data they currently merely display. According to Spivack (2007), this will result in data itself becoming smarter and the Web becoming more present, personalized and precise.

While Web 3.0 development is clearly underway, its concrete shape and extent can still not be grasped. Therefore, whether Web 3.0 technology will ultimately lead to a fundamentally different Web is not clear. Berners-Lee et al. (2001) define it as an extension of the current Web. Hendler (2009) defines it as an extension of Web 2.0 applications using Semantic Web technologies and graph-based, open data. Boyd (2007) envisions it as a profoundly different Web in which browsers and documents become obsolete and the lines between application and information become blurred. Markoff (2006) describes Web 3.0 development efforts as either involving the creation of new structures to replace the existing Web or the design of pragmatic tools to extract meaning from the current Web. Spivack (2007) subscribes to a more comprehensive view of Web 3.0, assuming that it will include mobile computing and sensors and therefore be a far broader, pervasive or ambient Web.

2 Economic aspects

Web 2.0 is often also linked to changes in the production of value and distribution of resources. While economics is traditionally concerned with the exchange of scarce resources, economics 2.0 has to deal with abundance and sharing. It is also characterized by network effects, power laws and the long tail (Anderson 2006). Tapscott and Williams (2008) termed it Wikinomics, stressing the democratization of value creation and mass collaboration. Crowdsourcing, collective intelligence, crowdfunding, open source, open data/APIs, social commerce etc. are only some of the terms that have emerged in relation to Web 2.0 and suggest that resources, value creation models and resource exchanges have been transformed. One specific emerging economic dimension of Web 2.0 is sometimes referred to as the sharing or peer economy (The Economist 2013), referring to consumers renting/sharing assets when they are not needed (e.g. rooms through Airbnb). Web 2.0 technology facilitates the sharing economy by reducing transaction costs and helping establish credibility through social networking and/or review mechanisms.

While many of the Web 2.0 technologies and applications emerged from grassroots efforts and often in direct response/protest to economic exploitation (e.g. music file sharing platforms, couchsurfing), the fact that Web 2.0 has become big business cannot be ignored. Facebook and Twitter are highly valued corporations and many smaller platforms have been successfully bought out by giants such as Google. As Gehl (2010) suggests, it is also important to recognize that many Web 2.0 website owners are companies that effectively take advantage of users' desires to create content and exploit their willingness to provide services for free that were traditionally paid labour. New advertising models have emerged based on Web 2.0

technology, which fuel the commercial success of Web 2.0, e.g. location-based promotions in Foursquare and the social graph-based behavioural targeting of Facebook. The business equivalent of the Web as a platform principle is service dominant logic (SDL) (Vargo and Lusch 2008). Value co-creation is one of its core principles and such co-creation is increasingly facilitated by Web 2.0 technology.

Entire industries have been transformed because of Web 2.0 technology. For instance, TripAdvisor realized a level of transparency in the hospitality industry that forced hoteliers to completely rethink the way they manage quality control and customer complaints. Crowdfunding and social media marketing have profoundly influenced the way start-ups, creative industries and charities operate. The music industry is struggling to survive in the new world of file sharing and micro-celebrities (Marwick 2011). The opportunities to organize online through Web 2.0 technologies have changed politics, including election campaigns. Another sector that was completely transformed is traditional media, with news being increasingly sourced from social media, putting citizens and consumers at the centre of media production. Gehl (2010) further points out that there is a lot of utopian discourse around Web 2.0, especially with relation to everyone being able to innovate, create and be an entrepreneur, suggesting entrepreneur-worship and an illusion of a complete democratization of resources and production means. It seems that the economic revolution brought about by Web 2.0 technology is one of new business models, aka new forms of capitalizing on free labour and free data, including social connections made visible. Consequently, Web 2.0 companies are increasingly transforming into big data companies taking advantage of the digital traces and contents supplied by users.

Although some discourse exists around the economic implications of Web 3.0 technology (see for example Almeida et al. 2013), the concepts are currently underdeveloped. There is a general idea that greater personalization can be economically exploited, but what economic developments go hand in hand with advances in Web 3.0 technology remains unclear.

3 The social and cultural superstructure

Safko and Brake (2009: 6) describe the Social Web as "activities, practices, and behaviours among communities of people who gather online to share information, knowledge, and opinions". The notion of the Social Web therefore emphasizes aspects of the Web that make it a networked conversation space in which social dynamics play an important role. Many authors have argued that Web 2.0 technologies have led to the emergence of new communication and sociability patterns. Miller (2008) suggests that we see an increasing flattening of social bonds as we move towards networked sociality and also points out a trend towards non-dialogic and non-informational communication, which he coins "phatic culture"

(p. 388). Indeed, the Facebook "Like" is a great example of phatic communication that allows us to establish connections without having to engage in real conversations. Similarly, Licoppe and Smoreda (2005) propose that Web 2.0 technologies blur the lines between absence and presence of social connections, leading us to be constantly contactable or in a state of connected presence. Rainie and Wellman (2012) conceptualised this phenomenon as "networked individualism", advocating that greater connectivity and greater empowerment through Web 2.0 supported media liberate us in many ways and lead to new forms of learning, problem solving and self-realization. This stands in stark contrast to Turkle's notion of "alone together" (2012), who argues that the new forms of communication and connectivity afforded by new media are poor substitutes for real community and intimacy. Knorr-Cretina (1997) describes such a state as post-social. According to Miller (2008), social networking, for instance, is no longer about communication but rather about establishing impressive profiles and engaging in collecting connections (termed "whoring" in social networking contexts) that are then displayed on these profiles.

The concepts of micro-celebrity and self-branding (Marwick 2011; Page 2012) are prominently discussed in relation to Web 2.0, suggesting that it facilitated the emergence of media and practices that fuel narcissism and exhibitionism as well as voyeurism. Social media stars have in some instances become more powerful than traditional celebrities (e.g. in the case of fashion bloggers). In fact, influence is the new currency in the new media world. As far as identity is concerned, the Social Web supports both the opportunity to display one's identity in incredible detail and easy ways to hide behind fake profiles or steal somebody else's identity. Oinas-Kukkonen and Oinas-Kukkonen (2013) speak of an identity paradox. They also describe a privacy paradox, pointing out that Web 2.0 has simultaneously increased privacy concerns and willingness to openly publish personal information. Web 2.0 platforms are at the centre of privacy discussions and scandals. Knowing one's rights with respect to privacy and understanding what information is appropriate to post online are essential components of Web 2.0 media literacy. They also refer to a credibility paradox, establishing that Web 2.0 leads to the creation of more authentic information but at the same time makes it easy for opinions to be presented as facts.

The influence on language is also very obvious. A plethora of new terms have emerged (tweets, defriend, to facebook message someone, hashtag, etc.) or new meanings have been added to existing words such as "like", "tag" and "profile". Technological limitations of Web 2.0 media have fostered the use of abbreviations and emoticons even more so than early Internet technologies such as email. There is also an ever greater move towards multimedia messages. Finally, the spreading of messages has evolved. While digital one-to-many and many-to-many communication models were established through Web 1.0, Web 2.0 adds new layers of modes and speed. Posting, reposting/retweeting/sharing, pinning, etc. are all aimed at spreading messages throughout one's personal social networks and beyond. Also, the Internet Meme as a new communication form is a concept that is intrinsically connected with Web 2.0 (Shifman 2012).

Since Web 3.0 technologies mostly focus on data integration and communication between databases and systems, the societal and cultural implications are not so obvious. Impact has been mostly articulated in the context of learning (Kurilovas et al. 2014). It has also been discussed in relation to specific industries such as tourism (Eftekhari et al. 2011). Hendler and Berners-Lee (2010) envision it leading to new ways in which humans and machines cooperate and the development of what they call "social machines" that will make it possible for users to interact with large datasets in unprecedented forms.

4 Conclusion

While one might argue that technologically, Web 2.0 and 3.0 are logical evolutions of the Web that are intricately embedded in the technological and philosophical origins of the Web and the Internet, socially, culturally and economically they represent revolutions that have brought and are expected to continue to bring about profound change in the way we communicate, conduct business, establish social connections, define relationships, etc. Following this imaginary trajectory, the immediate question is of course what Web 4.0 will look like. For many it denotes the age of intelligent electronic agents. Similar to Web 2.0 and 3.0, many of the technological foundations for such a Web are already available; however, the social, cultural and economic changes Web 2.0 has brought about and Web 3.0 is believed to encourage have yet to be envisioned and experienced for such a new wave.

Reference

Allen, Matthew. 2013. What was Web 2.0? Versions as dominant mode of internet history. *New Media & Society* 15(2). 260–275.

Almeida, Fernando, José Duarte Santos & José A. Monteiro. 2013. e-commerce business models in the context of web3. 0 paradigm. *International Journal of Advanced Information Technology* 3(6). 1–12.

Anderson, Chris. 2006. *The long tail: Why the future of business is selling less of more*. Hachette Digital, Inc.

Berners-Lee, Tim, James Hendler & Ora Lassila. 2001. The Semantic Web: A new form of Web content that is meaningful to computers will unleash a revolution of new possibilities. Scientific American, May 17, 2001. Accessed online (October 15, 2013) at: http://www.cs.umd.edu/~golbeck/LBSC690/SemanticWeb.html

Bijker, Wiebe E., Thomas P. Hughes, Trevor Pinch & Deborah G. Douglas. 2012. *The social construction of technological systems: New directions in the sociology and history of technology*. Cambridge, MA: MIT Press.

Boyd, Stowe. 2007. Jason Calacanis on Web 3.0. Blog post. Accessed online (December 5, 2013) at: http://stoweboyd.com/post/960708441/jason-calacanis-on-web-3-0.

Cabage, Neal & Sonya Zhang. 2013. Web 3.0 Has Begun. *Interactions,* September/October. 26–31.

Eftekhari, M. Hossein, Barzegar Zeynab & M. T. Isaai, Mohammad T. 2011. Web 1.0 to web 3.0 evolution: reviewing the impacts on tourism development and opportunities. *Human-Computer Interaction, Tourism and Cultural Heritage,* 184–193. Springer Berlin Heidelberg.

Flew, Terry. 2008. *New Media: An Introduction*. 3rd Ed. Melbourne: Oxford University Press.

Gehl, Robert W. 2010. *A Cultural and Political Economy of Web 2.0*. Unpublished dissertation. George Mason University.

Gillin, Paul. 2007. *The new influencers: A marketer's guide to the new social media*. Sanger, CA: Quill Driver Books.

Hendler, James. 2009. Web 3.0 Emerging. *Computer* 42(1). 111–113.

Hendler, James & Tim Berners-Lee. 2010. From the semantic Web to social machines. A research challenge for AI on the World Wide Web. *Artificial Intelligence* 174. 156–161.

Herman, Ivan. 2009. An Introduction to the Semantic Web. Presentation. Accessed online January 2, 2014 at: http://www.w3.org/2009/Talks/1030-Philadelphia-IH/

Knorr-Cetina, Karin. 1997. Sociality with Objects. *Theory, Culture & Society* 14(4). 1–30.

Kurilovas, Eugenijus, Svetlana Kubilinskiene & Valentina Dagiene. 2014. Web 3.0–Based personalisation of learning objects in virtual learning environments. *Computers in Human Behavior* 30. 654–662.

Licoppe, Christian & Zbigniew Smoreda. 2005. Are Social Networks Technologically Embedded? *Social Networks* 27(4). 317–335.

Manovich, Lev. 2001. *The language of new media*. Cambridge, MA: MIT press.

Markoff, John. 2006. Entrepreneurs See a Web Guided by Common Sense. The New York Times, November 12, 2006. Accessed online (November 15, 2013) at: http://www.nytimes.com/2006/11/12/business/12web.html?pagewanted=all & _r=0

Marwick, Alice E. 2011. I tweet honestly, I tweet passionately: Twitter users, context collapse, and the imagined audience. *New media & Society* 13(1). 114–133.

Miller, Vincent. 2008. New Media, Networking and Phatic Culture. *Convergence: The International Journal of Research into New Media Technologies* 14(4). 387–400.

Nielsen, 2012. State of the Media – The Social Media Report 2012. Accessed online (November 2, 2013) at: http://www.nielsen.com/us/en/insights/reports/2012/state-of-the-media-the-social-media-report-2012.html

Oinas-Kukkonen, Harri & Henry Oinas-Kukkonen. 2013. *Humanizing the Web: Change and Social Innovation*. New York: Palgrave-MacMillan.

O'Reilly, Tim. 2005. What is Web 2.0? Design patterns and business models for the next generation of software. Accessed online (August 14, 2014) at: http://mediaedu.typepad.com/info_society/files/web2.pdf

Overdrive Interactive. 2014. Social Media Map 2014. Accessed August 18, 2014 at: http://www.ovrdrv.com/files/knowledge/Social-Media-Map.pdf

Page, Ruth. 2012. The linguistics of self-branding and micro-celebrity in Twitter: The role of hashtags. *Discourse & Communication* 6(2). 181–201.

Rainie, Lee & Barry Wellman. 2012. *Networked: The New Social Operating System*. Cambridge, MA: MIT Press.

Safko, Lon & David Brake. 2009. *The Social Media Bible*. Hoboken, NJ: Wiley.

Shifman, Limor. 2012. An anatomy of a YouTube meme. *New Media & Society* 14(2). 187–203.

Spivack, Nova. 2007. Understanding the Semantic Web: A Response to Tim O'Reilly's Recent Defense of Web 2.0. Blog post. Accessed online (January 2, 2014) at: http://www.novaspivack.com/technology/understanding-the-semantic-web-a-response-to-tim-oreillys-recent-defense-of-web-2-0.

Tapscott, Don & Anthony D. Williams. 2008. *Wikinomics: How mass collaboration changes everything*. Penguin.

The Economist. 2013. The rise of the sharing economy: On the internet, everything is for hire. Accessed online (August 18, 2014) at: http://www.economist.com/news/leaders/21573104-internet-everything-hire-rise-sharing-economy.

Turkle, Sherry. 2012. *Alone Together: Why We Expect More from Technology and Less from Each Other*. Cambridge, MA: MIT Press.

Vargo, Stephen L., & Robert F. Lusch. 2008. Service-dominant logic: continuing the evolution. *Journal of the Academy of Marketing Science* 36(1). 1–10.

Webopedia. 2014. Web 2.0. Accessed online, August 15, 2014 at http://www.webopedia.com/TERM/W/Web_2_point_0.html.

II. Communication technologies and their enviroment

Tim Unwin

10 ICTs and the dialectics of development

Abstract: This chapter is an exploration of recent research and practice in the use of ICTs for 'development'. It begins with an overview of some of the challenges that need to be considered in defining the notions of both ICTs and 'development', arguing that both must be seen as contested terms that serve specific interests. In particular, the notion of 'development' depends very much on the political and social agendas of those espousing its usage; the interests behind development seen as 'economic growth' or 'social equity' are thus fundamentally different. The chapter adopts an overtly dialectical approach that first seeks to identify the main grounds for a thesis of the 'good' in the use of ICTs in development practice. It then develops an antithesis that proposes that the use of ICTs has actually increased inequality at a range of scales, and has thus worked against a definition of 'development' based on social equity. The conclusion seeks to explore what a synthesis of these two diametrically opposed positions might look like, and suggests first that the role of the state is crucial in ensuring effective and appropriate development, and second that well crafted multi-stakeholder approaches might indeed offer one way through which the poorest and most marginalised might indeed make effective use of ICTs to enhance their life experiences.

Keywords: ICTs, development, information, communication, technology, dialectic, inequality, mobile devices, multi-stakeholder partnerships

This chapter provides an overview of research and practice at the interface between "information and communication technologies" (ICTs) and "development". In line with the other contributions in this handbook, it explores some of the more interesting recent literature in the field, but it seeks to do so from a very specific theoretical standpoint that adopts an overtly dialectic and "critical" stance concerned with improving practice.

Over the last decade there has been a very rapid increase in the amount of research in the field of ICTs and development, closely matching the increasing pervasiveness of digital technologies in everyone's lives (see for example, Castells et al. 2009; Dodson et al. 2013; Dutton 2013a; Gomez 2013). Most of this literature adopts a positive stance, seeing and applauding the potential of ICTs to contribute to economic growth and better governance in poorer countries of the world (see for example, OECD 2009; Pélissié du Rausas et al. 2011; Dalberg 2013). Such arguments undoubtedly have some traction (Walsham 2001) and hence are taken in this chapter as the dominant *thesis*. However, as the middle of the second decade of this century approaches, increasing numbers of scholars and practitioners are questioning the value of many of the interventions that have sought to use ICTs to contribute to development (see Nisbet et al. 2012; Unwin 2013). As Dodson et al. (2013: 29) have recently argued, for example, "top-down, technology-centric, goal diffuse approaches to ICTD contribute to unsatisfactory development results". Such views are represented as the *antithesis*, and by seeking to engage with these two very different perspectives, the aim of the chapter is to develop a more nuanced understanding of how ICTs and "development" intersect. In so doing the chapter concludes with a synthesis of practical recommendations for how the current use of ICTs in "development" may be restructured in the interests of the poorest and most marginalised communities.

Both of the terms "ICTs" and "development" are commonly used, yet neither are simple to understand or agree on. Indeed, the very different ways in which these concepts are perceived represents one of the greatest challenges in grappling with the complexity of their relationship. Hence, the chapter begins with a brief introduction to the contrasting meanings attributed to these concepts, before elements of the thesis and the antithesis are each addressed in turn.

1 Boundaries of the dialectic

1.1 Information and Communication Technologies (ICTs)

There is nothing new about ICTs. In one sense, prehistoric people used the products of their empirical practices to identify the best ways to communicate through carvings and rock art. More recently, the development of movable type in China in the 11th century, and then the evolution of mechanical printing technology in Europe in the 15th century transformed communication, and for the first time created opportunities for a much more rapid expansion in the dissemination of ideas. Indeed, Francis Bacon in the early 17th century commented that along with gunpowder and the magnets used in compasses, printing was one of the three mechanical discoveries unknown to the "ancients" that had transformed "the whole face of things throughout the world" (Bacon 1620: Book I, CXXIX). The rapid expansion of modern ICTs, and particularly the Internet has often likewise been seen as transforming the whole face of the world (Castells 1996).

The creation of the first modern computers in the 20th century laid the foundations for the emergence of modern ICTs, but it is important to recall that the idea of a physical calculating machine can be traced back in origin some 4,500 years to the earliest abacus developed in ancient Sumer. This historical context is important, because it emphasises that technologies are not simply autonomous material products with some kind of inherent power within themselves, but are rather developed by individuals and communities to solve particular problems. They are inherently bound up with the societies that produce them, and this is as true today as it was in the past. Very specific interests underlie the development of new technologies. It is critically important to think of modern ICTs in this light if their interaction with "development" is to be understood. An instrumental view that ICTs are somehow autonomous, value free, "things" that can automatically do good, or be seen as a "silver bullet" to "fight" poverty, is fundamentally problematic. Moreover, the whole gamut of processes associated with the notion of globalisation (Wallerstein 1983, 2000; Harvey 2000; Stiglitz 2002) have always been closely related to technology, be it with the development of the nautical compass in the medieval period, or the use of the Internet for financial transfers instantaneously across the world today.

Most definitions of modern ICTs have tended to concentrate primarily on the material products that enable new forms of information sharing and communication to take place, dominated by computers and, more recently, mobile-'phones ('phones is an abbreviation for telephones, and thus the use of <'> indicates the loss of the <tele> in a contraction). A decade ago, Weigel and Waldburger (2004: 19) thus used the term to refer to "technologies designed to access, process and transmit information. ICT encompass a full rage of technologies – from traditional, widely used devices such as radios, telephones or TV, to more sophisticated tools like computers of the Internet". Their book, entitled *ICT4D – Connecting People for a Better World*, captured the very essence of the exciting modern millennial vision that technology could indeed be utilised to make the world a better place for everyone to live in. As this chapter goes on to highlight, though, this dream is far from yet being realised. Indeed, depending on how "development" is seen, the vision could turn instead be a nightmare where ICTs may actually lead to the exact opposite of what development as human betterment might be conceived as.

Recent definitions of ICTs have tended to be more nuanced and reflect a much more complex world that includes not only the hardware and devices, such as laptops, tablets, and televisions, but also the software needed to run them, the content that they enable, the infrastructure that connects them over wireless or fixed networks, and even the regulatory environments that allocate spectra. In his helpful overview of Internet studies, Dutton (2013b) thus separates out three distinct objects of study: the technology, its use in different contexts, and the laws and policies that shape and design its use.

The very rapid growth and spread of ICTs during the 21st century, as well as the burgeoning amount of funding available to researchers to work in this area, has also meant that academics from many different backgrounds have tended to approach the field from their own particular disciplinary contexts (Walsham 2012; Dutton 2013a). This not only means that there has been much duplication and overlap, with academics from one discipline often being blithely unaware of what those in other disciplines are doing, but also that it is extremely difficult to map out the entire field. The same is also very true of notions of "development" to which attention now turns.

In essence, those who tend to see technologies as being neutral instruments which that can be used for "good" support the dominant *thesis*; those who see them as being constructed primarily to serve the interest of the powerful support the *antithesis*.

1.2 Understanding and delivering "development"

The notion of "development" has long been contested and controversial, not only in terms of academic *understandings* of the concept (see for example Easterley 2006; Kothari 2006; Pieterse 2010), but also in terms of what is seen at any one time as being good *practice* in its implementation (Sachs 2005). Interestingly, there is often a marked divide between those who write about development, frequently from a critical perspective, and those who actually engage in development practices (Karlan and Appel 2012). Indeed, there are those building on the work of Escobar (2005) who reject the entire development project and focus instead on post-developmentalism.

Most definitions of "development" imply some notion of "progress" and "growth", and are usually seen as being derived from the European Enlightenment of the 18th century (Gay 1996; Bronner 2004; Sachs 2005). Whilst such a view has undoubtedly been challenged (Easterley 2006; Unwin 2009), it has had lasting impact. Ever since the 17th century, technology and science have generally been used by those in power to implement a particular kind of progress, be it in the industrial "revolution" of the 19th century, or the information and communication "revolution" of the late 20th century. Most frequently, this progress is measured in terms of economic growth, which despite numerous criticisms remains the dominant leitmotif of development practice across the world today. Accordingly, development has largely been seen as the reduction or elimination of poverty through economic growth. Extreme poverty, defined by the World Bank since 2005 as being those living on less than $ 1.25 a day (equivalent to $ 1 a day in 1996 US prices) is thus the general marker used by the international development community, as in the Millennium Development Goals (MDGs), and economic growth is normally seen as the main vehicle through which people will be lifted out of poverty (Sachs 2005; see also www.un.org/millenniumgoals).

Such arguments are usually premised on an absolute definition of poverty that in essence implies that economic growth in a given place will create more to be shared amongst everyone living there, thereby raising them all above whatever absolute level one cares to define as extreme poverty. However, critics of such arguments have long argued that there is no guarantee that the benefits of economic growth will be spread equally, and that the greater the economic growth, the greater will be the resultant inequalities in its distribution, unless very specific redistributional policies are put into practice (O'Boyle 1999; Unwin 2004). Hence, such critics favour a focus on *relative* rather than *absolute* definitions of poverty. There is now strong evidence that the economic growth encountered in the first decade of the 21st century was very closely associated with dramatically increased inequalities across the world. As Bauman (2013, unpaginated) comments, "The stubborn persistence of poverty on a planet in the throes of economic-growth fundamentalism is enough to make thoughtful people to pause and reflect on the direct as much as the collateral casualties of that redistribution of wealth". Significantly, much of this economic growth and increasing inequality has taken place alongside the dramatic expansion of ICTs, the Internet and mobile telephony. As later sections of this chapter highlight, the expansion of ICTs has actually occurred at the same time as an increase in global inequality rather than its diminution. Making the causal link between these processes, although intuitively compelling, remains challenging to prove in practice.

Whilst economic growth models of development as poverty elimination dominate, particularly in development practice but also in terms of much of the neoliberal academic discourse, there have always been alternative interpretations of "development" that have placed greater emphasis on other social, cultural and political agendas (see for example Sen 1985, 2002). One extreme alternative has been the notion of gross national happiness, coined in 1972 by the fourth Dragon King of Bhutan (www.grossnationalhappiness.com, accessed 15th October 2013; Ura et al. 2012). As yet, few academics have addressed the interface between ICTs and Gross National Happiness, but Heeks (2012) provides a brief exploration of some of the possible interconnections.

More mainstream has been the Human Development Index developed in 1990 largely by Mahbub ul Haq and Amartya Sen, which draws on life expectancy, education and income indices. This has subsequently formed the basis of the UNDP's *Human Development Reports*, with changes in the mode of calculation of the index being introduced in the report for 2010, which also included the introduction of an inequality adjustment (UNDP 2010). Although remaining largely focused on economic agendas, this does nevertheless seek to widen interpretations of development.

In part building on this work for the UNDP, Sen (1999) has however offered an alternative view of development, which does not see it so much focused on poverty elimination, but rather on the enabling of freedoms (for a critique, see O'Hearn 2009). For Sen, "freedom" is both the ultimate goal of societies and also the means of realising general welfare. In essence, he claims that development consists of the unlocking of "unfreedoms" that include a lack of political rights, vulnerability to coercive relations, and exclusion from economic choices and protections. Sen's arguments are therefore still embedded fundamentally within an economic tradition, but he does so by seeking to address the social basis of individual well-being and freedom.

Whilst intellectuals have struggled over understanding how "development" should be conceived and implemented (Kothari 2005), practitioners have also grappled with the main challenges of actually delivering an impact on the lives of poor people (McMichael 1996; Easterley 2006). Many different strands within the dominant economic growth agenda have therefore been adopted by development agencies over time. The neo-liberal agenda of the 1980s is thus widely seen as leading into the so-called Washington Consensus and budget support mechanisms of the 1990s (Williamson 1990; Burnell 2002). These became associated with the Structural Adjustment Programmes (SAPs) required by the major international donors, criticisms of which then led to the Poverty Reduction Strategy Papers of the late 1990s and early 2000s (Mohan et al. 2000). Gradually, project or programme based development assistance came to be seen by bilateral donors as failing to deliver on the systemic transformation of societies that had been expected, and consequently during the 2000s greater attention turned to the provision of budget support mechanisms (Unwin 2004), whereby donors agreed to provide substantial sums of money to recipient governments, providing they had a strategy in place explicitly to reduce poverty.

The Paris Declaration of 2005 and the Accra Agenda for Action in 2008 then sought to place the relationships between aid donor and recipient countries on a new footing that would focus much more on ownerships, alignment, harmonisation, results, mutual accountability and partnership (www.oecd.org/dac/effectiveness/parisdeclarationandaccraagendaforaction.htm). Interestingly, although human rights, democracy and good governance were included in Section V of the Millennium Declaration of 2000 (www.un.org/millennium/declaration/ares552e.htm) they were not explicitly mentioned among the MDGs in 2000. Nevertheless, despite a lack of conclusive empirical evidence to support the argument, a coalescence of interests between the corporate sector and governments in north American and European countries strongly argued that democracy and good governance are essential for economic growth, and hence by the middle of the decade this had become yet another important theme within the international development community, reflected for example in the UK's 2006 White Paper on international development, entitled *Eliminating World Poverty: Making Governance Work for the Poor* (DFID 2006).

Significantly, many civil society organisations challenge this move to closer partnerships between bilateral donors and the governments of poor countries, preferring instead to continue with project based "aid" designed to help individuals. This is largely on the grounds that governments often do not in practice serve the interests of the poorest of their citizens, and indeed should not necessarily always be expected to do so. This is important, because it has often been these civil society organisations that have developed ICT-based projects that specifically offer alternative perspectives to the grand strategies and policies of governments.

There are many other approaches to development thinking and practice, but what this short introduction has sought to do is to highlight three main things:

- definitions of development have changed considerably over time, and that interpretations of the value of ICTs in development practice will depend very largely on the notion of "development" that is adopted;

- the emergence of ICTs in development theory and practice has occurred at a time when economic growth and "development partnerships" have become dominant discourses in the overall development rhetoric; and
- there is increasing acceptance of the idea that development as economic growth has not only failed to deliver on poverty elimination, but that it has also led to greater inequalities in the world, which have potentially grave consequences for global peace and stability, and thus the very "development" that economic growth is intended to foster.

Having established some of the contours of debate around notions of development and ICTs, there now follows a summary of the contrasting positions that have been articulated at their intersection. These are deliberately written in rather different styles, to reflect the project-based enthusiasm of the *thesis*, and the more reflective and conceptual arguments of the *antithesis*.

2 ICTs for good: the *thesis*

As well as the difficulty of defining exactly what "good development" might be, another significant challenge in reaching any conclusive agreement on the relationship between ICTs and development is that remarkably few high quality comparative monitoring and evaluation studies have yet been undertaken. Much reporting on the impact of ICTs on development has thus been anecdotal, or based on a research design that is highly likely to produce positive results. This is hardly surprising given the interests underlying the use of technology for development. Both academics and companies who have developed a new technological "solution" are nearly always eager to show that it is successful, either for academics to enhance their careers, or for companies to increase sales of their products. The high costs, for example, of introducing computers into one set of schools, while also contributing an equal amount of money and resources to another educational intervention, and then comparing the differences in learning achievements of children in both environments, usually prevent this sort of comparative study that might yield really interesting results!

Despite this caveat, there is much evidence of the positive impact that ICTs have indeed had on development, however it is defined (Heeks 2009). This is not surprising given the increasingly integral part that the ICT sector has played in the global economy over the last two decades. Bilbao-Osorio et al. (2013: xi) comment in the World Economic Forum's *Global Information Economy Report 2013* on growth and jobs in a hyperconnected world, that "it remains challenging to isolate the impact of ICTs as their economic impacts have often occurred when combined with other broad social and business changes". Nevertheless, they also emphasise that both the pull factors, such as the need for developed economies to reinvent themselves to maintain their competitiveness, and the push factors whereby technological progress continues apace, combine together to reveal a close connection between ICTs and economic growth.

There is much research to support this notion of a strong relationship between the expansion of ICTs and economic growth. Indeed, this is almost a tautology, since the ICT sector is an integral part of the global economy, contributing some €2500 billion in 2010 (Net!Works 2012). Nevertheless, significant claims have been made about the contribution of ICTs across all sectors, as evidenced for example by the considerable interest in e-health, e-learning, e-agriculture and e-finance, as well as across all scales from the global to the local.

2.1 ICTs and development at a national and international level

Some of the strongest evidence of the contribution of ICTs to development is found in global generalisations by international agencies. A recent report by UNCTAD (2011), for example, argues that most empirical research suggests that there are indeed positive impacts in the use of ICTs for development

> (...) for economies, businesses, poor communities and individuals. Impacts are direct and indirect, and include impacts across the economic, social and environmental realms. There is case study and some macro-level evidence that ICT may contribute to poverty alleviation. Mechanisms include trickle-down effects from overall economic growth, employment and self-employment opportunities, establishment of microbusinesses that are in the ICT sector or related to it, such as the retailing of mobile 'phone cards, and the use of ICTs, such as mobile 'phones by small businesses (UNCTAD 2011: 17).

Yet, as the report goes on to note, there are also negative impacts, although much less research has been undertaken on them, and that which has been done tends to be anecdotal.

Similarly, Pélissié du Rausas et al. (2011) have shown that the total estimated worldwide contribution of the Internet is $1,672 billion, representing 2.9 % of global GDP. Indeed, in large and developed economies, this figure rises to 3.4 % of GDP. Likewise, the World Bank (2009) has claimed that in low- and middle-income economies, a 10 % increase in broadband penetration alone can accelerate growth in GDP by 1.38 %, and in high-income economies by 1.21 % (see also Dalberg 2013). These generalised figures remain, though, fraught with difficulty, because as Bilbao-Osorio et al. (2013) comment in a different context, much depends on local context; these average figures are made up of considerable variations in particular national circumstances.

It is not just in the economic context, though, that claims have been made that ICTs can contribute significantly to development at global and national scales. Thus, in an interesting report entitled *ICTs and Human Rights: an Ecosystem Approach*, Ericsson (2013: 4) have argued that as well as contributing to economic growth, “ICT also promotes greater transparency and enhances many fundamental human rights – such as the right to health, education, freedom of assembly and freedom of expression”. Despite advocating a strongly commercial approach to the benefits of a free Internet, they do note the increasing risks that “misuse of ICT poses to human rights”, emphasising in particular challenges around “freedom of expression and assembly, data privacy and security, and the relationship with law enforcement agencies” (Ericsson 2013: 14). Some have gone much further, and argued that access to the Internet is itself a human right. The UN Special Rapporteur for Human Rights, Frank La Rue, thus argued that “The Internet has become a key means by which individuals can exercise their right to freedom of opinion and expression” (UN 2011: 7), and as Cisco and the ITU (2013) have reported, courts or parliaments in Estonia (in 2000), France (in 2009) and Costa Rica (in 2010) have all declared access to the Internet as a human right. As with the economic growth agenda, ICTs are becoming completely intertwined with debates on human rights, which as noted above have become a central element of development discourse.

2.2 ICTs and development at a local and sectoral scale: successful projects and practices

There are countless examples of the purported value of ICT across almost every sector, and in almost all countries of the world (Weigel and Waldburger 2004; Unwin 2009). Most of these are pilot projects, and one of the real challenges is that all too often they have not been effectively scaled up or made sustainable.

One of the most high profile, but controversial, uses of ICTs in education in recent years has been the One Laptop per Child initiative, whose mission is to empower the world’s poorest children through education. Its website (http://one.laptop.org), for example, states that “We aim to provide each child with a rugged, low-cost, low-power, connected laptop. To this end, we have designed hardware, content and software for collaborative, joyful, and self-empowered learning. With access to this type of tool, children are engaged in their own education, and learn, share, and create together. They become connected to each other, to the world and to a brighter future”. Many reports typified by those for the Solomon Islands (Australian Council for Educational Research 2010) and Ethiopia (Hansen et al. 2009) do indeed show that this initiative has been successful (although for a critical perspective, see Krstić 2008; and Hollow 2008), and the initiative claims that over 2 million children and teachers in 42 countries are learning with their laptops today (http://one.laptop.org/about/countries).

Another, now classic, digital learning initiative was the Hole in the Wall concept initially developed by the charismatic Sugata Mitra’s as a joint venture between NIIT (India) and the International Finance Corporation (www.hole-in-the-wall.com/abouthiwel.html). The combination of the man and the project led to Mitra winning the 2013 TED prize, and it was also part of the inspiration behind the film Slumdog Millionaire. This model of digital learning too, though, has not been without its critics (Clark 2013).

In the finance sector, one of the most lauded recent initiatives has been M-PESA in Kenya (www.safaricom.co.ke/?id=257), which is widely seen as having transformed banking in Kenya. Launched in 2007 by Safaricom, it is lauded as a popular, affordable payment service that only requires a minimal engagement with traditional banking services. Mas and Ng'weno (2010: 1) thus note that within two-and-a-half years of its launch, "surveys of users show it is a highly valued service, and Safaricom continues to expand the range of applications it can be used for". They highlight three key reasons why it was so successful: "creating awareness and building trust through branding", "creating a consistent user experience while building an extensive channel of retail agents", and "a customer pricing and agent commission structure that focus on key drivers of customer willingness to pay and incentivized early adoption" (Mas and Ng'weno 2010: 1). To these, the important role of initial pump-priming funding from DFID, an international donor, and the high transactional costs of traditional banking in Kenya must also be added, in part as explanations for why other countries have not yet seen the very dramatic take-off of mobile banking witnessed in Kenya. Figures vary substantially on the percentage of Kenya's population that now uses mobile 'phones, with some estimates suggesting that it is as high as 93 % (Demombynes and Thegeya 2012), but what must be remembered about these figures is that mobile banking is not necessarily used by the poorest of the poor.

In agriculture and health there are likewise numerous initiatives that show how ICTs can have life changing development impacts (Yunkap Kwankam et al. 2009; Day and Greenwood 2009). For example, the e-agriculture community launched by the FAO in 2007 (www.e-agriculture.org), with a mission to share knowledge, learn from others and improve decision making about the use of ICTs in agriculture, had more than 10,000 members drawn from 160 countries and territories by 2013 (see also World Bank 2011). One of the most frequently cited benefits of ICTs for farmers is the use of mobile 'phones to enable them to have greater knowledge about market prices. The *Times of India* (2013: 1) captured this well in reporting in June 2013, for example, that "Chief Minister Naveen Patnaik on Tuesday distributed 5,000 mobile 'phones to farmers saying that the instrument would help them plan farming, track market price of various agricultural produce and weather". The evidence, though, is not always as straightforward as is sometimes claimed (Aker and Fafchamps 2013), with such market information only being of real benefit to those farmers who are actually able to get their produce to the markets with the best prevailing prices.

As with agriculture, there have been great strides in the use of digital technologies for health purposes, ranging from telemedicine initiatives that permit remote diagnosis and treatment, to the use of Geographical Information Systems in Demographic Surveillance Systems for the enhanced planning of health service provision (Mwageni et al. 2005), and the creation of a plethora of e-health networks both globally and nationally. Additionally, and entirely separately from any health issues, people with disabilities across the world have vast amounts to gain from the use of ICTs, particularly through the use of assistive technologies and text-to-speech and speech-to-text software on mobile devices (G3ICT and ITU 2012).

A very different kind of success story is represented by initiatives that use social media and crowd sourcing. One of the most successful of the latter is Ushahidi (www.ushahidi.com), initially established to share information about the political violence following the Kenyan elections in 2008. Since then, it has grown considerably, drawing on three key aspects of the use of ICTs: the development of free and open source software; the ability to connect together the many different members of the team, even though they work in different physical locations; and the crowdsourcing potential of collecting and mapping information from large numbers of people who can upload information through the Internet. Deliberately intended as a disruptive organisation to tackle the traditional ways in which information flows, Ushahidi builds tools for democratising information, increasing transparency and lowering the barriers for individuals to share their stories.

Likewise, social media more generally has often been seen as a substantial force for enhancing democracy, and thus "development" in large part on the grounds that "everyone" has access to the Internet, and that the vast majority of people have mobile 'phones. This agenda was euphorically lauded in terms of the role of social media in the so-called Arab Spring of 2011, sometimes also termed the Facebook Revolution (Reardon 2012). More nuanced reflections nevertheless indicate that this is far too simplistic a view, and that although blogging and interaction on Facebook are indeed strong organising tools, they were not what brought the people out onto the streets of Cairo or Tunis. More recent events in Turkey again illustrate this tension, with the mass media emphasising the potential of social media for political protest. Kantrowitz (2013), for example, comments that "thousands of protesters have taken to the streets across Turkey, using social media with great skill to propel their rebuke of Prime Minister Recep Tayyip Erdogan forward". Interestingly, he grounds their success in a traditional form of Turkish social media called *sozluks* (user generated dictionaries) that existed long before Facebook and Twitter were created.

These examples represent just some of the many claims that have been made to illustrate that ICTs have a very positive impact on development, not only in terms of their contribution to economic growth, but also through the very real practical ways through which they can help individuals and communities transform their lives. However, as the above accounts indicate, even in writing this it was difficult not to challenge and question some of the all too often taken for granted assumptions about such interventions. The next section therefore builds on these challenges to construct a completely different approach to ICTs and development.

3 The construction of digital inequality: the *antithesis*

Much of the rhetoric concerning the use of ICTs for development has emphasised the contributions that ICTs can make to economic growth (ITU 2011), and until recently rather less has focused on inequalities and other dimensions of development, including its social, cultural and political aspects (although see Unwin 2009). The purpose of this section is therefore to argue that instead of being the positive force for good, ICTs have actually had a strongly negative impact on development. The core argument is based on the observation that ICTs have led to much greater inequality in the world, but it also draws on questions concerning power, privacy, sustainability and the symbolic value of technology.

The fundamental point that is all too often ignored is that if it is argued that ICTs can have a dramatically positive impact on people's lives, then those without access to the technologies, or who cannot afford them, will increasingly be left behind; inequality will increase. Thus, if development is defined in relative terms, then the expansion of ICTs can be seen as actually leading to the exact opposite of what could righty be deemed to be "development". Figures 1 and 2 highlight this in terms of mobile 'phones and the Internet.

Figure 1 traces the expansion in mobile 'phone subscriptions in developed, developing and least developed countries. The trends, though, could just as easily represent urban, peri-urban and rural areas, or the differences between various

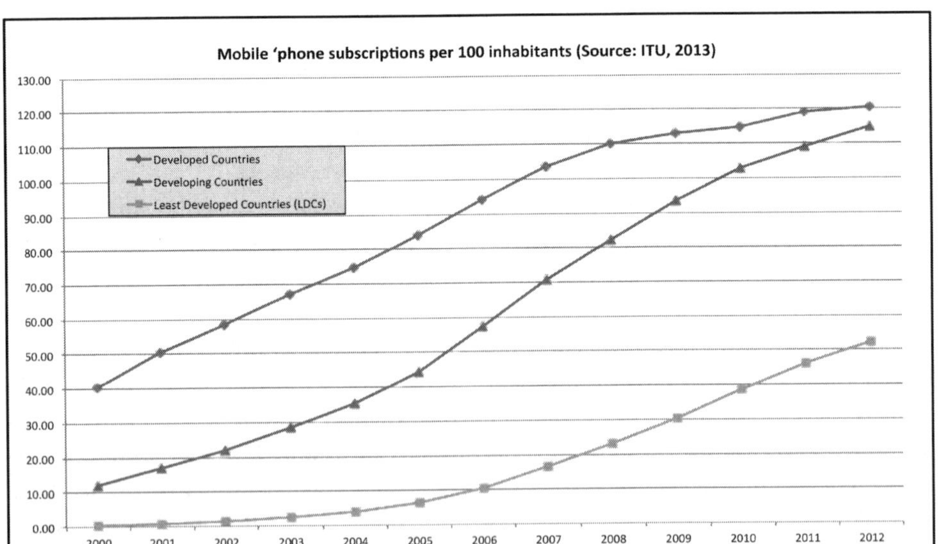

Fig. 1: Mobile 'phone subscriptions per 100 inhabitants.
Source: Rachel Strobel based on ITU data (2013).

ethnic groups in some countries. Moreover, in patriarchal societies, the different curves for developed and least developed countries in Figure 1 could also represent the difference between subscription numbers for men and women. Such data are extremely difficult to interpret, not least because in many poor countries a single 'phone is used by many people. Moreover, some individuals have multiple subscriptions, and others have inactive subscriptions. There is thus increasingly strong evidence that certain interests in the mobile telephony industry have significantly exaggerated the extent of the penetration of mobile telephony. When 2.4 million unregistered SIM cards in Kenya were switched off in January 2013, some 900,000 "users" failed to file their personal details with the respective operators; by March 2013 the registered number of users was thus only 29.8 million compared with a figure of 30.7 million in December 2012 (Kamau 2013).

The broad trends are nevertheless clear. On the one hand, the level of subscriptions is going up for all countries, but the critical point is not so much this, but rather that the difference between developed and least developed countries is now much greater that it was in 2000. Indeed, it appears that the curve for least developed countries is actually levelling off, and not continuing to increase in the way that more developed countries have done. Moreover, the business model for companies driving such expansion varies categorically between the rich and the poor: for the rich, most companies are eager to develop new high value devices and services that offer much greater functionality, whereas the poor are generally left with cheaper, less functional devices and services. Hence, even this image of a divided world of mobile subscriptions, hides the even greater functional diversity between what more affluent users of mobile broadband can do, and what those who only have "regular mobile 'phones" or other such devices are able to gain from the technology. To be sure, there are indeed initiatives such as FrontlineSMS (www.frontlinesms.com), which have sought to use basic mobile technologies to promote positive social change, but this does not disguise the observation that what and how one can learn using a regular 'phone is very different from what is possible with a smartphone.

The picture is even starker for data on Internet access (Figure 2). The difference between the percentage of people using the Internet in developed and least developed countries has thus risen from 12 % in 2000 to some 63 % in 2012. Again, put quite simply, if the Internet is a vehicle for development as growth, then it is the already strong economies that are benefiting most from it, and pulling yet further away from the least developed economies.

Furthermore, this is not just a question of "development" in different categories of country. Inequalities exist in every country, with a report by the Office for National Statistics (2013) in the UK for example emphasising that some 4 million, or 17 % of all, households still did not have Internet access in 2013. At a time when the UK government, as are many others, is increasingly making services available online, purportedly to save money while also providing a better service, those

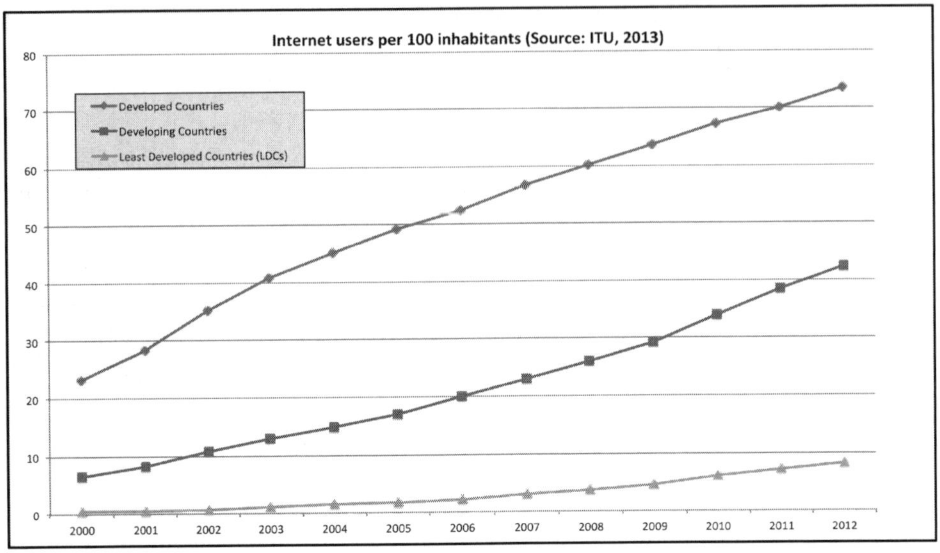

Fig. 2: Internet users per 100 inhabitants.
Source: Rachel Strobel based on ITU date (2013).

without such access are therefore becoming yet further disadvantaged. This is true across the world, as most countries are being encouraged to adopt e-government initiatives. Two-thirds of the world's poor actually live in middle income countries (Alkire et al. 2013), and if such poor people do not have access to the benefits of ICTs then the potential for violent social and political change will increase dramatically. Hence, initiatives such as the ITU and UNESCO's Broadband Commission for Digital Development (www.broadbandcommission.org/) and the Alliance for Affordable Internet (www.a4ai.org/) are becoming increasingly important in trying to ensure that the poor can actually benefit wherever they are found.

The expansion of the Internet, though, also carries with it great challenges concerning the relationships between citizens on the one hand, and states and corporations on the other. During the first decade of the 21st century, it was widely believed that the Internet could be truly liberating and even anarchic, enabling new more equitable social formations to emerge. Thus, as Thelwall (2013: 71) has commented, "The Web has been widely heralded for its potential to democratize access to, and the provision of, information". Yet, as he goes on to say, "This theoretical equality in information provision does not occur in practice" (Thelwall 2013: 71). Two incidents in 2013 highlight the significance of this. First, Edward Snowden's leaks relating to the extensive Internet- and 'phone-surveillance by US intelligence services indicate just how far states have gone to use the Internet to monitor their citizens (BBC 2013). Second, Google's business model is well known, whereby it offers people something for free, such as an Internet search engine or an e-mail account, but then uses the information that it can accumulate from such usage to make very substantial profits (Vise 2005). The scandal that broke in 2012 surrounding its accessing of information about personal web activity from domestic WiFi networks through its Street View cars, and more recently the acknowledgement by Google in August 2013 that people sending e-mails to any of its 425 million Gmail users have no reasonable expectation that their communications are confidential, both raise enormous questions about the company's claim that it does no evil.

One of the most remarkable things about the Internet, is the way that it has enabled companies such as Google and Facebook to encourage people to transform their views of privacy. What in the 20th century was seen as being private and confidential, is now more often than not seen as being something that people are willing to disclose. This raises very fundamental questions about the nature of privacy. On the one hand, Etzioni (2005) has argued that privacy is a good that can be weighed up against other goods. Accordingly, citizens are willing to give up some rights to privacy, if this is necessary for governments, for example, to protect them from so-called "terrorism". This would indeed seem to be what has largely happened with the use of the Internet (Barnes 2006). In contrast, however, it can also be argued that privacy is actually a means through which individuals retain power over their own lives, and that it is therefore something fundamentally more important (Friedman 2005). Power relationships are asymmetric, with states having vastly more power than any one individual usually has, and as Friedman (2005: 265) again comments, "Reducing government's ability to do bad things to us, at the cost of limiting its ability to protect us from bad things done to us by ourselves or by other people, may not be such a bad deal". This raises fascinating issues about the relationships between ICTs, freedom and development (Unwin 2010). Indeed, where citizens do not trust their governments, they are usually much more willing to share personal information with companies than they are with any organ of government. Paradoxically, therefore, a coherent argument can be constructed that ICTs are actually reducing our privacy and freedom, through the actions of states and corporations that ever more carefully monitor our every digital action. Certainly, the uses to which freedom on the Internet is controlled vary significantly between countries (Freedom House 2009), but the key point that must be made here is that it is crucially important to challenge the simple association that relates development as freedom, the Internet as necessarily being democratic and liberating, and thereby reaching the conclusion that the Internet encourages development as freedom.

Two further important elements on the *antithesis* remain to be noted. The first relates to so-called sustainable development (Mansell and Wehn 1998), and the second to the symbolic nature of ICTs. ICTs are often seen as being valuable because they contribute to sustainable development by helping to reduce carbon emissions. The Global e-Sustainability Initiative (GeSI – http://gesi.org/), which boasts more than 30 of the world's leading ICT sector companies as its members, is thus intended to help create a sustainable world through responsible, ICT-enabled transformation. However, most analyses of the contribution of the ICT sector as a whole to sustainability are only partial, and fail sufficiently to take into consideration many of the wider factors that need to be considered (Fujitsu 2012). Moreover, much of the sector specifically focuses on creating products that are deliberately not designed to last, so that consumers have to purchase newer and more expensive models, thereby increasing corporate profits. This is scarcely creating an environmentally sustainable economy. Two examples of such practices will suffice:

- Many mobile 'phones and laptops are built not to be robust so that they only last two or three years. Even where this is not the case, customers are encouraged to "upgrade" their 'phones, often after merely a two-year contract, again leading to replacement of hardware. Old 'phones are either passed on through recycling schemes to poorer people, or all too often contribute to landfill.
- Technological advances in hardware design are often such that "old" software will no longer run on newer machines, and software providers collaborate in this process by not providing software upgrades for versions that run on older machines. Hence, customers have to buy new hardware to run new software, again leading to perfectly usable devices, albeit slower and less functional, being "retired" as waste.

Inherent within the very conceptualisation of the "modern" ICT sector and its thirst for profit is thus an inherent drive to replace "old technology" with "new", which is generally not conducive to sustainability.

The use of the word "modern" also highlights another fundamental dimension of ICTs that is problematic for development when defined in non-economic ways. At one level, development is indeed about creating new things. It is about modernity. As Lushaba (2009) from an African perspective notes, there is a very close correspondence between the constructions of both "modernity" and "development". Hence, the introduction of "modern technologies" such as computers and mobile 'phones can simply by its very presence create a sense of development in poor countries. Such aspirational aspects of ICTs are too often ignored in purely economic analyses of their impact, and are perhaps one of the most significant reasons why ICTs have become such a symbol of contemporary development. However, for fundamental changes to take place in economically poor societies, much more is needed than mere symbols. Indeed, such new technological symbols are all too swiftly replacing many of the traditionally valued symbols of African and Asian society. The technologically "modern" world is not necessarily a world that celebrates the diversity and richness of these traditional cultures. Furthermore, just as factories dehumanised labour in the 19th century, the increasingly digital world of production in the 21st century is also further dehumanising labour, as computers increasingly control their office workers who communicate through e-mail, and rarely have that physical, and indeed real, human experiences of physical communication.

4 In conclusion: towards a *synthesis* – markets and partnerships

The above summaries remain brief and incomplete, but they do capture enough examples to indicate that there are indeed diverse views concerning the relationship between ICTs and "development". The challenge that remains is therefore to explore whether there are any grounds whereby some kind of synthesis can be achieved between what appear to be diametrically opposed ideas. The argument presented here in the conclusion is that it is indeed possible to do so, by arguing that we need to combine both the role of the market and the role of the state together at two different levels to ensure that all people have as equal an opportunity as possible to benefit from the potential of modern ICTs to transform their lives.

First, it seems unlikely, at least in the short term, that the contemporary capitalist world order will be transformed at all dramatically, and therefore it is probable that most countries will be locked into a generally low tax, low state intervention ethos, designed to enhance economic growth, for the next decade or so. As has been argued convincingly by those supporting the positive thesis outlined above, ICTs can indeed contribute substantially to such economic growth, and it is therefore very important that the market functions as effectively as possible, both nationally and globally, so as to facilitate such development. Hence, much work has been done to design, implement and support the creation of effective regulatory mechanisms that not only permit, but also actively encourage, competition in the ICT market place (www.ictregulationtoolkit.org). The effects of this have been witnessed particularly through the dramatic increase in the number of mobile 'phones and broadband connectivity across the world over the last decade (Pélissié du Rausas et al. 2011; Cisco and ITU 2013). Getting the balance right, though, is very far from easy, especially when in many of the world's poorer countries telecommunications regulation is seen as being a key contributor to the overall national GDP, with governments using such revenue to drive a wide range of other development activities. Senior managers in the private sector, who generally see themselves as being much more efficient than governments, frequently object to such intervention, pointing out that less regulation would actually enable them to get on and deliver the intended development outcomes more efficiently and at lower cost. Getting regulators and operators to work together supportively and effectively can thus be a very considerable challenge. Getting the balance right in terms of the level of government taxation and intervention is even more challenging.

Second, it must be recognised that the basic interests of states and of the private sector are fundamentally different. Companies exist to make profit for their shareholders, and will therefore tend towards investing first in the most profitable markets; they cannot survive simply on investing in loss-making ventures. States, on the other hand, are ultimately the guarantors of stability for all of their citizens, and regardless of their different political shades, politicians usually recognise that this requires them to enable all of their citizens to benefit from a standard set of services and utilities. With regard to ICTs, and especially telecommunications, the market alone has not yet proven capable of delivering universal access, let alone affordable access, to everyone. It is generally very costly to deliver affordable broadband Internet access to people living in isolated areas of poor countries, and it is here that states that aspire to offer such services, be it for economic growth or for more social and cultural reasons, have to find a way of intervening.

Sometimes this is attempted through Universal Service or Access Funds, which generally derive their funding from levies on companies in the ICT sector. Again, though, such companies often argue that they could deliver such services much more cheaply, if governments would let them do so. LADCOMM/GSMA (2013) thus note that less than one in eight of the universal service/access funds they surveyed had actually met their targets, and that some US $11 billion was languishing unused in such funds at the time the report was published.

Another alternative that has therefore been promoted is the use of multi-stakeholder partnerships for implementing effective ICT solutions for people for whom the market will not deliver cost effective solutions. These too have been controversial (Unwin 2014), with many claimed such partnerships often failing to deliver on their intended outcomes (Geldof et al. 2011). Much of such failure has been due to a lack of understanding about the enormous complexity, and thus cost, of delivering such partnerships effectively, and also because they have too often been conceived of as just involving governments and the private sector. Multi-stakeholder partnerships, that explicitly engage civil society and international donors alongside governments and the private sector, building on the expertise and resources of all involved, can though offer one way of resolving the very differing interests underlying the *thesis* and *antithesis* represented in this chapter. Such a synthesis recognises that a market-driven approach to economic growth will indeed remain prominent in delivering development through the use of ICTs, but that inherent within such an approach is its tendency to lead to enhanced inequality, the *antithesis* of "development". By crafting and supporting sustainable and effective multi-stakeholder partnerships, governments can nevertheless seek to address such inequalities, and thereby work to enable all of their citizens to derive benefits from the wider roll-out of the Internet.

This, nevertheless, raises the challenging moral question that if the only way to lift the poor out of poverty is for the rich to get even richer, is this right?

Note: All links were checked on 15th August 2013.

References

Aker, Jenny C. & Marcel Fafchamps. 2013. Mobile 'phone coverage and producer markets: evidence from West Africa. CEPR Discussion paper DP9491, http://www.cepr.org/active/publications/discussion_papers/dp.php?dpno=9491

Alkire, Sabina, José Manuel Roche & Suman Seth. 2013. Identifying the "bottom billion": beyond national averages. Oxford: OPHI. http://www.ophi.org.uk/wp-content/uploads/Identifying-the-Bottom-Billion-Beyond-National-Averages.pdf?7 ff332

Australian Council for Educational Research. 2010. Evaluation of One Laptop per child (OLPC) Trial Project in the Solomon Islands, Australian Council for Educational Research for Ministry of Education and human Resources Development, Solomon Islands Government (http://wiki.laptop.org/images/0/0b/SolomonIslandsOLPCTrialsEvaluationByACER2010.pdf)

Bacon, Francis. 1620. *Novum Organum*, Part 2 of *Istauratio Magna*. Translated and edited by James Spedding, Robert L. Ellis and Douglas D. Heath. 1857–74. London: Longmans.

Barnes, Susan B. 2006. A privacy paradox: social networking in the United States. *First Monday* 11(9), 4 September 2006. http://firstmonday.org/ojs/index.php/fm/article/viewArticle/1394/1312 %2523

Bauman, Zigmunt. 2013. Does the richness of the few benefit us all? *Social Europe Journal*, 28/01/2013 http://www.social-europe.eu/2013/01/does-the-richness-of-the-few-benefit-us-all/.

BBC. 2013. Profile: Edward Snowden, 7 August 2013, http://www.bbc.co.uk/news/world-us-canada-22837100.

Bilbao-Osorio, Beñat, Soumtira Dutta and Bruno Lanvin (eds). 2013. *Global Information Economy Report 2013: growth and jobs in a hyperconnected world*. Geneva: World Economic Forum.

Bronner, Stephen Eric. 2004. *Reclaiming the Enlightenment: Toward a Politics of Radical Engagement*. New York: Columbia University Press.

Burnell, Peter. 2002. Foreign aid in a changing world. In Vandana Desai and Rob B. Potter (eds) *The Companion to Development Studies*, 473–477. London: Arnold.

Castells, Manuel. 1996. *The Rise of the Network Society. The Information Age: Economy, Society and Culture, Volume 1*. Oxford: Blackwell.

Castells, Manuel, Mireia Fernández-Ardèvol, Jack Linchuan Qiu & Araba Sey. 2006 *Mobile Communication and Society: a Global Perspective*. Cambridge, MA: MIT Press.

Cisco and ITU. 2013. *Planning for Progress: Why National Broadband Plans Matter*. Geneva: Broadband Commission.

Clark, Donald. 2013. Sugata Mitra: Slum chic? 7 reasons for doubt, http://donaldclarkplanb.blogspot.co.uk/2013/03/sugata-mitra-slum-chic-7-reasons-for.html.

Dalberg, X. 2013. *Impact of the Internet in Africa. Establishing Condition for Success and Catalysing Inclusive Growth in Ghana, Kenya, Nigeria and Senegal*. Washington DC and London: Dalberg http://www.impactoftheInternet.com/pdf/Dalberg_Impact_of_Internet_Africa_Full_Report_April2013_vENG_Final.pdf .

Day, Bob and Peter Greenwood. 2009. Information and communication technologies for rural development. In Tim Unwin (ed.) *Information and Communication Technology for Development*, 321–59. Cambridge: Cambridge University Press.

Demombynes, Gabriel & Aaron Thegeya. 2012. Kenya's mobile revolution and the promise of mobile savings. World Bank Policy Research Working Paper 5988, Washington DC: World Bank http://elibrary.worldbank.org/deliver/5988.pdf;jsessionid=5f0jf60m5ba6e.z-wb-live-01?itemId=/content/workingpaper/10.1596/1813-9450-5988 & mimeType=pdf

DFID. 2006. *Eliminating World Poverty: Making Governance Work for the Poor*. Norwich: HMSO. (http://www.official-documents.gov.uk/document/cm68/6876/6876.pdf).

Dodson, Leslie, S. Revi Sterling, John K. Bennett. 2013. Considering failure: eight years of ITID research. *Information Technologies and International Development* 9(2). 19–34.

Dutton, William H. (ed.). 2013a. *The Oxford Handbook of Internet Studies*. Oxford: Oxford University Press.

Dutton, William H. 2013b. Internet Studies: the foundations of a formative field. In William H. Dutton (ed.), 2013a. *The Oxford Handbook of Internet Studies*, 1–23. Oxford: Oxford University Press.

Easterley, William. 2006 *The White Man's Burden: Why the West's Efforts to Aid the Rest Have Done so Much Ill*. New York: Penguin.
Ericsson. 2013. *ICTs and Human Rights: an Ecosystem Approach*. Stockholm: Ericsson.
Escobar, Arturo. 1995. *Encountering Development: the Making and Unmaking of the Third World*. Princeton: Princeton University Press.
Etzioni, Amitai. 2005. The limits of privacy. In A. Cohen and C. H. Wellman (eds.), *Contemporary Debates in Applied Ethics*, 253–262. Oxford: Blackwell.
Freedom House. 2009. *Freedom on the Net: a Global Assessment of Internet and Digital Media*, Washington, DC: Freedom House.
Friedman, David D. 2005. The case for freedom. In Andrew Cohen and Chrisotpher H. Wellman (eds.), *Contemporary Debates in Applied Ethics*, 262–275. Oxford: Blackwell.
Fujitsu. 2012. *ICT Sustainability: the Global Benchmark 2012*. No place: Fujitsu. https://www-s.fujitsu.com/global/solutions/sustainability/Fujitsu-Sustainability.html
G3ICT and ITU. 2012. *Making Mobile Phones and Services Accessible for Persons with Disabilities*. Geneva: ITU.
Gay, Peter. 1996. *The Enlightenment: an Interpretation*. New York: W. W. Norton.
Geldof, Marije, David J. Grimshaw, Dorothea Kleine & Tim Unwin. 2011. What are the key lessons of ICT4D partnerships for poverty reduction? London: Department for International Development.
Gomez, Ricardo. 2013 The changing field of ICT4D: growth and maturation of the field, 2000–2010. *The Electronic Journal on Information Systems in Developing Countries* 58(1). 1–21.
Hansen, Nina, Tom Postmes, Annemarie Bos & Annika Tovote. 2009. Does technology dirve social change? Psychological, social and cultural effects of OLPC among Ethiopian children. University of Groningen, Engineering Capacity Building Programme (http://www.gg.rhul.ac.uk/ict4d/NinaandTom.pdf).
Harvey, David. 2000. *Spaces of Hope*. Oxford: Blackwell.
Heeks, Richard. 2009. Where next for ICTs and international development? In OECD (2009) *ICTs for Development: Improving Policy Coherence*, 29–74. Paris: OECD and *info*Dev.
Heeks, Richard. 2012. Information technology and gross national happiness. *Communications of the ACM* 55(4). 24–26.
Hollow, David. 2008. Low-cost laptops for education in Ethiopia. ICT4D working Paper (http://www.gg.rhul.ac.uk/ict4d/workingpapers/Hollowlaptops.pdf)
ITU. 2011. *The role of ICT in Advancing Growth in Least Developed Countries*. Geneva: ITU.
Kamau, M. 2013. Telecoms regulator notes mobile 'phone penetration decreased by 900,000 after December exercise. *Standard Digital*, 31 July 2013, http://www.standardmedia.co.ke/?articleID=2000089774&story_title=sim-card-switch-off-sees-subscriber-numbers-drop
Kantrowitz, Alex. 2013. The secret behind the Turkish protesters' social media mastery, *Mediashit*, 1 July 2013, http://www.pbs.org/mediashift/2013/07/the-secret-behind-the-turkish-protesters-social-media-mastery
Karlan, Dean and Jacob Appel. 2012. *More than Good Intentions: Improving the Ways the World's Poor Borrow, Save, Farm, Learn and Stay Healthy*. New York: Penguin.
Kothari, Uma. 2005. *A Radical History of Development Studies: Individuals, Institutions and Ideologies*. London: Zed Press.
Krstić, Ivan. 2008. Sic transit Gloria laptopi, http://radian.org/notebook/sic-transit-gloria-laptopi
LADCOMM/GSMA. 2013. *Universal Service Fund Study*. London: LADCOMM for GSMA.
Lushaba, Lwazi Siyabonga. 2009. *Development as Modernity, Modernity as Development*. Dakar: Codesria.
Mansell, Robin & Uta Wehn. 1998. *Knowledge Societies: Information Technology for Sustainable Development*. Oxford: Oxford University Press.
Mas, Ignatio & Amolo Ng'weno. 2010. Three keys to M-PESA's success: branding, channel management and pricing. *Journal of Payments Strategy and Systems* 4(4). Available: http://

www.gsma.com/mobilefordevelopment/wp-content/uploads/2012/03/keystompesassuccess4jan69.pdf

McMichael, Philip. 1996. *Development and Social Change: a Global Perspective*. Thousand Oaks: Pine Forge.

Mohan, Giles, Ed Brown, Bob Milward & Alfred B. Zack-Williams. 2000. *Structural Adjustment: Theory, Practice and Impacts*. London: Routledge.

Mwageni, Eleuther, Honorati Masanja, Zaharani Juma, Devota Momburi, Yahya Mkilindi, Conrad Mbuya, Harun Kasale, Graham Reid & Don de Savigny. 2005. Socio-economic status and health inequalities in rural Tanzania: evidence from the Rufiji demographic surveillance system. In Indepth network (eds), *Measuring Health Equity in Small Areas*, 19–32. London: Ashgate Publishing Ltd. Net!works (2012).

Net!Works. White Paper on: Economic impact of the ICT sector. http://www.networks-etp.eu/fileadmin/user_upload/Publications/Position_White_Papers/Net_Works_White_Paper_on_economic_impact_final.pdf. Accessed 15th August 2013.

Nisbet, Erik C., Elizabeth Stoycheff & Katy E. Pearce. 2012. Internet use and democratic demands: a multinational, multilevel model of Internet use and citizen attitudes about democracy. *Journal of Communication* 62(2). 249–65.

O'Boyle, Edward J. 1999. Toward an improved definition of poverty. *Review of Social Economy* 57(3). 281–301.

OECD. 2009. *ICTs for Development: Improving Policy Coherence*. Paris: OECD and *info*Dev. http://www.keepeek.com/Digital-Asset-Management/oecd/development/icts-for-development_9789264077409-en

Office for National Statistics. 2013. *Statistical Bulleting: Internet Access – Households and Individuals, 2013*. London: Office for National Statistics. http://www.ons.gov.uk/ons/dcp171778_322713.pdf

O'Hearn, Denis. 2009. Amartya Sen's Development as Freedom: ten years later. *Policy and Practice: a Development Education Review* 8. 9–15. (http://www.developmenteducationreview.com/issue8-focus1).

Pélissié du Rausas, Mathiew, James Manyika, Eric Hazan, Jacques Bughin, Michael Chui, & Rémi Said. 2011. *Internet Matters: the Net's Sweeping Impact on Growth, Jobs and Prosperity*. McKinsey Global Institute, Insights and Publications, http://www.mckinsey.com/insights/high_tech_telecoms_Internet/Internet_matters

Pieterse, Jan Nederveen. 2010. *Development* Theory. 2nd edn. London: Sage.

Reardon, Sara. 2012. Was the Arab Spring really a Facebook Revolution? *New Scientist*, 13 April 2012. http://www.newscientist.com/article/mg21428596.400-was-the-arab-spring-really-a-facebook-revolution.html#.Uy7NZf3oluY

Sachs, Jeffrey. 2005. *The End of Poverty: How we can make it Happen in our Lifetime*. London: Penguin.

Sen, Amartya. 1985. A sociological approach to the measurement of poverty: a reply to Professor Peter Townsend. *Oxford Economic Papers* 37(4). 669–676.

Sen, Amartya. 1999. *Development as Freedom*. Oxford: Oxford University Press.

Sen, Amartya. 2002. Globalisation, inequality and global protest. *Development* 45(2). 11–16

Stiglitz, Joseph E. 2002. *Globalization and its Discontents*. London: Penguin.

Thelwall, Mike. 2013. Society on the Web. In William H. Dutton (ed.), *The Oxford Handbook of Internet Studies*, 69–85. Oxford: Oxford University Press.

Times of India. 2013. Naveen Patnaik gifts mobile 'phones to 5,000 farmers. June 19th. http://articles.timesofindia.indiatimes.com/2013-06-19/india/40069215_1_farmers-mobile-phones-market

UN. 2011. Report of the Special Rapporteur on the Promotion and Protection of the Right to Freedom of Opinion and Expression, Frank La Rue. A/HRC/17/27 (http://www2.ohchr.org/english/bodies/hrcouncil/docs/17session/A.HRC.17.27_en.pdf)

UNCTAD. 2011. *Measuring the Impacts of Information and Communication Technology for Development*. Geneva: UNCTAD Current Studies on Science, Technology and Innovation, No 3.

UNDP. 2010. *Human Development Report 2010 – 20th Anniversary Edition. The Real Wealth of Nations: Pathways to Human Development*, New York: UNDP.

Unwin, Tim. 2004. Beyond budgetary support: pro-poor development agendas for Africa. *Third World Quarterly* 25(8). 1501–1523.

Unwin, Tim. 2007. No end to poverty. *Journal of Development Studies* 43(5). 929–953.

Unwin, Tim (ed.). 2009. *Information and Communication Technology for Development*. Cambridge: Cambridge University Press.

Unwin, Tim. 2010. ICTs, citizens, and the state: moral philosophy and development practices. *The Electronic Journal of Information Systems in Developing Countries* 44. https://www.ejisdc.org/ojs2/index.php/ejisdc/article/view/744

Unwin, Tim. 2013. The Internet and development: a critical perspective. In William H. Dutton (ed.), *The Oxford Handbook of Internet Studies*, 531–554. Oxford: Oxford University Press.

Unwin, Tim. 2014 forthcoming. Multi-Stakeholder Partnerships in Information and Communication for Development Interventions. In *International Encyclopedia of Digital Communication and Society*. London: Wiley

Ura, Karma, Sabina Alkire, Tshoki Zangmo & Karma Wangdi. 2012. *A Short Guide to Gross National Happiness Index*. Thimphu: The Centre for Bhutan Studies. (http://www.grossnationalhappiness.com/wp-content/uploads/2012/04/Short-GNH-Index-edited.pdf).

Vise, David A. 2005 *The Google Story*. New York: Bantam Dell.

Wallerstein, Immanuel. 1983. *Historical Capitalism*. London: Verso Books.

Wallerstein, Immanuel. 2000. Globalization or the age of transition? A long-term view of the trajectory of the world system. *International Sociology* 15(2). 251–267

Walsham, Geoff. 2001. *Making a World of Difference: IT in a Global Context*. Chichester: Wiley.

Walsham, Geoff. 2012. Are we making a better world with ICTs? Reflections on a future agenda for the IS field. *Journal of Information Technology* 27(2). 87–93.

Weigel, Gerolf & Daniele Waldburger (eds). 2004. *ICT4D – Connecting People for a Better World*. Berne and Kuala Lumpur: Swiss Agency for Development and Cooperation and Global Knowledge Partnership.

Williamson, John. (ed.). 1990. *Latin American Adjustment: How Much has Happened?* Washington DC: Institute for International Economics.

World Bank. 2009. *Information and Communications for Development 2009: Extending Reach and Increasing Impact*. Washington DC: World Bank.

World Bank. 2011. *ICT in Agriculture e-Sourcebook: Connecting Smallholders to Knowledge, Networks and Institutions*. Washington DC: infoDev, ARB, and IBRD of World Bank.

World Economic Forum. 2013. *Global Information Technology Report 2013*. Geneva: World Economic Forum.

Yunkap Kwankam, S., Ariel Pablos-Mendez & Misha Kay. 2009. e-Health: information and communication technologies for health. In Tim Unwin (ed.), *Information and Communication Technology for Development*, 249–282. Cambridge: Cambridge University Press.

Martin J. Eppler

11 Information quality and information overload: The promises and perils of the information age

Abstract: In this contribution we present two key concepts in the realm of modern day communication infrastructures: the prescriptive notion of information quality – as the fitness for use of information products, services, and infrastructures for its various stakeholders – and the descriptive concept of information overload, i.e., not being able to process the provided quantity of information adequately. We outline why these two issues are highly relevant for the field of communication, where and how they have been addressed in research, how they can be conceptualized, how they inter-relate (i.e., how information quality can help to prevent information overload), and how this line of research may evolve in the future.

Keywords: information quality, data quality, information overload, cognitive overload, communication overload, communication quality, information processing, management, information technology

> *Where is the Life we have lost in living? Where is the wisdom we have lost in knowledge? Where is the knowledge we have lost in information?* (T. S. Eliot)

It has become a truism to state that we are living in an information age, where information has become not only a decisive production factor of a knowledge-based economy, but also a key facet (and at times nuisance) of everyday life. Digital, printed and visualized information is all around us and we at times struggle to make sense of it all, or filter out what is really relevant to our context. As more communication means are available to more people than ever before, more information is generated and disseminated in the process. This creates a twofold problem: It creates information stress in the sense that it strains our ability to process information adequately, and it makes it more difficult to pay attention to the (right) information that we really require (and can rely on) to solve problems, make decisions, perform certain tasks, or build up new knowledge.

In this contribution we address these challenges by discussing two influential concepts that both have a long-standing research tradition: information overload and information quality. We argue that these two notions are key constructs of the information age and should be known by anyone active in the communication field. The concepts' cash values, as the pragmatist philosopher Charles S. Peirce would say, seem particularly high in a time when communicators, such as journalists, analysts, scientists, consultants and other (information-driven) professional groups claim their legitimacy by stressing their ability to produce high-quality information and to cut through the clutter of information overload.

The questions that we consequently address in this contribution are the following:

1. Why are information quality and information overload relevant and instructive concepts to comprehend communication in the 21st century?
2. What do these two concepts entail and how can they be systematically conceptualized and used?
3. How are the two concepts related?
4. What kind of research has been undertaken to study the two concepts and which results have been achieved?
5. What are the current and future research needs in this area?

Our main focus in this contribution will be on question 2 or on how to systematically conceptualize the two notions. First, however, we will briefly discuss the relevance of the two concepts for the communication field in general and for communication infrastructures in particular. After section 2 on their relevance, we will present the two concepts in detail in sections 3 and 4. We begin with the problem of information overload in section 3 and then show one possible response to this challenge in section 4: the quest for information quality. In sections 3 and 4 we thus also discuss how the two constructs are related. Both sections 3 and 4 contain an informative research synopsis on the respective constructs. In section 5 we derive research avenues from these reviews of literature and from current challenges in the realm of communication and its infrastructures.

1 The Relevance of information quality and information overload for communication

In order to understand the timely relevance of the concept of information overload and the pressing need for high-quality information, it is sufficient to screen the current statistics on the information explosion that has been taking place around us and is affecting us on a daily basis:

About every five minutes, the typical individual employee is interrupted by email (according to Jackson et al. 2003 cited in: Sumeckia et al. 2011). Another statistics claims that every minute of the day e-mail users send more than 204,000 messages to one another (Jackson et al. 2012). Hair et al. surveyed a group of academics and creative workers, and found that 34 % felt stressed by the volume of emails, 50 % checked their email every hour and 35 % checked their email every 15 minutes; one in three workers suffered from email stress. But e-mail is not the only symptom of an overloaded society: the average US teen is assailed by 3,000 advertisements per day; a staggering 10 million by age 18 according to Hayward (2013). The average American also watches 5.1 hours of television per day according to a recent study (Joo et al. 2013).

In addition to these quantifiable trends, we also see an ever-growing threat of misinformation. Social media have been accused of making it very easy and practically seamless for anyone to spread lies, falsified information, rumors, outdated insights, or distorted facts. This further increases the need for understanding the overload problem and the promise of (detecting and managing) information quality.

These developments have led to a constant over-supply of information and an under-supply of time and attention. This ever-growing offer is straining our ability to screen, filter and assess information. It requires technology and infrastructures that help us deal with the data deluge and assist us in identifying truly high quality information – and this without usability problems that further increase the cognitive load on users.

The trends described above also lead to what could be called a *commoditization* of information, making it a seemingly free or cheap item. To fight against this perception information providers (i.e., journalists, analysts, researchers, commentators, or consultants) need to distinguish their offer through higher quality information. They need to understand what makes information valuable to their audiences.

But what characteristics constitute quality information and how can these traits be systematically ensured? Does quality really address some of the root causes of the overload challenge and if so which ones?

In light of these questions, we need to clarify how information overload comes about, how it can be reduced and how a systematic management of information quality can reduce this challenge. This clarification can be achieved by conceptualizing the two phenomena from a management perspective, which we attempt in the following two sections.

2 The concept of information overload

In this section we briefly provide an overview on the facets of information overload and offer a diagrammatic approach to this complex phenomenon. In doing so, we highlight the main causes, consequences and countermeasures of information overload.

In every day communication, the notion of information overload simply describes situations in which we receive too much information to sensibly deal with it all in our available time frame. We may express this feeling by complaining about being inundated by e-mail and text messages, constantly interrupted by phone calls and meetings, or flooded with reports, blogs, tweets or facebook updates. We may experience what Alvin Toffler called information overload in the early 1970s (York 2013: 3) or Richard Wurman had labeled information anxiety (Wurman 1990), namely the feeling that there is always another memo, magazine article or book that we just have to read to be in the know. There are various synonyms and related constructs that have been suggested for this phenomenon in the literature: cognitive overload, sensory overload, communication overload, knowledge overload, or information fatigue syndrome are a few of the more prominent ones. These constructs have been applied to a variety of overload situations, ranging from auditing, strategizing, business consulting, management meetings to supermarket-shopping. The following table provides a synoptic view of some of the typical definition of such situations (taken from Eppler and Mengis 2004: 328).

Tab. 1: A compilation of definitions of the information overload concept.

Definitions	Components/Dimensions	References
The decision maker is considered to have experienced information overload at the point where the amount of information actually integrated into the decision begins to decline. Beyond this point, the individual's decisions reflect a lesser utilization of the available information.	– inverted u-curve: relationship between amount of information provided and amount of information integrated by decision maker – information utilization	Chewning and Harrell (1990) Cook (1993) Griffeth et al. (1988) Swain and Haka (2000)
Information overload occurs when the volume of the information supply exceeds the limited human information processing capacity. Dysfunctional effects such as stress or confusion are the result.	– volume of information supply (information items versus – chunks) – information processing capacity – dysfunctional consequences	Jacoby (1974) Malhotra (1982) Meyer (1998)
Information overload occurs when the information processing requirements (information needed to complete a task) exceed the information processing capacity (the quantity of information one can integrate into the decision making process).	– Information processing capacity – Information processing requirements	Galbraith (1974) Tushman and Nadler (1978)
Information overload occurs when the information processing demands on time to perform interactions and internal calculations exceed the supply or capacity of time available for such processing.	– Time demands of information processing; Available time versus invested time – Number of interactions (with subordinates, colleagues, superiors) – Internal calculations (i.e., thinking time)	Schick, Gordon and Haka (1990)

Definitions	Components/Dimensions	References
Information overload has occurred when the information-processing requirements exceed the information-processing capacity. Not only the amount of information (quantitative aspect) that has to be integrated is crucial, but also the characteristics (qualitative aspect) of information.	– Information-processing requirements – Information processing-capacity – Quantitative and qualitative dimensions of information (multidimensional approach)	Keller and Staelin (1987) Schneider (1987) Owen (1992) Iselin (1993)
Information overload occurs when the decision maker estimates to have to handle more information than he/she can efficiently use.	– Subjective component: opinion, job- & communication-satisfaction – situational factors and personal factors	Abdel-Khalik (1973) Iselin (1993) O'Reilly (1980) Haksever and Fisher (1996)
Amount of reading matter ingested exceeds amount of energy available for digestion, the surplus accumulates and is converted by stress and over-stimulation into the unhealthy state known as information overload anxiety.	– Subjective cause component: energy – Symptom: stress, over-simulation – Subjective effect: information overload anxiety	Wurman (1990) Wurman (2001) Shenk (1997)

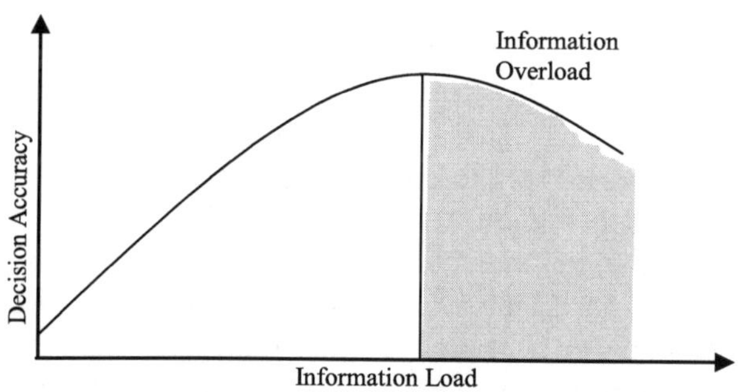

Fig. 1: Information overload as the inverted u-curve.

Next to these verbal definitions, there is also a diagrammatic representation of the phenomenon that has been validated in several research contexts and is widely used. It depicts information overload as the negative effect of increasing information quantity or cognitive load on decision accuracy and is often referred to as the "inverted u-curve" view of overload.

This visual description of information overload emphasizes the (for some counter-intuitive) fact that more information (or more communication for that matter) does not always lead to greater clarity or better decisions. Rather, there is a threshold point beyond which every additional piece of information may actually reduce

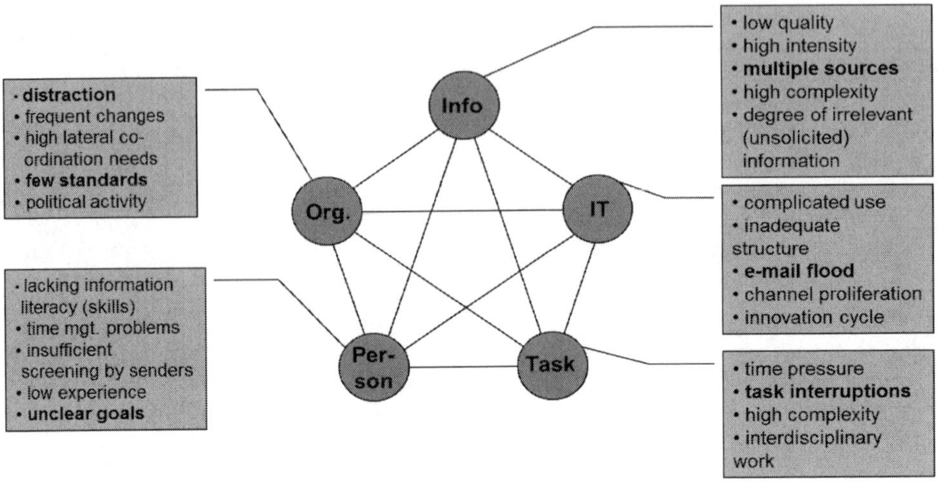

Fig. 2: Five main causes (and remedies) of information overload.

our decision accuracy. A main reason for this sudden decline of decision accuracy as information increase lies in what Herbert Simon called "bounded rationality" or simply our limited ability to process information. As our cognitive load increases, we begin to take mental shortcuts to ease the burden of excessive complexity. We begin to filter out information (for example facts that run counter to our point of view). We pay more attention to information that is attractive (rather than influential) or give more weight to the last pieces of information rather than to information that was processed earlier. These and other mental shortcuts decrease our decision accuracy. It may also lead to not making any decision at all, the so-called paralysis-by-analysis phenomenon. We, as individuals, are not the only ones to blame for overload. Next to the individual receiver or sender of information, there is also the information itself which can contribute to an excessive cognitive load, as well as our task design and the way that the organization in which we operate reduces or aggravates the problem. Last but not least, information technology infrastructures have a key role to play in creating or reducing information overload. In the following we describe these five causes of information overload in more detail (taken from Eppler and Mengis 2004: 330–332). Usually, information overload emerges not because of one of these factors, but because of a mix of all five causes.

An important factor influencing the occurrence of information overload is the *organizational design* of a company. Changes in the organizational design, for instance due to disintermediation or centralization (Schneider 1987) or because of a move to interdisciplinary teams (Bawden 2001), can lead to greater information processing requirements because they create the need for more intensive communication and coordination. On the other hand, a better coordination through standards, common procedures, rules or dedicated coordination centers (Galbraith 1974) can reduce the processing needs and positively influence our ability to process information (Galbraith 1974; Schick et al. 1990; Tushman and Nadler 1978).

Next to the organizational design, another important overload cause is the *nature of information* itself. Schneider (1987) stresses the fact that it is not only the amount of information that determines information overload, but also the specific characteristics of information. Such characteristics are the level of uncertainty associated with information as well as the level of ambiguity, novelty, complexity or intensity (Schneider 1987). Simpson and Prusak (1995) argue that modifying the quality of information can have great effects on the likelihood of information overload (see the next section for more on this). Improving the quality (e.g., conciseness, consistency, comprehensibility etc.) of information can improve the information processing capacity of the individual, as he or she is able to use high-quality information quicker and better than ill-structured, unclear information.

The *person* and his or her attitude, qualification or experience is another important element to determine at which point information overload may occur. While earlier studies simply state that a person's capacity to process information is limited (Jacoby et al. 1974; Galbraith 1974; Malhotra 1982; Simon 1979; Tushman and Nadler 1978), more recent studies include specific limitation factors such as personal skills (Owen 1992), the level of experience (Swain and Haka 2000), or the motivation of a person (Muller 1984).

Another influential cause is constituted by the *tasks and processes* which need to be completed with the help of information. The less a process is based on reoccurring routines (Tushman and Nadler 1978) and the more complex it is in terms of the configuration of its steps (Bawden 2001; Grise and Gallupe 1999), the higher the information load and the greater the time pressure on the individual (Schick et al. 1990). The combination of these two factors can lead to information overload. Information overload is especially likely if the process is frequently interrupted and the concentration of the individual suffers as a consequence (Speier et al. 1999).

Finally, *information technology* and its use or misuse is a major reason why information overload has become a critical issue in the 1980s and 1990s within many organizations. The development and deployment of new information and communication technologies, such as the Internet, intranets and extranets, but especially e-mail are universally seen as one major source of information overload (Bawden 2001). Closely related to the problem of e-mail overload is the discussion of pull- versus push-technologies and whether they have a positive or negative impact on an individual's processing capacity. To push selected pieces of information to specific groups reduces on the one hand their information retrieval time, but increases on the other the amount of potentially useless (or low-quality) information that a person has to deal with (Edmunds and Morris 2000). In addition, it causes more frequent interruptions (Speier et al. 1999). Information technology can thus potentially reduce overload or further increase it. The figure below summarizes these key drivers of the overload challenge. They are also the areas that need to be addressed to reduce information overload: We must optimize the organizational design to reduce unnecessary communication (for example through adequate e-mail policies and efficient communication routines). We must qualify people to adequately filter, aggregate, interpret and convey information. We must help them in aligning their information needs to their goals to avoid the trap of excessive information consumption. We should design tasks in a way so that interruptions are eliminated and unnecessary time pressure can be avoided. We should offer communication infrastructures that consolidate or aggregate information channels rather than multiply them. We should, in other words, design communication infrastructures that assist people in screening, organizing or reviewing information effectively. As a final, and particularly important countermeasure (because each one of us can directly contribute to it) we should increase (whenever possible) the quality of the communicated information, so that it can be more efficiently interpreted, used or sent on.

Having examined the key challenge of information overload we can now look at one possible remedy to this challenge in more depth, namely the promise of high-quality information.

3 The concept of information quality

Everything that can be said, can be said clearly (Ludwig Wittgenstein).

One of the key countermeasures against information overload, as outlined above and stated in the relevant literature streams ranging from (news) consumption, management, to corporate communication, is to address the overload-causing information attributes and raise the quality of information to that it can be more easily found or accessed, interpreted and used. Many of the countermeasures against overload that regard information attributes in fact can be related to augmenting information quality. Before examining the individual attributes that constitute high-quality information, it is useful to define the term. In the review of existing literature on information quality, we have found at least seven kinds of definitions of information quality of which we give one example each below:

1. Information quality can be defined as information that is fit for use by information consumers (Huang et al. 1999).
2. Information quality is the characteristic of information to meet or exceed customer expectations (Kahn and Strong 1998).
3. Quality information is information that meets specifications or requirements (Kahn and Strong 1998)
4. Information quality is the characteristic of information to meet the functional, technical, cognitive, and aesthetic requirements of information producers, administrators, consumers, and experts (Eppler and Muenzenmayer 1999).

5. Information quality is the characteristic of information to be of high value to its users (Lesca and Lesca 1995).
6. The degree to which information has content, form, and time characteristics which give it value to specific end users (O'Brien 1991).
7. Quality of information can be defined as a difference between the required information determined by a goal and the obtained information. In an ideal situation there will be no difference between the required and obtained information. A qualitative measure for information quality is expressed by the following tendency: the smaller this difference, the greater the quality of information (Gerkes 1997).

These definitions stress the relative dimension of information quality, that is to say the notion that information has to fit the needs of certain groups and provide value to them (beyond intrinsic quality issues as accuracy or correctness). Such definitions however, are easier to state than to achieve or implement. For the purpose of assuring and managing information quality, researchers and practitioners in the field have gone beyond simple definitions and created a systematic set of information attributes that can be used to specify and subsequently manage information quality. In the next section we propose such a systematic framework to better comprehend and assure information quality in various contexts. It is a generic framework in the sense that it should be applicable across a wide range of contexts. Typical application areas in which information quality issues have been so far studied are:
- data warehouses, client data bases, or panel / househould address data (see Zhu et al. 2012 for a short overview)
- online information (such as Wikipedia entries)
- journalism (such as the quality of news items or press agency statements)
- auditing and financial reporting (i.e., the quality of an annual report)
- management information (the quality of a management information system)
- contents of knowledge management systems (i.e., the quality of documented experiences)
- patient health records
- publicly available statistical information

While there are several domain-specific information quality frameworks, there are very few proposals for a comprehensive approach to conceptualizing information quality. One such approach is described below.

3.1 A Framework for information quality

The basic premise behind any information quality framework is that you cannot achieve what you do not specify. If the traits that lead to high quality information are not made explicit in a systematic manner, then there will be no way to coordinate the efforts to improve the quality of information. An information quality framework thus provides a blueprint for everybody involved in the creation, dissemination and use of information and acts as a sort of boundary transgressing object to align information producers, administrators, infrastructure managers, and communicators. It answers the following question: what do we need to address in order to ensure that information is of high value?

The overall structure of the present framework is derived from an implicit convention which most recent information quality frameworks seem to follow, whereby they are divided into two sections: a category section, and a criteria (or dimensional) section. Thus, in most IQ-frameworks, the individual quality criteria are grouped into a few information quality categories. In the current framework this twofold structure is also used, but with qualifying category names, which already include a quality criterion on a higher level. The four IQ-categories or views are:

1. *Relevant information*: This category relates to whether the information is comprehensive, accurate, and clear enough for the intended use, and whether it is easily applicable to the problem at hand. This category is also called the community view, since the relevance of a piece of information depends on the expectations and needs of a certain (writer, administrator, or user) community.
2. *Sound information*: This second category contains criteria which describe the intrinsic or product characteristics of information, such as whether it is concise or not, consistent or not, correct or not, and current or not. Whereas the criteria in the first category (relevance) are subjective (indicated through the term "enough") the criteria in this category should be relatively independent of the targeted community (indicated through the term "or not").
3. *Optimized Process*: The third category contains criteria which relate to the content management process by which the information is created and distributed and whether that process (or information service) is convenient (for writers, administrators, and users), and whether it provides the information in a timely, traceable (attributable), and interactive manner.
4. *Reliable Infrastructure*: The fourth and final level contains criteria which relate to the infrastructure on which the content management process runs and through which the information is actually provided. Reliability in this context refers to a system's easy and continuous accessibility, its security, its maintainability over time and at reasonable costs, and its high speed or performance. As the figure below shows, the upper two levels of the framework are labeled content quality, while the lower two are referred to as media quality.

The first two categories, relevance and soundness, relate to the actual information itself, hence the term content quality. The second two categories, process and infrastructure, relate to the management of that information, and whether the

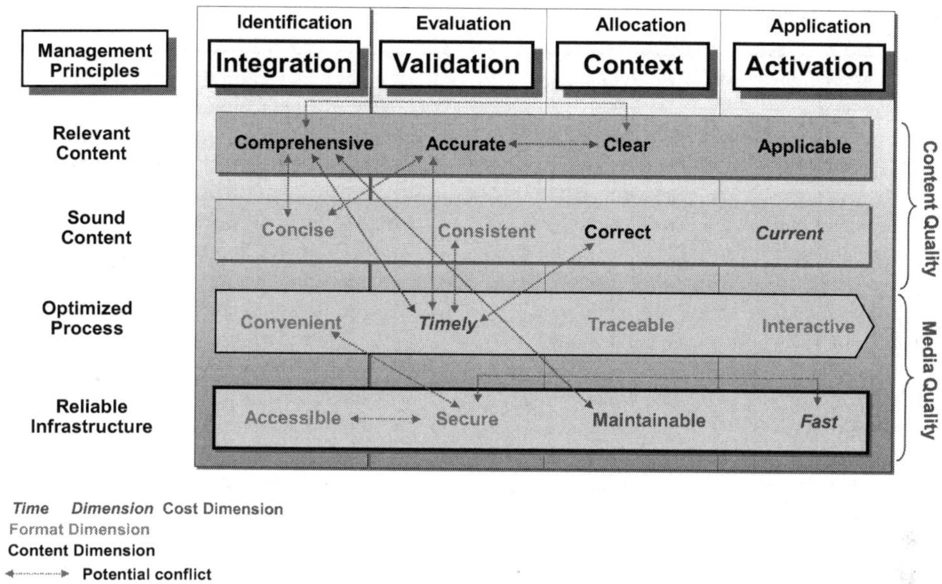

Fig. 3: A systematic framework for information quality management.

delivery process and infrastructure are of adequate quality, hence the term media quality, which stresses the channel by which information is transported. For the end-user, both segments, media and content quality, may be perceived as one final product – the information and its various characteristics. For the information producers and administrators however, this difference is crucial, since the authors usually cannot influence the media quality, and the administrators only have limited possibilities of influencing the content quality. The framework separates these areas because they fall into separate areas of responsibility. In order to be of practical value, the framework has to distinguish between these responsibilities and indicate which areas are the responsibility of management or the information producers (i.e., the area labeled as content quality), and which domain is the responsibility of the support or IT-staff (the two levels labeled media quality).

Besides the four levels of responsibility, the framework also incorporates different *traits* of information that can and should be distinguished. By color-coding the 16 key information attributes that lead to high-quality information, we can highlight the facets of information that give it value, namely its content dimension (the what of information), its time dimension (the when), and its format dimension (the how in terms of representation or delivery). The framework also highlights possible goal conflicts or trade-offs among individual criteria. To give one example: there is usually a trade-off between accuracy and timeliness, as checking for (and if needed improving) accuracy involves considerable delays in providing the information, thus reducing its timeliness. Such trade-offs must be addressed by prioritizing or weighing information quality attributes or by anticipating conflicts

Tab. 2: Seventy typical information quality criteria.

1. Comprehensiveness	25. Reliability	48. Response time
2. Accuracy	26. Price	49. Believability
3. Clarity	27. Verifiability	50. Availability
4. Applicability	28. Testability	51. Consistent Representation
5. Conciseness	29. Provability	52. Ability to represent null values
6. Consistency	30. Performance	53. Semantic Consistency
7. Correctness	31. Ethics/ ethical	54. Concise Representation
8. Currency	32. Privacy	55. Obtainability
9. Convenience	33. Helpfulness	56. Stimulating
10. Timeliness	34. Neutrality	57. Attribute granularity
11. Traceability	35. Ease of Manipulation	58. Flexibility
12. Interactivity	36. Validity	59. Reflexivity
13. Accessibility	37. Relevance	60. Robustness
14. Security	38. Coherence	61. Equivalence of redundant or
15. Maintainability	39. Interpretability	distributed data
16. Speed	40. Completeness	62. Concurrency of redundant or
17. Objectivity	41. Learnability	distributed data
18. Attributability	42. Exclusivity	63. Nonduplication
19. Value-added	43. Right Amount	64. Essentialness
20. Reputation (source)	44. Existence of meta	65. Rightness
21. Ease-of-use	information	66. Usability
22. Precision	45. Appropriateness of meta	67. Cost
23. Comprehensibility	information	68. Ordering
24. Trustworthiness	46. Target group orientation	69. Browsing
(source)	47. Reduction of complexity	70. Error rate

through adequate escalation mechanisms. The main elements of the framework are the individual information quality criteria. They have been carefully selected from the information quality literature and systematically organized. Eppler (2006) discusses their selection in detail, as well as the reason for in- and exclusion of certain criteria. The sixteen criteria have been chosen from a myriad of attributes suggested in the literature. Below we summarize seventy of the most frequently mentioned criteria that cover aspects of information quality.

The sixteen criteria in the framework are structured along four *phases*. These four vertical sections or steps in the framework are the result of an amalgamation of phase models on information behavior found in relevant academic literature. These models typically distinguish between three to six phases of information processing, starting with an information searching activity and finishing with the actual use of the allocated and evaluated information. All of these models stress the identification or selection, evaluation or assessment, transformation or allocation, and application or use of information. The horizontal structure of the framework reflects a chronological sequence (or phases) from the user's point of view. For him (or her) information may be the answer he needs to find, understand and evaluate, adapt to his context and apply in the correct manner. Thus, an information system (based on a general understanding of the term) should assist him in identifying relevant and sound information. It should help him to evaluate whether the information is adequate for his purpose. It should assist him in re-contextualizing the information, that is to say understand its background and adapt it according to the new situation. Finally, the communication infrastructure should provide assistance in making the found, evaluated, and allocated information actionable, e.g., use it effectively. The key questions (of an information consumer) to be answered in each phase are the following ones: 1. Where is the information I need? (identification) 2. Can I trust it? (evaluation) 3. Can I adapt it to my current situation? (allocation) 4. How can I best use it? (application).

The last part of the framework consists of four management *principles* to address the respective information attributes in the four phases. As such, they provide a coordinative mechanism that helps in collaboratively ensuring the traits that give value to information. They provide pragmatic help in implementing the framework and achieving the quality criteria contained in it. The principles are also placed horizontally along the framework since they follow the same step-by-step logic as the four phases discussed above. Every principle relates to the criteria that are found in the same column as the principle. The integration principle, for example, aims to improve the criteria of comprehensiveness, conciseness, convenience, and accessibility. There are various ways to apply the principles in order to improve the quality of information. The figure below gives typical examples from communication infrastructures for each principle.

The principles are especially useful to provide a synoptic overview of what needs to be done to assure the information quality of an information product such as a news portal, a knowledge management system, or even a library catalogue. Below we summarize each principle briefly:

In order to enable information consumers to identify relevant information effectively, information must be *integrated*, i.e., it must be aggregated and cleansed from unnecessary elements to reveal its most pertinent insights. Dispersed sources

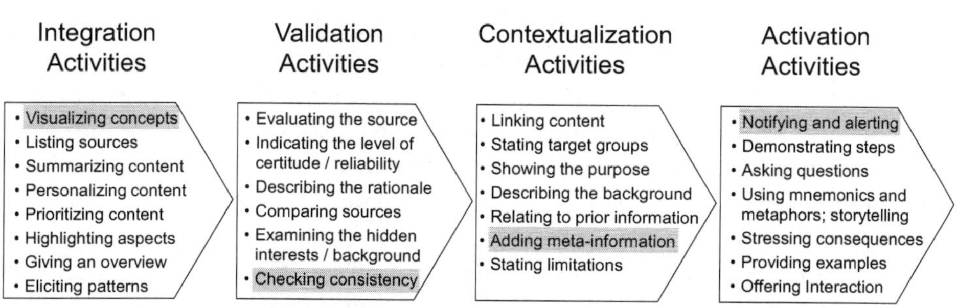

Fig. 4: Value-adding activities as background for the information quality principles.

of information must be compiled to provide unified access and to provide the information user with a convenient way to identify and access relevant information. Thus, the integration principle states that information has to be provided in a comprehensive and concise, convenient and accessible manner to be of high quality.

In order to enable information consumers to evaluate information effectively, it must be *validated,* i.e., it must be checked for accuracy, consistency, timeliness and security. Indicators of a piece of information's validity must be provided with the information to facilitate this task.

In order to enable information consumers to understand and adapt information correctly, i.e., to allocate it adequately, they need to be provided with the *context* of that information. The provided context should increase the clarity of the information (what it is about), the perceived correctness (where it applies and where it leads to false results), the traceability (where it comes from and how it originated), and its maintainability (where and how it can be updated).

In order to enable information consumers to use information effectively, it has to be provided in a directly applicable format, in its most current version, in a highly interactive process, and on a fast infrastructure. This assures that information is *applied* because it is available when needed and because it is easily noticed and remembered.

In this way the framework defines the requirements of high-quality information on four levels with the help of information quality attributes and categorizes them into four phases that can be supported through four corresponding principles. The framework has been successfully applied to diagnose, improve, and sustain the information quality of various communication infrastructures ranging from corporate intranets, newspapers, to e-government initiatives (see for example Cantoni 2006 and Cantoni et al. 2005).

4 Outlook: future research avenues

We have now given a brief overview on two sides of the information society in terms of its perils (i.e., loosing oneself in information) and promises (having the right information, at the right time, in the right format). These two influential and informative concepts have several implications for communication infrastructures: Sustainable communication infrastructures should on the one hand *ease the cognitive burden on information consumers* by assisting them in filtering and aggregating information, as well as contextualizing it and making it easy to use and apply. On the other hand a communication infrastructure should be designed in a way as to foster high quality information, by ensuring that any such platform has a reliable infrastructure, a (continually) optimized information provision process, and contains sound and relevant information. We have also seen, however, that such technical and process-oriented solutions are only a part of a comprehensive solution. The quest for information quality and the risk of overload are also personal challenges that must be addressed by each and every individual knowledge worker. There are in fact many challenges that remain in this ever-changing field both from a practice and a research perspective. Below we provide a few pointers to future research avenues in the context of information overload and quality.

Future research on the concepts of information overload and information quality can be broadly separated into three vectors: *new challenges* brought about by novel formats of information, *new solutions* to these challenges, and *novel research approaches* to study these phenomena. Regarding the first area, novel information formats and corresponding challenges, the following research questions are emerging:

- What is the impact of social media on people's information overload and on their information quality perceptions?
- Does citizen journalism affect the very notion of quality information and if so how?
- In what ways do relatively novel channels such as Facebook or Twitter contribute to information overload?
- Do mobile phone users experience overload differently than desktop users and if so how?
- Is the quality of information provided by private twitter accounts higher or lower than that of corporate accounts?
- Which overload and information quality topics emerge in new big data applications (see, for example, Redman, 2013?

Regarding the second area, novel solutions, research questions like the following can be envisioned:

- In which ways can novel (social) filter mechanisms like tagging, liking or retweeting reduce the overload and misinformation problem?
- How can low-quality information be detected in tweets?
- How can visual tools and methods reduce information overload and improve the fitness for use of information?
- What role can technology (such as artificial intelligence applications) play in reducing the overload challenge and improving information quality?

Besides these two substantial research streams, there is also a need to investigate new methods of studying both information overload and information quality. Key questions in this direction are the following:

- How can information overload be studied in a more authentic, close-up manner?
- How can the trade-offs between information quality attributes be researched? How can the research on information quality become more differentiated, for

example by examining which information quality attributes matter most to today's information consumers and in which channels or situations?
- How can we design research that simultaneously addresses information overload and information quality issues (like Keller and Staelin 1987 or Simpson and Prusak 1995)?
- In what way can research be used to educate consumer about the dangers of information overload or low information quality?

Whatever route future research will take, one thing will remain: The insight that information is not always a blessing, but needs to be actively managed to provide value, and this management effort must address both, the quantity and the quality of information. Only in this way can we regain the knowledge that we have lost in information.

References

Abdel-Khalik, A. Rashad. 1973. The effect of aggregating accounting reports on the quality of the lending decision: An empirical investigation. *Journal of Accounting Research* 11. 104–138.

Bawden, David. 2001. Information overload. *Library & Information Briefings*, 92. http://litc.sbu.ac.uk/publications/lframe.html (Accessed 15 April 2015)

Cantoni, Lorenzo & Martin J. Eppler. 2005. eGovernment: The Role of Information Quality and Training (in Russian). In Sergey Yuri Naumov (ed.), *Globalization: Issues of International Cooperation and Solution of Panhuman Tasks,* 109–120. Saratov: Stolypin Volga Region Academy for Civil Service.

Cantoni, Lorenzo. 2006. Information Quality in Electronic Government Websites: a Case Study from Italy's Ministry for Public Administration. In Viktor Mayer-Schönberger and David Lazer (eds.), *Information Government*, 257–260. Cambridge (MA): MIT Press.

Chewning, Eugene C. Jr. & Adrian M. Harrell 1990. The effect of information load on decision makers' cue utilization levels and decision quality in a financial distress decision task. *Accounting, Organizations and Society* 15(6). 527–542.

Cook, Gary J. 1993. An empirical investigation of information search strategies with implications for decision support system design. *Decision Sciences* 24(3). 683–699.

Edmunds, Angela & Anne Morris. 2000. The problem of information overload in business organizations: A review on the literature. *International Journal of Information Management* 20. 17–28.

Eppler, Martin J. 2006. *Managing Information Quality: Increasing the Value of Information in knowledge-intensive Processes* (2nd Edn). Berlin: Springer.

Eppler, Martin J. & Jeanne Mengis. 2004. The Concept of Information Overload: A Review of Literature from Organization Science, Accounting, Marketing, MIS, and Related Disciplines. *The Information Society. An International Journal* 20(5). 1–20.

Eppler, Martin J. & Peter Muenzenmayer. 1999. Information Quality on Corporate Intranets: Conceptualization and Measurement. In *Proceedings of the 1999 Conference on Information Quality*, Massachusetts: MIT.

Galbraith, Jay R. 1974. Organization design: An information processing view. *Interfaces* 3. 28–36.

Gerkes, M. 1997. *Information Quality Paradox of the Web*, online manuscript, retrieved from http://izumw.izum.si/~max/paper.htm, last accessed in 2005.

Griffeth, Rodger W., Karry D. Carson & Daniel B. Marin. 1988. Information overload: A test of an inverted U hypothesis with hourly and salaried employees. *Academy of Management Proceedings* 1988(1). 232–237.

Grise, Mary-Liz & R. Brent Gallupe. 1999. Information overload: Addressing the productivity paradox in face-to-face electronic meetings. *Journal of Management Information Systems* 16(3). 157–185.

Haksever, A. Mehmet & Norman Fisher. 1996. A Method of Measuring Information Overload In Construction Project Management. Paper presented at the *Beijing International Conference* in *Proceedings CIB W89*. 310–323.

Hayward, Keith. 2013. 'Life stage dissolution' in Anglo-American advertising and popular culture: Kidults, Lil' Britneys and Middle Youths. *The Sociological Review* 61(3). 525–548.

Huang, Kuan-Tsae, Yang W. Lee & Richard Y. Wang. 1999. *Quality Information and Knowledge*. New Jersey: Prentice Hall.

Iselin, Errol R. 1993. The effects of the information and data properties of financial ratios and statements on managerial decision quality. *Journal of Business Finance & Accounting* 20(2). 249–267.

Jacoby, Jacob, Donald E. Speller & Carol Kohn. Berning. 1974. Brand choice behavior as a function of information load: Replication and extension. *The Journal of Consumer Research* 1(1). 33–43.

Jackson, Thomas W. & Bart Van Den Hooff. 2012. Understanding the Factors that Effect Information Overload and Miscommunication within the Workplace. *Journal of Emerging Trends in Computing and Information Sciences* 3(8). 1240–1252.

Joo, Mingyu, Kenneth C. Wilbur, Bo Cowgill & Yi Zhu. 2013. Television Advertising and Online Search. *Management Science* 60(1). 56–73

Kahn, Berverly K. & Diane M. Strong. 1998 Product and Service Performance Model for Information Quality: An Update. In InduShobha N. Chengalur-Smith & Leo L. Pipino, *Proceedings of the 1998 Conference on Information Quality*, 103–115. Cambridge, MA: Massachusetts Institute of Technology.

Keller, Kevin Lane & Richard Staelin. 1987. Effects of quality and quantity of information on decision effectiveness. *The Journal of Consumer Research* 14(2). 200–213.

Lesca, Humbert & Elisabeth Lesca. 1995 *Gestion de l'information, qualité de l'information et performances de l'entreprise*. Paris: Litec.

Malhotra, Naresh K. 1982. Information load and consumer decision making. *The Journal of Consumer Research* 8(4). 419–431.

Meyer, Jörn-Axel. 1998. Information overload in marketing management. *Marketing Intelligence & Planning* 16(3). 200–209.

Muller, Thomas E. 1984. Buyer response to variations in product information load. *Journal of Applied Psychology* 69(2). 300–306.

O'Brien, James A. 1991. *Introduction to Information Systems in Business Management*, 6th ed. Boston: Irwin.

O'Reilly, Charles A. 1980. Individuals and information overload in organizations: Is more necessarily better? *Academy of Management Journal* 23(4). 684–696.

Owen, Robert S. 1992. Clarifying the simple assumption of the information load paradigm. In John F. Sherry, Jr. and Brian Sternthal (eds.), *NA – Advances in Consumer Research Vol. 19*, 770–776. Provo, UT: Association for Consumer Research

Redman, Tom. 2013. Data's Credibility Problem. *Harvard Business Review* 91(12). 84–88.

Schick, Allen G., Lawrence A. Gordon & Susan Haka. 1990. Information overload: A temporal approach. *Accounting Organizations and Society* 15(3). 199–220.

Schneider, Susan C. 1987. Information overload: Causes and consequences. *Human Systems Management* 7(2). 143–153.

Shenk, David. 1997. *Data smog. Surviving the information glut*. London: Abacus.

Simon, Herbert A. 1979. Information processing models of cognition. *Annual Review of Psychology* 30. 363–396.

Simpson, Chester W. & Laurence Prusak. 1995. Troubles with information overload – Moving from quantity to quality in information provision. *International Journal of Information Management* 15(6). 413–425.

Speier, Cheri, Joseph S. Valacich & Iris Vessey. 1999. The influence of task interruption on individual decision making: An information overload perspective. *Decision Sciences* 30(2). 337–359.

Sumeckia, David, Maxwell Chipulua & Udechukwu Ojiako. 2011. Email overload: Exploring the moderating role of the perception of email as a 'business critical' tool. *International Journal of Information Management* 3(5). 407–414.

Swain, Monte R. & Susan F. Haka. 2000. Effects of information load on capital budgeting decisions. *Behavioral Research in Accounting* 12. 171–199.

Tushman, Michael L. & David A. Nadler. 1978. Information Processing as an Integrating Concept in Organizational Design. *Academy of Management Review* 3(3). 613–625.

Wurman, Richard Saul. 1990. *Information anxiety. What to do when information doesn't tell you what you need to know*. New York: Bantam Books.

Wurman Richard Saul. 2001. *Information Anxiety 2*. Indiana: Macmillan Publishing.

York, Chance. 2013. Overloaded by the News: Effects of News Exposure and Enjoyment on Reporting Information Overload. *Communication Research Reports* 30(4). 1–11.

Zhu, Hongwei, Stuart E. Madnick, Yang W. Lee & Richard Wang, R. 2012–2013. *Data and Information Quality Research: Its Evolution and Future* (Working Paper CISL 2012–13). Cambridge, MA: MIT. http://web.mit.edu/smadnick/www/wp/2012-13.pdf (Accessed 15 April 2014)

Davide Bolchini

12 User experience and usability

Abstract: The goal of this chapter is to review key concepts related to the usability and user experience of interactive communication. The body of literature in this area (Law 2009) grew immensely over the last two decades, due to the pervasive expansion of information and communication technology in our everyday lives. As such, several disciplinary approaches came to bear onto these topics, from computer-mediated communication, computer science, information science, software engineering and human-computer interaction. Rather than a summary of all that have been said on these topics, this chapter seeks to synthesize a practical, integrated perspective across knowledge domains which mainly stems from usability engineering, the growing area of User Experience (UX) and interaction design. The contribution offered builds upon the research of the author in the context of the broader conversation about the fundamental notions, methods, techniques and approaches to usability and user experience design. The chapter is organized in two parts. The first part will investigate and review the boundaries of the concepts of user experience and usability, by engaging with the key dimensions characterizing these concepts that emerge from the academic and professional literature. Given the breadth of the conceptual grounds covered by these concepts, this is not aimed at being a comprehensive overview on the subject, but rather an original introduction to entice the reader to know more. The second part delves deeper into a specific perspective on user experience for interactive communication, namely proposing fundamental principles to systematically design and structure the conceptual, communicative "architecture" of the user experience.

Keywords: user experience, usability, user, design, tasks, goals, modelling

1 On scoping user experience and usability

> Ultimately, we are deluding ourselves if we think that the products that we design are the "things" that we sell, rather than the individual, social and cultural experience that they engender, and the value and impact that they have. Design that ignores this is not worthy of the name (Bill Buxton, from www.billbuxton.com).

Academic and professional books on "user experience" (commonly termed "UX") proliferate (Kuniavski 2002; Hartson 2012). As interactive information technology pervades more and more our lives and all sectors of the economy, the topic of understanding, designing and assessing the experience of users with interactive products (from websites to mobile applications) grew massively in importance over the last two decades. It is not uncommon to find full-time user experience consultants paid 6-figure salaries to execute research studies to re-design and optimize the user experience of multimillion dollars websites, on which entire businesses and organizations rely. A quick search on Google and Twitter for user experience jobs returns thousands and thousands of open positions worldwide. Similarly, hundreds of undergraduate, master and doctoral programs, as well dozens of international conferences worldwide, focus on advancing education and research exactly in this field.

But as the field is growing large and multifaceted, so is the concept of user experience. What is the user experience? First of all, as any experience, it is neither a property of the interactive system nor a property of the user, but rather a *quid* emerging from the encounter between the two. Oftentimes, we have to acknowledge that the term "user experience is simply misused as a more colorful definition of "web design". In these instances, who was focusing and evangelizing tools and wisdom about web design 20 years ago is today talking about the same concepts but applied to the design of the user experience. This stems from different reasons, all driven by formidable technology advancement. On the one hand, web design quickly evolved into mobile design, tablet design, large-display design. On the other hand, it became clear that the design was not just the graphics and layout of a webpage, but the manifestation of a complex communication process and a carefully crafted multi-purpose dialogue that goes on between stakeholders and the users.

To search for a definition of "user experience", let us start from our *own* experience. When we visit an ecommerce website for the first time to try to find and buy a sturdy but not too expensive cloth rack for our children's room, our interaction with the website can be characterized at various levels: how long it took to find and buy the desired product; how many clicks and keystroke we had to go through to get to the check out; how effort it took to figure out from the homepage if that website was the right one, and where to go to browse the product; how enjoyable were the high resolution pictures of the product, or how frustrating were the tiny and unclear pictures of some other products; how easy or difficult was to compare current products with previously browsed once; how flowing and engrossing was the sliding picture gallery of product categories for the children rooms; how relevant, engaging or annoying were the recommendations of similar products; how disconcerting was not to be able to re-find a product initially found but then lost during navigation; how unexpectedly long was the mandatory registration process to check out, and how expensive were the shipping costs. All these factors are examples of what can shape our user experience with that website in a given moment for a given scenario of use. What happens the next day, when we go again to that website and try to re-find other or similar products may be different: we may be already somehow familiar with the website, so we can reach faster key sections, we know what to expect, we become more efficient, we are glad to see that the site recognizes us and thus promotes recently visited categories of products front and center for easier access; or we may also be disappointed by the fact that we forgot our password and there is no way to retrieve it, or that the product we had seen and bookmarked yesterday is now out of stock.

Based on this simple example, and mindful of the many definitions existing in the professional and academic literature on the topic, we can attempt a definition of user experience:

> User experience is the multilayered set of user-perceived phenomena emerging from the encounter and situated dialogue between a user and the manifestation through a user interface of the designer's communication intents.

Such user-perceived phenomena can be described as cognitive reverberations during or post the experience (what users remember or think of his/her interaction with the system, and therefore what personal opinion s/he has developed about the overall "quality of use" of the system), or observed behavior (frustration, engagement, surprise, boredom, arousal, enjoyment, poor or good task performance). Other facets of the user experience pertain how much attached a user gets to the product or system, how "addictive" (or "sticky") its use becomes, and what social and individual traits the use of the product reflects on the user.

By encompassing some of these facets, the ISO Standard 9241-210 (ISO 2014, 2.5) provides a simpler definition of user experience: "a person's perceptions and responses that result from the use or anticipated use of a product, system or service". As annotations to this definition, the standard indicates that:

"User experience includes all the users' emotions, beliefs, preferences, perceptions, physical and psychological responses, behaviours and accomplishments that occur before, during and after use. User experience is a consequence of brand image, presentation, functionality, system performance, interactive behaviour and assistive capabilities of the interactive system, the user's internal and physical state resulting from prior experiences, attitudes, skills and personality, and the context of use." (ISO 9241-210:2010, 2.5) A collection of a broad set of definitions of user experience that summarizes input from both practitioners and researchers can be found at: http://www.allaboutux.org/ux-definitions.

A useful overview of the dimensions of user experience stems from considering the desired intents of user experience design in interactive communication. Rogers (2011) provides a rich set of examples of user experience goals that help clarify the nature of this multifaceted concept (Table 1).

Besides the aforementioned examples, let us consider a host of possible nuances and variations that our user experience with interactive product can assume, as witnessed – as a prominent example – by the not too recent introduction of the revolutionary Apple iPhone in 2007. Few days after the Apple event when the product was introduced, on (Grossman 2007, p. 35) an effective description of key user experience traits appeared:

Tab. 1: Examples of User Experience Goals.

Desirable Aspects	Undesirable Aspects
satisfying	boring
helpful	unpleasant
fun	frustrating
enjoyable	patronizing
emotionally fulfilling	making one feel guilty
motivating	making one feel stupid
provocative	annoying
engaging	cutesy
challenging	childish
surprising	gimmicky
pleasurable	...
enhancing sociability	...
rewarding	...

> The iPhone is a typical piece of Apple's design: an austere, abstract, Platonic-looking form that somehow also manages to feel warm, organic and ergonomic.

Clearly a set of apparently intangible but clearly perceivable attributes of the user experience caught the attention of the public and the critics, and soon defined a milestone in our understanding and expectations about what interactive product design can be. And these attributes are as (or even more) important for our successful interaction with the product as the level of the mechanical functionality, efficiency and speed of our interaction (which other mobile phones exhibited well before the iPhone).

By acknowledging the fundamental important of this concept for the welfare of the economy and all modern communications, the field of user experience design, user experience architecture, or more simply – UX – has developed by striving to conceive tools, techniques, guidelines, principles, models, languages, practices, and research methods to tease out and understand the nature of such phenomena, and drive the design of optimal user experiences across various channels, contexts and industries.

From our discussion of the concept of user experience, it is clear that it builds on basic, and more easily measurable aspects of the interaction with interactive products: can the user accomplish his/her task? If yes, can s/he do it in acceptable time and with reasonable effort? For example, if it takes 3 minutes just to locate the right category of products for the cloth rack (no matter how engaging, motivating, flowing and welcoming are the front page gallery images) the site has some issues regarding its usability, that is the property of the design that allows users to make use of the system in the most efficient and effective way.

The concept and field of usability (and usability engineering) long precedes the one of user experience. Since the late eighties, before the web, usability was systematically introduced by Jakob Nielsen (Nielsen 1994) and others as a fundamental trait of any user interface whose purpose is to serve information workers in accomplishing specific tasks as efficiently and as effectively as possible.

ISO Standard 9241 (ISO 2014) defines usability as "the extent to which a product can be used by specified users to achieve specified goals with effectiveness, efficiency, and satisfaction in a specified context of use." Traditionally, the concept of usability or "ease of use" has been broken down in several components, including:

- *Learnability*: How easy is it for users to accomplish basic tasks the first time they use the application?
- *Effectiveness:* are all tasks relevant to targeted user profile and context of use supported and doable?
- *Efficiency*: Once users have learned the design, how quickly can they perform tasks?
- *Memorability*: When users return to the design after a period of not using it, how easily can they re-establish proficiency?
- *Errors*: How many errors do users make, how severe are these errors, and how easily can they recover from the errors?
- *Satisfaction*: How pleasant is it to use the design?

Complementary to the discourse on the user experience, specific examples of goals related to usability can be found in Table 2.

Tab. 2: Examples of Usability Goals.

Effective to use
Efficient to use
Safe to use
Have good utility
Easy to learn
Easy to remember how to use
Reliable
Helps recover from error
...

Nowadays, the concept of user experience – given its broader scope – is often used as more general used label that captures usability as part of it.

By building on the notions presented so far, the rest of the chapter deepens our exploration by approaching these concepts from the perspective of the conceptual activity of the designers in charge of "architecting" user experiences.

2 Design Reasoning for User Experience Architectures

As web applications increase in size and complexity, the details of the user interface no longer constitute the chief design problem. When interactive applications incorporate a vast array of interconnected information and are designed for multiple delivery channels (tablet, mobile, desktop), the organization of the overall user experience presents a new set of challenges: macro-organization of content, user navigation paths, and exploration of high-level design alternatives. Surprisingly, human-computer interaction (HCI) design methods today remain focused on the *surface* layer of the user experience (the user interface) (Stolterman 2010), while not addressing the representation of the *deeper, organizing structure*. For example, best practice in HCI encourages sketching the "pages" of the user interface (Buxton 2007), but does not provide a comprehensive blueprint for understanding the *types* of pages to design, their functions, or how navigation paths and content fit together in a coherent whole. In the world of structural architecture, this would be analogous to focusing on the optimal design of the doors while failing to consider the whole building. Addressing this critical gap requires that user interface designers approach their task in a fresh, new way: they must consider the design of the web user experience holistically – that is, architectonically – so that user interface decisions can emerge within a unifying strategy.

In both structural and software systems architecture (Garlan 1993), design tradition has evolved from focusing on elementary components to framing design decisions into an envisioned, overarching blueprint of the system-to-be. HCI design methods, however, have yet to make this critical evolution; they concentrate on superficial, *page-level design* and fail to capture the overall, organizing architecture of the user experience. Recent efforts in "interaction design" (Rogers 2011) broaden this view, but they do so by focusing on elements external to the system (e.g., user context and activities) instead of modelling a unifying composition of the design elements (Stolterman 2010). As a result, a critical gap exists in our current knowledge: how to represent a *coherent vision* for the user experience during the design process of large-scale web applications.

This gap results in design processes that focus on page-level features (e.g., menus and layout) rather than on a *conceptual model* that keeps navigation structures, pages and content together in a unifying whole. This situation creates considerable practical problems in HCI design and design education: What principles should guide the transition from user requirements to interface design? What is the role of each page in the overall architecture of the user experience? How do we identify the core of the design in order to prioritize the design resources? Usability expert J. Nielsen echoes these concerns, stating that designers "treat a site like one big swamp with *no organizing principle* for individual items" (Nielsen 2009: page). This lack of conceptual modelling beyond the page level results in such negative consequences as lack of overview structures, confused nesting of sub-sites poorly integrated with the main site, and user disorientation.

2.1 Web design practice

The practice of web design employs a variety of informal and manual techniques to represent site prototypes and user interface specifications (Rosenfeld 2002). Page sketches, user interface mock-ups, wireframes, site maps, and storyboards are among the basic components of a designer's toolkit. These techniques enable designers to first produce and share their rough ideas and then iteratively refine their prototypes with off-the-shelf software tools (Bowles 2011). The literature in this area has grown steadily over the last decade, demonstrating the relevance of methodological support for web design (Young 2008); as well as the need for effective tools to enhance both design communication and the quality of the design deliverables for web projects (Lund 2011).

The strength of existing tools is their agility: designers can quickly generate and improve multiple user interface design ideas. However, when web designers reason only on the user interface (pages and links), they focus their view on optimizing each single page without a blueprint of the larger design. The field of *information architecture* has proposed techniques, tools and notation systems for representing page flows, as well as recognizing and designing recurrent patterns and high-level structures connecting site pages (e.g., hierarchical, linear, or web-like structures) (Van Duyne 2007). Visually engaging, high-quality site maps for website planning and design have also been proposed and published, leveraging some of the visual and projection techniques used in urban planning and architecture (Kahn 2001). However, such approaches suffer two major limitations. First, producing architecture maps remains an "ad-hoc" process which lacks a common, reusable, and stable vocabulary that can be employed across design contexts. Second, existing notations still narrowly focus on page-by-page reasoning when designing large-scale structures.

This *lack of high-level principles* for understanding complex designs is evident when existing methods (site maps, user workflows, etc.) try to operate on a large, non-trivial scale. Large information architectures for hundreds of pages quickly become unmanageable and unusable due to visual clutter: the designer's mental model becomes overwhelmed, resulting in a sub-optimal end-user experience. Current efforts to create a coherent "conceptual model" as a foundation for a better user experience are either too unstructured (Johnson 2002), thus providing little guidance for the rest of the development; or too flattened on the visual appearance of a single web page (Hartson 2012), thus not capturing high-level structures beyond the page level. As such, the modelling of the overarching organization of content and navigation paths to support optimal usability remains a largely unexplored area.

2.2 Model-driven Design in Web Engineering

Over the last two decades, web engineering research has proposed several methodologies for designing hypermedia and web systems from a conceptual standpoint (Garzotto 1993; Gomez 2001; Eisenstein 2001). Such design models and frameworks recognize the importance of odeling the structure of the system-to-be in terms of conceptual entities, domain relationships, and business operations and transactions; and provide concepts and processes for moving from a high-level representation of the system to the engineering of the user interface.

Traditionally, these models have always operated under the assumption that web engineers would create a conceptual model of the application, which would then be used to guide the implementation of the system. However, *interaction designers* – also called *user experience designers* – are typically not considered by these approaches, because the proposed notations and conceptual tools are geared towards engineering the system instead of quickly exploring high-level design alternatives. Yet, user experience designers are key stakeholders who impact major decisions, including the high-level design, content structure, navigation architecture, and browsing patterns. As such, they must be equipped with the proper conceptual tools for making their voices heard during the strategic, fundamental aspects of the design process.

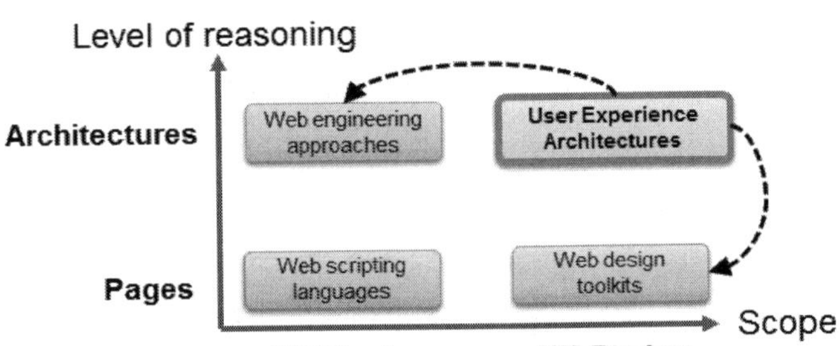

Fig. 1: User experience architectures elevate the level of reasoning in web design and complement web engineering methodologies by focusing on the user experience (UX).

Indeed, existing web design toolkits do focus on crafting the user experience, but limit reasoning to the page-level; while web engineering approaches do rise to architectonic level, but only for odeling the software (SW) infrastructure of the system. A critical piece of this puzzle is missing: mapping the *architecture* of the *user experience* (Figure 1). Such mapping will not only frame page-level toolkits, but provide a more structured input to the system implementation as well.

2.3 Capturing "the whole": the importance of holistic reasoning in UX design

The importance of architectonic reasoning in information systems has been theoretically elaborated in the tradition of design thinking. Whereas a tectonic way of design primarily operates at the analytical level on the parts of a system, an architectonic approach conceives and understands the system "as a whole" (Arnheim 1995), or coherent unit. Working architectonically should be understood as a process in which every detail is as important as the overall framework (Nelson 2003). On the one hand, a disaggregated, tectonic perspective (and design training) emphasizes tactical, rather than long-term strategic decisions and can result in a lost sense of purpose in complex designs. On the other, explicitly imprinting a unifying principle, or "format", on the design of a system conveys a sense of wholeness that helps sustain the design process *and* enables users to understand and effectively use the system.

2.4 Principles for reasoning on user experience architectures

Adopting an architectural perspective of the design of the user experience is fundamental because it expands the current approach of designing large-scale web applications by focusing design activities on a unifying view of the envisioned user experience. User experience architecture will also serve as *linchpins* in the design process, because they enable designers to maintain coherence among the fragmented, scattered pieces of design artifacts generated in a complex project (e.g., storyboards, mock-ups, page prototypes, wireframes, and sketches), and thus help compose a shareable mental model (Young 2008). In this way, the critical function of each single page can find its place in the rationale of the total user experience architecture.

In what follows, we review some fundamental principles that are necessary to understand and design "user experience architectures", bird's-eye views of the deeper, organizing structure of complex interactive communication designs.

2.4.1 UX architectures mediating the shift from requirements to user interface design

A critical yet ill-supported aspect of the HCI design process is the transition from requirements to early design. This gap is evident, not only in the published literature on HCI design methods, but also in several HCI curricula. In structural architecture, the process of envisioning and giving a high-level shape to the mass and space comprising the building's overarching structure is known as "massing" (Akin 1993). This fundamental step of early-stage design helps architects manage

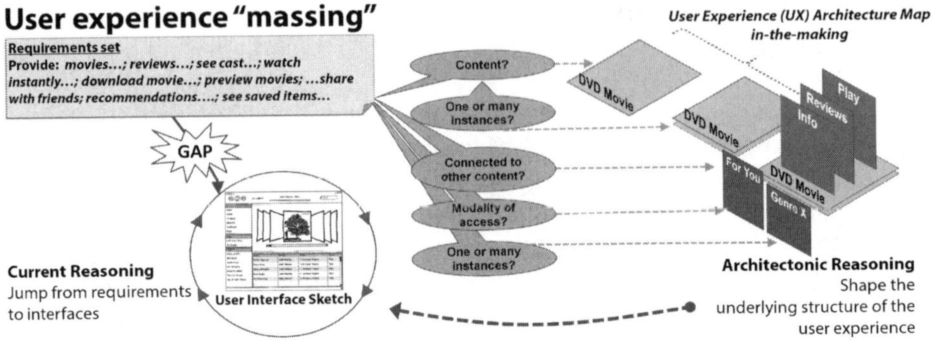

Fig. 2: Bridging the gap between requirements and user interface sketches through UX architectures.

part–whole relationships, control hierarchies of elements, scaffold the design process, and structure sub-problems. Such high-level composition provides the conceptual foundation that drives the rest of the design process. Existing HCI design methods have not evolved in this respect: from a list of user requirements, designers explore alternative user interface elements and iteratively refine and evaluate them (Figure 2). Although the sketching of alternative designs is crucial for achieving better designs (Dow 2011), the unsolved problem with current methods is an unproductive "leap of abstraction" from requirements specifications to interface design. For example, discussing the graphical layout of a page can easily lead to premature fixation (Jansson 1991) on such secondary aspects of the design as button appearance and position, if this is done without having a broader view of the function, content, and role of that page within the user experience. In structural architecture, this would be like discussing the location of the doors without first having established the building's traffic flow.

By adopting an architectonic-level perspective on the design, designers can examine requirements specifications to *gradually drive the exploration of the conceptual infrastructure of the design* (similar to "massing" in structural architecture). This conceptual structuring defines the all-supporting framework for the user experience: it includes content structures, access structures, and navigation patterns (Garzotto 1999). Specific questions guide this reasoning. From a list of requirements for a movie rental website, for instance (Figure 2), designers can reason such questions as: what is the key content of interest for the user experience? How can the information about a movie be organized for optimal exploration? How can users discover and navigate to more movies? Which alternatives should I consider? The answers to these questions will help designers gradually structure the overarching architecture of the user experience, which in turn helps them identify additional problems and sub-problems for further investigation (Bolchini, 2011). For example, in a movie rental site, requirements are examined to identify the site's core content topics (e.g., movie, actor), access criteria, and navigation paths. Such overarching conceptualization is captured in user experience architecture maps that provide a framework for sketching and exploring page-level decisions (Figure 4). Whereas emerging HCI practice emphasizes interface sketching (Buxton, 2007), architectonic reasoning promotes *conceptual sketching* as a key activity for structuring the user experience. Such activity is not unstructured, but can follow existing conceptual design models (Bolchini, 2008).

2.4.2 Emerging dominance

Deciding upon *which part of the user experience* design to focus in order to generate multiple interface sketches is currently left to intuition or "best guess." This problem is also reflected in HCI students' tendency to spend an inordinate amount of time getting the homepage right without first having considered the *core* of the user experience (i.e., *what do users do past the homepage?*).

To combat this, user experience architecture maps may be used as frames of reference for such critical design decisions. Designers can be guided to easily and proactively recognize "emerging" parts of a user experience architecture map. Emerging features can be visually spotted on a map by noting, for example, the "size" of the topic and the "number" of access structures. The larger a topic, the more content, proportionately speaking, it will entail; similarly, a higher number of access structures indicates the relative importance of the content to which access must be provided. For example, a user experience architecture map could clearly manifest "movie" as the emerging, richest, and most complex topic (Figure 3). It would be critical to start sketching and carefully evaluating interface solutions for the "movie" information and its surrounding access structures (e.g., movie by genre, recommended movies) prior to discussing the homepage, the pages for the news, the registration, or the studios. Based on this principle, the *emerging dominance* of the architectural elements help prioritize page-level design and evaluation. With UX maps, designers are able to pre-attentively recognize the

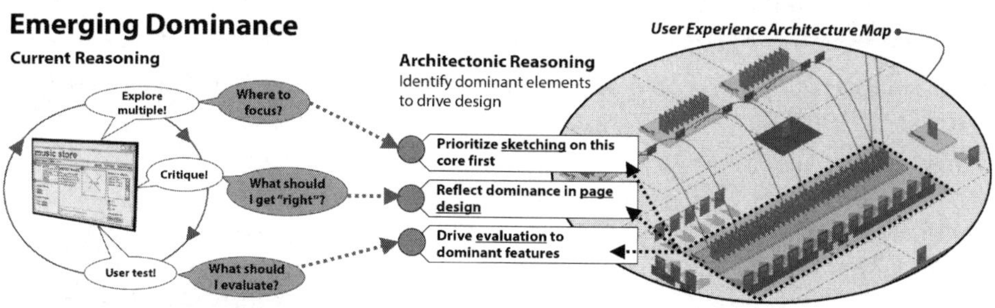

Fig. 3: User experience architectures elevate the level of reasoning in web design and complement web engi

emerging parts of the user experience from the map, and more effectively *drive lower-level design decisions* focusing on these emerging elements.

2.4.3 Generating Design Alternatives and Elevating the Nature of Feedback

Architectonic reasoning can bring transformative changes for those designers and students who employ a structured approach to interactive design. Here is a typical scenario: imagine a team of designers in the early phases of website development. The site is aimed at connecting talent scouts to local musicians. By following existing techniques (Kalbach 2007), designers would sketch a site map (Figure 4) based on an initial set of requirements. Starting from the homepage, they would map out further pages around an intuitive, hierarchical structure. These pages could include a *concerts section*, an *artists' section*, and each *artist's detail page*. Such a map would lead designers to focus on each page for designing interface elements and layout. Although straightforward and popular, this approach cannot answer such fundamental questions as: How can a talent scout pinpoint those artists specializing in jazz, find the most popular acts, or identify emerging talent? How can users decide what to look for? A tree-like map modelled in terms of sections and pages (Figure 4) does not allow for exploration of such questions, because current methods implicitly encourage designers to *conceive* structures by following the *same steps* that users will follow to *use* those structures. Reasoning on what the user does at a page-by-page level is a valid approach for small, simple sites, but it can inhibit understanding of what users *could conceivably do*.

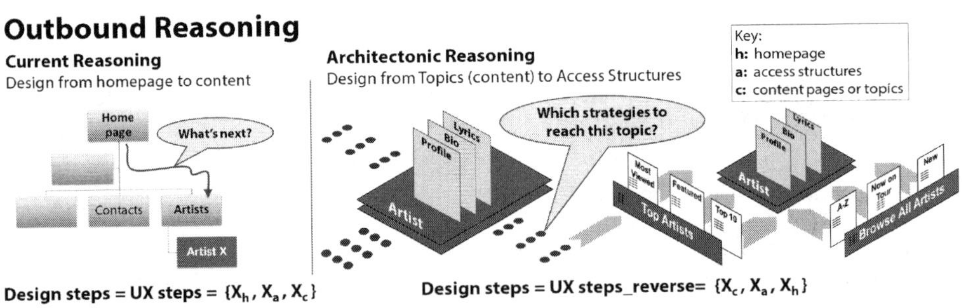

Fig. 4: Designing from content to its access paths is counterintuitive but amplifies the design space (Bolchini 2006).

2.4.4 Outbound Reasoning

To address these limitations, user experience architect can amplify the traditional way of conceiving navigation by following a more fruitful, *outbound trajectory* from the content to its access structures. For example, designers can establish that the *artist* (with detailed profile, biography and lyrics information) is one of the core contents of the application. They will then creatively explore all the relevant alternative navigation paths for enabling access to this content (see Figure 4). Artists can be accessed by genre, style, or years of performance. They can be grouped together for any number of reasons: they are currently on tour, they are the most popular, or they have been hand-picked by editors as emerging artists. Artists can also be assembled into larger groups, for example, to support browsing of “all” or “top” artists (Figure 4). Although counterintuitive, outbound reasoning can greatly *broaden the design space*, as shown in this example. Exploring a variety of “local” access structures around identified *topics* unlocks design fixation and provides a key reasoning process for structuring richer navigation architectures. By decoupling the processes by which users navigate, from those which designers follow to conceive the architecture (Figure 4), browsing structures emerge that can meet a broader range of user’s goals.

2.4.5 Disruptive feedback – beyond the page level

Interaction design progresses through frequent cycles of evaluative feedback accomplished through design critiques, expert reviews, usability testing, and consultation with libraries of usability heuristics (Nielsen 1994). Improving label and button visibility, restructuring page layout for better readability, and abiding by interface standards are a few examples of recurrent feedback themes among HCI professionals and students (Figure 5). Yet, because designers seek feedback to advance their work, the most useful would be that which helps them *rethink* the design rather than suggests minor refinements. Unfortunately, feedback at the level of the user interface focuses on making the “design right” (Buxton 2007).

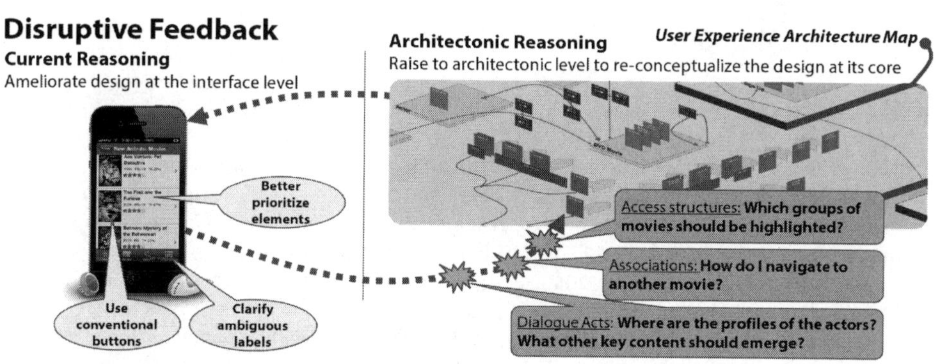

Fig. 5: User Experience Architectures give designers full access to a critical, new level of design feedback.

Making the "right design" is more difficult, requiring deeper, more *disruptive feedback* that is directed to the conceptual architecture of the user experience.

The example in Figure 5 illustrates how architecture-level feedback can push designers to re-consider such *structural issues* as: How can the user explore additional movies or easily pick a different one? How can the user browse actor profiles or movie reviews? Would this movie be accessible to non-registered users? How? How can I improve visibility and access to this movie? Supported by user experience architecture maps, disruptive feedback addresses the conceptual model underlying the user experience. When reasoning from an architectural perspective, designers can raise substantially the quality of feedback sought from and provided to peers.

3 Conclusions

This chapter reviewed key traits of the concepts of user experience and usability particularly as related to the design of interactive communication systems such as websites and mobile applications – systems currently mediating a large part of our interaction with the world of information. In the first part of the chapter, we investigated the notion of user experience as the set of multifaceted, perceived phenomena emerging from the interaction with interactive products, and then connected it to the notion of usability, which captures the effectiveness and efficiency by which the application supports users in accomplishing their tasks. In the second part, we elucidated in depth some essential traits that can characterize our reasoning around the design of user experience in terms of "user experience architectures", conceptually rich representation of the complex structure of the interactive communication design that can drive our conception of user experience and usability beyond the superficial level of the user interface to introduce us to the underlying composition of the intents of communication designers.

References

Akin, Omer. 1993. Architects' Reasoning with Structures and Functions. *Environment and Planning B: Planning and Design* 20(3). 273–294.

Arnheim, Rudolf. 1995. Sketching and the psychology of design. In V. Margolin and R. Buchanan (eds), *The Idea of Design*, 15–19. The MIT Press, Cambridge, MA.

Bolchini, Davide & France Garzotto. 2008. Designing Multichannel Web Applications as "Dialogue Systems": the IDM Model. In G. Rossi, O. Pastor, D. Schwabe & L. Olsina (eds.), *Web Engineering: Modeling and Implementing Web Applications, Human-Computer Interaction Series*. 193–219. New York, NY: Springer.

Bolchini, Davide, & Adam Neddo. 2011. Beyond Interfaces and Flows: Abstractions for Mapping Organic Architectures. *ACM Interactions* 18(1). 56–61.

Bolchini, Davide & Paolo Paolini. 2006. Interaction Dialogue Model: A Design Technique for Multichannel Applications. *IEEE Transactions on Multimedia* 8(3). 529–541.

Bowles, Cennydd & James Box. 2011. *Undercover User Experience Design (Voices That Matter)*. Berkeley, CA: New Riders.

Buxton, B. 2007. *Sketching User Experiences: Getting the Design Right and the Right Design*. San Francisco, CA: Morgan Kaufmann Publishers Inc.

Dow, Stephen P., Julie Fortuna, Dan Schwartz, Beth Altringer, Daniel L. Schwartz & Scott R. Klemmer. 2011. Prototyping Dynamics: Sharing Multiple Designs Improves Exploration, Group Rapport, and Results. *Proceedings of the SIGCHI conference on Human factors in computing systems*. 2807–2816.

Eisenstein, Jacob, Jean Vanderdonckt & Angela Puerta. 2001. Applying Model-Based Techniques to the Development of UIs for Mobile Computers. *Proceedings of the 6th International Conference on Intelligent User Interfaces*. 69–76.

Garlan, David & Mary Shaw. 1993. An Introduction to Software Architecture. In V. Ambriola & G. Tortora (eds.), *Advances in Software Engineering and Knowledge Engineering*, Volume I, 1–40. World Scientific Publishing Co. Pte. Ltd.

Garzotto, Franca & Paolo Paolini. 1993. A Model-Based Approach to Hypertext Application Design. *ACM Transactions on Information System* 11(1). 1–26.

Garzotto, Franca, Paolo Paolini, Davide Bolchini & Sara Valenti. 1999. Modeling-by-Patterns of Web Applications. *Proceedings of Advances in Conceptual Modeling Lecture Notes in Computer Science*, 1727/1999. 293–306.

Gómez, Jaime, Cristina Cachero & Oscar Pastor. 2001. Conceptual Modeling of Device-Independent Web Applications. *IEEE Multimedia* 8(2). 26–39.

Grossman Lev. 2007. The Apple of Your Ear. *Time Magazine* (Europe Edition), 169 (2), Jan. 22, 2007. 35–38.

Hartson, Rex. 2012. *The UX Book*. CA: Morgan Kauffman.

ISO 9241-151:2008, retrieved on August 1, 2014 at: http://www.iso.org/iso/home/store/catalogue_tc/catalogue_detail.htm?csnumber=37031

ISO 9241-210:2010, retrieved on August 1, 2014 at: http://www.iso.org/iso/home/store/catalogue_tc/catalogue_detail.htm?csnumber=52075

Jansson, David & Steven Smith. 1991. Design Fixation. *Design Studies* 12(1). 3–11.

Johnson, Jeff & Austin Henderson. 2002. Conceptual Models: Begin by Designing What to Design. *Interactions* 9(1). 25–32.

Kahn, Paul, Lenk Krzysztof & Piotr Kaczmarek. 2001. Applications of Isometric Projection for Visualizing Web Sites. *Information Design Journal* 10(3). 221–228.

Kalbach, James. 2007. *Designing Web Navigation: Optimizing the User Experience*. Sebastopol, CA : O'Reilly Media, Inc.

Kuniavsky, Mike. 2003. *Observing the User Experience: A Practitioner's Guide to User Research*. San Francisco, CA: Morgan Kaufmann Publishers Inc.

Law, Effie L., Virpi Roto, Marc Hassenzahl, Arnold P. O. S. Vermeeren & Joke Kort. 2009. Understanding, scoping and defining user experience: a survey approach. *Proceedings of the 27th international Conference on Human Factors in Computing Systems* (Boston, MA, USA, April 04–09, 2009). CHI '09. ACM, New York, NY, 719–728. DOI= http://doi.acm.org/10.1145/1518701.1518813.

Lund, Arnie. 2011. *User Experience Management: Essential Skills for Leading Effective UX Teams*. San Francisco, CA: Morgan Kaufmann Publishers Inc.

Nelson, Harnold G. & Erik Stolterman. 2003. *Design Way: Intentional Change in an Unpredictable World – Foundations and Fundamentals of Design Competence*. Englewood Cliffs, NJ: Educational Technology Publications.

Nielsen, Jakob. 1994. *Usability Inspection Methods*. Chichester, UK: John Wiley & Sons, Ltd.

Nielsen, Jakob. 2009. Top 10 Information Architecture Mistakes. Jakob Nielsen's Alertbox. Retrieved from http://www.useit.com/alertbox/ia-mistakes.html

Rogers, Yvonne, Helen Sharp & Jennifer Preece. 2011. *Interaction Design: Beyond Human – Computer Interaction* (3rd ed.). Chichester, UK: John Wiley & Sons, Ltd.

Rosenfeld, Louis & Peter Morville. 2002. *Information Architecture for the World Wide Web* (2nd ed.). Sebastopol, CA: O'Reilly Media, Inc.

Stolterman, Erik. 2010. Complex Interaction. *ACM Transactions on Computer-Human Interaction* 17(2). 8:1–8:32.

Tohidi, Maryam, William Buxton, Ron Baecker & Abigail Sellen. 2006. Getting the Right Design and the Design right: Testing Many Is Better than One. *Proceedings of the SIGCHI conference on Human Factors in computing systems*. 1243–1252.

Van Duyne, Douglas K., James A. Landay & Jason I. Hong. 2007. *The Design of Sites: Patterns for Creating Winning Web Sites (2nd ed.)*. Upper Saddle River, NJ: Prentice Hall.

Young, Indi. 2008. *Mental Models: Aligning Design Strategy with Human Behavior*. Brooklyn, NY: Rosenfeld Media.

Brian Winston

13 Impact of new media: A corrective

Abstract: The dominance of an insistence in public discourse and policy making in communications on the radical affordances of the digital disguises the continued cultural vitality and economic significance of "old media." Their persistence speaks to the overall social shaping of technology which, of itself, suggests a limit to the technicist hyperbolic rhetoric of the "information revolution." The hyperbole now not only insists on deep technological changes (a "lesser" technicism) but claims, in the name of cybernetics, even more profound alterations to the human condition (a "greater" technicism). Now, supposedly, we face not only interactivity but "post humanity" with lesser "technologies of freedom", "networks of outrage and hope" and greater cyborgian envisionings. It can be argued that much of this is illusory and that this is well reflected in the persistence of old media. There is little evidence of any supervening social necessities for such revolutionary change and much – from resource depletion to the rise of religiosity – to suggest the opposite. In fact, the usual suppression of any new communication technology's radical potential is still in play.

Keywords: old media, technicism, cybernetics, cyberology, "post-humanity", interactivity, diffusion theories, social-shaping of technology (SST), digital impact, "new" economy, supervening social necessity, suppression of radical potential, "technologies of freedom"

> *"The report of my death was an exaggeration"*
> (Mark Twain – on reading his own obituary in a newspaper)[1]

Twain's famous correction should not be forgotten when considering the impact of new media. In the first decade of the 21st century, something in the order $ 3.6 billion was taken at the cinema box-offices of the world's top 20 film-going countries. Recorded music formats, although repeatedly transformed by technological innovation, still yielded +/– £ 1 billion a year in the UK alone (BPI 2010: 4). Despite the supposed locust-like destructive power of unauthorized downloading, by 2009 paid-for downloads worldwide were worth some $ 3.7 billion to the music industry (IFPI cited in Robinson 2010: 17). Broadcasting was reaching vast audiences, one way or another. Nine of every 10 Britons, 46 million plus, were still listening to radios (RAJAR 2011). And television: in fact, thanks to judicious exploitation of the

1 *New York Journal*, 2 June 1897, <http://oupacademic.tumblr.com/post/48310773463/misquotation-reports-of-my-death-have-been-greatly> [accessed 1 November 2013].

new platforms and time-shifting capacities, viewership was actually rising. In the UK, it increased by half-an-hour a day in the twenty years to 2010 (BARB 2011: 1); Plunkett 2010: 6). Moreover, all over the world, these viewers were watching almost completely unchanged forms of content (that is, to take the Western example, material recognizable, say, to ancient Greeks as drama or pubic spectacle *etc*). The world's book publishers were producing two million new or reprinted titles every year. At least (an audited) 114,000,000+ copies of daily newspapers were being sold world wide in 2011 (IFABC 2012).

That all these industries constantly cry foul at the new media is more an expression of corporate greed than pain (except perhaps with the yet powerful and wealthy newspapers which, against absolute falling circulations for over 50 years, are more truly struggling to monetise their web presence). Otherwise, the expressions of largely *faux* distress are enthusiastically trumpeted by all who would see the impact of new media as revolutionary.

1 The persistence of the old

1.1 The analogue watch face

Despite the largely ignored evidence of the persistence of old media, it is obviously not true to say that the media environment is in a state of stasis; that is not the point. Of course, it is constantly developing. The point is that it does not aid understanding to claim seismic shifts in this process. So to do privileges changes in technology in isolation from social contexts, in ignorance of historical developments and in defiance of old-media persistence with its time-honored modes and topoi. Such a privileging of the claim of radical change may be termed *technicism*. It sees technology as the over-arching determinant of social realities.

However, sustaining such a view requires considerable hyperbole, underpinned by highly selective evidence, as well as the historical grasp of an amnesiac. Rather than technicist assertions of 'revolution', it is the case that old technologies have a surprising resilience, a capacity to absorb changes and developments. Evolution rather than revolution is the constant and it is underpinned by factors beyond, and as important as, technology. Technicism proposes that technology is, *de facto*, the sphere containing the social when – in fact and self-evidently – it is itself an expression of the social. Technicism's limitations as an approach to understanding the nature of change and the impact of the new are well illustrated by the media. To emphasize this point, though, I would first propose a clearer and less ambiguous example than that afforded the socio-cultural complexities of contemporary media: consider the wristwatch.

Nothing better illustrates the follies of assuming old technologies are replaced by new in some inevitable (and rapid) march of "progress" than do horological developments in the late 20th century. Technicism cannot explain why the vast majority of wristwatches shipped in 2012, 1.2 billion pieces, had analogue faces of a pattern finally established half a millennium ago (FHS 2012). Watches had acquired fully electronic mechanisms from the late 1980s on but the interface between machine and user was in the majority of cases, three decades later, unchanged. What is even more at odds with technicist assumptions is that not only the faces but also the majority of the mechanisms of these watches, contrary to popular understanding, were mechanical (including self-winding models).

Technicist hyperbole, however, ignores this history as it (typically) simply moves on to the "next big thing" – in this case the smart watch: "App-enabled smart watch shipments will reach 36M per year by 2018, up from 1M in 2013" (Bhas 2013). And, indeed, these devices, in so far as they do not duplicate smart phones, smart tablets *etc ect*, will likely have functions unmatched and of self-evident usefulness – unlike those of the digital watch that were either irrelevant or that duplicated advanced mechanical timepieces. Think GPS. However, it will have taken (at least) 40 years for the development to reach the market and even then (as is usually the case with technicist statistics) the apparent exponential take-up of the new is somewhat illusory. On present showing, 36 million pieces will be around 3% of the market.

Ditto, then, the media. The introduction of the new, technicism's focus, requires also consideration of the persistence of the old, and technicism is myopic about that. The analogy of the mismatch between analogue face and the clockwork or digital chip beneath is exactly reflected by the media. Capacities might well be increased and modalities be enhanced but the surface modes of representation – as with language itself – change far more slowly. For one thing, modes of communication are inherently limited by the capacities of the human sensorium. Producing a sound reproduction system, say, with a range greater than 20–20,000 Hz is pointless for humans as that is the audible range for the ear. (It would be of value to dogs, of course). Technicism tends to ignore such limitations.

1.2 The two technicisms

In fact, there are two technicisms, a "greater" and a "lesser". At its most strident, greater technicist rhetoric about the digital does actually claim an evolutionary impact on our very humanity. There are those who would suggest that exceeding the limitations of the human sensorium indeed causes it to change; to be augmented, for example, with implants. Or, even without surgical intervention, the overall impact will force – is forcing – our minds to adapt. However, assertions that the digital is remaking our brains are without much compelling foundation. Nevertheless, they can be retailed to the point where they earn, as in the following example, a Pulitzer Prize nomination.

There was a contention as the century turned that the Internet was rewiring the brain. A central "proof" of this, apart from some barely pertinent anecdotes and misreadings of Turing *etc*, was reports such as the following from the refereed pages of *The American Journal of Geriatric Psychiatry* (Small et al. 2009). This was on "patterns of cerebral activation during Internet searching", a study of the electrical activity in 24 geriatric brains responding as their owners surfed the net. Results suggested to the psychiatrists that: "Internet searching may engage a greater extent of neural circuitry not activated while reading text pages but only in people without prior computer and Internet search experience" (Small et al. 2009: 116; emphasis added). Many would call this evidence of the process of learning: "temporary synaptic rewiring happens whenever anybody learns anything" (Harris 2011: 9). Such prosaic explanations do not bestseller popular-science books make, however, and so were eschewed in favour of sexier causality: *What the Internet Is Doing to Our Brains* (Carr 2010)[2]. In the 1880s one might as well have written on what the typewriter is doing to our brains; or the automobile; or any technology requiring learning to achieve mastery.

The transformative-impact trope can be traced to the wide-spread reception of the technicism of Marshall McLuhan, the leading proto media "guru" of the 1960s. Whatever the insightfulness of his cultural criticism, his influence on debates about culture's technological expression was less than helpful:

> The mosaic form of the TV image demands participation and involvement in depth, of the whole being, as does the sense of touch ... It is not a photo in any sense, but a ceaselessly forming contour of things lined by the scanning finger. The resultant plastic contour appears by light *through* not light *on* and the image so formed has the quality of sculpture and icon, rather than of picture (McLuhan 1964: 313–334; italics in original)[3].

This reflects the oxmoronic concept of "haptic visuality" (Marks 2002: 2), apparently grounding it in some notion that the eye/brain perceives the television image in ways different from general perception. In fact, though, the interlaced raster of the conventional analogue television tube exactly confuses the watcher because of the usual critical fusion limitation of the brain. In no meaningful scientific sense do TV screens massage eye-balls anymore than web search engines deform brains.

Nevertheless, never-mind the popular: cutting-edge scholarly thinking about the digital in the greater technicist mode also insists on profound, indeed quasi-evolutionary, change.

---

2 Nicholas Carr cashed in on this to the point of a Pulitzer nomination in 2011 (Anon, 2011). Previously he had discussed the impact as more of a question: 'Is Google Making Us Stupid? What the Internet is doing to our brains" was the title of his six-page cover story for the July/August 2008 edition of *The Atlantic* magazine. However, the science of studies such as that of Small *et al* allowed him to drop the query when expanding his article into a book.

3 I have always felt that this extraordinary description of the TV image was not a little a consequence of the fact that McLuhan kept his television set in the basement of his house in Wychwood Park, Toronto. It was incapable of receiving a clear signal.

The problematics of the human/machine interface had been first described as issues of control mechanisms – christened as a new science of "cybernetics". It was the control of anti-aircraft weapons that inspired Norbert Wiener, who was responsible for the concept. He had spent World War II grappling with the design of gun sighting-mechanisms – a compelling issue as the speed of the object to be attacked (a warplane at some 700 kilometers an hour) rendered conventional ranging systems useless. Post-war, his thinking extended to what he saw as the commonalities between automatic machines (*ie* the computers then emerging in response to the challenges of thermo-nuclear bomb design) and the human nervous system. Both swim against the entropic deterioration of the universe and in Wiener's view such resistance defines "life". A machine which produced order out of chaos was, therefore, as "alive" as a human-being (Wiener 1954).

Prosaically this was to lead to the design considerations of ergonomics, but the elegance of Wiener's conceit resonated beyond that. It meshed well with the popular reception of the earliest computers as "electronic brains" and, indeed, it was eventually suggested that: "No natural or human science has been unaffected by these technical and theoretic transformations" (Harraway 1991: 59). It was also to feed into a wide-spread emerging *trahison des clercs* questioning the continued viability of the concepts of the 18th century European Enlightenment. However much impact was, in reality, more theoretic than actual ("technical"), cybernetics was to have definitional consequences for the very idea of the "human". It leads to the melded human/machine figure of the cyborg and the assertion that: "there are no essential differences or absolute demarcations between bodily existence and computer simulation" because "embodiment in a biological substrate [ie the human being] is seen as an accident of history rather than an inevitability of life" (Hayles 1999: 2–3). By this reckoning, whether held to have utopian or dystopian outcomes, we are now deemed to be, in fact, in a technology-based "post-human' condition.

The failure to acknowledge the constraints of the sensorium (when coupled, as it often is, with technological naivety) also infuses the concept of "convergence" as a radical reality. To suggest that electronic encoding of audio and visual information by the same digital techniques means that the resultant signals are in some ways "converged" ignores the difference between the eyes and ears as receivers of the said encodings. At this level "convergence" is meaningless. Better by far, as a basis for social policy and general understanding, to acknowledge that significant media convergence is an expression of the monopolistic tendencies of late capital. It is a function of moguls not machines, speaking to industrial infrastructure, not electronic encodings. However, such is the seductiveness of the greater technicism that "convergence", by ignoring the sensorium, also feeds into the "post-human" implicit critique of the values of liberal humanism.

As with all such anti-Enlightenment attacks, the "post-human" position needs careful evaluation – especially when it is grounded as much in literary theory as actual verifiable data on real, unambiguous impacts. In critiquing McLuhan, Dwight Macdonald had written: "If I have inadvertently suggested that [McLuhan's thinking is] pure nonsense, let me correct that impression. It is impure, nonsense adulterated by sense" (Macdonald 1967: 205). I am not sure that even this much can be said for the current fad for the "cyberology" of the greater technicism. One can see the attractions of such thinking as a rebuttal, for example, of the very bases for patriarchal assumptions about gender. Yet I fear that, for me, essentially un-evidenced talk of human beings as "biological substrate" far too much, however inadvertently, echoes the rhetoric of the ramp at Auschwitz. I find it hard to avoid hearing in the concept of the "post-human" implicitly *Übermenschen*-echoes of an obverse – one that too easily sees sub-human ("sub-strate"?) creatures – mere machines – that can readily become a genocide's victims.

It is not an accident, perhaps, that the most overt response to the supposition that the Internet has had profound mental effects comes from totalitarian China. There the authorities have recognized "addictive" Internet use as a mental illness and have moved to imprison, in goodly numbers, those (mainly young men) deemed to be suffering from it in boot camps for "re-education"[4].

In general, the assertions of greater technicist rhetoric are less fact-based than the musings (or, as McLuhan himself might say, the bemusings) of the scientifically illiterate[5]. Utopianist or distopianist, such technicism ignores the realities – a world of sectarian violence and rising religiosity, collapsed economic verities, bigotry and greed, environmental abuse, finite resources. Greater technicism tells us little about concrete impacts of real media on the actually-existing social sphere.

1.3 The lesser technicism

Nor does the lesser technicism do so either, for all that its claims are of somewhat less draconian – albeit still profound – impact. It does not – cannot – acknowledge the implications that, at a fundamental level, no new communications mode has *entirely* replaced any pre-existing medium (although, of course, it has often displaced their primacy). The question of impact is therefore not the self-evident matter the lesser technicism suggests it is. Claims as to impact's extent are grounded in judgment as to whether or not such displacement as occurs can be characterized, as it overwhelmingly tends to be, as revolutionary and radical. Rather, however, it could be that even where technological developments offer previously unavailable capacities, such as the interactive property of the web, widely supposed societal effects are actually overstated. (This is possible even

4 As documented for example in *Electronic Heroin* (Shosh Shlam and Hila Madalia 2013) Israel/China.

5 In Hayles case, given her training as a chemist, this is particularly surprising.

when lesser consequences than the remaking of eyeballs and brains is not being proposed).

Interactivity, for example, is properly seen as a critical new mass-media capacity but the extent to which its attractions are an alternative to older media cannot be assumed. For one thing, the persistence of older media as reflected in the statistical record suggests that their uses and gratifications might still have potency. Interactivity's superiorities can be overstated. Interactive narratives, for example, deny the deeply human attractions of time-honored non-interactive narrativity. "Tell me a story" is not the same as "I want to intervene in this story"; to hear a story is not the same as, in effect, playing a game and the former has universal appeal whatever the attractions of the latter. Change, as ever, is of course underway, but radically disruptive dynamics – despite the hyperbole of those selling the new media – are not so obvious.

Technicism's history of technology, ignoring such socio-cultural factors as the attractions of story-telling, becomes an easily tabulated succession of technological "breakthrough" dates. The most widely received accounts of these phenomena (*eg Wikipedia passim*) offer, in essence, a series of overlapping biographies of "inventors" – Gutenburg and Daguerre, Morse and Bell, the Lumières and Marconi, Baird and Turing *etc etc*– until, by the mid-20th century, the work of "invention" is wholly subsumed by the research and development laboratories of great corporations whence emerge, as the products of usually anonymous technicians, the transistor and the pocket calculator, the digital watch and the videotape recorder, the mobile phone and the TV satellite, the i-this and the i-that – in an unending stream, usually perceived as ever more life-enhancing (or, more rarely, culture-threatening).

There are severe limitations to such an account of media technological history as the "progress of great men" – aside, of course, from the fact that women and non-Caucasians figure little, if at all, in it. It grossly simplifies the complexities of technological creativity; it pays scant regard to social contexts and realities; and it is frequently distorted by national biases and basic error[6].

---

6 Johannes Gutenberg (c. 1400–1468) is conventionally credited with the first printed press from moveable type in the West; but he was not the only person at work on this technology at the time and moveable type was anyway long established in, at a minimum, Korea and China. Charles Daguerre (1787–1851) is conventionally credited with the development of photography but he was a showman working to an agenda determined by his partner Nicéphore Nièpce (1765–1833); and anyway his system produced unique images unlike the photographic process that was to prevail (for example Fox-Talbot's) which used negatives capable of producing multiple copies. Samuel Morse (1791–1872) was a portrait painter who relied on other scientists and technologically more sophisticated minds, for example, Joseph Henry, a physics professor at the institution that was to become Princeton. Morse too had many legitimate rival claimants to the title "inventor of the telegraph". Alexander Bell (1847–1922) was an elocution teacher with an interest in deafness. He, like Morse, relied on assistants, notably Thomas Watson (1854–1934), an able young electrical engineer. It is also probable that Bell's first effective telephone was the result of his being given a sight (illegally) of the patent application of his rival Elisha Gray (1835–1901). The Lumière brothers

Even at its most refined, when the lesser technicist approach does tackle the impact of the new with more nuanced sociological and economic insight, it still has an inherently linear sense of technological developments. It is seen as a species of "progress" as that term is commonly understood. Rogers' "diffusion theory" (1962), for example, shares such progressivist linearity with the "great man" approach, albeit with anonymous "innovators" as the instigators of the process rather than great-man "inventors". Diffusion theory takes greater cognizance of societal factors – the timeliness of the innovation, the efficacy whereby knowledge of it is communicated *etc etc* – than does crude technicism. Rogers' research suggested that technologies diffuse from innovators (2.5 per cent), through early adopters (13.5 per cent) to early majority (34 per cent), late majority (34 per cent) and laggards (16 per cent). However, the research was initially based on mid-Western farmers' adoption of a new seed. What is forgotten is that this group was very much conditioned by an activist federal agriculture department to have a certain attitude to new techniques. Washington had, since the mid-19th century, been conditioning farmers and farming educators to adopt an open-minded approach – an applied science, intensive, interventionist mind-set. Rogers's ideas about "early adopters" *etc* forces us to extrapolate from a group formally exposed to constant innovation and encouraged to adopt it by direct government policy. Clearly, doing so is not as straightforward as the wide application of his model over the past half-century suggests it is. For example, it is less likely to provide strategically viable information about the consumerist behavior of other less so-conditioned groups – teenagers consuming music, *eg*. Moreover, technological creativity is a given as, successively, these "innovators" give way to later "adopters". These phases are not contextualized in a broader historical and cultural setting.

(August [1862–1954] and Louis [1864 to 1948]) were not, as is commonly supposed, responsible for the first cinema show for a paying audience in December 1895. Theirs was the fourth and they were among a host of entrepreneurs and technologists working on movie image systems at that time (see below, footnote 10). Guglielmo Marconi (1874 –1937) was more of a scientist than these others being a physics graduate from the University of Bologna – which is why he knew about the devices he used for his radio experiments – the coherer. His contribution – no small thing – was to realise that the higher the aerial the further the signal could be sent. Needless to say, others were at work on wireless telegraphy at the same time. (He was also the only one among this group to receive a Noble prize – for physics in 1909.) John Logie Baird (1888–1946) was a failed entrepreneur who became obsessed with mechanically scanned television and never fully grasped the principles of electronic scanning which were to prevail. Nevertheless, the British popularly persist in regarding him as "the father of television". Alan Turing (1912–1954) was, undoubtedly, one of the most influential mathematicians of the 20th century, responsible, in 1936, for a breakthrough paper which lies at the very foundation of computing science; but he was not, as is increasingly being popularly claimed, a father of computing in any practical sense. In fact, he had trouble changing light bulbs. The popular understanding of the nature of technological innovation in the media is completely inadequate. (Winston 1995; 1998)

Similar critiques can be offered of other "adopter-focused" (or, better, consumerist) models for technological impact assessments[7]. Consumerist analysis still privileges the "progress" of technology as having a determining influence on the broader society. They display an advertiser's faith in the malleability of human beings not a compelling account of how they actually behave.

1.4 Social shaping theory

There is a critique of technological determinism, however, which offers an alternative to its flawed conceptual approaches, whether greater or lesser. A second, less popularly understood, conceptualization of media technology's impacts on society denies technology as the prime driver of social change. Instead, society is conceived of as the major factor determining the technological agenda itself and conditioning the diffusion of the technologies it produces. This "social shaping of technology" – *SST* (Noble 1984; Bijker et al. 1987; Mackay and Gillespie 1992; Winston 1998) – is also deterministic and can therefore be termed "cultural determinism"[8].

The cultural determinist account, although still prone to the reductionist simplicities of linear causality, seeks to place the work of the technologist within the broader social sphere. It suggests that the technological agenda is constrained by social needs and that the successful diffusion of any given technology depends on its social acceptability, its "fit" (as it were). As it denies technology a determining role in society, it tends to be less judgmental as to technology's effects, seeing them rather as consequences of other social factors. However, given the wide reception of technicist explanations of technological change, SST can often seem counter-intuitive, rejecting technology as an engine of social change. It resists arguments that technology is either "out of (social) control" or that it is a prime force materially altering social – much less (as is sometimes claimed – *vide supra*) human sensory/cognitive – realities.

Cultural determinism accepts that Western society is dynamic, forever changing and that technology is an element effecting those changes; but that the changes in media technology *per se* presage any fundamental, determining of society can be denied nevertheless. I see no support in Western everyday realities for an "information revolution" which is having, of itself, fundamental rapid effects.

7 *E.g.*, Gartner Inc's "hype cycle"; Moore's "Crossing the Chasm"; or the (mis?)-application of the "hockey-stick analogy" (Moore 1998; Fenn and Raskino 2008; Essex and McKitrick 2012: 154–17). *The Hype Cycle*, for example, is actually a product created and exploited by a consultancy; *Crossing the Chasm* is the title of a neo-liberal marketing "bible"; the *Hockey-Stick Controversy* is about how to read metrological data.

8 The term SST was originally SCOT, the "Social Construction of Technology". Neither SST nor SCOT is in wide use, even academically.

(Hence that the reports of the death of old media are an exaggeration.) This, though, is not to say that the media environment is static and the old media of print and live (including live broadcasting) and recorded performances (acoustic or video) in their rich variety of forms are in rude health. Twain might have been living when he rebutted *The New York Journal* but that is not to say he was well[9]. The old media face new competition in an ever-weakening state. But to overstate their frailty obscures the reality of such impacts as are really underway. It privileges the technological surface over society's deep socio-cultural roots. It misreads history to stress speed and significance when both of these factors are exaggerated.

We must note that these two positions – technicist and SST – do not reflect a difference between those enthusiastic for new media technologies and those, cynical or hostile, who – as E. P. Thompson indicated – it is erroneous to call "Luddite" (Thompson 1968: 600). Positive and negative technicsm, whether of the greater or lesser variety, share an assumption that technological change in media has significant, profound and fundamental social impact. Both agree that technologies, as Raymond Williams once put it, "are discovered, by an essentially internal process of research and development, which then sets the conditions of social change and progress" (Williams 1974: 13). The new media technology "then changes the society or the sector into which it has emerged" (Williams 1989: 120). Williams would dispute this account as would all cleaving towards the alternative SST approach. For SST, society is conceived of as the major factor conditioning the technological agenda itself and shaping the diffusion of the technologies it produces. It holds that the dynamics in play, even though they can be disruptive in certain regards, should be not seen as "revolutionary". They usually indicate more gradually, containable change.

SST's antecedents lie with the French *Annaliste* school of historians which dates back to the 1920s. For example, Marc Bloch's classic essay on the diffusion of the watermill in Medieval Europe focuses on the social and legal structures pushing or inhibiting its introduction and says little about the technical knowledge leading to its development (Bloch 1967 [1935]: 136–138). The most immediate consequence of SST is that the impact of new technologies is conceived of happening within existing societal frames. Evolution, therefore, is far more logically the likely consequence of diffusion than is revolution.

Fernand Braudel discusses the common phenomenon that new technology, contrary to popular assumptions, is not dependent on advances in fundamental science. Innovation is far more often the result of system-engineering applying existing understanding (aka "science") and techniques in new conjunctions and applications than it is of "*eureka*" moments. For the *Annaliste*s the question is always why does the innovation occur when it does since it so seldom relies on

9 In fact, he was. A decade after the erroneous obituary, he received a D. Litt. from Oxford. He died two years after that in 1910.

new knowledge. Braudel considered this in the context of the most profound technological change in modern Western history – the application of steam power to manufacturing (Braudel 1982: 183). The "Industrial Revolution" was dependant on a physical phenomenon known in antiquity but it was the social circumstances of the 18th century which conditioned its application and diffusion. Even then, however, the term "revolution" can be queried. It certainly transformed mass labor but took the considerable length of the "long" 18th century (1688–1815) to do it; and it left class/gender relationships intact, arguably (in the eyes of today's radicals) into the present.

But to return to communications: the cinema, for instance. Why 1895? André Bazin asked,

> How was it that the invention took so long to emerge, since all the prerequisites had been assembled ... The photographic cinema could just as well have grafted itself onto a phenakistoscope foreseen as long ago as the 16th century. The delay in the invention of the latter is as disturbing a phenomenon as the existence of the precursors of the former (Bazin 1967: 19).

Projecting lanterns were known for centuries as was the *camera obscrua* and the *camera obscura portabilis*. Illusionist animated toys, as a species of mass-produced consumer product, were a fad from the 1840s on. Photography – itself a system-engineering response to a new rising social demand for personal images – dates from the same time (Winston 1996). Celluloid, an early plastic, was introduced to make billiard balls, piano keys and, more pertinently, in 1846, as a base for wound dressings. In 1861, 1700 people in a Philadelphia theater watched, as an orchestra played live, a fleeting repetitive projected photographic moving image of a couple dancing. A series of plates were mounted on a wheel to pass through the projector at a speed sufficient for the crucial fusion factor to create the illusion (Mannoni 2000). So, why 1895? Why not 1861, or indeed 1761 *ect*?

In short, in 1895 the American vaudeville industry was selling one million tickets a week and the term "show business" (as opposed to "shows") was a neologism. By the end of the century the force of urbanization was reaching a critical dominant position which it did not have previously. This created the economic basis for the culturally-determined growing social need for entertainment. Mechanization of the live popular theatre industry (*ie* the cinema) met this need. As with the steam engine, the parts of the system needed to enable this were all to hand. Hence the usual phenomenon with these innovations of multiple "inventors"[10]

10 *I.e.*, at a minimum, Edison and Dickson, Auguste and Louis Lumière, Max and Émile Skladanowsky, C. F. Jenkins and Thomas Arnat. The roll-call of "great men" has many other names, their cinematographic "inventions" lost or ignored: Friese-Green, for example, was experimenting in the late 1880s and at one point had his patents confirmed as the master ones. Edison however had more money and Friese-Green's victory was pyrrhic. The "inventors" individually and even collectively "invented" (in the *eureka* sense) nothing. They were systems-engineers. The Lumières even acquired the name *cinématographie* – from one Léon Bouly who coined it for a patent of 1892.

responding to this, as it might be, "supervening social necessity" (urban mass entertainment) at the same moment of time (Winston 1998).

This ensured the diffusion of the device but also that its radical potentials were contained. It "fitted" the culture. Cinema still required paying audiences to sit in the dark and watch theatrical narratives (or factual images arranged in culturally-satisfying narrative forms)[11]. It quickly replaced the theater as the dominant site of popular entertainment for the masses. However, it did not remove the social needs which the live theater had been satisfying for centuries. The bourgeoisie, especially, hung onto theater as culture, eschewing the cinema as an unrespectable form of entertainment for decades in the early 20th century. The live theater survived, survives, in the West. In 2011/12, over a century after the cinema swept across the globe, Broadway took more than $ 1.1 billion at the box office (Anonymous 2013). In London's West End, in 2011, the take was £ 525 m (SOLT 2012).

2 Digital exceptionalism?

2.1 Digital's unexceptional history

For the technicist at both greater and lesser levels of thinking, however, all these accounts of evolutionary change and coexistence, even if acknowledged, do not begin to address the supposed exceptional impacts of the digital. The digital is, allegedly, no mere matter of substituting one cultural form for another; one system of communication for a replacement; one signal modulation method for an alternative. The digital is of far greater moment than this: it is the basis, at the greater level, for "post-humanity". At the lesser: "The digital microchip is the Gothic Cathedral of our time ... It will transform business, education and art. It can renew our entire culture" (Gilder 1985: 15–16). Of course, the failure of the technology to do any such thing in a structurally significant way in the past quarter century can be simply ignored in favor of continued assertion: "The Internet, like the steam engine, is a technological breakthrough that has changed the world" (Singer 2010: 27). And this is self-evident: 300,000,000 *Twitter* texts and 356,000,000 *Wikepedia* "visitors" a day. In 2012, 1.19 billion, one of every 13 human beings, was "on" *Facebook*; there were 1,873,910,000,000 *Google* searches; and *Amazon* shipped

11 There is an argument that the first post-1895 motion pictures more reflected popular non-narrative 19th century spectacles (*e.g.*, the *Diorama)* than they did the theatre. This "cinema of attractions" was "less a way of telling stories than [] a way of presenting a series of views to the audience" (Gunning 1990: 57). Scopophilia certainly fueled Victorian society, but nevertheless the concept of a "cinema of attractions" can be disputed. The examples given (*e.g.*, a – tame – striptease) clearly have narrative arcs and, without question, narrative triumphed within that first decade.

306 items every second of everyday of in that year. –35.5 million packages. In the last quarter of 2012, its revenues falling, *Apple* still earned $ 36.0 billion with a net profit of $ 8.2 billion. The company had greater reserves than the US government[12]. The existence of a global economy, underpinned by a 24-hour global share market, is crucially enabled by the web. Obviously, then, even if the world's continuities – pope and queen, banker and terrorist, man and woman, oppressor and oppressed – are acknowledged, their continued existence is no bar to the digital, the Internet and the i-Machines fulfilling their promised world-changing transformative potential. The assumption is that the digital is exceptional.

SST, however, takes an opposite view of digital's development and impact. Instead, it argues that, just as with the steam engine, these current advances exactly do not change the world so much as emerge from it and fit into it. The statistics are overwhelming, but their significance is rather more obscured. One can apply the withering interjection – "so what?" – too much of this evidence of impact. Remember the philosopher Henry Thoreau facing the pioneering electrical communication technology, the telegraph, in the early 1850s: "We are in great haste to construct a magnetic telegraph from Maine to Texas; but Maine and Texas, it may be, have nothing to communicate" (Thoreau 1995 [1854]: 34). SST will acknowledge the efficacy of networked computers-as-typewriter/calculator/telephone/telegraph and entertainment devices but it is not beglamoured by this. Cultural determinism disputes digital exceptionalism in terms of development and diffusion, seeing instead a certain conformity to previous patterns. First, then, it rejects the amnesiac technicist history which misreads or ignores *la longue durée* of the new technology's development.

The conceptualizations underpinning the diffusion of digital modulation techniques include Leibniz's *Explication de l'Arithmétique Binaire* (1703); Babbage's theorizations for a "difference engine" in 1820s; Nyqvist's work on digitization formulae in the 1920s; Turing's 1936 conceptualization of the human computer confronting the *Entscheidungsproblem in mathematics;* and crucially, in the following decade, in the midst of war, the first design of a machine to alter its procedure in the light of its own calculations, the definition of the (non-human) computer: Goldstine and Mauchly's 1943 "Report on an Electronic Diff.* {sic} Analyzer"[13]. In

12 These un-triangulated statistics (except for the *Apple* company results) are drawn from the net, *viz.*: <https://www.google.com/search?q=facebook<http://en.wikipedia.org/wiki/Wikipedia: Statistics>; <http://expandedramblings.com/index.php/by-the-numbers-a-gigantic-list-of-google-stats-and-facts/>; <http://www.statisticbrain.com/google-searches/>; <http://venturebeat.com/2012/12/27/at-peak-amazon-sold-a-whopping-306-items-per-second-in-2012/>; <http://www.apple.com/uk/pr/library/2012/10/25Apple-Reports-Fourth-Quarter-Results.html>; <http://www.bbc.co.uk/news/technology-14340470> [accessed 1 November 2013].

13 *E.g.*, Nyquist, Harry (1928) "Certain topics in telegraph transmission theory", Trans. AIEE, vol. 47, pp. 617–644, April); Turing, Alan (1936/7) "On Computable Numbers, with an application to the *Entscheidungsproblem*" (Proceedings of the London Mathematical Society, Ser. 2, Vol. 42, 1937; Ser. 2, Vol. 43, 1937, pp. 230–265). [The same year, 1936, an alternative solution was published by Emil Post (Post, E.I. (1936). "Finite combinatory processes – Formulation 1", *Journal of Symbolic Logic,* the metal, discounting calculating machines from the 17th century on[14], digital devices were being essayed from the 1930s. Crucially, a *P*ulse *C*ode *M*odulation (PCM) digital electronic signal sampling device, to be used as part of a telephone exchange, was patented and built by Alec Reeves in 1938. Solid state has an equally protracted history, the first diffused devices to utilize the phenomenon dating from the World War I (as "cat's whiskers" wireless receivers). The transistor and the development of integrated circuits dates from the 1950s. The first widely-diffused commercially-marketed computer, as opposed to the early Cold-War one-off valved (tubed) machines, appears 20 years after the Goldstine and Mauchly paper: the IBM 360 series of 1964.

This history is, in terms of its pace, of a piece with digital predecessors' pre-diffusion incubation periods. Whatever the digital's ultimate impact on society, and whatever start points are chosen, the pattern of its development as a technology is not significantly different from the other communications technologies. We are, by any measure of the actual technology involved, in, at least, the eighth decade of the "information revolution". And, secondly: as for development, so for diffusion.

2.2 Digital's unexceptional diffusion

The history of the diffusion of digital devices perfectly illustrates the balancing pressures whereby new media technologies infiltrate the social sphere. The initial supervening social necessity for the digital, bringing together pre-existing scientific understanding and techniques, was the atomic bomb. The needs of big business (for, say, inventory or payroll purposes) were being met by advanced electrical calculators, following Hollerith's pioneering tabulator of 1890, increasingly from the 1920s on[15]. There was no obvious demand there and no immediate pressure to miniaturize the computing machines either although, post-transistor, this was increasingly possible. Not only was the possibility not exploited, on the contrary, small devices – developed to test peripherals such a memory stores – were

vol 1, no. 3, September); Goldstine, Herman & Mauchley, John (1943) "Report on an Electronic Diff.* [sic] Analyser" (in Burks 1980: 314).

14 *E.g.*, from Napier's Bones (1617) through Pascal's device (the "Pascaline", 1643), refined by Leibnitz's wheels (1694), to the diffused Baldwin desk-adding machine of the 1870s, and Hollerith's Jacquard-card controlled and electrically-driven tabulator of 1889 (the foundational technology upon which IBM was to be established) (Winston, 1998).

15 The supervening social necessity for Hollerith was the census of 1890. This is perhaps the clearest example of social need driving innovation in the record. The results of the 1880 US census took seven years to produce and were by then outdated. As the census was constitutionally mandated, a public competition was organized to find a quicker system to be used for the 1890 census. Herman Hollerith won with an electo-mechanical device using punch-cards to input data. The cards had been introduced as a control for pattern weaving on looms in 1802 (Noble 1984: 147).

discarded. The potential of the small computer was contained and disruption was suppressed. Social stability required that the digital initially coexist with all the other machines which had emerged in the 19th century to service developed capitalism – from the typewriter and the desk calculator through the telephone to the office building's elevator and Hollerith's "business machine"; from the cinema to broadcasting.

With the digital, the push – which Braudel terms the "accelerator" which can be thought of as supervening social necessity – is as much matched by the "brake" (which I would term the "suppression of radical potential") as is the case with other communication innovations (Braudel 1981: 430; Winston 1998). The process can be seen clearly at work in, for example, the current controversy about copyright.

Digitalization permits the easy destruction of copyright but this potential is by no means an inevitability. *Google* has been digitizing the world's print libraries and has scanned in excess of 10 million titles. This though, as suppression of potential would suggest would be the case, is not a direct challenge to copyright. The titles are in legal limbo – the books are either in the public domain or are out-of-print although not, technically, out of copyright. *Google* has not been digitizing live copyrighted material and digitizing these "orphan" works is presented as an altruistic gesture by the company to "expand the frontiers of human knowledge" (Skidelsky 2009: 20). However, its beneficent intentions have not been readily believed. *De facto*, the *Google* copies create a basis for exploitation; a template for further raids on intellectual property. And as such that has been resisted for the better part of a decade.

Technicism takes this threat to copyright as being clear evidence of digital impact: copyright cannot survive it. For SST, on the other hand, it is rather that copyright is in something of a crisis irrespective of the impact of new technology. Copyright terms have been repeatedly extended in international law at the behest of greedy rights holders until, today, they can persist for more than a century. However, 85 % of titles are out of print within 28 years (Boyle 2008: 9).

> Once upon a time, three things held true. Copyrights were relatively short. You had to renew them (most people did not). You didn't get one unless you asked. Now none of those holds true and copyright can last for over 100 years. Copyright has exhausted its function yet the works remain trapped in our cultural black hole (Boyle 2009).

At the outset (the first Western copyright act passed in the UK in 1710[16]), the protection of the economic interests of publishers (not authors, *NB*) were balanced against the monopoly of the copyright license by limiting its term to 14 years. (Capitalism protects such interests but is hostile to monopolies. As for "intellectual

16 8 Ann, c. 19 (1710) 'An Act for the Encouragement of Learning, by Vesting the Copies of Printed Books in the Authors or Purchasers of such Copies'.

property", as a legal concept to protect writers and other creatives, the Anglo-Saxon common law waited – tardily – on post-1789 French legal developments.)

Google, then, has merely forced a confrontation on this matter. The supervening necessity for the project – increased accessibility to otherwise unobtainable texts – is clear. Capitalism's instinct, however, is not readily to countenance any threat to property, even intellectual. This is not to say that digitizing the world's libraries will have no impact. It will, and it will be significant. But it is taking a not unusual time to be absorbed. *Google's* initiative will result in a reform of the law of copyright, not its removal. The social need for such a reform is overwhelming and this, as much as the exploitation of the duplicating capacities of the digital is what is at stake. The point is that, hyperbole aside, *Google* has not been free to ride rough-shod over publishing. The radical potential of the technology is being contained, suppressed.

2.3 Illusions of virtuality and the persistence of irrationality

The result of the diffusion of digital techniques protects society's deeper structures. "Revolution" is illusory. Consider a technicist manifesto of 1995, "A Declaration of the Independence of Cyberspace":

> Governments of the Industrial World, you weary giants of flesh and steel, I come from Cyberspace, the new home of Mind ... You have no sovereignty where we gather ... Cyberspace consists of transactions, relationships, and thought itself, arrayed like a standing wave in the web of our communications. Ours is a world that is both everywhere and nowhere, but it is not where bodies live ... Your legal concepts of property, expression, identity, movement, and context do not apply to us. They are all based on matter, and there is no matter here (Barlow 1996).

The realities of life and work are not so easily set-aside and glib denials of materiality are merely bemusing hyperbole.

The physical realities of cyberspace in fact constitute a further element of digital conformity (as it were). For example, search engines, in a world by no means secure in its energy provisions, require vast supplies of electricity. Simply, overall, it is an absurdity to suggest that "there is no matter here". The "matter" is in the chips, screens, wires, transmitters, metal and plastic boxes. It is in vast profits of the service providers and the manufacturers. It is in the dead bodies of those workers driven to suicide by the conditions of labour in which these materials and services are produced (Tam 2010). It is in the (very material) profits of this "virtual" world.

This materiality extends, finally, into other economic realities. A "new economy" was being widely proclaimed in the 1980s as the developed world moved its manufacturing to cheaper off-shore venues. This "globalization" was reflected in a concomitant, new technology based, integrated 24-hour global market. The ever-open trading of shares, unconstrained by national boundaries, is an undeniable consequence of the computer and the Internet. However, at the deeper level, there is little "new" about it. It is still a market and it behaves just as its more constricted predecessors did. Capitalism's greatest post-World War II financial crisis in 2008 well illustrated the ambiguous advantages of the digital "revolution" in this regard.

The persistence of traditional market instabilities is not only a feature of the application of the digital to general trading; it infects the new technologies themselves. The irrationality of the first "dot.com boom" of 1995 on echoed the "Tulip Mania" of the Netherlands in 1634–7; and, by 2000, the market in new technology stocks had collapsed. The irrationality continued/continues: before it was destroyed by *Facebook*, *My Space* had 100 million users and Rupert Murdoch bought it for $580 million in 2002. A decade later he sold it at a loss of $545 million. *Bebo*, fed by teenage faddism, had 40 million users and was sold to AOL for $850 million within three years of its launch in 2005. AOL off-loaded it two years later when its user numbers were down by 75% (Kiss 2010: 3). The impact of these phenomena, for all the millions of people involved, is more like that of the hula-hoop than, say, the contraceptive pill.

The oral contraceptive pill, introduced in 1960, can be taken as the model of a technology which has had real, indisputable and deep structural effects on society. SST would note, of course, that its development was by men to be used by women, the technology of male oral conception not therefore unexpectedly languishing. However, meshing with economic changes, the pill has contributed to transforming gender relations in the West, fuelling feminism, reforming the concept of morals and impacting on basic demographics (Tryler 1999). Leave aside the unsubstantiated claims of the "post-human": can anything like be said of the "information revolution"?

3 "Technologies of Freedom"?

3.1 The collapse of the Wall

In my view, it cannot – for all that contraceptive-pill style consequences are implicit in the claims of the digital's transformative power. Profound impacts on the social sphere are central to the technicist view of the "Information Revolution". From the outset, it has been represented not as merely a somewhat obscure matter of developments in electrical modulation systems, nor even as an expression of late capital. Rather, the digital has been presented as a tool for reform, change and social upheaval. It was not only to transform "business, education and art ... renew our entire culture". By the early 1980s, even before the full-diffusion of personal computers and well before the Internet, hyperbolic "new economy" rhetoric was already in play; and such was technicist enthusiasm that it embraced not only a "new" sustaining of capitalism but also its obverse social revolution. The digital was producing "technologies of freedom" (Sola Pool 1983) – quite how, though, was less obvious. De Sola Pole, for example, based his utopianism on nothing more than cable television and electronic publishing. Technicism was providing a frame through which to see the world even when evidence of a transformation of human affairs – technologically-based or not – was not readily apparent.

At the time, and irrespective of technology, a desire for "freedom", political freedom conventionally understood, was (as ever) fuelling a series of significant popular protests. Now, though, these socio-political "revolutions" were to be emphatically claimed as consequences of the digital, of the i-phone, of *Facebook*. Irrespectively of the effectiveness and durability of their outcomes, any upheavals are presented as being dependent on the "Information Revolution". And this, according to the lesser technicsm, is the digital's most profound immediate social impact. An SST reading, however, suggests otherwise.

The most significant and (thus fare) durable of the social upheavals of the digital age, drawing a line under the "short" (1918–1989) 20th century, was the collapse of the Soviet Union at the end of the 1980s. However, here was a world-event in which, without question, the "technologies of freedom" played little or no part.

The popular revolts that led to the destruction of the Berlin Wall and the collapse of state communism in Eastern Europe took place in a situation of, exactly, extremely restricted and controlled modes of communication. Rather, they depended on the phenomenon of "informational cascades" created by the face-to-face interactions of human beings (Bikhchandani et al. 1992). These, of course, can involve technologized communication modes but, equally well and in fact more effectively on most occasions, they can occur without them. An "informational cascade" explains the individual's tendency, in any given circumstance, to follow the crowd. The concept addresses not only the question of political behavior but the phenomenon of fads and street fashion, of peer pressure and other consequences of non-consumerist influences. This has long been a concern of media sociology. The classic concept of the "two-step [or multi-step] flow of information" suggested, contrary to assumptions about media power, at best a certain diffusion of media impact occurs (Lazarsfeld et al. 1944: 151). Media messages have the greatest impact on "opinion leaders" who in turn impact on their circles face to face. Media messages are thereby opinion-forming catalysts, not sole determinants of it.

Media did not, then, figure in initiating the protests leading to the Soviet collapse. The "informational cascade" began with the spontaneous occupancy of the West German Embassy in Prague by hordes of East Germans being allowed to cross the border unimpeded in 1989. This provoked sustained weekly *Friedensgebet* (prayer for peace) gatherings back-home in the Nikolaikirche in Leipzig – prayers, note, in that most hallowed of Western public sites, the church as a social sanctuary. At the most, the technology involved was the telephone: and that would not appear to be crucial either.

We can note that a previous upheaval in the Soviet sphere, the Solidarity movement in Poland at the beginning of the decade, was not halted when the authorities imposed martial law and closed down the telephone system. US money underwriting continued trade-union protests was a far more pertinent factor (Judt 2005). As for East Germany a few years later: "It was commonly known that each Monday at about 6 p.m., a large number of people would come streaming out of both the Nikolai Church and other nearby churches" (Lohmann 1994: 67; emphasis added). They congregated in the adjacent Karl Marx Platz and, by so doing, established (or rather re-established) the square – in fact, democracy's ancient *agora* – as the prime physical site of the public sphere. Lohmann's full analysis of this process points out that the only media involved – state newspaper reports – constantly downplayed the numbers involved until the scope of the civil unrest reached such a mass that editorial distortion could not be continued.

This is not to deny technology a role in civil disturbances. It is, rather, to insist that other factors be taken into account: as with Solidarity and foreign support, so the impact of *Friedensgebet* in Leipzig without *perestroika* in Moscow would have been negligible. It was a weakened and unprotected East German regime's opening of its borders that triggered the revolution, not the media. And it was essentially word of mouth, not the media, which sustained it. That a contrary lesson as to the limited role of media could be learned from the Soviet collapse was ignored.

3.2 The twittered revolution

Instead, technicism's pervasiveness, fuelled by the later stages of the diffusion of networked micro-computers, was to seize in its name the succession of social upheavals across the globe after the Soviet collapse as evidence of digital impact. This was despite the conspicuous failure of the new technology actually to deliver on its promise of "freedom" to any salient effect. The digital had, however, enabled the appearance of collective enterprises in the new century such as *Wikipedia* and the exponential growth of the social networks and this was sufficient to sustain and further encourage the utopianist vision of the web's possibilities. For example, Clay Shirky became "the prime booster of the better world order being ushered in by the Internet" (Appleyard 2011: 42) with his 2008 best-seller, *Here Comes Everybody: The Power of Organizing Without Organizations*. This was yet another *vade mecum*, to the "new" digital world, with a monocausal insistence now on the latest stage of digital diffusion as a prime agent of social change (Shirky 2008).

It is not surprising that, in such a context, the seductiveness of a utopianist lesser-technicist vision proved irresistible as a tool to explain the civil disorders which continued in the new century. The connection between the web and social unrest was too obvious to ignore. Technological applications designed, say, to establish social contact between economically privileged young men and women in the West (*eg Facebook*) could equally serve to aggregate revolutionary leaders elsewhere. Shortly it became a given that: "the revolution will be twittered". A series of protests following a disputed election in Iran in 2009 were supposedly facilitated by digital communication modes outwith the control of the authorities. An American blogger came up with the twitter phrase out of whole cloth. Amplified by traditional media (in the usual way), the connection between social media and social revolution was secured as received opinion. Never-mind that at the time less than 1,000 Iranians were actually connected to *Twitter* (Morozov 2011: 1); and never-mind that the protests were suppressed and no revolution actually occurred. This is not to say that informational cascades never involve new media. That would be absurd since they self-evidently can. It is, rather, to say that the claim of their primacy as a tool of revolt can be easily disputed.

A most egregious example of the overwhelming power of the vision of net-based revolt can be found in the explanation of the UK riots of 2011. As in Iran, the authorities contained the disturbances; but, unlike Iran, there was in fact rather more substantial evidence of the use of social networks to announce protests and direct people to the sites of disturbances (Ball and Brown 2011: 12–14). Nevertheless, this does not prove impact because news of the riots was also a consequence, *de facto*, of all-pervasive live old-media transmissions, such as radio, as well as word of mouth (Adegoke and Ball 2011: 15). Informational cascades, as I have said, can of course involve new media. The point is that, by the end of the 21st century's first decade, so bemusing had the power of technicist rhetoric become that the mere use of social media to propose civil disturbance was treated, by the hysterical and grossly unprepared British authorities, as, straightforwardly, actual. The virtuality of the riots proposed or reported in texts send by several impish – and easily traced – young men got them the reality of several years of imprisonment each. In each instance there had been no response to their messages. These caused no disturbances; no riots were traced to them or their texts. In effect, the incidents far, far more indicated the limited impact of new media than its power; but technicism had no more fervent followers than the judges sentencing these perpetrators to outrageously disproportionate, positively 18th century prison terms (Davies 2011: 17; Carrell 2011: 16). The propaganda of the UK government, suddenly seized by technicist verities, obscured the causes of the *jacquerie* – economic decline, austerity, sustained attacks on the welfare state – by insisting the riots were the fault of "gangs" toting Blackberries.

3.2 A false spring

If the role of new media can be so misrepresented in the West, how much more obscure is it elsewhere? Arguably, nowhere more than of the Middle East does the

West need understanding; and nowhere had it, collectively, less. In Islam, the term *jahiliyya* is used to describe the state of pagan ignorance of the inhabitants of Mekkah prior to their acceptance of Mohhamed's teaching. For modern Islamists, *jahilliyya* well describes the West and it was with profound ignorance that the West reacted to the revolts across the Arab world that began in 2011 in Tunisia – the so-called "Arab Spring". The events were shoehorned into Western frames of understanding, their religio-political and social roots ignored. The millennial-long stresses within the *Umma* of the schism between Sunni and Shi'a, and the modern reflection of that as competition between the Sunni Saudis and the Shi'ite Iranians, were but dimly perceived, as were other time-honoured distinctions. Cognizance of the pattern of acceptionist and rejectionist responses to the West was disregarded in favor of a simplistic belief that Arabs, naturally as it were, were playing out a Western revolutionary script by overthrowing autocrats in the name of the people.

Seized with an oxymoronic misunderstanding of Islamism, the emerging politics were deemed to be "moderate", again in the Western sense. Seemingly knowing nothing of 20th century Islamism's roots as reaction to the Western presence in the post-Ottoman British and French spheres of influence, the distinctive Islamic rejection of the separation of church and state befuddled Western reaction. Of the Muslim Brotherhood in Egypt, Manuel Castells, for example, writes: "In sum for the Muslim Brotherhood, Islam and democracy are fully compatible" (Castells 2012: 75). This is absolutely not the case in any Western sense. Castells has been confused by the party's election campaign rhetoric.

In fact, in Islamist theory, no mere democratic election can be allowed to remove a government bound by *shariah* law to substitute a secular regime which might have threaten that law – indeed, have as its primary objective its abandonment. As Afshin Ellian points out: "A liberal democracy cannot survive when divine laws trump those made by man" (Baruma 2006: 24–5).

And so it is turning out. As these *jaqueries* ran into the sand (as it were) by 2013, and the extremely limited (or non-existent) social changes caused by the protests became ever clearer, Western policy was reduced to impotence. *Vide* Syria. Citizens – those wealthy enough to possess the technology – can "twitter" away but as the regional powers fought out a proxy war and the great powers fatuously meddled on one side or another, this was of little real moment. For SST, all these factors need to be considered if technological impact is to be assessed. Technicist *jahilliyya*, by contrast, is untouched by such sociopolitical realities; it is uninformed by history.

Castells insists: "Most of the Arab uprisings started with organization, debate and calls to rise up on the Internet" (Castells 2012: 103). But such an interpretation requires a real measure of cognitive dissonance. Castells, whose own personal history speaks to a brave and sophisticated oppositional political activism, knows full well that insisting on the impact of costly Western technologies in societies which he himself describes as suffering from dire poverty and underdevelopment, *prima facie* begs many questions.

New technology penetration can be admitted as a factor in creating informational cascades among the educated, better off urban population of these countries, especially the younger generation. Disturbances limited to this demographic, however, pose little threat to the authorities. Witness Iran in 2009. The "Arab Spring" was significantly different exactly because of the involvement of proletarian masses. The role of the new media, for all the self-serving hype of Western technicist interests preening as agents of the people's revolutionary struggle, has necessarily been quite limited. The economic realities of underdevelopment, including the fragility of power-supply and communications infrastructures ensure this.

In Egypt, handbills were more important as an agent of revolution than was *Facebook*: "The changes in Tunisia and Egypt were not driven by technology. These were revolutions driven by people" (Fisher 2011: 150). One can add: driven by people assembling in the mass; moreover, poor people with limited access to Western technologies of any kind. The sites of the "Arab Spring" were not in cyberspace but the city square. Informational cascades used every possible mode – with word of mouth paramount. It is a "trick of the light" to believe otherwise and what is being demonstrated is not the power of the digital but the reverse (Appleyard 2011: 42). Twittering against tanks is of limited efficacy.

3.3 The end of secrecy

As is the wholesale exposure of state "secrets" against the realities of (aptly-termed) *realpolitik*. It is suggested that the impact of electronic communications has meant that, in the name of democratic freedom, no state secret can be safe any longer. Without question, the Internet enables wide distribution of whistle-blower disclosures; but, again, it is bemusing to believe revelations require this new technology. Newspapers have been engaged in exposures since (at least) Nellie Bly (Elizabeth Cochrane) spent 10-days undercover as an inmate of the New York Women's Asylum in 1887 (Kroeger 1994: 86–97). In the West, the whistle-blower and the press have been in symbiotic relationship ever since. *Wikileaks* was founded in 2006 to provide a web-based alternative to print and broadcast platforms, but its purpose was time-honored. After a measure of success exposing the kleptocratic nature of Kenyan politics and the names and assets of a Swiss bank's customers, in 2010 it achieved world-wide impact with the release of many thousands of secret American diplomatic cables – now sent, supposedly encrypted, electronically.

In the light of hyperbolic claims as to the significance of this event, a number of factors need be remembered. The breakthrough followed previous revelations which had had limited impact and was achieved, significantly, because the release was redacted for simultaneous publication in four old-fashioned press outlets of high reputation: *The Guardian, The New York Times, Der Spiegel* and *Le Monde.*

Secondly, the revelations in the documents were startlingly banal. For the veteran reporter, "there are few facts or observations in these leaks that an American official would not confide to a respected journalist" (Frankel 2010: 36). There was little "shock" in the revelation that American diplomats were rude, dismissive, ignorant and bigoted about the countries in which they were serving. Ambassadors, famously people "sent to lie abroad for the good of their country", were ever thus[17].

Thirdly, there was little or no *realpolitik* threat involved. The *frisson* of the revelations was more a consequence of the official mania for declaring everything secret than it was of the actual importance of the communication. In 2013, another set of whistle-blower revelations had somewhat greater impact. The National Security Agency of the US secret service, drunk on the possibilities of the new technology, was exposed as listening to millions upon millions of conversations, including the private mobile-phones of the leaders of nations supposedly allied to America. But here again, the essence of the revelation was that states are amoral which is scarce news – much less news dependent on new technology.

Of course these events were significant but not in ways that technicism suggests. The impact of *Wikileaks* was real enough on a number of persons exposed as being informants for, or agents of, the Americans[18]. In the longer term neither the *Wikileaks* nor the NSA exposés have had any obvious impact on American foreign policy. Foreign entanglements continue as before. One consequence for governments in general, though, is likely: they will no doubt "finally acknowledge that secrets shared with millions of 'cleared' officials, including lowly army clerks are not secret. They must decide that the random stamping of millions of papers and computer files ["secret"] each year does not a security system make" (Frankel 2010: 32). Look, then, to a return to paper, safes and couriers (typewriters even) as evidence of the impact of the digital.

Predictions along these lines might well be disputed; but we should not be confused by the rabid persecution of these whistle-blowers or the threats to the editors who publish their leaks, attacks on the press of a severity barely seen in the West in modern times. This vengefulness reflects official annoyance, anger, and embarrassment. It has nothing to do with technology and what is incontrovertible is that the state will no more "wither away" in the face of the digital than it did in the face of the "dictatorship of the proletariat".

---

17 An ambassador was defined in the 17th century as: "*Legatus vir bonus peregre missus ad mentiendum Respublicae causa*' / 'an honest man sent to lie abroad for the good of his country' (Wotton 1672 [1604]: xii).

18 These *ad hominem* exposures were caused by the subsequent release of all the unredacted documents by *Wikileaks*. The papers and magazine involved in publishing the initial story had been at pains to protect identities.

What is occurring here is parallel to the confusions around copyright. The containment is underway of a supervening social necessity – in this case, the belief of the authorities in the bourgeois democracies that they need to maintain restricted state communications. The containment is in a suppression of the radical potential of this technology arising from another need – the need to prevent it destroying the privacy of the individual citizen. Individual privacy as a concept (in contradistinction to the state predilection for secrecy) only emerges over the centuries as the European bourgeoisie slowly divide their living spaces into specialised closed-off areas accessible by corridors. Windows acquired glass and with that curtains (Rybczynski 1987). But from the architecture of the bourgeois home emerges the powerful Enlightenment concept of a right to a private life as an essential human right. We are not facing an "Information Revolution" so much as a conflict between an insistence on an individual's private rights and the state's claimed prerogatives. "Revolution" is in the process of being contained.

4 Impact

Despite it all-pervasiveness, the digital is actually still in a process of diffusion which these conflicts – copyright, secrecy, privacy *etc etc*– well illustrate. As with the social implications of new media technologies generally, the most powerful supervening necessity, that which conditions the research and development agenda, can produce unwanted side-effects which need to be suppressed. Moreover the supervening necessity itself can be obscured. Communicating with ships at sea was the necessity pushing for the development of wireless telegraph. Its more general use on land was deemed to be unlikely as messages could be eavesdropped by anybody with a coherer. It was only after World War I that this disadvantage emerged as "radio" to meet another necessity: home entertainment.

Confusion typically characterizes the phase between introduction and full diffusion. Edison, for example, thought the phonograph, as a recording device, would primarily be used to make records of telephone conversations (Chanan 1995: 3). Not only did this not happen until after World War II (and then only if the receiving-phone was unattended); in fact, the development of wire and tape recording technology was actively suppressed before World War I to contain fears of creating, in an unwanted fashion, legally binding "documents" with such audio records (Nmungwun 1989: 43–8). Social impacts are ever likely subject to misinterpretations, especially by conservative dystopian technicists. With the telephone, there were worries that house-maids would be seduced by lothario telephone callers; that the lower orders would use the device to speak to each other to cheek their betters; even that colds and other communicable diseases would be transmitted down the wire (Marvin 1988: 81–97). With the Internet, the confusion turns exactly on naïf misunderstandings of its impact on the received notion of the distinction between private and public. The simple truth – that Cyberspace is as public as a city square – is only slowly being realized.

The claims for digital's impact have been naïf from the outset. Far from being "technologies of freedom" the supervening necessities that determined its development and applications were all concerned with, exactly, the reverse: from nuclear ignition to the 24-hour electronic market; from new lines of consumer durables (i-Machines) to previously unimaginable levels of surveillance. Technicism might see "networks of outrage and hope"; SST, on the other hand, sees an authoritarian's dream technology. The most obvious digital potential is that – like the reverse TV systems of Orwell's 1984 – it can permit constant monitoring of the citizenry. Any single word in any communication – site, blog, text, email, phone call – is now traceable, its use recordable and storable (Shah 2011: 37). Such is the power of this possibility that the prevalence of pornography, hate speech or "trolling" are deemed to be a price worth paying for control. So to is the existence of radical web-based campaign sites trading on the illusion that change can be effected by merely clicking a computer screen. "Clicktivism", however, is not a sustitute for activism but only one element within it, one factor among many. As a result, this illusion of political action is but a minor inconvenience for authority.

This dystopian big-brother vision of constant surveillance, intrusion and control, however, is as unlikely as the hyperbolic utopian one of freedom and revolution. The negatives are contained, just as the positives will be unachieved – at least by the impact of the technologies alone. Negative necessities are balanced in some measure by constraint. Positive potentials are also constrained. The technology will not destroy society in any revolutionary upheaval; it serves the evolutionary norm. It is diffused only to the degree it will be made to "fit", to meet deep-seated cultural needs, to conform to agreed social an economic norms. Old media might (eventually) disappear but the new will meet the social needs they were serving. It is the need which will determine diffusion, not the capacities of the technology.

As Marx wrote: "Men {sic} make their own history, but they do not make it as they please; they do not make it under self-selected circumstances, but under circumstances existing already, given and transmitted from the past" (Marx 1972 [1854]: 437).

References

Adegoke, Yemisi & James Ball. 2011. I just listened to the radio to find a riot. Simple as that. *Guardian,* 8 December.

Anonymous. 2011. The 2011 Pulitzer Prize Winners: General Nonfiction. <http://www.pulitzer.org/citation/2011-General-Nonfiction> (accessed 11 April 2014).

Anonymous. 2013. Statistics and Facts on Theater & Broadway. *The Statistics Portal.* http://www.statista.com/topics/1299/theater-und-broadway/ (accessed 1 November 2013).

Appleyard, Bryan. 2011. The Net Delusion. *New Statesman*, 20 January. <http://bryanappleyard.com/the-net-delusion/> (accessed 1 November 2013).

Ball, James & Symeon Brown. 2011. Cheap, secure and efficient: the gadget of choice for rioters. *Guardian*. 8 December.

BARB – Broadcasters' Audience Research Board. 2011. Viewing Data. London: BARB. <http://www.barb.co.uk/resources/tv-facts/annual-share-of-viewing?_s=4 (accessed 1 November 2013).

Barlow, John Perry. 1996. Declaration of the Independence of Cyberspace. Davos, Switzerland, 8 February. https://projects.eff.org/~barlow/Declaration-Final.html (accessed 31 October, 2011).

Buruma, Ian. 2007. Murder in Amsterdam :The Death of Theo van Gogh and the Limits of Tolerance. London: The Penguin Press.

Bazin, André. 1967. *What is the Cinema? Volume One*. Berkeley: University of California Press.

Bhas, Nitin. 2013. App-enabled Smart Watch Shipments to Reach 36 Million by 2018, Driven by New Multi-function Segment. *Juniper Research*. 28 August 2013, Hampshire, UK. http://www.factbrowser.com/facts/13428/> (accessed 1 November 2013).

Bijker, Wiebe E., Thomas P. Hughes & Trevor Pinch. 1987. *The Social Construction of Technological Systems: New Directions in the Sociology and History of Technology*. Cambridge, MA: MIT Presslate.

Bikhchandani, Sushil, David Hirshleifer & Ivo Welch. 1992. A Theory of Fads, Fashion, Custom and Cultural Change as Informational Cascades. *Journal of Poltical Economy* 100(5). 992–1026.

Bloch, Marc. 1935. The Advent and Triumph of the Watermills. *Land and Work in Medieval Europe*. Rrans. J. E. Anderson (1967). Berkeley: University of California Press.

Boyle, James. 2008. *The Public Domain: Enclosing the Commons of the Mind*, New Haven, CT: Yale.

Boyle, James. 2009. A copyright black hole swallows our culture. *Financial Times*, 6 September. http//www.thepublicdomain.org/2009/09/06/google-books-and-the- escape-from-the-black-hole (accessed 29 September 2009).

BPI – British Phonograph Industries. 2010. *Statistical Handbook 2010*. London: BPI.

Braudel, Fernand. 1981. *Civilisation and Capitalism 15th–18th century Volume 1: The Structures of Everyday Life, The Limits of the Possible*. New York: Harper Row.

Braudel, Fernand. 1982. *Civilisation and Capitalism 15th–18th century Volume 2: The Wheels of Commerce*. New York: Harper Row.

Burks, Arthur Walter. 1980. From ENIAC to the stored programme computer: two revolutions in computers. In Nicholas Metropolis, Jack Howlett and Gian-Carlo Rota (eds), *A History of Computing in the Twentieth Century*. 311–344. New York: Academic Press.

Carrell, Severin. 2011. Dundee teenagers jailed for trying to start a riot in Dundee via Facebook. *The Guardian*, 12 December. http://www.theguardian.com/technology/2011/dec/12/dundee-teenagers-jailed-riots-facebook (accessed 11 April 2014).

Carr, Nicholas G. 2010. *The Shallows: How the Internet is Changing the Way We Think, Read and Remember*. New York: W. W. Norton.

Castells, Manuel. 2012. *Networks of Outrage and Hope: Social Movements in the Internet Age*. Cambridge: Polity.

Chanan, Michael. 1995. *Repeated Takes: A Short History of Recording and its Effects on Music*, London: Verso.

Davies, Caroline. 2011. Riot appeal hearings to start next week. *The Guardian*, 21 September. http://www.theguardian.com/uk/2011/sep/21/riot-appeal-hearings-start-next-week (accessed 11 April 14).

Essex, Christopher & Ross McKitrick. 2002. *Taken By Storm: the troubled science, policy, and politics of global warming*. Toronto, Ont.: Key Porter Books. 154–17.

Fenn, Jackie and Mark Raskino. 2008. *Mastering the Hype Cycle: How to Adopt the Right Innovation at the Right Time.* Cambridge Mass.: The Harvard Business Press.

Fisher, Alan. 2011. The "Arab Spring", social media and Al-Jazeera'. In John Mair and Richard Keeble (eds.), *Investigating journalism: Dead or Alive?* 150–164. Bury St. Edmunds: Abramis.

Frankel, Max. 2010. Shared with millions, it is not secret. *The Guardian*, 1 December.

Gilder, George. 1985. *Life after Television*, Nashville, TN: Whittle Books.

Gunning, Tom. 1990. The Cinema of Attractions: Early Film, Its Spectator and the Avant-Garde. In Thomas Elsaesser (ed.), *Early Cinema: Space, Frame, Narrative.* 56–63 London: British Film Institute.

FHS – Fédération de industrie Horologic Suisse. 2012. The Swiss and world watchmaking industry in 2012. http://www.fhs.ch/statistics/watchmaking_2012.pdf (accessed 1 November 2013).

Harraway, Donna J. 1991. *Simians, Cyborgs and Women: the Reinvention of Nature*, New York, NY: Routledge.

Harris, John. 2011. How the Internet is altering your mind. *The Guardian*, 20 August. http://www.theguardian.com/technology/2010/aug/20/internet-altering-your-mind (accessed 11 April 2014).

Hayles, Katherine. 1999. *How We Became Posthuman: Virtual Bodies in Cybernetics, Literature and Informatics.* Chicago: University of Chicago Press.

IFABC – International Federation of Audit Bureaux of Circulations. 2012. National Newspapers Total Circulation 2011. http://www.ifabc.org/docs/circulation_report.pdf (accessed 1 November 2013).

Judt, Tony. 2005. *Post War: A History of Europe Since 1945.* London: Penguin.

Kiss, Jemima. 2010. From teen favourite to web graveyard, Bebo faces closure as members log off. *Guardian*, 8 April.

Kroeger, Brooke. 1994. *Nellie Bly: daredevil, reporter, feminist.* New York: Times Books.

Lazarsfeld, Paul Felix, Bernard Berelson & Hazel Gaudet. 1944. *The people's choice: how the voter makes up his mind in a presidential campaign.* New York, Columbia University Press.

Lohmann, Susann. 1994. The dynamics of informational cascades: the Monday Demonstrations in Leipzig, East Germany, 1989–1991. *World Politics* 47(1). 42–101.

McLuhan, Marshall. 1964. *Understanding Media: The Extensions of Man*, New York, NY: McGraw Hill.

Macdonald, D. 1967. He has looted all culture ... In Gerald Emanuel Stearn (ed.), *McLuhan Hot & Cold*, 203–211. New York: Signet.

Mannoni, Laurent. 2000. *The Great Art of Light and Shadow: Archaeology of the Cinema.* Exeter: University of Exeter Press.

Marks, Laura. 2002. *Touch: Sensuous Theory and Multisensory Media.* Minneapolis: University of Minnesota Press 2002.

Marvin, Carolyn. 1988. *When Old Technologies Were New: Thinking about Communication in the Late Nineteenth Century*, New York: Oxford University Press.

Marx, Karl. 1972 [1852]. *The eighteenth Brumaire of Louis Napoleon* in *The Marx-Engles Reader.* New York: Norton.

MacKay, Hughie & Gareth Gillespie. 1992. Extending the Social Shaping of Technology Approach: Ideology and Appropriation. *Social Studies of Science* 22(4). 685–716.

Moore, Geoffrey. 1998. *Crossing the Chasm: Marketing and Selling High-Tech Products to Mainstream Customers.* North Mankato, MN: Capstone

Morozov, Evgeny. 2011. *The Net Delusion: How Not to Liberate the World.* London: Allen Lane.

Nmungwun, Aaron. 1989. *Video Recording Technology: Its Impact on Media and Home Entertainment* New York: Routledge.

Noble, David. 1984. *Forces of Production: A Social History of Industrial Automation.* New York: Knopf.

Plunkett, James. 2010. Long live TV – time spent in front of the box grows despite the internet. *Guardian*, 19 August.

RAJAR – Radio Joint Audience Research. 2011. Quarterly Listening Figures. *Reports*. London: RAJAR.

Robinson, James. 2010. Three quarters of music tracks "downloaded illegally". *Guardian*, 17 December.

Rogers, Everett M. 1962. *Diffusion of Innovations*. Glencoe: Free Press.

Rybczynski, Witold. 1989. *Home: A Short History of an Idea*. Harmondsworth: Middlesex: Penguin.

Shah, Saeed. 2011. Butt out! Rude words banned from Pakistani text messages. *Guardian*, 18 November.

Shirky, Clay. 2008. *Here Comes Everybody: The Power of Organizing Without Organizations*. New York: Penguin.

Singer, Peter. 2010. The promise of the web. *Guardian*, 5 April.

Skidelsky, William. 2009. 'Google's plan for world's online library: philanthropy or act of piracy?'. *The guardian*, 30 August. http://www.theguardian.com/technology/2009/aug/30/google-library-project-books-settlement (accessed 11 April 2014).

Small, Gary W., Teena D. Moody, Prabha Siddarth & Susan Y. Bookheimer 2009. "Your brain on Google: patterns of cerebral activation during internet searching", *American Journal of Geriatric Psychiatry*, 17(2). 116–126.

Sola Pool, Ithiel de. 1983. *Technologies of Freedom*. Cambridge, MA: Harvard University Press.

SOLT – Society of London Theatre. 2012. SOLT 2011 Box Office Data Report on sale now. Press Release. Thursday 5 July 2012. <http://www.solt.co.uk/downloads/pdfs/pressroom/box-office-data-report-2011.pdf> (accessed 1 November 2013).

Tam, Fiona. 2010. Foxconn factories are labour camps: report. *South China Morning Post*, 11 October, 2010. <http://www.scmp.com/article/727143/foxconn-factories-are-labour-camps-report> (accessed 1 November 2011).

Thompson, E. P. 1968. *The Making of the English Working Class*. Harmondsworth, Middx: Penguin.

Thoreau, Henry David. 1995 [1854]. *Walden; or, Life in the Woods*. New York: Dover.

Tryler, L. 1999. Introduction of the pill and its impact. In *Contraceptions*, 59(1 Suppl):11S–16S. NIH – U.S. National Institutes of Health. http://www.ncbi.nlm.nih.gov/pubmed/10342090 (accessed 1 November 2013).

Wiener, Norbert. 1954. *The Human Use of Human Beings*. Boston: Houghton Mifflin.

Williams, Raymond. 1974. *Television: Technology and Cultural Form*. London: Fontana.

Williams, Raymond. 1989. *The Politics of Modernism*. London and New York, NY: Verso.

Winston, B. 1996. *Technologies of Seeing: Photography, Cinematography and Television*. London: BFI.

Winston, Brian. 1998. *Media Technology and Society: A History from the telegraph to the Internet*. London: Routledge.

Wotton, Henry. 1672 [1604]. *Reliquiae Wottoniana*. London: Marriott, Tyton, Collins and Ford.

Claire Hewson

14 Research methods on the Internet

Abstract: This chapter considers how Internet technologies can be used to support primary research on the internet, or *internet-mediated research* (IMR). First, a brief history of IMR is offered, and then the key methods that have been implemented in IMR are outlined, including: surveys and questionnaires; interviews and focus groups; experiments; observation and document analysis. Illustrations of the key methodological approaches are provided, and the main tools and technologies which can support them are described. Advantages and disadvantages of IMR are highlighted, and caveats noted. The key issues and debates which have emerged in IMR are discussed, including issues of data quality, sampling and sample bias, and ethics. Finally, some reflections on future possibilities and promising directions in IMR are offered.

Keywords: internet research methods, internet-mediated research, online research methods, online surveys, big data, digital methods

This chapter considers how the various internet communication technologies discussed in Section I can and have been used to support, and arguably transform, social and behavioural research methods. Focus here is on *primary research*, that is, "the acquisition and analysis of data to produce novel evidence and research findings" (Hewson, Laurent and Vogel, forthcoming). Primary research on the internet has been referred to as internet-mediated research (IMR) (Hewson et al. 2003) and this term will be adopted throughout the present chapter. The chapter will not consider secondary research on the internet, such as conducting a literature review or creating a bibliography; for useful guides on the latter see Hewson et al. (2003); Ó Dochartaigh (2012). The structure of the chapter is as follows: overview and history of IMR; scope and range of IMR methods, with illustrations; key issues and debates in IMR (data quality, sampling, ethics); future possibilities and directions.

1 Overview and history of IMR

The birth of IMR can be traced back to the early 1990s when researchers from a variety of disciplines and backgrounds started to pilot a range of techniques, most notably survey and interview approaches, experiments, and observational research. Over the last two decades the growth and diversification of IMR has been quite astounding. Supported by the parallel growth of internet technologies and the internet user population, the present scope, use and potential of IMR methods is now vast. To illustrate the extent of these developments, consider the early examples which used technologies such as email to administer surveys (Dillman 1991; Dommeyer and Moriarty 1999), interviews (Murray and Sixsmith 1998), and experiments (Hewson 1994). Email approaches were attractive at the time due to not requiring sophisticated computer programming and server configuration skills. Researchers wishing to use more sophisticated techniques, such as web-based surveys and experiments, were required to engage directly with programming code, or have access to support teams which could carry out these implementations. That involved working with HTML code, javascript, server-side scripts, and so on. Although a number of early guides emerged to assist the interested researcher in getting to grips with such programming techniques (e.g. Birnbaum 2000; Hewson et al. 2003), there was clearly a bias in these early days towards the more computer- literate and interested researcher amongst those engaging in piloting IMR methods. One major development since then has been the impact of technologies in expanding both the range and accessibility of IMR methods. For example, there now exists a wide range of online survey software solutions which offer relatively easy-to-use options for administering surveys via the world wide web (e.g. SurveyMonkey: surveymonkey.com; Qualtrics: qualtrics.com); some of these packages now also embody optimisations for mobile delivery (e.g. for delivery via smart phones). Similar packages for assisting in implementing web-based experiments exist, though these are far fewer and certainly more technically challenging to use, with different solutions differing on their ease of implementation. For example, WEXTOR offers a fairly friendly web interface for designing and creating web-based experimental designs (see: http://wextor.org/wextor/en/), whereas WEBEXP (see: www.webexp.info) requires more advanced computing skills to set up and use, including hosting a web server. Technological changes have also facilitated the accessibility of other methods, such as unobtrusive observation approaches which gather data from existing online sources such as discussion group archives; many discussion groups these days can be accessed by a user-friendly web interface, rather than requiring a newsreader application and subscription to this and downloading of messages. Some of the state-of-the-art technologies presently available for supporting a range of IMR methods will be discussed further below, as each of these methods is considered in more detail.

A second major development in the history of IMR is the recent expansion and explosion of unobtrusive data collection approaches, variously referred to as 'data-mining', 'data-scraping', and 'data harvesting'. Many of these approaches may be described as 'unobtrusive observation techniques', others as document analysis. The distinction between observation and document analysis in this context can become blurred, but a useful working definition is that observation involves sourcing data from people's *interactive* behaviours online (intra- and inter- personal), whereas document analysis involves sourcing data from static, finished, published products. Examples of both types of data sources are presented in the next section, to illustrate. These techniques can lead to the creation of 'big data' sets, i.e. data sets which are so large that they are difficult to store, manage and analyse. Techniques for presenting the results of analyses of such data sets often rely on graphical methods, where complex patterns connecting a large array of 'nodes' (which act as data points) are offered.

Thirdly, developments in the growth and diversity of the internet-user population (IUP) over the last several decades have played an important role in expanding the scope of the internet as a site for conducting social and behavioural research. Sample bias was an early major concern in IMR (e.g. Schmidt 1997; see further discussion below), but many recent studies have indicated that data generated from samples recruited online can produce high quality, reliable data comparable (or even superior) to that generated offline. Statistics on the expansion and diversity worldwide of the IUP support the idea that the internet now enables arguably unprecedented access to a vast, diverse pool of potential participants (some recent figures are presented below under sampling).

2 Scope and range of IMR methods

Internet-mediated research methods have now expanded to create a diverse, cross-disciplinary, comprehensive array of techniques and procedures from different research traditions (notably, both qualitative and quantitative) spanning a variety of research domains. Most popular, accessible and widely used is the online survey or questionnaire (e.g. psychometric scale). Looking at a few of the online survey clearing houses presently residing on the web indicates the breadth and scope of this mode of data collection in IMR, though there is a clear bias towards psychological research in the existing resources available, which are often hosted by psychologists or psychological organisations (e.g. see: Online Psychology Research: www.onlinepsychresearch.co.uk). The IMR survey literature clearly indicates, however, that other disciplines have used online survey methods (e.g. sociology: Valliant and Dever 2011; politics: Malhotra and Krosnick 2007; nursing studies: AbuAlRub 2006). This expansion in the use of online surveys has been noted by some as a reflection of the 'democratisation' of the survey (e.g. Couper 2000), for example allowing researchers who may otherwise have lacked the necessary resources (i.e. time, funding, research assistance) to conduct survey-based research (Carter-Pokras, McClellan and Zambrano 2006). Others, on the other hand, have raised concerns regarding the potential for numerous poorly designed surveys to appear online which could damage the reputation of online surveys, similarly to the impact that telemarketing had on the reputation of telephone surveys (Fielding, Lee and Blank 2008).

Other key methods which have now become established in IMR are: experiments, interviews, observational approaches, and (though there are fewer examples) document analysis. Examples of each of these approaches are given in this section. Again, these methods have spanned a range of disciplines and research orientations, with some methods appearing more prominently in certain areas and domains, e.g. experiments have been particularly common in cognitive psychology research. A useful framework for thinking about the array of IMR methods presently being used, and for classifying these, is proposed by Hewson, Laurent and Vogel (forthcoming). They consider the 'obtrusive-unobtrusive' dimension as effective for distinguishing different types of IMR methods, noting that this distinction can help clarify how different issues, e.g. the ethics concerns that can arise, may map on to this dimension. Essentially, obtrusive methods involve participants actively engaging in research procedures in the knowledge that they are taking part in and contributing data for a research study, while unobtrusive methods gather data without such awareness by participants. Naturally, in considering this distinction as a dimension rather than a dichotomy, particular methods will tend to be 'more' or 'less' obtrusive or unobtrusive, rather than being clear cut instances of either category. Hewson et al. (forthcoming) position different methods along the obtrusive-unobtrusive dimension based on how they have typically been implemented, each method spanning a range of conceivable possible positions (depending on the particular design and implementation choices made). For example, interview and survey methods are (necessarily) obtrusive, though perhaps more or less so depending on whether e.g. a synchronous or asynchronous interview is conducted. Document analysis tends to be unobtrusive (unless, perhaps, documents are solicited), and observation methods have spanned most broadly the whole range of the dimension (more recently, data mining techniques have expanded the scope and prevalence of approaches very much towards the unobtrusive end).

Now, illustrations of each of the key methods in IMR mentioned above will be considered, along with an outline of the tools and techniques which have enabled their implementation. In the next section, some of the key issues which have emerged in IMR more generally will be considered, with reflections on how these issues might relate in particular to the methods that have been discussed in this section.

2.1 Surveys and questionnaires

The majority of surveys and questionnaires administered online are now web-based, facilitated by the many cost-effective, user-friendly software packages now available to support this method. Different packages offer different ranges of functionality, and vary in price and levels of support offered; see the WebSurveyMethodology (WebSM) resource (websm.org) for a useful, searchable database of the many packages presently available. SurveyMonkey has been a particularly popular option with UK-based academic researchers, though Qualtrics is now starting to reach the UK market, having become well-established as a popular option at US universities. Qualtrics is appealing in offering competitively priced department, faculty and university-wide licenses (prices for the last two options negotiated individually with universities, depending on their needs). Both packages are good examples of user-friendly interfaces offering reliability, and a reasonable range of functionality for low to mid prices options. Email-based surveys are still used (e.g. Bigelsen and Schupak 2011), though are not very common these days. One of the key advantages of web-based (as opposed to e.g. email) approaches in IMR survey methods is the scope for maintaining greater control over presentation and other procedural parameters, as well as the potential enhanced functionalities that can be integrated. Thus, maintaining consistency of factors such as font size, page layout, order of presentation of questions, and so on is easier to achieve using web-based survey methods. Now, most packages are careful to implement measures that maximise consistency of display across different browsers. Additional features which web-based methods can readily support, compared with email, include skip logic (moving to particular subsequent questions depending on participants' answers to previous questions), question piping (customising later questions using text entered by participants in earlier responses), and checking for response completeness and correct format. This gives web-based approaches a big advantage when it comes to considerations of flexibility, consistency, reliability and validity (see further discussion of data validity below).

Examples of web-based survey research have illustrated how both extremely large sample sizes, and specialist difficult-to-reach populations can be obtained fairly time and cost effectively. Thus Reece et al. (2010) report obtaining a sample of 25,294 gay and bisexual men who completed a survey on sexual behaviours. Their recruitment methods involved sending emails directly to members of a popular internet social and sexual networking site for men seeking men (see below for further discussion of sampling in IMR). In this case both a very large, and relatively specialist sample was achieved. Bigelsen & Schupak (2011) report successfully obtaining a particularly specialist, hard-to-reach population, which they note would have been difficult to obtain using offline methods, in their study of people who identified as excessive or maladaptive fantasisers. They managed to recruit a sample of 90 self-identified excessive fantasisers. In this case, a questionnaire was sent for completion and return by email. Another example in which a specialist population was obtained is reported by Anderson et al. (2003) who recruited 157 tinnitus patients, who after a screening test were administered questionnaires either offline or online; the online recruitment methods involved posting requests to the webpages of the Swedish Hard of Hearing Association (offline newspaper recruitment adverts were also used).

A very common use of online survey and questionnaire methods has been to administer psychometric test instruments. Often, these implementations are carried out in the context of 'validating' a particular existing psychometric scale for use online, since porting such tests to an online medium alters the testing context and conditions (e.g. participants are unsupervised, procedural and presentation details can differ slightly, etc.), and thus psychometric test properties and norms established offline cannot be assumed (see below for more discussion of IMR validation studies). The aforementioned study by Anderson et al. (2003) adopted this approach, reporting comparability of the Hospital Anxiety and Depression Scale (HADS) for administration both online and offline to tinnitus sufferers. Another example is presented by Hewson and Charlton (2005) who administered the well-utilised Multidimensional Health Locus of Control Scale (MHLC) (Wallston and Wallston 1981) online, noting that while many studies had demonstrated the validity of unidimensional tests for online administration, few had reported results for multidimensional scales. They found the properties of this scale to be at least as good as when administered offline (which they did concurrently in the same study). Still, as Buchanan and Smith (1999) have pointed out, finding that one psychometric scale seems able to be validly administered online does not allow generalisations to other scales; each scale needs to be validated independently for porting to an IMR setting.

Marketing and consumer research have also enjoyed a flurry of activity in the online survey arena (Grandcolas et al. 2003), the method being a quick, effective way to obtain commercially valuable data from a potentially large number of participants. This area is one where the aforementioned concerns regarding the quality, integrity and 'image' of online survey research may be most pertinent (many readers will be familiar with the marketing survey pop-up boxes that can sometimes become quite annoying when trying to use the world wide web). Crucial to minimising such possible influences on academic social and behavioural research is to implement and present well-designed surveys which adhere to principles of good practice, including ethics procedures (see more on ethics below). Fortunately there are plenty of resources now available on good practice in online survey design and implementations. A particularly helpful resource for researchers wishing to explore possibilities for implementing IMR surveys is the *Web Survey Methodology* website (WebSM: www.websm.org). There are also now several books dedicated to the topic (e.g. Couper 2008; Sue and Ritter 2012), and the more generic online research web resource *Exploring Online Research Methods* (www.restore.ac.uk/orm/) which provides some very useful training and teaching materials on web and other online (e.g. email) survey approaches.

2.2 Interviews and focus groups

Researchers have been conducting interviews and focus groups in IMR since the very early days (e.g. Gaiser 1997; Murray and Sixsmith 1998). However, this IMR method has not taken off in quite the same way as others, particularly the online survey. It is useful to distinguish between *asynchronous* and *synchronous* methods for conducting IMR interview-based research. Asynchronous methods typically use email (or perhaps discussion group technologies) and involve a longer timescale, conversants (interviewer and interviewee) sending exchanges back and forth as and when it is convenient. A number of researchers have reported success in using this approach, and noted several benefits, including the extended timescale (e.g. compared with an offline face-to-face [ftf] interview) enabling more time for reflection and 'fact-checking' (compared with relying on memory). Also, it has been suggested that such approaches can help balance out power relationships between researcher and interviewee, due to the greater control over the timescale of the interview delegated to the interviewee (Madge and O'Connor 2002). Synchronous approaches have been less well-used, on the whole, perhaps hindered by less successful reports from some of the early attempts at implementing these, but also possibly due to the requirement to engage with more advanced, less widely used 'chat' technologies which are required to enable these approaches (O'Connor et al. 2008).

Compared with traditional offline ftf interviews, the nature of the online communication medium – which typically supports *text-based* exchanges (as in the above described approaches) – has been noted as presenting both possible benefits, and possible drawbacks, in IMR approaches. Thus, a key consideration has been how the lack of normally present visual and extralinguistic cues available offline may impact upon the quality of the data obtained in an online interview, particularly the 'richness' of the data, and the levels of detail, accuracy and reflexivity that can be achieved. These issues have been linked to the lack of proximal contact with participants in IMR contexts making it more difficult to establish good levels of rapport (Jowett et al. 2011), as well as the possible ambiguities that may arise where normally available cues such as intonation, body language, facial expression, etc are absent. There have been mixed reports to date from online interview researchers regarding the effects of the online medium on data quality, richness, reflexivity, clarity, and so on. Notably, in relation to the aforementioned synchronous-asynchronous distinction, the majority of successful reports to date seem to have come from researchers using asynchronous methods (Bowker and Tuffin 2004; Kenny 2005; McDermott and Roen 2012). Synchronous approaches have more often been reported as tending to generate less detailed, flippant, or playful responses (Davis et al. 2004; Gaiser 1997). Having said this, there have been suggestions that synchronous approaches may have the advantage of being better able to maintain 'conversational flow' (Hewson 2008). That is, the potentially lengthy breaks between question and response in an asynchronous email interview, for example (which could involve days), could make it difficult for conversational partners to maintain continuity and coherence in the themes of the conversation, particularly where larger groups are involved (such as in a focus group). In this sense, synchronous approaches could offer an advantage. However, the reliance on all participants being together at the same time creates an obvious restriction for such methods. Asynchronous approaches offer more scope for bringing together participants from a diverse range of geographical and time zone locations (though Gaiser 2008, has also noted the potential difficulty for a researcher in managing asynchronous focus groups where participants are widely geographically dispersed and in different time zones).

The lack of proximal contact in the online interview, and resulting effect of reduced availability of cues which may normally be available (i.e. in ftf contexts) to help enrich and disambiguate data, may on the other hand have possible benefits. Thus the potentially enhanced levels of anonymity in an online interview context, alongside the scope for engaging in quite elaborate and intimate exchanges, has been noted as opening possibilities for strategic, deliberate attempts to try and reduce the impact of biosocial factors (sex, class, race, etc.), which could otherwise lead to unwanted biases (Hewson et al. 2003). However, in general, interview researchers have tended not to emphasise and test such possibilities, striving rather to employ strategies and procedures for reducing levels of perceived anonymity as far as is possible. For example, several researchers have emphasised the importance of using carefully constructed 'rapport-building' exercises, for example involving personal self-disclosure on the part of the researcher, so as to create a 'bond' with participants and encourage rich and candid exchanges (Madge and O'Connor 2002). Indeed, some researchers have reported very successful outcomes when adopting such techniques, even using synchronous approaches (which as noted earlier, have generally been less successful) (Barratt 2012; Madge and O'Connor 2002). Nevertheless, there is some evidence that enhanced levels of anonymity in online communications can encourage more candid, less socially desirable responses (Joinson 2005). The extent to which striving to capitalise on such possible advantages of enhanced levels of anonymity in online contexts, or adopting strategies to reduce anonymity and enhance rapport, will prove most beneficial in IMR interview approaches requires further exploration. To date, it seems that rapport-building strategies have been shown to be especially important and beneficial. Other methods (e.g. surveys) may find higher levels of anonymity especially advantageous.

One possibility for creating an online interview context that perhaps most closely approximates offline ftf approaches, and allows extralinguistic information to be utilised, is to use multimedia technologies, such as Skype. Hanna (2012) has piloted this approach, conducting one-to-one interviews with participants using Skype. However, presently such approaches are hampered by issues of low reliability in audio and video quality, and risks of losing connections, due to present technological constraints. The emergence of better and faster internet technologies and connection (download and upload) speeds may well improve these issues in the near future. Though it is worth also noting here that there is some evidence that the aforementioned benefits of online communication modes for enhancing candour and disclosure may apply only to visually anonymous computer-mediated communication (CMC) contexts (Joinson 2001).

2.3 Experiments

Like surveys, IMR experiments concur many advantages from being implemented via the world wide web, most notably due to greater levels of control over a range of parameters, compared with other possible methods. In many cases the more complex requirements of experimental designs, compared with online surveys, e.g. implementing timings, maintaining control over variables that must crucially remain constant, and so on, demand web-based implementations. However, in simple designs other approaches are possible. For example, Hewson (1994) reports an early experiment in which the experimental manipulation involved presenting participants with different versions of a vignette, and associated questions which elicited responses. It was thus possible to send the experimental materials in the body of an email, and ask participants to send their responses back by email. While a web-based implementation would certainly have had the advantage of allowing greater control over various parameters (see Hewson 2003, for a more detailed 'case study' discussion), the email-based procedure did generate valuable data which allowed the research questions to be answered. Now that there exist software packages to assist in web-based experiment implementations (see above), it is easier to construct these than in the early days when first-hand programming techniques and web server management were necessary. That said, these tools still require a level of computer-related expertise and/or commitment to use, and cannot be so easily adopted by anyone who has only modest levels of web browser and basic software use skills, as can online survey packages. However, in some cases survey packages might be used to implement relatively simple experimental designs, especially those packages which include random assignment of participants to conditions amongst their range of functionalities (e.g. SurveyMonkey 'Gold').

There now exists a good range of examples of web-based experiments, and several very useful guides on resources and design issues, as well as principles of good practice to bear in mind (Reips 2000; Reips and Krantz 2010). Much of the existing work using these methods has come from areas of psychological research, particularly cognitive psychology (as may be expected). Simple text-based materials are the most easy to implement (e.g. Pohl et al. 2002), whilst experiments incorporating graphics, audio, video, and precise timings introduce various increasing levels of complexity. Recent technological advances, including wider access to faster internet upload and download speeds, has facilitated the use and plausibility of these more complex implementations. For example, whereas several decades ago it seemed that implementing reaction time experiments in IMR which required precise timings (i.e. to the millisecond, as in much psycholinguistic research) was not feasible, researchers have since successfully carried out such studies (e.g. Corley and Scheepers 2002). Likewise, experiments involving the download of large-sized multimedia files were far less plausible in the early days, but have now been successfully demonstrated as possible. Thus Caro et al. (2012) conducted an experiment in which elderly participants were presented with video clip vignettes via the internet, in order to explore their views on residential options, including their choices when confronted with health and disability challenges. The study in this case was set up using bespoke software, and most participants took part in a supervised setting, though the researchers note that this would not have been necessary if the population of interest had not been relatively inexperienced in using computers. Though setting up the experiment for Internet delivery involved a number of technical challenges, it ran smoothly; the authors note that such video-delivery implementations in IMR should become more widely accessible to researchers in the future as appropriate technological solutions become more readily available.

Finally, experiments in IMR need not necessarily involve a single participant interacting with a computer program, but may potentially support designs allowing interactions between several participants simultaneously. Of course, this adds yet another layer of complexity, but there are already some examples of this type of approach, showing this it is indeed possible. For example, Horton, Rand and Zeckhause (2011) recruited participants from the online labour market 'Mechanical Turk' (www.mturk.com/mturk/) to take part in interactive experiments using the prisoner's dilemma paradigm, and replicated results previously found in offline contexts. They do, however, note that the present lack of availability of advanced, dedicated software for implementing these types of designs means bespoke solutions are still necessary. One set of technologies presently available which may provide a more accessible way of implementing interactive experiments in IMR is 'multi user virtual environments' (MUVEs). While there will clearly be limitations on what can be achieved using these environments, due to a lack of flexibility compared with what can be implemented using bespoke options, they may prove useful for some purposes, and potentially support some research designs (see Shroeder and Bailenson 2008, for a discussion of MUVE research; see also below for consideration of observational studies which use MUVEs and MUSEs [multi user simulated environments] and which overlap with experimental designs).

Given the inevitable greater complexity involved in implementing IMR experimental designs, as well as the enhanced need (compared with many other methods) for maintaining tight control over variables, this IMR method had generated a greater emphasis within the relevant literature on the details and intricacies of various design and implementation choices, particularly in terms of their potential effects on participants' responses, who volunteers to respond, and so on. For useful discussions of such methodological issues see Reips and Krantz (2010), Krantz and Williams (2010). Useful principles of good practice for IMR experimental design have been offered, such as using the lowest-tech solutions to achieve what is required (Reips and Krantz 2010) and making participants aware of any requirements for particular specialist software which cannot reasonably be assumed to be in widespread use (Hewson et al. 2003).

2.4 Observation and document analysis

Observation approaches in IMR have recently been of particular interest, largely in response to Web 2.0 developments and the accompanying increasing widespread recognition of the wealth of potential data now available in the form of traces of inter- and intra-person activity online (Hewson 2014). The range of approaches being conceptualised and implemented has expanded dramatically over the last decade or so. Both obtrusive and unobtrusive observational methods in IMR have now been applied across a broad array of research domains within different disciplines. An early example of an unobtrusive observational study which demonstrates some of the particular benefits of an IMR approach is presented by Bordia (1996). Bordia was able to locate and extract examples of rumour transmission from Usenet, Internet and Bitnet newsgroups, allowing investigation of this topic in a way that could not be easily achieved using offline methods; in particular the ability to effectively search and locate instances of rumours enabled more data to be generated than would typically have been possible offline, and the online medium facilitated the possibility of doing this unobtrusively (reducing data contamination from demand characteristics present in disclosed observational approaches). More recent examples involving the observation of traces from online discussion forums include Brady and Guerin (2010) who accessed archives of an online parenting support group website and subjected these to qualitative analysis. Being able to gain ready access to a vast, diverse collection of discussion archives online, and easily search these for specific topics and content, is a particular feature of IMR approaches which does not have an obvious parallel in offline observation research. Access to such discussion groups is possible using services such as the web-based interface Google Groups (groups.google.com).

Since the earliest text-based examples, observational IMR studies have expanded in scope, moving on to also incorporate multimedia sources (graphics, audio, video, etc.), drawing upon possibilities offered by Web 2.0 technologies. One relatively accessible approach is to use Social Networking Sites (SNSs), such as Facebook, and media sharing services (e.g. YouTube, FlickR, etc.) as platforms for gathering multi-media observational data (the latter perhaps being better suited to supporting document analysis approaches, depending on the nature of the particular sharing service). McCreanor et al. (2013) offer a review of studies which have explored the way in which alcohol marketing companies have used SNSs to promote their products. Such marketing via SNSs has raised concerns recently about the potential damaging effects on young people of exposure to alcohol marketing and pro-drinking messages on the Internet (e.g. see the recent 'Alcohol Concern' report on this topic: www.alcoholconcern.org.uk/publications/policy-reports/new-media-new-problem). Observational studies which interrogate the structure, function and content of these SNSs are thus of value in gaining insights into young people's alcohol-related attitudes and behaviours. A study by Yoo and Kim (2012) used YouTube (a media sharing website where people post up videos, and others can comment on these; see: youtube.com) in order to access and analyse videos which depicted obesity-related content, so as to explore the portrayal of obesity themes, and obese people, within this online medium.

Virtual Reality Environments (VREs), including MUVEs and MUSEs, have also been used in observational IMR. A very early example of this approach is presented by Givaty et al. (1998) who asked participants to navigate around a 3D virtual environment, created especially for the research study, in an investigation of aspects of visual cognition; these authors note the benefits of the IMR environment in allowing more participants to be obtained compared with offline supervised laboratory settings. Williams (2007), on the other hand, made use of an existing virtual environment (*Cyberworlds*) to conduct an ethnographic study in which he observed avatars, one of the reported benefits being the ability to readily record detailed field notes covertly. Using existing VRE resources in this way is generally likely to place lower technical skill demands on the researcher. Other well-known VREs that could potentially serve as a platform for such designs include Second-Life (secondlife.com). Schroeder and Bailenson (2008) review IMR approaches which have used VREs, many of which have involved computer science and educational applications, and utilised experimental designs. Experimental approaches (often overlapping with observation approaches) in VRE research hold particular promise due to the ability to manipulate and maintain close control over variables. Also, the facilitation of unobtrusive observation in such environments, as well as the ability to automatically record behaviours (of avatars), is of particular note.

The examples discussed above have for the most part involved observations of the 'content' of online interactions and behaviours, e.g. what people have said in discussion posts, in YouTube videos, while interacting in virtual environments, their responses recorded in experimental tasks, and so on. One interesting feature of people's interactions and behaviours online is the potential for (and often automatically occurring, e.g. in web log files) logging of details of the processes, structures and patterns of these interactions. For example, webpage navigations, search engine terms entered, Facebook status 'likes', and social network 'friend links' are all data forms involving the patterns and processes of online interactions. Such traces can be harvested (with a bit of technical know how) to be used in various imaginable research projects. Within the social and behavioural sciences, discussions of how such 'big data' sets (i.e. they often have scope to be extremely large) may be useful to a range of research domains and topics have only recently started to emerge. Marketing research is a domain where the approach has perhaps been most prevalent to date, companies finding data on customers' navigations of their webpages, e.g. searching and purchasing behaviour, to hold obvious commercial and marketing value. Indeed, Google Analytics (www.google.co.uk/analytics) provides a service to assist in collecting and analysing such data from webpage browsing activity. While social science applications remain largely to be explored, one domain in which online unobtrusive observation of this nature has had an impact is social network analysis (SNA). Essentially, SNA involves tracing and tracking the patterns, structures and movements of social groups and communities (and has been a thriving area of research offline). In an IMR context, the large volumes of data which can potentially be harvested from social networking sites has presented opportunities for bringing SNA techniques and methods to the online medium. Thus, data concerning who has 'friended' who, who has messaged who, which friends post on each other's walls, etc. may conceivably be of use. For a description of SNA in IMR see Hogan (2008).

Techniques such as SNA which involve collecting large volumes of process and pattern data do involve more technically complex procedures than many of the other IMR methods discussed above, typically requiring 'data scraping' algorithms to be created. Also, the analysis and presentation of such data can present technical challenges. However, some tools to assist in the collection of such data sources are starting to emerge. For example, Webometrics Analyst is a free software tool designed to help in producing network diagrams based on web searches of sources including Twitter and YouTube (see: http://lexiurl.wlv.ac.uk/index.html). Unfortunately this is a Windows-only resource (hence the author, a Mac-user, has not been able to try it out). Other potentially useful resources include Google's ngram viewer (see: https://books.google.com/ngrams). A dedicated tool for searching Twitter has been created by Ulf-Dietrich Reips, and can be located here: http://maps.iscience.deusto.es. The role of big data sets obtainable online in social science methods is a current thriving discussion topic. For further information see the current project on this topic being run by the Oxford Internet Institute: *Accessing and Using Big Data to Advance Social Science Knowledge* (www.oii.ox.ac.uk/research/projects/?id=98).

Document analysis approaches, as noted above, can be broadly distinguished from observational research in IMR in that they source data from relatively static individually-produced records published online, rather than using traces of (and potentially real-time) interactions and behaviours (such as web browsing activity, online discussion exchanges, etc.). Of course, this distinction can become blurred, as already noted. Though relatively few to date, compared with other methods, some examples of document analysis approaches in IMR do exist. Thus, several authors have reported carrying out analysis of blogs (Clarke and van Amerom 2008; Marcus et al. 2012). Others have carried out analysis of other types of websites (Thoreau 2006, conducted a qualitative analysis of the online magazine *Ouch!* produced by and for disabled people). One approach which has been successfully implemented is to solicit documents, rather than using existing online sources; thus, Hessler et al. (2003) in a study of adolescent risk behaviours recruited participants and asked them to submit daily diary entries online. This method was reported as being successful in generating valuable, rich data, in which participants gave candid and open responses, which the researchers comment can be hard to achieve with this population using offline ftf methods.

3 Key issues and debates in IMR

As has been apparent throughout the above discussion, there are a number of very appealing advantages of IMR methods, many of which have by now been well-demonstrated across a range of methods and research topics. Some of the more salient evidenced benefits of IMR approaches which have now been well-documented include: extended geographical reach, facilitating access to a large, diverse pool of potential participants worldwide; possibilities for implementing recruitment methods which can facilitate access to hard-to-reach, specialist populations; enhanced time and cost efficiency (e.g. due to automated data input, transcription, delivery of materials and procedures, etc.). Also, possible effects leading to enhanced candour and disclosure have been highlighted, as well as other advantages such as the enhanced scope for using unobtrusive approaches which may lead to more authentic data, and enhanced data quality due to the automated checks that can be implemented, e.g. in web-based surveys. Such benefits of IMR have been noted throughout the above discussion of specific methods. However, there are also issues of concern that have been raised in relation to IMR methods. In this section some of the key concerns that have been raised are considered, with an emphasis on their likely impact in terms of creating threats to the value, quality and reliability of IMR-generated data, and possible threats to being able to implement ethically sound research studies.

3.1 Data quality

One key threat to data quality in IMR which has been raised, particularly in relation to survey and experiment approaches, is the reduced levels of control over – and ability to monitor – the procedural aspects of a study. Thus due to the typical lack of physical proximity between researcher and participant(s), it can be difficult to control and monitor the participation context (e.g. are participants distracted, intoxicated, etc.), participant behaviours (e.g. are they multitasking, or not following procedures as instructed), and the details of presentation parameters and procedures (e.g. display variations, due to different hardware and software, and so on). The potential impact of all these factors, and particularly the perception that in online contexts people have a propensity to not be honest and authentic in their reports and behaviours, were initially a major concern in discussions of IMR methods. However, there is now a vast collection of research data – particularly from IMR validation studies (as mentioned above) – which strongly suggests that people do give genuine, honest responses when taking part in IMR studies, and that these studies can and do generate valid, reliable data comparable with that which can be achieved in more traditional offline contexts.

Validation studies which have demonstrated that psychometric tests administered online can display properties equivalent to their offline counterparts have been already mentioned above (Anderson et al. 2003; Hewson and Charlton 2005). There are numerous additional such examples confirming the validity of a diverse range of test instruments for online administration (e.g. Brock et al. 2012; Buchanan and Smith 1999; Kosinski et al. 2003). Only a few examples have found a lack of equivalence (e.g. Barbeite and Weiss 2004; Davis 1999), and in these cases it is not always clear whether the online or offline data is superior. Beyond psychometric scales, other types of questionnaires, and experimental designs, have also been found to show evidence of validity for online administration; for example, Voracek et al. (2001) replicated sex differences in sexual jealousy which had previously been established offline. This report is of interest in relation to the common perception that (because they can) people are more likely to lie about their personal characteristics, such as sex, age, and so on. Replication of established sex differences, or indeed the effects of other demographic and individual difference factors, would seem to indicate that, at least in these cases, people are accurate in their reporting of such personal characteristics. Other authors reporting replicating in IMR experiments findings established offline include Corley and Scheepers (2002); Linnman et al. (2006); see also McGraw, Tew and Williams (2000) for a review.

Interview and focus group methods are also potentially affected by reduced levels of control in IMR. Control over procedures (e.g. ensuring that messages can be sent and received, chat software doesn't stall mid-conversation, etc.) is important to ensure high quality data. Monitoring participants and what they are doing is generally easier in IMR interviews and focus groups than in online surveys and experiments, due to the greater presence of the researcher throughout the whole process; nevertheless, it would be possible for an interviewee to deliberately set out to deceive the researcher, for example by lying about their age, sex, ethnicity, and so on, should they be motivated to do so. The reported experiences of IMR interviewers to date does not suggest that such deception is likely to occur. Another possible threat to data quality in IMR interview research, as discussed above, is the reduction of interpersonal cues (due to the lack of direct physical proximity), possibly leading to ambiguities, misunderstandings, and generally impoverished data. However, a number of researchers have reported obtaining high quality, rich, reflective interview and/or focus group data in IMR. One factor which has been emphasised as important is the careful planning of rapport-building strategies (e.g. Jowett et al. 2011). Whilst there is some evidence that asynchronous interview approaches might be better able to generate high-quality rich, reflective, reflexive data than synchronous methods, both approaches have been reported to have been successful (e.g. Madge and O'Connor 2002).

Unobtrusive IMR methods are generally less affected by many of the factors discussed above. Control over participant behaviours becomes less relevant, for example, since unobtrusive methods are typically concerned with the observation and analysis of naturally occurring online behaviours, often using stored traces and archives of these behaviours. Also, since the focus in unobtrusive approaches is often on studying the idiosyncrasies of online interactions themselves, as compared with offline behaviours and interactions, the unusual nature of the online medium becomes itself the focus of enquiry, rather than a barrier to be overcome. However, the enhanced scope for individuals to manipulate their identities online, particularly where this may involve hiding aspects of their offline identities and attributes (e.g. biosocial characteristics), should be borne in mind by researchers. This is particularly relevant to approaches where the authenticity of the information provided is important, such as in an analysis of the types of posts made by males and females in an online forum. Researchers using these types of unobtrusive methods should remain aware of these caveats, and take them into account when considering interpretations of data derived from online sources. In other contexts, where accurate information about factors such as biosocial attributes is less crucial, ensuring that the computational procedures (algorithms) used for accessing, storing and analysing data sources are robust and reliable, and can extract and save the required information, remains important.

3.2 Sampling

Sample bias has been a longstanding concern in IMR, early researchers noting that the inherently biased nature of the internet user population (IUP) raised serious concerns regarding the quality of data that could be obtained using IMR recruitment procedures and study administration methods (Bordia 1996; Coomber, 1997). The perception in those days of the typical internet user being a well-educated, tech-savvy, middle class, western white male was prevalent (and not necessarily without good reason). Still, there were clearly also internet users who did not fit this stereotype, even in the very early days. Today, it seems that this conception of the typical internet user certainly needs revision, and while it is generally recognised that biases still exist (e.g. towards users in developed nations), many researchers now would find the early claims that IMR samples cannot provide useful, and reasonably representative, data untenable. Not least, the many services and functions the internet now provides to a broad, diverse population of users (as noted earlier, e.g. banking, shopping, socialising, etc.) indicates how it has permeated further and further into everyday lives over the last couple of decades.

Tab. 1: Number of recorded hosts, and estimates of the number of internet users, between 1995 and 2015.

Date	Hosts[1]	IUP (millions)[2]
1995	5	45
2000	72	430
2205	318	1094
2009	625	1825
2010	733	2030
2015	1013	2890

[1] Hosts figures taken from the Internet Systems Consortium 'ISC Domain Survey': www.isc.org/solutions/survey.
[2] IUP estimate figures taken from Computer Industry Almanac.

To consider the growth of the IUP over the last twenty or so years, refer to the figures presented in Table 1, which show the growth in the number of internet hosts (a host is an internet-connected computer that provides internet access to a larger number of connected computers, e.g. as in an 'internet service provider' or 'ISP', such as 'Zen Internet') and estimated numbers of internet users over the last twenty years.

Clearly, Table 1 shows the growth in the size of the IUP to have been quite dramatic over the last two decades. Evidence for the diversification of this population, e.g. in terms of penetration worldwide, has also been presented (e.g. see the summary figures presented by Hewson and Laurent 2008). Also, the comprehensive and systematic Oxford Internet Surveys (OxIS, see: http://microsites.oii.ox.ac.uk/OxIS/) have provided valuable data on the composition and habits of the UK IUP, as part of the World Internet Project (see: www.worldInternetproject.net/#news). For example, the 2011 report (available: http://microsites.oii.ox.ac.uk/OxIS/publications) refers to evidence of a new generation of internet users who access the internet from multiple locations and multiple devices.

Alongside the above evidence concerning the growth in size and diversity of the IUP, as well as emerging developments and trends in usage patterns, there is now also an emerging body of evidence indicating that internet-accessed samples recruited online can produce high quality data. Thus, the validation studies mentioned above provide such evidence, at least where (as in many cases) these used online recruitment methods to access internet samples. Other studies have set out in particular to compare various online and offline sampling methods. Early examples have indicated that internet-accessed samples can be more diverse than traditional offline samples on some variables (e.g. Smith and Leigh 1997). That said, often these studies (many being examples from psychology) have used undergraduate psychology students in their offline recruitment methods, which are not themselves especially diverse. More recently, however, systematic investigations drawing upon broader types of samples have been implemented, offering some very valuable insights. One such study, carried out by Malhotra and Krosnick (2007), compared a large scale, offline (face-to-face) probability sample with a non-probability volunteer internet sample; the internet sample was recruited from an online panel, using stratification to match to population parameters. Administering a survey on voting behaviour and attitudes, these researchers found evidence that the offline sample data was superior, and that the offline sample better represented the (US) population at large on key parameters. Other researchers, however, have reported finding that online volunteer samples can generate data comparable to that obtained from offline probability samples (e.g. Miller et al. 2010), while others still have shown that online probability samples can be obtained which display good levels of broader generalisability (e.g. Chang and Krosnick 2009). It seems that contexts where generalisability to a broad (e.g. national) population is crucial may be most at risk from issues related to biases in internet-accessed samples. However, where access to probability samples can be achieved, e.g. by use of a panel recruited using offline probability sampling methods, these risks may be lowest. Studies comparing internet-recruited volunteer samples with offline non-probability samples have typically found the online data to be at least as good quality as the offline data (e.g. AbuAlRub 2006; Gosling et al. 2004).

Of course, as indicated in the above discussion, the types of online recruitment methods used to access internet samples will play a key role in determining the types of samples obtained, including in terms of their demographics and other characteristics. It is beyond the scope of this chapter to review the range of methods available in further detail, but a comprehensive account can be found in Hewson, Laurent and Vogel (forthcoming).

3.3 Ethics

IMR methods and approaches, particularly some of the more recent unobtrusive methods that take advantage of the wealth of traces of online activity now available, have raised some tricky ethics issues. On the one hand, there can be issues involved in properly porting existing ethics principles and practices to an IMR context; on the other, new issues and challenges may emerge. Some of the issues which IMR procedures raise can be non-obvious to the researcher accustomed primarily to standard ethical practice in the context of traditional offline research (BPS 2013). Though that is not to suggest that the same ethics principles and practices should not also apply in IMR. The recent BPS guidelines on ethics in IMR (BPS 2013) outline some of the key issues to emerge. As well as issues such as how to properly implement informed consent, debrief and withdrawal procedures (e.g. how can you verify online that a participant consenting to answer a web survey is over 16, or whatever the recommended required minimum age is in order to be able to give valid informed consent in other ethics guides), new issues which arise are discussed. One such key issue which has been debated in the IMR literature is that of the nature of the public-private domain distinction online. In particular, leading to (yet unresolved) questions concerning when it may or may not be acceptable to undertake observation (particularly non-disclosed) of people's online activities which have become readily accessible through the ways in which traces are recorded, stored and archived online. Thus, for example, the ready access to many archives of online discussion forum posts via Google Groups was mentioned previously. It is not always clear in what circumstances it is ethically justifiable for a researcher to access and use such posts as sources of research data. Indeed, there has been widespread disagreement and contrasting practice on this issue in the history of IMR. The BPS (2013) guidelines offer some useful advice on this and other key ethics issues related to IMR methods and procedures.

To illustrate the ongoing disparities in both opinions and practices relating to the issue of the use of readily accessible data traces online, as well as the related issue of disclosure and whether usage of such traces should be overt or covert, consider the following contrasting examples. Fox, Ward and O'Rourke (2005) adopted a strategy of full disclosure in a study in which they followed discussions on a pro-anorexia website; they disclosed their intentions as researchers, and participated in discussions on the website's message boards. Brotsky and Giles (2007), on the other hand, also carried out a similar participant observation study involving pro-anorexia discussion and support websites, but without disclosure. In this case, one researcher posted as a plausible persona on the website, and engaged in interactions with it's users, in order to gather research data. In each case the researchers presented arguments to back up their design choices, and justified these in terms of their ethical appropriateness; also each study was approved by a research ethics committee (REC). It is the similarity of these two studies, alongside their very different stances on whether disclosure was appropriate, which is striking. Another interesting illustration comes from Tackett-Gibson (2008) who intended to engage in full disclosure of a research study which set out to observe online communities engaged in exchanging drug use information; however, this intention was blocked by the group moderators who were only prepared to give permission to lurk and carry out observations unobtrusively (as well as have access to stored discussion archives). Arguments that disclosing research intentions might sometimes serve to disrupt or damage a group are relevant here, and moderators may often be in a good position to offer advice on this point (so contacting them first is generally good practice).

One study which directly addressed the issue of disclosure in online observational research is presented by Hudson and Bruckman (2004), who carried out an experimental study in which they observed the responses of chat room participants to disclosures of observation intent by researchers (recording the discussion), compared with undisclosed entry and lurking. In both cases hostile responses were encountered, but the researchers report being kicked out less often when merely entering and lurking. While this result is interesting, it should be noted that the researchers engaged in no attempt to build any type of rapport with participants (as has been found to be a good strategy in interview approaches, as discussed above). Indeed, their observation announcement ("We are researchers recording this chatroom for a study on language use in online environments. For questions or comments, email study@mail.chatstudy.cc.gatech.edu. Thank you!") may rather have been perceived as intrusive and abrupt.

In relation to the above discussion, and the difficulties that can sometimes emerge when trying to make ethically sound decisions about appropriate IMR procedures and practices, perhaps a useful principle to bear in mind is that from the BPS (2013) guidelines, which recommend that a researcher should "ensure that ethics procedures and safeguards are implemented so as to be proportional to the level of risk and potential harm to participants." (BPS 2013: 8). For other useful guidelines on ethics in IMR see: the 2012 report of the Association of Internet Researchers (AoIR: http://aoir.org/reports/ethics2.pdf); Buchanan and Ess (2009). These guides offer advice on important issues, such as how to maximise the effectiveness and reliability of informed consent, debrief and withdrawal procedures in IMR; how to ensure confidentiality and anonymity of data; legal issues, including copyright and data protection laws; considerations relating to sensitive research topics, and conducting research with minors.

4 Future possibilities and directions in IMR

This is an exciting period in the development of IMR methods. As noted above, changes in the way people are using the internet (e.g. the new generation of mobile users) and the integration of internet technologies into our everyday lives (e.g. in learning, socialising, travelling, exercising, shopping, etc.) are creating all sorts of imaginable opportunities for incorporating these developments to inspire creative, new IMR techniques and approaches. Thus, we might imagine incorporating multimedia information (pictures, videos, audio recordings) into survey and interview methods, not just in presenting materials to participants (as has already been done), but also in determining the way participants can respond. Geolocation data could also be incorporated, allowing tracking and recording of a participant's movements during a study procedure. Such approaches could allow richer and more detailed data to be collected (perhaps supporting visual methods, which have been of recent interest in social research, Bagnoli 2009, or 'walking interviews', see Jones et al 2008). The scope for implementing in-situ procedures, in which participants are immersed in particular contexts relevant to the research study, expand with such possibilities. Early innovators have already started to explore such creative applications, using a range of internet technologies, including social media sites, and mobile 'apps' (a software application designed to run on smart phones, tablets, and such devices). One example is the project 'mappiness', based at the London School of Economics (homepage: http://www.mappiness.org.uk/) which aims to map happiness across the UK, and examine how people's happiness may be affected by their local environment. Participants are provided with a smart phone app to download (iphone only), which bleeps every day to ask how they are feeling, as well as record their location, and some other information (such as who they are with). The app also allows a photograph to be uploaded, to show the current location. The findings from this project, so far, are reported as indicating that happiness is greater in natural environments (blog.mappiness.org.uk, accessed 22nd July 2014).

Observational research in IMR was noted above as one of the major areas of development and expansion in recent years. The anticipated increasing integration of the internet – and the 'internet of things' (internet-connected everyday objects, such as cars, fridges, televisions, running shoes, hospital beds, and so on) – into a vast number of people's everyday lives opens scope for further enhancing observational IMR in all sorts of ways. Of course, as already highlighted, ethics considerations (relating in particular to the aforementioned public-private domain distinction) emerge in relation to such possibilities. One illustrative example of observational IMR is the "we feel fine" project (Jonathan Harris and Sep Kamvar; homepage: http://wefeelfine.org). This project collects large volumes of observational data, unobtrusively, from blog posts worldwide. These data provide a live feed (updated every minute) of human emotional expressions, available to view in various summary presentation formats from the project homepage. Other examples were offered above. As the internet-connectedness of everyday objects continues to expand, and the population of users interacting on social media and other internet sites also expands, such approaches will no doubt be facilitated and opportunities enhanced. Observational data, where ethically viable, could also be combined with other sources (e.g. from surveys, or interviews) in a mixed- or multi- method approach. The possibilities for obtaining extremely large data sets, facilitated by the ever-expanding data trail of people's online interactions and communications, is of particular intrigue. One existing example of a project which has generated an extremely large data set, in an obtrusive questionnaire-based research design, is the 'mypersonality' project (http://mypersonality.org/wiki/doku.php?id=start). This project uses a Facebook application which allows users to take psychometric tests, repeated at regular intervals, so that the data can be used to examine how responses change, or remain consistent, over time. At the time of writing the project website reported that nearly 7.5 million people had used the application to complete a questionnaire.

The rapid pace of development of internet technologies will continue to open up opportunities in IMR. Already mentioned above was the likely facilitation in the very near future of approaches involving real-time multimedia interactions online, as internet connection speeds and software and hardware technologies continue to improve (for example, fibreoptic broadband is now becoming widely available, at a reasonable cost, in the UK). Skype-type interviews may very well be a real, reliable possibility within a few years time. Other technologies which hold intriguing potential, and have already started to be used in IMR, include virtual reality environments (VREs), as discussed above. It is now around two decades since the first IMR studies started to appear. It is intriguing to imagine all the possible ways in which IMR methods, and the rapidly evolving technologies supporting them, might develop over the next two decades. Some speculations about some likely future directions have been offered above. The work of researchers who continue to pilot innovative new techniques and methods will shape what the future holds.

References

Anderson, Gerhard, Viktor Kaldo-Sandstrom, Lärs Strom, & Tryggve Stromgren. 2003. Internet administration of the Hospital Anxiety and Depression Scale in a sample of tinnitus patients. *Journal of Psychosomatic Research* 55(3). 259–262.

AbuAlRub, Raeda Fawzi. 2006. Replication and examination of research data on job stress and coworker social support with Internet and traditional samples. *Journal of Nursing Scholarship* 38(2). 200–204.

Barratt, Monica J. 2012. The efficacy of interviewing young drug users through online chat. *Drug and Alcohol Review* 31(4). 566–572.

Bagnoli, Anna. 2009. Beyond the standard interview: The use of graphic elicitation and arts-based methods. *Qualitative Research* 9(5). 547–570.

Barbeite, Francisco G. & Elisabeth M. Weiss. 2004. Computer self-efficacy and anxiety scales for an Internet sample: Testing measurement equivalence of existing measures and development of new scales. *Computers in Human Behavior* 20(1). 1 – 15.

Bigelsen, Jayne & Cynthia Schupak. 2011. Compulsive fantasy: Proposed evidence of an under-reported syndrome through a systematic study of 90 self-identified non-normative fantasizers. *Consciousness and Cognition* 20(4). 1634–1648.

Birnbaum, Michael H. (ed.). 2000. *Psychological Experiments on the Internet*. Academic Press, San Diego.

Bordia, Prashant. 1996. Studying verbal interaction on the internet: The case of rumour transmission research. *Behavior Research Methods, Instruments, & Computers* 28(2). 149–151.

Bowker, Natilene & Keith Tuffin. 2004. Using the online medium for discursive research about people with disabilities. *Social Science Computer Review* 22(2). 228–241.

BPS – British Psychological Society. 2013. British Psychological Society: Ethics guidelines for internet-mediated research. INF206/1.2013. Leicester.

Brady, Ellen & Suzan Guerin. 2010. "Not the romantic, all happy, coochy coo experience": A qualitative analysis of interactions on an Irish parenting web site. *Family Relations* 59(1). 14–27.

Brock, Rebecca L., Robin A. Barry, Erika Lawrence, Jodi Dey & Jaci Rolffs. 2012. Internet administration of paper-and-pencil questionnaires used in couple research: Assessing psychometric equivalence. *Assessment* 19(2). 226–242.

Brotsky, Sarah R. & David Giles. 2007. Inside the "Pro-ana" community: A covert online participant observation. *Eating Disorders: The Journal of Treatment & Prevention* 15(2). 93–109.

Buchanan, Tom & John Smith. 1999. Research on the Internet: Validation of a world-wide web mediated personality scale. *Behavior Research Methods, Instruments, & Computers* 31(4). 565–571.

Buchanan, Elisabeth & Charles Ess. 2009. Internet research ethics and the Institutional Review Board: Current practices and issues. *Computers and Society* 39(3). 43–49.

Caro, Fracis G., Teck Ho, Daniel McFadden, Alison S. Gottlieb, Chrisitine Yee, Taizan Chan & Joachim Winter. 2012. Using the Internet to administer more realistic vignette experiments. *Social Science Computer Review* 30(2). 184–201.

Carter-Pokras, Olivia, Leah McClellan & Ruth E. Zambrana. 2006. Surveying free and low-cost survey software. *Journal of the National Medical Association* 98(6). 881–886.

Chang, Linchiat & Jon A. Krosnick. 2009. National surveys via RDD telephone interviewing versus the Internet: Comparing sample representativeness and response quality. *Public Opinion Quarterly* 73(4). 641–678.

Clarke, Juanne & Gudrun Van Amerom. 2008. A Comparison of Blogs by Depressed Men and Women. *Issues in Mental Health Nursing* 29(3). 243–264.

Coomber, Ross. 1997. Dangerous drug adulteration – An international survey of drug dealers using the internet and world wide web (www). *International Journal of Drug Policy* 8(2). 71–81.

Corley, Martin & Christoph Scheepers. 2002. Syntactic priming in English sentence production: Categorical and latency evidence from an Internet-based study. *Psychonomic Bulletin Review* 9(1). 126–131.

Couper, Mick P. 2000. Web surveys: A review of issues and approaches. *Public Opinion Quarterly* 64(4). 464–494.

Couper, Mick P. 2008. *Designing Effective Web Surveys*. CUP: New York.

Davis, Robert N. 1999. Web-based administration of a personality questionnaire: Comparison with traditional methods. *Behavior Research Methods, Instruments and Computers* 31(4). 572–577.

Davis, M. Bolding, G. Hart, G. Sherr, L. & Elford, J. 2004. Reflecting on the experience of interviewing online: Perspectives from the internet and HIV study in London. *Aids Care* 16(8). 944–952.

Dillman, Don A. 1991. The design and administration of e-mail surveys. *Annual Review of Sociology* 17. 225–249.

Dochartaigh, Niall Ó. 2012. *Internet Research Skills*. 3rd Edition. Sage.

Dommeyer, Cout J. & Eleanor Moriarty. 1999. Comparing two forms of an email survey: Embedded vs. attached. *Journal of the Market Research Society* 42(1). 39–50.

Fielding, Nigel, Raymond M. Lee & Grant Blank (eds.). 2008. *The Sage Handbook of Online Research Methods*. Sage: London.

Fox, Nick, Katie Ward & Alan O'Rourke. 2005. Pro-anorexia, weight-loss drugs and the internet: an "anti-recovery" explanatory model of anorexia. *Sociology of Health & Illness* 27(7). 944–971.

Gaiser, Ted. 1997. Conducting online focus groups: A methodological discussion. *Social Science Computer Review* 15(2). 135–44.

Gaiser, Ted. 2008. Online focus groups. In Nigel Fielding, Raimond M. Lee & Grant Blank (eds.), *The sage handbook of online research methods*. Sage: London.

Givaty, Galia, Hendrik A. H. C. van Vaan, Chris Christou & Henrick H. Bülthoff. 1998. Tele-Experiments – Experiments on Spatial Cognition using VRML-based Multimedia. *http://www.is.tuebingen.mpg.de/fileadmin/user_upload/files/publications/pdf233.pdf* (Accessed 23rd July, 2014).

Gosling, Samuel D., Simine Vazire, Sanjay Srivastava & Oliver P. John. 2004. Should we trust web-based studies? A comparative analysis of six preconceptions about internet questionnaires. *American Psychologist* 59(2). 93–104.

Gosling Samuel D. & John A. Johnson (eds.). 2010. *Advanced Methods for Conducting Online Behavioral Research*. American Psychological Association.

Grandcolas, Ursula, Ruth Rettie & Kira Marusenko. 2003. Web survey bias: Sample or mode effect?. *Journal of Marketing Management* 19(5). 541–561.

Hanna, Paul. 2012. Using internet technologies (such as Skype) as a research medium: A research note. *Qualitative Research* 12(2). 239–242.

Hessler, R. M., Downing J., Beltz C., Pelliccio A., Powell M., & Vale W. (2003). Qualitative research on adolescent risk using email: A methodological assessment. *Qualitative Sociology* 26(1). 111–124.

Hewson, Claire M. 1994. Empirical evidence regarding the folk psychological concept of belief. In *Proceedings of the Sixteenth Annual Conference of the Cognitive Science Society*, 403–408. Georgia, USA.

Hewson, Claire. 2003. Case Study 1: Empirical evidence regarding the folk psychological concept of belief. In Claire M. Hewson, Peter Yule, Dianna Laurent & Carl M. Vogel (eds.), *Internet research methods: A practical guide for the social and behavioural sciences*, 125–130. New technologies for social research. London, UK: Sage.

Hewson, Claire, M. Peter Yule, Dianna Laurent & Carl M. Vogel (eds). 2003. *Internet research methods: A practical guide for the social and behavioural sciences*. New technologies for social research. London, UK: Sage.

Hewson, Claire, M. Peter Yule, Dianna Laurent & Carl M. Vogel (eds). 2003. *Internet research methods: A practical guide for the social and behavioural sciences*. 2nd ed. London, UK: Sage.

Hewson, Claire. 2008. Internet-mediated research as an emergent method and its potential role in facilitating mixed methods research. In Sharlene Nagi Hesse-Biber and Patricia Leavy (eds.), *Handbook of emergent methods*, 543–570. New York: Guilford Press.

Hewson, Claire & John P. Charlton. 2005. Measuring health beliefs on the internet: A comparison of paper and internet administrations of the Multidimensional Health Locus of Control Scale. Behavior Research Methods, Instruments & Computers, 37(4). 691–702.

Hewson, Claire & Dianna Laurent. 2008. Research design and tools for Internet research. In Nigel Fielding, Raimond M. Lee & Grant Blank 2008 (eds.), *The Sage Handbook of Online Research Methods* 58–78. Sage: London.

Hewson, Claire. 2014. Conducting research on the internet: Updates in a web 2.0 era. To Appear in *The Psychologist* 27. 946–951.

Hesse-Biber, Sharlene Nagi & Patricia Leavy (eds.) *Handbook of emergent methods*. New York: Guilford Press.

Hine, Christine (ed.). 2005.*Virtual Methods: Issues in Social Research on the Internet*. Oxford: Berg.

Hogan, B. 2008. Analysing social networks via the internet. In Nigel Fielding, Raymond M. Lee & Grant Blank (eds.), *The handbook of online research methods*. London: Sage.

Horton, J. J., Rand, D. G. & Zeckhause, R. J. 2011. The online laboratory: Conducting experiments in a real labor market. *Experimental Economics* 14. 399–425.

Hudson, J. & Bruckman, A. 2004. "Go Away": Participant objections to being studied and the ethics of chatroom research. *The Information Society* 20(2). 127–139.

Joinson, Adam N. 2001. Self-disclosure in computer-mediated communication: the role of self-awareness and visual anonymity. *European Journal of Social Psychology* 31. 177–192.

Joinson, Adam N. 2005. Internet behaviour and the design of virtual methods. In Christine Hine (ed.), *Virtual Methods: Issues in Social Research on the Internet* 21–34. Oxford: Berg.

Jowett, Adam, Elizabeth Peel & Rachel Shaw. 2011. Online interviewing in psychology: Reflections on the process. *Qualitative Research in Psychology* 8(4). 354–369.

Kenny, Amanda J. 2005. Interaction in cyberspace: An online focus group. *Journal of Advanced Nursing* 49(4). 414–422.

Kosinski, M. Bayliss, M. S. Bjorner, J. B. Ware, Jr. J. E. Garber, W. H. Batenhorst, A. Cady, R. Dahlöf, C. G. H. Dowson, A. & Tepper, S. 2003. A six-item short-form survey for measuring headache impact: The HIT-6™. *Quality of Life Research* 12(8). 963–974.

Krantz, John H. & John E. Williams. 2010. Using graphics, photographs, and dynamic media. In Samuel D. Gosling & John A. Johnson (eds.), *Advanced Methods for Conducting Online Behavioral Research*. American Psychological Association.

Lefcourt, Hebert M. (ed.). 1981. *Research with the Locus of Control Construct: Vol. 1. Assessment Methods*. New York: Academic Press.

Linnman, Clas, Per Carlbring, Asa Ahman, Hakan Andersson & Gerhard Andersson. 2006. The Stroop Effect on the internet. *Computers in Human Behavior* 22(3). 448–455.

Madge, Clare & Henrietta O'Connor. 2002. On-line with e-mums: Exploring the Internet as a medium for research. *Area* 34(1). 92–102.

Malhotra, Neil & Krosnick, Jon A. 2007. The effect of survey mode and sampling on inferences about political attitudes and behavior: Comparing the 200 and 2004 ANES to Internet surveys with non-probability samples. *Political Analysis* 15(3). 286–323.

Marcus, Madalyn A., Herry A. Westra, John D. Eastwood & Kirsten L. Barnes. 2012. What are young adults saying about mental health? An analysis of Internet blogs. *Journal of Medical Internet Research* 14(1). http://www.ncbi.nlm.nih.gov/pmc/articles/PMC3374526/ (Accessed 29th April, 2013).

McCreanor, Tim, Antonia Lyons, Christine Griffin, Ian Goodwin, Helen Moewaka Barnes & Fiona Hutton. 2013. Youth drinking cultures, social networking and alcohol marketing: Implications for public health. *Critical Public Health* 23(1). 110–120.

McDermott, Elisabeth & Katrina Roen. 2012. Youth on the virtual edge. *Qualitative Health Research* 22(4). 560–570. doi: 10.1177/1049732311425052.

McGraw, Kenneth O., Mark D. Tew & John E. Williams. 2000. The Integrity of web-delivered experiments: Can you trust the data? *Psychological Science* 11(6). 502–506.

Miller, P. G. Johnston, J. Dunn, M. Fry, C. L. & Degenhardt, L. 2010. Comparing probability and non-probability sampling methods in ecstasy research: Implications for the Internet as a research tool. *Substance Use & Misuse* 45. 437–450.

Murray, Craig D. & John Sixsmith. 1998. E-mail: A qualitative research medium for interviewing? *International Journal of Social Research Methodology: Theory & Practice* 1(2). 103–121.

O'Connor, Henrietta, Clare Madge, Robert Shaw & Jane Wellens. 2008. Internet-based interviewing. In Nigel Fielding, Raimond M. Lee & Grant Blank 2008 (eds.), *The Sage Handbook of Online Research Methods*. Sage: London.

Pohl, F. Bender, M. & Lachmann, G. 2002. Hindsight bias around the world. *Experimental Psychology* 49(4). 270–282.

Reece, Michael, Joshua G. Rosenberger, Vanessa Schick, Debby Herbenick, Brian Dodge & David S. Novak. 2010. Characteristics of vibrator use by gay and bisexually identified men in the united states. *Journal of Sexual Medicine* 7(10). 3467–3476.

Reips, Ulf-Dietrich. 2000. The web experiment method: Advantages, disadvantages, and solutions. In Michael H. Birnbaum (ed.), *Psychological Experiments on the Internet*, 98–117. San Diego: Academic Press.

Reips, Ulf-Dietrich & John H. Krantz. 2010. Conducting true experiments on the web. In Gosling Samuel D. & John A. Johnson (eds.), *Advanced Methods for Conducting Online Behavioral Research*, 193–216. American Psychological Association.

Schmidt, William C. 1997. World-wide web survey research: Benefits, potential, problems, and solutions. *Behavior Research Methods, Instruments, & Computers* 29(2). 274–279.

Smith, Michael A. & Brant Leigh. 1997. Virtual subjects: Using the Internet as an alternative source of subjects and research environment. *Behavior Research Methods, Instruments, & Computers* 29(4). 496–505.

Schroeder, R., & Bailenson, J. 2008. Research uses of multi-user virtual environments. In Nigel Fielding, Raymond M. Lee & Grant Blank (eds.), *The handbook of online research methods*. London: Sage.

Sue, Valerie M. & Lois A. Ritter. 2012. *Conducting Online Surveys*. Thousand Oaks, CA: Sage.

Tackett-Gibson, M. 2008. Constructions of risk and harm in online discussions of ketamine use. *Addiction Research & Theory* 16(3). 245–257. doi:10.1080/16066350801983699.

Thoreau, Estelle. 2006. Ouch! An Examination of the Self-Representation of Disabled People on the Internet. *Journal of Computer-Mediated Communication* 11(2). 442–468. doi:10.1111/j.1083-6101.2006.00021.x.

Valliant, Richard & Jill A. Dever. 2011. Estimating propensity adjustments for volunteer web surveys. *Sociological Methods and Research* 40(1). 105–137.

Voracek, M. Steiger, S. & Gindl, A. 2001. Online replication of evolutionary psychology evidence: Sex differences in sexual jealousy in imagined scenarios of mate's sexual vs. emotional infidelity. In Ulf-Dietrich Reips and Michael Bosnjak (eds.), *Dimensions of Internet Science*. Pabst Science Publishers, Lengerich, Germany.

Wallston, Kenneth A. & Barbara Strudler Wallston. S. 1981. Health locus of control scales. In Hebert M. Lefcourt (ed.), *Research with the Locus of Control Construct: Vol. 1. Assessment Methods*, 189–243. New York: Academic Press.

Williams, Matthew. 2007. Avatar watching: Participant observation in graphical online environments. *Qualitative Research* 7(1). 5–24.

Yoo, Jina H. & Junghyun Kim. 2012. Obesity in the new media: A content analysis of obesity videos on YouTube. *Health Communication* 27(1). 86–97

Emanuele Rapetti and Francesc Pedró

15 Digital Natives, New Millennium Learners and Generation Y, does age matter? Data and reflection from the higher education context

Abstract: This contribution faces the controversial and high-debated question whether it exists a *generation of digital(ized) learners* and which are its expected traits, shaped during years of reflection around this topic.The first paragraph introduces the discussion, outlining how digital media and connectedness change the life of today's learners and explaining why suggestive images, like the "digital natives" one, have emerged to evoke and summarise in an intuitive form a given set of expectations. The second and the third paragraphs examine in depth the debate in its complexity at the theoretical level. From the analysis of the literature, three different views to approach this topic seem to emerge (the "enthusiasts", the "concerned ones", the "critics") and, consequently, the need to systematize the characteristics attributable to the *Learners of Digital Era (LoDE).* Therefore, a perspective of analysis and a theoretical framework are suggested and detailed. Then, data coming from a research run at the higher education context in Switzerland put in evidence how the age variable is inappropriate to depict a cohort of learners with common traits. In the fifth paragraph authors point up some research implications of the above; and, finally, outlooks and perspectives for the future are offered.

Keywords: Learners of digital era, digital natives and immigrants, new millennium learners, generation Y, digital learners, learners and new media, eLearning, media education, connectedness, pedagogy for the 21st century

1 Digital natives, new millennium learners, generation y, what are we talking about?

In all countries digital media and connectedness are an integral part of the lives of today's learners. It is often claimed that the current cohorts of students are so adept at digital technologies that their lifestyles are, as a matter of fact, determined by the use they make of these technologies in areas such as personal communication, entertainment, and social interaction. In many countries this is not so surprising as it applies also to a growing percentage of the adult population.

The debates about the implications of this phenomenon for education have been already taking place for some time. Powerful and suggestive images, like the "digital natives" one, have emerged to evoke and summarise in an intuitive form a given set of expectations about today's learners. Whether or not the level of technology adoption or dependence is having an impact on the way students manage knowledge and learn and, therefore, on their expectations about teaching and learning has been open to discussion. Such a discussion has often devolved into an irreconcilable confrontation between the advocates of educational change and those who look at technology in teaching merely as a tool to complement what teachers do. The former see in the new generations of technology-adept students an opportunity to radically transform teaching and learning in formal education. The latter claim that technology should be used to enhance and improve current practices and that the level of technology adoption should be a function of two criteria: convenience and productivity.

Nevertheless, rarely are these discussions backed by empirical evidence. Many works on this topic can be seen as stimulating and challenging essays that expand the horizon of the debate and drive it by anticipating plausible hypotheses. However, they often fail to provide the empirical evidence that could contribute to inform the policy debate at the institutional level and to help teachers to make professional choices about technology adoption in teaching on a sound basis.

To contribute to bridge this gap, this chapter is organized as follows: the next paragraph will outline the debate on the topic, while the following one suggests a more balanced approach, able to integrate pros and cons offered in the debate, it also presents a brief inter-disciplinary itinerary to root such proposal. The fourth paragraph summarizes results of an empirical research, which has shown that age does not matter when it comes to preferences of tools by higher education students (or even it does sometimes, but in a counter-intuitive way); the fifth paragraph and the conclusion discuss some implications of the proposed reflections.

2 Facing the debate around the topic in its complexity

Looking at the past years, it is possible to find an almost never-ending list of contributions trying to describe our times, with a crucial split among the commentators who are enthusiast about the *digital* present and the ones who are nostalgic of an *analogic* past. The debate still lacks a comprehensive and mature reflection about the future (Skiba, 2010), and not in terms of what (and how) ICTs can offer us for a better life, but rather in terms of anthropological reflection. Only very recently, the topic has started to be addressed in a more founded way (Oh & Reeves, 2014).

It is enough to count how many articles turn around labels such as Digital Natives, to realize the entity of the debate. For instance, in Google Scholar “digital natives” tag appears more than 5,000 times excluding citations and just observing 2012–half0013 period; and it is worth to be noted that some contributions assume digital natives do exist, while other articles are more cautious, andothers are even sceptic.

Evidently, such discussion has a great importance in education, especially for what concerns the relationship between communication, technologies, and instruction.

Moreover, it is observable a lack of comprehensive perspective on learning in the digital era, and it seems that the most relevant pieces of knowledge can be understood as best practices or good intuitions. Indeed, two key-assumptions lead the narrative behind our idea of Knowledge Society – or Information Age (Castells 2000): living nowadays means living in a digital environment; this environment impacts on our everyday life, then on learning experience as well. Even if, *a priori*, this reasoning allows us to expect young learners (grown up in the digital era) to be “digital learners”, we must recognize, *a posteriori*, that such expectations very often do not appear. Just to offer two examples: the high dropout rates of eLearning in vocational training (Succi and Cantoni 2008), and the non-enthusiastic acceptance in universities by students (Bullen et al. 2011).

A review of the literature about the dyad “learners and new media” shows a great level of uncertainty. Thinkers and scholars debate whether a generation of “digitalized” learners has started to populate educational institutions in north-western world, and strong oppositions are registered.

In previous works, the gap between noteworthy intuitions and solid evidence-based data was already put in evidence: if the first ones are widespread, the second ones are rare, unveil very little certainties, and a number of controversial points (OECD 2012).

Since the beginning of the debate, voices obtaining the bigger audience and fame were the ones providing a catchy label/expression,effective to concentrate all the implied reflections. Labels are so important in our analysis because they are powerful and answer to an anthropological need: naming unknown things and phenomena.Though, labels are not neutral: they limit and tune the observation around the ideas carried by the expression; and might even drive the attention away from other issues.

“Generation Y”, “Millennials” (Strauss and Howe 1991), “Digital natives” (Prensky 2001a, 2001b), “Grasshopper Minds” (Papert 1993), or “Net Generation” (Tapscott1998; 2009) are likely to be the most known labels; not to mention the “IPod generation” (Akande2008). Each of those focus rhetorically on one (or more) of the following points:

1. Effects of digital familiarity/behaviours: changes in learning-related behaviours due to the presence of digital in social activities.

2. Effects of digital routines/behaviours: changes in cognitive processes and behaviours due to the presence of digital in everyday life (psycho-cognitive approach).
3. Generational characteristics: socio-cultural differences among generations.
4. Impact of digital environment: societal changes due to digital context affecting everyone.
5. Impact of digital familiarity: societal changes due to digital experiences affecting particularly young people.
6. Impact of internet: specific focus on spread of the "network logic" and internet experiences.
7. Millennium turn: specific focus on the historic-cultural aspects of the millennium turn.
8. Technological development: specific focus on the overall impact of technological development over society.

Previous eight points are anything but an arbitrary abstraction, but they might be helpful to understand implications and consequences of such labels.

2.1 Three (possible) views on the topic

Our research in this complex field leads to the elaboration of a schematic approach to systematize the literature review on the topic. This step became necessary in order to overcome the limits of adopting a specific label – and the implied theoretical assumptions.

We identified (at least) three potential views, namely the "evangelists", the "catastrophists", and the "skeptics" (OECD 2012: 75–88), elsewhere defined as the "enthusiasts", the "concerned ones" and the "critics" (Rapetti 2012: 123–140).

This is our understanding of the abovementioned groups:

1. The "enthusiasts" (about the impact of ICTs on learners' skills and behaviours) are firmly convinced that digital technologies are making the generation of younger learners a very skilled one. Within them it is possible to identify three different approaches, depending on the observed ICTs' effects on learners behaviours and attitudes:
   a. the historic-sociological approach, moving from all the differences between the current generation and the previous ones (Howe and Strauss 2000);
   b. the psycho-cognitive approach, claiming that the everyday usages of ICTs change the cognitive abilities of young people (Prensky 2001a, 2001b; Small and Vorgan 2008);
   c. the socio-pedagogical approach, based on the paradox "everywhere ICTs, except at schools", and asking for a reform/revolution in school and uni-

versity systems (Oblinger and Oblinger 2005; Junco and Mastrodicasa 2007).
2. The “concerned ones”, almost “reactionaries”, accepting as well this idea of a digitalized generation of learners, but concerned about potentially dangerous effects, such as violence, dumbness, harassment, addiction, etc. (Bauerlein 2008).
3. The “critics”, who question the idea of characterising the set of skills of the young generation simply based on ICTs’ usages, criticizing all the unrequested generalizations, and calling for deeper studies and localized analyses (Selwyn 2009).

3 Looking for a theoretically wise re-framing of the topic

It has to be noticed that a constantly growing number of studies approaches the topic in a descriptive way, basically stressing statistical data (Ghait 2010; Joiner et al. 2013; Kubiatko 2013; Kurt et al. 2013; Lapa & Cardoso 2013); but – and this is a crucial point – the main question (“are digital natives a myth or a reality?”) is still to be answered (Margaryanet al. 2011).

A major contribution has been provided by the Organisation for Economic Co-operation and Development (OECD) within the New Millennium Learners (hereinafter NML) (oproject (Pedró 2007; 2009a; 2009b; OECD 2010a; 2012). The results obtained in this project suggest a mixed and far more complex picture than it is often presented in most of the well-known essays about this topic. To begin with, although an increasing percentage of students can be said to be adept in technology, it is misleading to consider that all of them fit equally well into the image of new millennium learners. As it happens with learning styles, there are different student profiles regarding technology adoption and uses, and in many respects clear digital divides still exist. The use of concepts such as the new millennium learners can be helpful in so far as it evokes a clear and powerful image, but misleading if used as a cliché or stereotype. For the purposes of improving teaching and learning in higher education, the diversity of students and situations matters the most.

While higher education institutions and teachers are increasingly adopting technology in teaching, there is an urgent need to address this issue in a systemic way. This means to identify what policies and practices can best suit the objective of providing students with a rich learning environment while improving their satisfaction and boosting learning gains. More has to be done to improve the knowledge base about technology use in higher education so as to inform the debates. This requires not only more experimental research but also increased efforts to disseminate existing findings and to avoid the "reinvention of the wheel".

3.1 A suggested perspective to observe the topic

It is offered here the "Learners of Digital Era" (hereinafter LoDE) perspective; moreover, a brief presentation of the theoretical framework adopted in the research is presented. It was built at the crossing point of four involved disciplines in the debate: Communication Sciences, Pedagogy, Sociology, and Anthropology (Rapetti and Cantoni 2013).

In order to support a cautious perspective, a "not-labelling label" seems to be necessary, namely an expression that does not provide, *a priori*,the answer to the question. Such a research attitude is aimed to seek as much neutrality and objectivity as possible. Moreover, using a brand new label, the research is free of all the (both positive and negative) implications and biases involved in other expressions.

The proposed one is the label LoDE, which would like to stress the following facets (Rapetti 2012: 191):
- The focus is on persons, so the first word refers to them.
- The perspective is anthropological-pedagogical, so the chosen word is about learning.
- Not only young people learn though ICTs in the Knowledge Society, and relevant effects on digital learners can be observed as well on adults.
- The lesson learned from "Digital natives" label, i.e.: the pervasion of digital technologies in everyday life has a great impact on learning experiences, the "digital" adjective it is not applied to people, but to the context they live in.
- The lesson learned from "generation Y" and "NetGeneration" labels, i.e.: the use of digital technologies (especially the portable ones) has impacted our way of living, it is something new in history – therefore, a new "era", but it does not mean, again, that it is a matter of age.

3.2 Characteristics attributable to LoDE

The generation of "digital learners" has been characterized by a number of traits, features, expectations, and assumptions. All of them can be taken into consideration to elaborate the LoDE perspective, according to the following criteria: reasoning about the legitimacy of the characteristics pointed out by enthusiasts and the alarms expressed by the concerned ones, subtracting/taking away all the fallacies highlighted by critics.

3.2.1 Synthesis of characteristics pointed out by enthusiasts

What works as a turning point to describe current learners is a claimed "difference": because of the permeation of ICTs since their early childhood, they are said to have developed a different way of thinking (Prensky 2001a).

In thevery famous book *Connecting to the netgeneration. What higher education professionals need to know about today's students* (Junco and Mastrodicasa 2007), young learners are defined – in their socio-cognitive attitudes – as: special, sheltered, confident, conventional, team-oriented,achieving, pressured.

And considered according to the following educational characteristics (Junco and Mastrodicasa 2007: 138–144):

- Driven to success: goal-oriented, they want high-rates and prefer to get the best mark instead of criticizing a teacher.
- Social: they love to interact in learning, fewer lectures and more discussions.
- Experiential learners: by learning with PCs, they developed the trial-and-error way of thinking, dislike instructional manuals and like to gather information surfing freely on the internet.
- Multitasking: because of the "development of their learning processes using such technologies, they developed cognitive processing styles that can be described as 'hypertext' in nature".

Other authors offer the following description of the Net Generation (Oblinger and Oblinger 2005):

- They are intuitive visual communicators.
- They are better able to integrate visual spatial skills (possibly because of computer games).
- They learn better by discovery than being told.
- They can shift their attention easily from one thing to another.
- They have a fast response time and demand fast turnaround time as well.

These differences as described above become important to facilitation as they impact the learning styles of the millennials. Some of the important learning styles presented are:

- They prefer to work in teams.
- They are achievement oriented and like structure as opposed to ambiguity.
- They like interactivity and a rapid pace. They may need to be encouraged to stop and reflect.
- They are more comfortable with visuals than with text.
- They like to be involved in community activities and believe they can make a difference, especially using science and technology.

The points by Junco and Mastrodicasa are close to the ones articulated by Oblinger and Oblinger, so it's possible to affirm that the points above constitute the core of the educational differences indicated by enthusiasts (see also: Holt 2009).

3.2.2 Synthesis of alarms expressed by concerned ones

The negative effects of digital life inspired many worried voices; most relevant concerns related to learning experiences are summarised in the following list:

- No time management in using digital devices.
- Serious problem of concentration.
- Ignorance-enhancement, when processes of learning and knowledge development are misled with information retrieval ability.
- Net-addiction, PC-addiction, mobile-addiction (etc.).
- Loss of sense of reality and disconnectedness from the educative context.
- Brain laziness.

3.2.3 Synthesis of fallacies highlighted by critics

The last list informing LoDE perspective presents two important aspects: the fallacies in the assumptions about a digitalized generation of learners, and the research warnings highlighted by critics:
- Do not provide a discourse, which replicates the divide and the oppositions between generations.
- There is not enough scientific evidence to state a different brain functioning.
- Do not approach the topic with determinism or recipe-style.
- Do not offer worldwide generalisation, especially if moving from the observation of local or regional trends.
- Do not take for unquestionable the stereotype of tech-savvy young people, since often they use extensively ICTs, but their technological skills seem to be rather limited.
- Do not rely in terms of scientific evidence on the main claims of the enthusiasts, when they are just speculative or anecdotal.
- Consider that skills developed in informal experiences of learning (e.g.. online social networking) do not transfer easily to formal contexts, like schools and universities.

3.3 A suggested theoretical framework to set the perspective on the topic

Aiming to order and frame the subject, the suggested theoretical framework is enriched by the contribution of four different disciplines, namely Communication Sciences, Pedagogy, Sociology, and Anthropology. Such interdisciplinary theoretical framework is meant to overcome a *naïf* view of the discourse: the goal is to understand why and how the diffuseness of digital is affecting learners' everyday life.

For each discipline one (or more) authors and/or approach have been chosen and, in the following text, some essential key-messages coming from the chosen references are briefly summarised.

3.3.1 The contribution of Communication Sciences

Three voices inspire the suggested theoretical framework, namely: the concept of "second orality" by Ong; the Adoption theory proposed by Rogers, enriched by the idea of media appropriation; the NewMinE Lab's triangle of instructional design.

In details, first, we can recognize that the existence of a new form of communication requires to develop new forms of didactics, able to respond to new dynamics in communicating. The objective is not to substitute, rather to integrate, the classical concepts of orality and literacy (Ong 1982).

In addition, it has to be considered the rate of adoption of a certain technology, individually analysing the different communities of potential adopters. It is necessary not to superimpose the diffusion of innovations with the speed of development of human habits (especially in education) (Rogers. 1995). Furthermore, it has been studied that adoption is a social phenomenon but what influences communication and learning styles are the individual processes of appropriation (MEDIAPPRO 2006).

Finally, all the above must be framed into an approach that strikes a right balance among the three vertices of the triangle of instructional design, that are (i) persons, (ii) strategies & tools, (iii) contents & teaching/learning goals. In light of that, we should refuse the faith in means and every deterministic approach (Cantoni et al. 2007).

3.3.2 The contribution of Pedagogy

From the field of Pedagogy, the approach proposed by Media Education, the work of Vygotskij, and some of the lessons learned by constructivism are useful to inform this framework.

The approach of media education (McLuhan 1964) is a useful way to overcome the opposition between education and communication mediated by technologies, and it can inspire efficient pedagogical strategies (Bevort & deSmedt 1999; Rivoltella 2001). The digital context of learning can be understood thanks to the concept of "media convergence" (Rivoltella 2006); this must also become a focus in didactic.To explore effective media usages and their relevance in learning, one can exploit tools like media diet diaries (Ridehout et al. 2010).

Second, the concept of Zone of Proximal Development is a main reference for what concerns eLearning and digital pedagogy (Schmoelz 2010); learners are, nowadays, in condition to receive significant support to their performance, thanks to ICTs (Vygotskij 1978).

3.3.3 The contribution of Sociology

It is necessary to consider which modifications in society are registered. In particular, two social phenomena must be taken into consideration: the development of networks, permeating any aspect of relationship, knowledge and power; and the existence of digital divides, splitting and promoting the digital-adapted over the others.

The reflections suggested by Castells (2000) put in light that, in the digital environment, the learner has to learn how the "network logic" works. This is both a physical and a theoretical concept behind the organization of knowledge and power in the knowledge society, and it is not a neutral approach to knowledge (who is more able to retrieve/create nodes, is the more powerful one).

Consequently, it emerges the issue of the so-called digital divide. In the net, only the ones who are connected do exist. Educational processes have to be implemented in order to fight any kind of digital unplugging and divide. The educational experiences of technologies have to offer to the eLearner the appropriate set of skills to face job market and, even more important, to exercise the citizenship (Marshall, Kinuthia & Taylor 2009).

3.3.4 The contribution of Anthropology

To complete the framework, in order to develop an idea of education that respects the anthropological condition of learners as persons, also contributions from Anthropology are to be adopted. In particular, suggestions coming from the work of Maritain Baumann.

First, humans cannot be understood within simplistic, deterministic or instrumentalistic visions (Rosmini 1964). Even if they experience the condition of *homo technologicus*, their anthropological freedom must guide the concept we have of them (Maritain 1973).

Therefore, we have to be aware that "liquidity" fragments the social action and reduces the integrity and identity of humans. Technologies, which "pack in boxes" interpersonal communication, can result in alienation. The *homo consumens* no longer recognizes her/himself for what s/he is, but for what s/he has or would have; hyper technologization of learning can reduce a person to a consumer of digital information (Bauman 2007).

4 Does age matters? Data and reflections

From the review of theliterature, a very complex set of issues has emerged; the most controversial to be explored here is "does age matter?".

Intuitions (and the "enthusiasts") would say yes, but the reality is more complex and scattered than this; and also authors' experience leads to cautious considerations. For example, from a research conducted at the higher education level (Rapetti and Cantoni 2012), non-univocal results emerged. LoDE, certainly, are massive users of ICTs in everyday life, consequently also in their learning practices. Nonetheless, data do not depict current university learners as particularly skilled exploiters of new technologies.

Furthermore, the concept of "media convergence" (Rivoltella 2006) appears to be the most consistent explanation of reality, that is: learners adopt in studies the tools they already manage in their private life, and compose a learning-environment suitable to their needs. Results of a research conducted at the Open Campus of the University of the West Indies (Learners' Voices at UWIOC) confirm this attitude of "media convergence" (Rapetti and Marshall 2010: 125).

4.1 Some results concerning the age variable

In this subparagraph results concerning the age variable – coming from the research "Learners' voices @ USI-SUPSI" – will be shown. The research was run at the university institutions of Ticino (Switzerland): Università della Svizzeraitaliana (USI) and Scuola Universitaria Professionale della Svizzeraitaliana (SUPSI), and involved 562 respondents. It was run between 2009 and 2011, and designed to combine a quantitative phase with a qualitative (quasi ethnographic) one. The adopted questionnaire was structured in 25 questions, organized as follows: socio-demographic data (age, gender, course, country of origin); owned digital technologies; access to the internet; online activities and frequency of usage; most used applications; the role of ICTs everyday life; learning preferences (in general and concerning ICTs). The mean age of respondents were 24.5 years (median: 23; the age ranging from 17 to 75 years). Respondents were divided into three "age classes": 17 to 23 years (58.5 % of the sample), from 24 to 29 (28.1 %), and 30 and over (13.3 %). This is primarily aiming to highlight any possible differences between LoDE belonging to Gen Y – namely, born after 1980 – and the others, who had in 2009 more than 30 years. Furthermore, it was interesting to offer a further comparison within the Gen Y itself (for more details see E. Rapetti PhD thesis, free readable at http://amala.rero.ch/record/30474).

Crosstabs procedure was run for all possible crossings between items about ICTs and learning and "age classes".

In order to verify any statistical influence, 81 tabs were analysed applying Pearson's Chi-Square to check the assumed relationship; while to determine its nature Cramer's V value (converted in %) was used. Such a procedure makes it possible to answer the question: *does Age make any statistically relevant difference?* Out of 81 crossings, Pearson's values resulted significant in 8 cases, meaning "age classes"[2] variable was proofed to have a statistical influence. Nonetheless, this influence is interesting only in two cases (highlighted in bold in the following tables), and quite counter-intuitive, if we consider the "digital natives" *vulgata*.

Tab. 1: Crosstabs' synthetic results "age classes" question 4.4.

The fact of being older ...	
... increases of ...	**... the likelihood to consider that ICTs improved significantly ...**
0,8 %	"the way you practice your hobbies or interests"
0,5 %	"the way you do your students' tasks"
0,1 %	"the way you learn"
3,9 %	"the way you collaborate with your peers"

Tab. 2: Crosstabs' synthetic results "age classes" question 5.2.

The fact of being older ...	
... increases of ...	**... the likelihood to be more in favour of ...**
0,2 %	"lectures in classroom"
0,6 %	"printed dictionary/encyclopaedia"
0,3 %	"online platforms (eLearning)"

Tab. 3: Crosstabs' synthetic results "age classes" question 8.1.

The fact of being older ...	
... increases of ...	**... the likelihood to answer that ...**
4,0 %	"It would be good if there was more eLearning in my courses"

It must be underlined that, in particular, age-issues – and socio-demographic variables in general – do not provide unidirectional or resolving interpretations, despite they are the leading point behind the idea of a generation of digital learners. High familiarity with, and massive adoption of ICTs seem to be the unique universal traits attributable to LoDE. ICTs' familiarity explains LoDE learning behaviours more accurately than age variableis, as if to say that the massive exposure to digital devices must be considered like an environmental factor and not a direct cause (neither an effect) for ICTs predilection in learning.

The image emerging from such results suggests that LoDE do prefer a quite rich learning diet, encompassing both face to face, established media and new media; only encyclopaedias and dictionaries appear to be outdated, and clearly substituted by their online counterparts, which play a major role for (quick) information search and retrieval. A very little preference for educational usages of social networks suggests, moreover, that LoDE are not ready/interested in adopting such applications for their learning at the university, maybe keeping them just for informal learning (Livingstone 2001; Madge et al. 2009).

In conclusion, such results suggest to deepen studies about the media convergence issue. Learners of digital era seem to be much more complex than dominant/ *à la page* descriptions about them. They are not ICTs-addicted, neither techno-luddites (Cantoni and Tardini 2010), rather, they arrange the best learning environment, adopting new and old media, in order to fit and respond to their educational needs and interests (Rapetti and Cantoni 2012).

In synthesis, within the studied community of learners, age does not matter at all, or explains very little, when it comes to preferences and beliefs connected with ICTs in learning, no gap exists between younger generations and their 30+ colleagues.

Evidently, such data do not explain the problem in all its facets, but put in light the weaknesses of adopting the "generational divide" theory. Converging results are readable also in other recent publications (Romero et al. 2013; Bullen et al. 2012), observing and comparing this issue in international contexts.

5 Some research implications

If we come back to the origin of the idea of a *generation of digital learners*, we substantially have to deal with the bigger educators' problem: how to engage learners in *boring* teaching, considered that they have an *exciting* "digital" life. At a first look, the greatest difference between a teacher and a student is the familiarity with new media; consequently, the first upcoming solution can not help to be "let's put some ICTs in education; let's go in internet! ICTs are everywhere, except in school ...". Unfortunately (or luckily), technology in itself is not the solution to didactical troubles.

What has been presented in this chapter depicts a description of LoDEthat allows to infer also some useful points, which might be of interest also for educational practitioners and instructional designers.

(i) Dealing with the dichotomy "education and ICTs", we must not confuse: educational needs, didactical constraints, vogues of innovation, and educational-anthropological implications.
(ii) ICTs have to be integrated in education because needed; and not because students ask for them, teachers hope to solve methodological troubles, or policy officers are following some technological trends.
(iii) Pedagogy, communication processes, and technological development proceed on different paths, at different rhythms.
(iv) One thing is the technical competence, another thing is the media competence, which implies the ability of managing information to produce knowledge.

(v) Anthropological “horizons of meaning” (Degani and Mozzanica 2009) are not simply reachable through methodological turns.
(vi) The challenge is to educate critical thinking so to be able to find more information than the teacher (who is *magis-ter*, in Latin: “three times more”). The learner can be media competent, the educator has to be pedagogically competent.

The above list opens to three great issues, which need to be faced within the broader discourse about the pedagogy for the 21st century: the problem of “digital illiteracy”; the learning transfer from informal to formal experiences; the skill transfer from individually massive ICTs’ usages to educational contexts.

The problem of “digital illiteracy”(Gavin 2011) represents – probably – the biggest question about education in the 21st century: nowadays learners are, willy-nilly, eLearners; universal schooling – pushed by a society structured around the industrial revolution of 19th century – was conceived to universalise literacy and basic numeracy, which are either contents and methods considered to be fundamental; so, given that times are changed, the question is whether the notion of literacy has to be enlarged to digital. The perception is that digital development is driving too much our pedagogical horizons, while it would be better being “inspired by technology, driven by pedagogy” (OECD 2010b), embracing a systemic approach.The misalignment between the time of innovation in learning and the time of diffusion of ICTs (Rogers 1995) makes it possible a reality in which, at the same time, someone claims a tool or a facility – for instance Twitter – isalready dead (Scheper 2010), while someone else calls for a didactic approach to History by tweets (Burchia 2012). While “digital detoxing” is becoming an issue (Sieberg 2011), governments are discussing whether to create a “school 2.0” accepting the Negroponte prophecy of one-laptop-per-student (Avvisati et al.2013). Digital technologies have to become part of teaching and learning experiences; but the educational-psychological-anthropological stages of development come first.

Connected to the above issues, there is the reflection about how to make it possible the learning transfer from everything individuals can learn in informal experiences with media. The theme is much less theoretical, and it concerns the need to capitalise a big amount of information; hopefully transforming it in knowledge. But the difficulty regards the selection process. The power of media is intrinsic to their characteristics: for instance, Facebook (or any other social network) is great to make people share their vacation’s pictures because it is especially aimed to that scope, and – not at all irrelevant – people do like to share their vacations’ pictures; when teachers try to use social networking for learning purposes they struggle with the fact that such online facilities are not conceived to make students share their mathematic homework, and – clearly – this latter is very much less attractive ... According to Schulmeister, the turning point is dealing with motivation and behaviours:

> In deconstructing media use, one of the most important arguments is that doing something similar does not mean that the actors share the same intention. Therefore it is necessary [...] a step to deconstruct the motives for computer and Internet use by uncovering observable behaviour or asking subjects for their intention (Schulmeister 2010: 43).

A further open question is, somehow, the other facet of the just discussed theme: the skill transfer from individually massive ICTs' usages to educational contexts. The issue behind such idea is how to capitalise the eLearners' experience for what concern their repeated actions with digital tools, and how to transfer this skill into formal contexts. For instance, the speediness in retrieving information and the mental protocol people put in place to get the searched data. All in all, such skill transfer is harder to be evaluated. Additionally, it has been remarked that:

> A transfer of competences acquired by using media to learning does not seem to take place or at least not in the expected amount and direction. The use of computers and other media for learning is unemotionally taken as a means to an end. [... and we must take into account]many of the down sides of these technologies – the undemocratic tendencies of many online 'communities', the limited nature of much so-called digital learning and the grinding tedium of much technologically-driven work. It also tends to romanticize young people, offering a wholly positive view of their critical intelligence and social responsibility (Schulmeister 2010: 55).

6 Conclusions: outlooks and perspectives for the future

Although an increasing percentage of young people can be said to be adept in technology, it is misleading to assume that all of them fit equally well into the image of digital natives (Valtonen et al. 2010). As it is the case with learning styles, there are different student profiles regarding technology adoption and use, and in many respects clear digital divides still exist. The use of concepts such as that of "digital natives" can be helpful in so far as such concepts evoke a clear and powerful image, but misleading if they are used as clichés or stereotypes. For the purposes of improving teaching and learning in formal education, it is the diversity of students practices, preferences, and situations in relation to technology, that matters the most.

Second, there is not enough empirical evidence yet to support the idea that students' use of technology and digital media is transforming the way in which they learn, their social values and lifestyles, and, finally, their expectations about teaching and learning. In particular, students' attitudes towards technology use in teaching and learning appear to be far from what many would guess to be the dominant patterns that would emerge. Rather, students tend to be more reluctant in this respect than the image of the new millennium learner might suggest. Most of them do not want technology to bring a radical transformation in teaching and learning but would like to benefit more from their added convenience and increased productivity gains in academic work. If those gains do not become apparent to students, then reluctance emerges.

Inevitably, a couple of questions emerge: What will the future bring? How should education institutions prepare for that future?

If anything is clear is that technology will continue to evolve as fast as it has been doing in the past decade, if not faster. Digital devices that are considered to be indispensable by today's higher education students were not accessible to a majority of them only five years ago, or even less. As a number of reports outline (Johnson et al. 2009), the future will bring also new applications and environments that may have, once again, an impact on the way young people communicate, get entertained, socialize, and deal with their coursework.

It is unclear however if and how the new technology developments will impact students' learning expectations and demands.

Until now, educational institutions have done a lot to support technology adoption in teaching, with important investments in infrastructure as well as in services both for students and teachers. They have to keep up with emerging technology developments, equipments and applications, and contribute to support innovations intended to explore the value and possible benefits of adoption for teaching.

Yet, there is a need to invest more in empirical research to elucidate in which ways technology can provide more than convenience and productivity, in particular learning benefits either by providing a more rewarding experience or better learning outcomes, or both at the same time. As Dede (2007) has outlined, one starting point for fruitfully locating technology in higher education pedagogy is to observe how students are using technology in other aspects of their lives, "sifting out the dross of behaviours adopted just because they are novel and stylish from the core of transformational approaches to creating, sharing, and mastering knowledge" (Dede 2007: 4). What is at least as important as the research effort is the ability to share the results in fora where they can be translated into recommendations for better practice. This should not be an individual task, but a commitment of the whole educational community.

References

Akande, Benjamin Ola. 2008. The IPOD generation. *Diverse: Issues in Higher Educatio* 25(15). 20.

Avvisati, Francesco, Sara Hennessy, Robert B. Kozma & Stéphan Vincent-Lancrin. 2013. *Review of the Italian Strategy for Digital Schools* (No. 90). OECD Publishing.

Bauerlein, Mark. 2008. *The dumbest generation: How the digital age stupefies young Americans and jeopardizes our future (or, don't trust anyone under 30)*. London: Penguin Books.

Baumann, Zygmunt. 2007. *Homo consumens. lo sciame inquieto dei consumatori e la miseria degli esclusi.* Trento: Centro Studi Erickson.

Bevort, Eveline & Thierry DeSmedt. 1999. *Évaluation des pratiques en éducation aux medias, leurs effets sur les enseignants et les élèves*. Paris: Editions du CLEMI.

Bullen, Mark, Tannis Morgan & Adnan Qayyum. 2011. Digital learners in higher education: Generation is not the issue. *Canadian Journal of Learning and Technology / La revue canadienne de l'apprentissage et de la technologie* 37(1).

Bullen, Mark, Albert Sangrà Morer, Montse Guitert Catasus & Marc Romero Carbonel. 2012. Digital learners at the Open University of Catalonia: a skeptical view of the phenomenon of the net generation. In *Conference proceedings of "eLearning and Software for Education"* No. 01. 294–299.

Burchia, Elmar. 2012. *Ora twitter sale in cattedra. Alle elementari si insegna con la lingua da 140 caratteri.* Retrieved from: http://www.corriere.it/cronache/12_gennaio_09/tweetscuola_7ad1cd8c-3ac3-11e1-8a43-34573d1838c1.shtml (Accessed 14 January 2014).

Calvani, Antonio & Mario Rotta. 1999. *Comunicazione e apprendimento in internet. Didattica costruttivista in rete*. Trento: Erickson.

Cantoni, Lorenzo, Luca Botturi, Chiara Succi & NewMinE Lab. 2007. *eLearning capire, progettare, comunicare*. Milano: FrancoAngeli.

Cantoni, Lorenzo, Emmanuele Rapetti & Stefano Tardini. 2010. Generation Y and "glocal" working.In Barbara Bertagni, Michele La Rosa & Fernando Salvetti (eds.), *"Glocal" working, living and working across the world with cultural intelligence*, 252–272. Milano: FrancoAngeli.

Cantoni, Lorenzo & Stefano Tardini. 2010. Generation Y, digital learners, and other dangerous things. In *QWERTY Journal of Technology, Culture, and Education* 5(2) [Special Issue on "Generation Y, Digital Learners and Other Dangerous Things"]. 11–25.

Casati, Roberto. 2013. *Contro il colonialismo digitale.* Roma: Laterza

Castells, Manuel. 2000. *The information age. The rise of the network society,* 2nd edn. Oxford: Blackwell Publishers Ltd.

Dede, Chris. 2007. *Foreword.* In Gail Salaway, Judy B. Caruso & Mark R. Nelson (eds.), *The ECAR Study of Undergraduate Students and Information Technology*, 43–62. Boulder, CO: Educause.

Degani, Luca & Raffaele Mozzanica. 2009. *Integrazione socio-sanitaria. Le ragioni, le regioni, gli interventi.* Rimini: Maggioli Editore.

Gavin, Dudeney. 2011. No place in class for digital illiterate. *The Guardian Weekly,* 6 December. Retrieved from: http://www.guardian.co.uk/education/2011/dec/06/teaching-digital-literacy (Accessed 11 April 2014).

Ghaith, Ghazi. 2010. An exploratory study of the achievement of the twenty-first century skills in higher education. *Education + Training* 52(6/7). 489–498.

Holt, Lori. 2009. Educating the net generation – article review. *IThreads,* (Fall 2009)

Howe, Neil & William Strauss. 2000. *Millennials rising. The next great generation.* New York: Vintage Books.

Johnson, Larry, Alan Levine & Rachel Smith. 2009. *The 2009 Horizon Report.* Austin, Texas: The New Media Consortium.

Joiner, Richard, Jeff Gavin, Mark Brosnan, John Cromby, Helen Gregory, Jane Guiller, Pam Maras & Amy Moon. 2013. Comparing First and Second Generation Digital Natives' Internet Use, Internet Anxiety, and Internet Identification. *Cyberpsychology, Behavior, and Social Networking* 16(7). 549–552.

Junco, Reinol & Jeanna Mastrodicasa. 2007.*Connecting to the net.generation what higher education professionals need to know about today's students.* Washington, DC: NASPA, National Association of Student Personnel Administrators – Student Affairs Administrators in Higher Education.

Kubiatko, Milan. 2013. The Comparison of Different Age Groups on the Attitudes toward and the Use of ICT. *Educational sciences: Theory and Practice* 13(2). 1263–1272.

Kurt, Adile Aşkım,Selim Gunuc & Mehmet Ersoy. 2013. The current state of digitalization: Digital Native, Digital Immigrant and Digital Settler. *Ankara University, Journal of Faculty of Educational Sciences* 46(1). 1–22.

Lapa, Tiago & Gustavo Cardoso. 2013. What "Digital Divide" between Generations? A Cross-National Analysis Using Data from the World Internet Project. In Constantine Stephanidis and Margherita Antona (eds), *Universal Access in Human-Computer Interaction. User and Context Diversity*, 113–122. Springer Berlin Heidelberg.

Livingstone, David W. 2001. *Adults' informal learning: Definitions, findings, gaps, and future research*. NALL working paper #21.

Madge, Clare, Julia Meek, Jane Wellens & Tristram Hooley. 2009. *"Facebook," social integration and informal learning at university: "it is more for socialising and talking to friends about workthan for actually doing work"*. Learning, Media and Technology 34(2) [Special issue: Learning and social software – researching the realities]. 141–155.

Margaryan, Anoush, Allison Littlejohn & Gabrielle Vojt. 2011. Are digital natives a myth or reality? University students' use of digital technologies. *Computers & Education* 56(2). 429–440.

Maritain, Jacques. 1973. *Integral humanism*. Notre Dame, IN: University of Notre Dame Press.

Marshall, Stewart, Wanjira Kinuthia & Wallace Taylor. 2009. *Bridging the knowledge divide educational technology for development*. Charlotte, NC: Information Age Pub.

McLuhan, Marshall. 1964. *Understanding Media*. New York: Signet Books.

MEDIAPPRO. 2006. *An European research project. The appropriation of new media by youth*. Bruxelles: European Commission.

Oblinger, Diana & James Oblinger, 2005. Is It Age or IT: First Steps Toward Understanding the Net Generation. In Diana Oblinger & James Oblinger (eds.), *Educating the net generation*. www.educause.edu/educatingthenetgen/ (Accessed 11 April 2014).

OECD. 2010a. Foreword. In *Are the new millennium learners making the grade? Technology use and educational performance in PISA*. 3–4. Paris: OECD

OECD-CERI. 2010b. *Inspired by technology, driven by pedagogy*. Paris: OECD

OECD-CERI. 2012. *Connected minds: Technology and today's learners*. Paris: OECD-CERI.

Oh, Eunjung & Thomas C. Reeves. 2014. Generational Differences and the Integration of Technology in Learning, Instruction, and Performance. In J. Michael Spector, M. David Merrill, Jan Elen, M. J. Bishop (eds), *Handbook of Research on Educational Communications and Technology*, 819–828. Springer New York.

Ong, Walter J. 1982. *Orality and literacy. The technologizing of the word*. London, New York: Methuen.

Papert, Seymour. 1993. *The children's machine. Rethinking school in the age of the computer*. New York: Basic Books.

Pedró, Francesc. 2007. The new millennium learners. Challenging our views on digital technologies and learning. In *Nordic Journal of Digital Literacy* 2(4). 244–264.

Pedró, Francesc. 2009a. The new millennium learners. Main findings. Paper presented at the *New Millennium Learners Conference*, Bruxelles.

Pedró, Francesc. 2009b. New millennium learners in higher education: Evidence and policy implications. *Paris: Centre for Educational Research and Innovation (CERI)*. http://www.pgce.soton.ac.uk/ict/NewPGCE/PDFs10/NML-in-Higher-Education.pdf (accessed 14 April 2014).

Prensky, Marc. 2001a. Digital natives, digital immigrants part 1. *On the Horizon* 9(5). 1–6.

Prensky, Marc. 2001b. Digital natives, digital immigrants, part II: Do they really think differently?. *On the Horizon* 9(6). 1–9

Rapetti, Emanuele. 2012. *LoDE – Learners of Digital Era From the analysis of the current debate to an empirical study emphasising the learners' perspective*. Phd Dissertation. Università della Svizzera italiana, Lugano. Open accessat http://doc.rero.ch/record/30474.

Rapetti, Emanuele & Lorenzo Cantoni. 2010. "Digital Natives" and learning with the ICTs. The "GenY @ work" research in Ticino, Switzerland. *Journal of e-Learning and Knowledge Society* 6(1). 39 –49

Rapetti, Emanuele & Lorenzo Cantoni. 2012. Reconsidering "gen Y" & co. From minding the gap to overcoming it. Paper presented at *Open Learning Generations: Closing the Gap from Generation 'Y' to the Mature Lifelong Learners. EDEN 2012 Conference Proceedings,* Oporto. 6–9 June 2012.

Rapetti, Emanuele & Lorenzo Cantoni. 2013. Learners of digital era between data evidence and intuitions. *Proceedings of the international ATEE conference "Learning & teaching with Media & Technology"*, Genua, Italy, 7–9 March 2013.

Rapetti, Emanuele & Stewart Marshall. 2010. Getting over the 'generation Y' perspective. Observing ICTs in learners' experiences around the world. *QWERTY Journal of Technology, Culture, and Education* 5(2) [Special Issue: Generation Y, Digital Learners and Other Dangerous Things]. 58–85.

Rapetti, Emanuele, Sara Vannini & Anna Picco. 2011. Is mobile learning a resource in higher education? Data-evidence from an empirical research in Ticino (Switzerland). *Je-LKs, Journal of eLearning and Knowledge Society* 7(2). 47–57.

Renzi, Lorenzo. 2012. *Come cambia la lingua. L'italiano in movimento.* Bologna: Il Mulino

Rivoltella, Pier Cesare. 2001. *Media education modelli, esperienze, profilo disciplinare.* Roma: Carocci.

Rivoltella, Pier Cesare. 2006. *Screen generation, gli adolescenti e le prospettive dell'educazione nell'età dei media digitali.* Milano: Vita & Pensiero.

Rogers, Everett M. 1995. *Diffusion of innovations* (4th Ed.). New York: Free Press.

Romero, Marc, Montse Guitert, Albert Sangrà & Mark Bullen. 2013. Do UOC students fit in the Net Generation profile? An approach to their habits in ICT use. *The International Review Of Research In Open And Distance Learning* 14(3). 158–181.

Rosmini, Antonio. 1964. *Sull'unità dell'educazione.* Brescia: La Scuola.

Scheper, Scott. 2010. *How to get focused.* Sarasota (FL): First Edition

Schmoelz, Alexander, AndreaPayrhuber & Daniela Hahn. 2010. Instructional design and eDidactic. A student-centered approach for first-year students. *Proceedings of World Conference on Educational Multimedia, Hypermedia and Telecommunications 2010*, 374–379. Chesapeake, VA: AACE.

Schulmeister, Rolf. 2010. Deconstructing the media use of the net generation. In *QWERTY Journal of Technology, Culture, and Education* 5(2). [Special Issue: Generation Y, Digital Learners and Other Dangerous Things]. 24–57.

Selwyn, Neil. 2009. The digital native – myth and reality. *Aslib Proceedings* 61(4). 364–379.

Sieberg, Daniel. 2011. *The Digital Diet: the 4-step plan to break your tech addiction and regain balance in your life.* Souvenir Press

Scheper, Scott. 2010. *How to get focused* (3th Ed.). Faction.

Skiba, Diane J. 2010. Digital wisdom: a necessary faculty competency?. *Nursing education perspectives* 31(4). 251–253.

Small, Gari W. & Gigi Vorgan. 2008. *iBrain. Surviving the technological alteration of the modern mind* (1st Ed.). New York: Collins Living.

Strauss, William & Neil Howe. 1991. *Generations. The history of America's future, 1584 to 2069* (1 Quill Ed.). New York: Quill.

Succi, Chiara & Lorenzo Cantoni. 2008. A map of eLearning acceptance (MeLA) and a corporate eLearning readiness index (CeLeRI). *International Journal of Advanced Corporate Learning (iJAC)* 1(1). 39–47.

Tapscott, Don. 1998. *Growing up digital. The rise of the net generation.* New York, London: McGraw-Hill.

Tapscott, Don. 2009. *Grown up digital. How the net generation is changing the world.* New York, London: McGraw-Hill.

Valtonen, Teemu, Patrick Dillon, Stina Hacklin & Pertti Väisänen. 2010. Net generation at social software: Challenging assumptions, clarifying relationships and raising implications for learning. *International Journal of Educational Research* 49(6). 210–219.

Vygotskij, Lev S. 1978. *Mind in Society: The Development of Higher Psychological Processes.* Harward: University Press

* Note of authorial responsibility: the chapter was developed in collaboration, F. Pedrò wrote "Introduction: Digital Natives, New Millennium Learners, Generation Y, what are we talking about?" and "Conclusions: outlooks and perspectives for the future", while E. Rapetti wrote all the other paragraphs.

Wenhong Chen and Rich Ling

16 Mobile media and communication

Abstract: Mobile telephony is the most widespread mediation technology in the world. With access to the wireless internet, mobile devices have expanded from a tool of voice or text-based communication to devices and services for multimedia communication, consumption and production. More importantly, not only users and their communication but also the patterns and consequences of their action and interaction become mobile across time and space. In this chapter we focus on the social ramifications of mobile media and communication. We start with a brief history of mobile media and communication, highlighting the technological affordances of the mobile internet. We review theoretical issues and report major streams of empirical research. We conclude with a discussion of research agenda with both theoretical and practical implications and call for political, economic, infrastructural, and industrial analyses of mobile communication and media that contextualize and structure user experiences.

Keywords: mobile media, mobile communication, mobile internet, social implications, civic engagement, big data, privacy

Mobile telephony is the most widespread mediation technology in the world. It is estimated that there are 96 mobile subscriptions per each 100 inhabitants on planet Earth as of 2013 (ITU 2013a). At the moment there are 6.6 billion connections to approximately 3.2 billion unique mobile phones subscribers (GSMA 2013). In the U.S., more than 90 % of adults use mobile phone and 56 % of adults are smartphone owners (Smith 2013). Since introduced by Apple Inc. in 2008, apps have become fundamental features of mobile internet devices such as smartphones or tablets. Rapid technological advancements have expanded the definition of mobile media and communication from mobile telephones to mobile devices and services that enable "mediated social connectivity while the user is in physical motion" (Campbell 2013: 9).

As mobile communication has been seamlessly merged into the rhythms of everyday life, the literature has been inherently interdisciplinary. In this chapter we focus on the social ramifications of mobile media and communication. After a brief history of mobile media and communication, especially the development of cellular telephony and the rise of the mobile internet, we review theoretical issues such as digital divides, inequalities, the mobilization thesis, and the structuring of the mobile phone into social interaction. We examine how the adoption and diffusion of mobile technologies varies by socio-demographic groups and mobile communication in the developing world. We list a few promising research areas and conclude with a discussion of research agenda with both theoretical and practical implications.

1 History of mobile media and communication

1.1 The development of cellular telephony

Early mobile automobile-based communication systems, such as the one in St. Louis in 1946, had a single antenna covering a large city. This approach has several limitations including the need for powerful transmitters and receivers and a limited ability to handle many calls simultaneously. These limitations prompted the need for cellular based communication developed by Douglas Ring and W. Rae Young in 1947 at Bell Labs (Ring 1947). In this approach, an area is divided into adjacent radio "cells" that are much smaller than the city-wide areas used in the earliest mobile systems. The cells each cover a small area and are linked with adjacent cells such that a caller will be switched between the cells as he or she moves about. In this way, the transmission/receiver equipment can be much smaller and requires less electricity since they do not have to generate a powerful signal. In addition, this approach increases the number of people in, for example, a city that can use the system. Variations of this approach are currently the basis of the mobile telephony system.

The cellular (or mobile) telephony system allowed people to call, and somewhat later to text, one another. As we will discuss below, it made people individually addressable in a way that was not previously possible (Ling and Donner 2009). As we will discuss, this has changed how we coordinate interaction (Ling and Yttri 2002), how we maintain our social sphere (Ling 2008) and how we react to emergencies (Sundsøy et al. 2012).

1.2 From the Internet to the mobile internet

Roughly parallel to the popular diffusion of the cellular telephony network since the early 1990s, we have seen the development of the Internet. In the first years of the new century, there were various attempts to merge mobile telephony with internet-like services. The most prominent of these was Wireless Application Protocol (WAP). At approximately the same time, the iMode system in Japan also afforded access to mobile internet content (Lindmark 2002). Both of these were superseded in 2007 with the development of the Apple iPhone and later with the App Store. This was followed in 2008 by the commercialization of the competing Android operating system that is used on approximately two thirds of all smartphones as of 2013.

Mobile communication devices afford multiple modes of communication (oral, visual, and text-based). Smartphones are in fact computers with mobile interactivity and programmability through pre-installed or downloaded apps (Mosemghvdlishvili and Jansz 2013). They also accommodate pervasive, portable, personalized, and instant communication. Mobile devices have become increasingly powerful, serving as cameras, location-based mapping devices, messaging platforms, social networking terminals, archives and a seemingly endless number of other applications. Accordingly, besides placing a phone call, popular mobile phone activities include taking photos, text messaging, taking and watching video, playing games, accessing the Internet, emailing, downloading and listening to music, providing us with location based information (Humphreys 2007) and a seemingly endless number of other functions. As the tool of choice when users conduct many aspects of their daily lives, indeed, the mobile phone has become a hub and a symbol of a digital lifestyle (Horrigan 2009). Beyond traditional mobile telephones we are also seeing the diffusion of other mobile mediation devices. These include tablets and pads that often are more oriented towards computing and information retrieval than interpersonal voice communication. In the very sense that "media are the extensions" (McLuhan 2004) of our lives, mobile devices become not just portable but increasingly wearable including watches that replicate the functions of smartphones and also head-mounted gadgets such as Google Glass.

2 Theoretical issues and empirical studies

2.1 From descriptive to theoretical analysis of mobile communication

2.1.1 Descriptive analysis and domestication theory

Earlier research on mobile communication was often descriptive and examined issues such as coordination, its role in the lives of teens, use in emergency situations and use in inappropriate situations (Ling 2004). It examined the adoption by subgroups, drawing on digital divides and digital inequalities theories on how social inequalities structure individuals' access to, use of and gain from technologies and media (Rice and Katz 2003; Selwyn 2004). The low cost, low skill barriers and aggressive marketing strategies meant that the mobile phone diffused more quickly and more completely than the PC-based internet. Indeed the mobile phone quickly reached broader socioeconomic strata ranging from "bottom of the pyramid" farmers in remote African villages to technologists in Silicon Valley (Burrell 2010).

Mobile telephony has dramatically changed the communication infrastructure in many developed as well as many developing countries. Beyond the technical transition, this shift meant that many billions of people could, for the first time, use mediated communication. The fact that it made people directly accessible was a key link for many of these analyses. A common approach to understanding individuals' adoption and use of mobile communication has been the domestication approach (Haddon 2003; Silverstone et al. 1992). This line of research has examined how the technology has been adopted and how it finds a more or less stable position in the lives of users.

2.1.2 Bounded solidarity and taken for grantedness

A line of mobile-based theorizing has suggested that mobile communication supports the social cohesion of small groups. An extension of this is that it is also becoming a socially structured part of everyday life. Early studies of mobile communication indicated that it afforded increased interaction among small groups (Licoppe 2004). Unlike other forms of electronic mediation, it has been suggested that the mobile has the ability to increase social cohesion in the intimate sphere. Ling (2008) has drawn on classical sociology to clarify this and described it as bounded solidarity. Reid and Reid (2004), Ishii (2006), Igarashi (2005), Ling, Bertel and Sundsøy (2012), and Miritello et al. (2013) have reported that the mobile phone is disproportionally used to contact a small number of people, suggesting that it is an instrument of the intimate sphere.

Taking this line of thought one step further, Ling has suggested that mobile communication in its various forms is being structured into daily life (2012). That is, it is taken for granted. We increasingly expect one another to be available via mobile phones to the degree that planning and daily logistics rely on mutual access to this form of mediation. This suggests that the system has moved beyond simply serving the utility of an individual to where it is an assumed part of the social infrastructure that is inextricably intertwined into our daily activities.

The idea of being responsible to our social ties is also being considered as it feeds into the creation of automatic (habitual) use of the phone (Bayer et al. 2013). This line of thought suggests that we develop habitual or quasi-conscious behaviors when using the mobile phone. Giving users quick, real-time access to information, smartphones encourage users to form a checking habit (Oulasvirta et al. 2012). Just as a smoker develops habitual routines for getting out a cigarette and lighter as he/she goes out the door to the smoking area, we also are cued to "check our phones" in particular settings. Interestingly, use of the mobile phone in this way is driven by the weight of routine that is enforced by our internalized sense of reciprocal social expectations.

As mobile phones are so interwoven into the everyday life, people often take them for granted (Ling 2012). Following Vincent (2006), users are physically and psychologically attached to their mobile phones as a symbol of "me, my mobile and my identity" (2006: 42). The use of mobile phones may trigger changes in users' perception of the self as more networked and their perception of time as softening and fluid in a form of "present extensive", which "feeds a postmodern mentality" (García-Montes et al. 2006: 64; Ling 2004).

2.1.3 Apparatgeist

Offering a balanced view between technological determinism and social influence, Katz and Aakhus (2002) developed the theory of apparatgeist to capture "the spirit of the machine that influences both the designs of the technology as well as the initial and subsequent significance accorded them by users, non-users and anti-users" (2002: 305). They see technologies as "historically bound", which provide "a flexible menu of extensive, but not infinite, choices." (2002: 307). Early studies have shown certain levels of conformity and uniformity of mobile phone usage patterns across cultures (Campbell 2007; Katz and Aakhus 2002). More recently, interviews comparing American and German college students identify limited cross-cultural differences in mobile internet use (Humphreys et al. 2013).

2.1.4 The mobilization thesis

Thanks to the wider adoption and low barriers of use, mobile technologies have been considered as providing a more accessible and affordable venue for narrowing the digital divides than PC and PC-based internet (Chaudhuri 2012; Donner 2009). The mobilization thesis and the normalization thesis are developed in internet studies. The mobilization thesis argues that members of underprivileged social groups would be empowered as they gain access to previously unavailable technologies (Norris and Inglehart 2009), while the normalization thesis emphasizes that technologies primarily benefit those who are better off or already active (Hirzalla, van Zoonen and de Ridder 2010). For instance, research by Chen and Givens (2013) shows that seeking donations through mobile phones taps into publics that have not traditionally been engaged in donating and thus helps to bridge the significant age, race and socioeconomic status (SES) gaps in offline and internet donation.

2.2 The adoption and diffusion of mobile technologies

Mobile communication has been adopted by different groups for a variety of purposes. The first users were often business people and the extremely wealthy and privileged (Agar 2003). As the system developed, mobile telephony was used in commercial situations such as sales and distribution where the individual was away from a fixed location and the communication channel allowed for real-time microcoordination (Ling and Yttri 2002; Ling 2012). As prices fell and as the technology became more robust, it made the transition more squarely into the private sphere. Just as in commercial settings, use within the home was often focused on coordinating the ebb and flow of daily life. It is at this point that teens became users. As with older users, the fact that the device gave them control over an individualized communication channel meant that it was a vast improvement over the geographically tied landline system.

Beyond basic mobile phones, the web-enabled smartphones have been the next wave of development. Pew data suggest that in the U.S., men, people under 35, those who are better educated and urban and suburban residents are more likely to be smartphone owners (Smith 2013). There has also been a divide when considering app use along the familiar lines of social inequalities: Young, better educated, more affluent, urban or suburban residents are more likely to download apps to their smartphones or tablets, although the gender gap has been narrowed (Duggan and Rainie 2012). Yet, mobile divides are not simply mirroring internet divides, especially in terms of race and ethnicity. In what follows, we mainly discuss the variations and differences in mobile access and use by race, age, gender and socioeconomic status.

2.2.1 Mobile use among different groups

Considering the U.S., there are some groups that are more mobile "savvy" than others. As one would perhaps expect, young users lead older users in the breadth and depth of mobile phone use such as exchanging text messages (Smith 2011a; 2011b), sharing or recording video, accessing the internet, or playing music or games (Horrigan 2009; Ling et al. 2012). Young people are more likely than older people to adopt new mobile practices such as mobile donation (The mGive Foundation 2011). Young adults are also more likely to download apps to communicate with friends and family (Purcell 2011).

While racial minorities have been left behind in terms of internet access and use, African Americans and Hispanics led other racial groups in using the mobile internet (Horrigan 2009) and owning smartphones (Smith 2013). They are also more likely to pay for mobile local news and exchange more text messages than white Americans (Smith 2011a; 2011b) and use apps with which to communicate (Purcell 2011). Compared to the national average of 34 %, about half of Hispanics (47 %) and 38 % of African Americans live in a mobile-phone only household (Lopez et al. 2013). However, studies show that it is unlikely that the use of mobile internet devices would narrow inequalities in online participation and bridge digital divides in disadvantaged communities (Mossberger et al. 2012). In general, people with less economic wherewithal choose mobile access instead of investing in the relatively more expensive PCs, software and fixed-line internet access (Brown, Campbell and Ling 2011). In fact, low cost makes text message the most popular mobile phone function among low-income groups (Lin and Sun 2010). In comparison, people of high SES use mobile phone more frequently to interact with professional or personal contacts (Ji and Skoric 2013).

2.2.2 Mobile communication in the developing world

An important theme in the analysis of mobile communication has been its role in developing countries. As of 2013 there are three mobile phone subscriptions in the Global South for each one in the Global North (ITU 2013b). Not long ago the cost of establishing a copper network was prohibitively expensive. In their 1984 report Maitland et al. (1984) noted, for example, that in Tanzania there was one landline telephone for every 1060 persons in the rural areas and one for every 69 in the cities. As of 2011, there was about one mobile phone for every second person. Beyond the technical transition, this shift means that for many billions of people, they can, for the first time, use mediated communication. For instance, low-end "working-class ICTs" help to upgrade migrant workers in China from information have-nots to information have-less (Qiu 2009).

Mobile communication has resulted in major changes in the lives of the individuals, especially for women in the Global South (Ling and Horst 2011). It has facilitated more efficient logistics, engendered micro-entrepreneurship (Collings 2012; Donner and Escobari 2010; Jagun et al. 2008), affected the lives of women (GSMA 2012), and changed access to health systems (Chib and Chen 2011). At the community level, initiatives and programs have been carried out by governments as well as NGOs to draw on mobile technologies for socioeconomic development in disadvantaged communities such as promoting micro-businesses, public health, or women's rights. Indeed, an assessment of ICT for development projects across the world suggests that mobiles can better address users' needs than traditional PCs since "mobiles have emotional appeal because talking is a universal psycho-sociological propensity" (Chaudhuri 2012: 326).

2.3 Diverse mobile usage patterns

When seen somewhat abstractly the mobile can be characterized as a bundle of versatile technologies that can be used for instrumental and expressive actions such as seeking information (informational use), maintaining and developing social relations (relational or expressive use), relaxation and entertainment (recreational use), and micro coordination and logistical use (Campbell and Russo 2003; Licoppe 2004; Ling and Yttri 2002; Ling 2004).

2.3.1 Relational use

The core of relational use is in the maintenance of contacts with family and friends. Scholars have been concerned about the consequences of "absent presence" (Gergen 2008) as mobile users are actually physically in one place but socially elsewhere, interacting "side-by-side" rather than "face-to-face" with their co-present communication partners (Ito 2005: 14–15). Nonetheless, most studies have shown a positive relationship between expressive use of mobile phones and personal networks (Campbell and Russo 2003; Licoppe 2004). Some studies have investigated how mobile communication would affect users' network size and composition (de Zúñiga and Valenzuela 2011). Other work has suggested that mobile phone use is positively associated with the size and the diversity of core discussion networks in North America and that it may contribute to the pervasive awareness of one's close network contacts (Hampton et al. 2011) and the maintenance of existing social networks (McEwen 2009). Moving to the Japanese context, longitudinal analysis shows that freshmen's text message networks were built on their face-to-face networks and grew at a slower pace than face-to-face networks. Further, respondents considered the relationship with interlocutors stronger if they communicated by both face-to-face and by text message than if they only communicated via face-to-face interactions (Igarashi et al. 2005). Analysis of data from Scandinavia shows that the mobile phone was largely used to maintain strong ties (Ling et al. 2012).

2.3.2 Logistical use

Logistical mobile communication can include microcoordination or "nuanced instrumental coordination" that allows users to check out the accessibility and location of their family members, friends or colleagues and readjust meeting times and places as needed (Ling and Yttri 2002: 139). Hatching and changing plans via the mobile phone while on the go makes daily life more spontaneous and flexible. As microcoordination often involves people meeting up in public places, it increases the serendipity of meeting and talking with new people co-presented on the spot (Campbell and Kwak 2011; Hampton, Livio and Goulet 2010).

Another type of coordination activity with growing significance accompanied by the rise of smartphones is the use of location based services. These can include "check in" types of applications like Foursquare, but it can also include applications that guide users as they search for services and different types of locations

(Humphreys 2007; Sutko and De Souza e Silva 2011). Further, we can also use the mobile phone for what Bertel (2013) calls "flexible alignment". This is coordination, not just between individuals, but also between people and automatic scheduling systems. Bertel uses the idea of "flexible alignment" to examine how teens use smartphones to keep oriented with regards to shifts and changes in their daily schedules at school.

2.3.3 Mobile news distribution

Mobile devices have become an important distribution channel for news (Westlund 2011). Digital delivery of print news, in addition to radical changes in the structure of advertising, will result in fundamental changes in news production and consumption. In particular, mobile news primarily occupies interstices – small spaces or intervals of times – between users' scheduled activities (Dimmick et al. 2011). This in turn is likely to affect the way that people are informed about current issues and their ability to make choices when voting and when orienting themselves in relation to current events. As such, this area is worthy of analysis. As importantly mobile devices have been adopted as a useful tool for professional and citizen journalists, especially in conflict regions or during emergency, for news collection, transmission and publishing (Mills et al. 2012; Watkins et al. 2012).

2.3.4 Mobiles and entertainment

The mobile phone is a platform for gaming (Juul 2012) and the consumption of video and music. Mobile devices have become an increasingly important screen for media industries and advertisers for distributing content and message. Mobile media and technologies offer new venues and forms of cultural participation. It is possible to use a mobile device to view films, broadcast material and sporting events via services such as NetFlix, and apps for sporting channels and for traditional broadcast operators. On top of making cultural participation accessible anytime, anywhere, it makes the consumption of cultural content more instant, spontaneous, on-demand, interactive, participatory, and networked. Mobile apps allow cultural participants to generate and share content, showcasing their cultural appreciation and appropriation with friends and followers via Facebook, Twitter, Instagram, Tumblr, or locational-based service (i.e. Yelp, Foursquare etc.). Allowing participants to "engage with each other" and personally endorse a cultural event, such user-generated content is often encouraged or initiated by cultural organizations in the hope of engaging younger patrons from broader social strata in a time of dwindling visitors (Thompson et al. 2013).

Mobile games, especially pervasive games that integrate real world with the game world, expand spatio-temporal and social boundaries (Korhonen et al. 2008). Such mobile games can help players to engage with their local urban environment and maintain geographically distributed relationships, which can be mobilized for civic engagement or social support (Shklovski and de Souza e Silva 2013).

2.4 Implications of mobile communication

2.4.1 Interpersonal interaction, social bonds and perpetual contacts

Traditionally, the study of interpersonal interaction has been central to the study of mobile communication. Until the widespread development of the mobile internet and smartphones, mobile telephony was almost exclusively focused on people calling and texting one another on a one-to-one basis. Indeed, even after the diffusion of more advanced mobile devices, interpersonal interaction has remained one of the core functionalities. This thread of research has concentrated on how we use the mobile phone to carry out both expressive and instrumental activities. Research has examined how the device is used to facilitate logistics and how it is used to maintain expressive ties (Ling and Yttri 2002; Ling 2008).

With this foundation in place there was a focus on how mobile communication affected social structure. As noted in the discussion on bounded solidarity, one line of research examined the question of whether the mobile phone tightened or loosened social bonds. This question has a long tradition when considering other forms of mediation such as the television (Putnam 2000). Where broadcast technologies, especially TV, are often seen to wear at the social fabric of small groups, it was suggested that mobile communication provided a mediation form that allowed the cultivation of these ties regardless of physical proximity and indeed, when thinking of text, asynchronously (Licoppe 2004; Ling 2008). There is cause to question our mental "presence" in a particular setting if we are constantly interacting with non-present others via our mobile phones (Gergen 2003). Analysis also shows that our circle of mediated interlocutors is an exceedingly tight group. Indeed research indicates that the majority of our calls and texts go to only 3–5 other persons (Ling et al. 2012). Seen in this way, mobile communication, at least in its phase as a mediator of interpersonal communication, helps us to maintain our social lives. Moreover, in the case of emergency situations, this type of connectedness becomes extremely important (Cohen and Lemish 2005; Dutton and Naiona 2003; Katz and Rice 2002; Sundsøy et al. 2012).

2.4.2 Civic engagement

Another line of research in this area is the relationship between mobile phone use and civic engagement. While there are open research issues, there is work showing that decentralizing information and communication increases transparency and accountability in less democratic societies (Jun 2013). For instance, the levels of mobile phone penetration are significantly related with the levels of perceived corruption in Africa (Bailard 2009). Rheingold's work (2002) on the smart mobs and their role in toppling an unpopular president in the Philippines has inspired subsequent studies on the democratic potential of mobile devices. Comparing the People Power II movement in the Philippines and the Nosamo movement in South Korea, Qiu (2008) demonstrates that mobile communication is used to engage, compete with, or bypass traditional and digital media, which helps to speed up and scale up the formation of civil society based on existing political struggles. Research on the Occupy Movement shows that posting, sharing, or live-streaming photos or videos via mobile phones preserves and distributes eyewitness material of protest participants and police (Thorson et al. 2013).

Analysis has shown that informational and recreational mobile use contribute to civic and political participation (Campbell and Kwak 2010; Campbell and Kwak 2011b; Rojas and Puig-i-Abril 2009). Studies on the general American populations indicate that relational mobile phone use is not related to civic engagement, political participation, or talking with strangers in public place (Campbell and Kwak 2010; Campbell and Kwak 2011b). Yet, a study of college students in Asian cities reveals a positive link between relational mobile use and indicators of civic engagement such as social leisure activities, group involvement, and interpersonal trust (Chen et al. 2012).

The ubiquity of the mobile phone also makes mobilization and coordination for collective actions easier. In the context of the Haitian earthquake, more than 40 % of the donors in America to the Haiti Earthquake Relief contacted via text messages in turn encouraged their friends or family members to make a similar donation (Smith 2012). Practitioners speculate that text-to-give campaigns make donation more spontaneous and give donors instant satisfaction of contributing to a shared cause (Chen and Givens 2013; Heath 2010). Thus, further research can examine how mobile communication is used as a way to interact with voters (Campbell and Kwak 2012) and eventual donation providers (Chen and Givens 2013).

2.4.3 Mobile health

Mobile communication has been used to support and improve health delivery. This can be done allowing local healthcare workers to contact a doctor when meeting with a patient in a remote area (Chib and Chen 2011). Another dimension is using the mobile phone to develop patient administration systems to report in health information (Sanner et al. 2012). This approach speeds the collection of data and allows administrators to gain a broader overview of, for example, the spread of epidemics. Another type of mobile health is the use of the device to provide various types of interventions such as assistance in weight loss or dealing with alcohol issues (Campbell and Kelley 2006) or in disease management with, for example, diabetes (Piette et al. 2013).

Privacy, security and usability in using mobile phones remain important issues for wider adoption of mental health monitoring and management. As importantly, more evaluation research is needed to generate consistent evidence on the impacts of mHealth for health behavior changes (Gurman et al. 2012).

2.4.4 Mobile crowdsourcing

Two modes of mobile phone use include person to person interaction and also accessing information from remote servers by individuals. A third alternative is a type of "middle position" where dispersed individuals interact indirectly with one another through crowdsourced programs. It might be that this service is a type of crowdsourced utility, such as in the case of the GPS traffic program Waze, to pool information about, for example, road conditions, parking spaces, weather, or food price. Alternatively, crowdsourcing can be used to distribute the work (i.e. translation) to a wide group of eventual contract workers, as in the case of txteagle and mobile works (Eagle 2009).

3 Big data and privacy

Large-scale longitudinal behavioral data have great implications for many social science disciplines such as communication and media studies, urban planning, or public health. While many studies have been and remain relying on self-reported survey data, scholars have raised concerns on the validity of such data and found that they were only moderately related with server log data (Boase and Ling 2013; Eagle et al. 2009). Analysis of large scale log data has also provided insight into the structuring of society. Results on this count include the effect of strong and weak ties on social structure (Onnela et al. 2007), the size (Ling et al. 2012; Miritello et al. 2013) and structure (Eagle et al. 2009) of friendship networks, the regularity of mobility patterns (Song et al. 2010), and the reaction of people to emergency events (Sundsøy et al. 2012). Other applications that can be traced from mobile communication data patterns can include the analysis of social divisions (Bucicovschi et al. 2013), optimizing urban transportation systems (Berlingerio et al. 2013) and epidemiological spreading of disease (Lima et al. 2013).

The reliance on corporate social media means that users often have limited control over their data, let alone the "architectures of the spaces through which they communicate" (Poell 2013: 1). Both democratic and authoritarian governments have been actively monitoring and analyzing mobile traffic data. If mobile research on privacy used to focus on the prudence of individual users on what and what not to share, recent revelations on state surveillance and corporate monetization of mobile data adds more fuel to what Katz termed as "the perennial question", namely, "whether these technologies will fundamentally alter the practice of freedom and the level of anonymity that is the bedrock of contemporary sociopolitical life in democratic nations" (Katz 2007: 392).

4 Future research directions

The rapid evolving and diffusing mobile technologies are both a blessing and a curse for scholars. On the one hand, the literature on mobile media and communication has grown fast. On the other hand, many studies have been descriptive and tend to be blinded by a focus on the latest technology. While offering insight on how the latest technologies are appropriated is informative and interesting, the temptation of "running the research mill" can distract scholars from pushing for theoretical advancements (Sawhney 2007). In a similar vein, Hofheinz (2011) criticizes the obsession with Nextopia, i.e. focusing on lead users and the latest technologies. This is at the expense of assessing the large-scale, long-term structural forces that shape how individuals and groups use technologies for a variety of purposes and goals.

The debate is still ongoing whether mobile media and communication is an independent field or an important part of the larger field of information and communication technologies (ICTs), communication and media studies (Campbell 2013). Nonetheless, there has been a convergence with internet studies. According to Peng et al. (2012), internet study has focused on themes such as e-Health, e-Business, e-Society, and Human-Technology Interactions in internet research and relatively little research hasexamined mediated interpersonal communication as has been the case in mobile communication research.

More recently, with the diffusion of the smartphone and the mobile internet, we have seen the convergence of traditional mobile communication research and research on the internet. This development presents a challenge to researchers since cloud based services are in many cases terminal-agnostic. In this transition, a relevant research question becomes whether this matters and if so, how? In what follows, we discuss emerging research themes that have both theoretical and practical implications.

First, scholars of mobile communication and media need to shed more light on the patterns and implications of how people adopt and use mobile media and communication, especially their variations by social groups, cultural and institutional context as well as the social and psychological mechanisms that link the usage patterns and their personal and social consequences. It is important to push for more fine-tuned analysis on the specific types of mobile communication practices and their social implications (Campbell and Kwak 2010). Also, despite the dramatic growth, there has been a lack of research on apps (Goggin 2011; Mosemghvdlishvili and Jansz 2013).

Second, it is a perennial question whether mobile communication promotes or discourages diverse social networks. The adoption of the smartphone brings a new element into the mix, namely the ability to use the device as an entertainment/information terminal that eventually will diminish its function as a terminal for social interaction. On the one hand, mobile technologies enable anytime, anywhere, personalized, perpetual communication (Ling and Donner 2009) and ambient awareness of one's network contacts and physical place (Campbell 2013). For instance, users gain greater convenience and confidence by using location based service to explore new urban places (Sutko and de Souza e Silva 2011). On the other hand, mobile communication technologies enable users to withdraw from people and places to "telecocoons" (Habuchi 2005). They can distract and segment users' attention from their physical environment (Leyshon et al. 2013). An experiment suggests that the mere presence of mobile devices can negatively affect the quality of face-to-face conversation, especially when the topics of the conversation are perceived as personally meaningful (Przybylski and Weinstein 2013).

Third, the social shaping of mobile communication means that it can serve as "force multiplier" for the disadvantages and existing power holders (Katz 2007). Mobile technologies can empower the disadvantages such as female migrant workers in China and increase control from their employers (Wallis 2011) as well as facilitate illegal drug dealing and enhance law enforcement (McEwen 2011). More studies need to examine the multidirectional implications of mobile technologies as agents for social change and new practices and reinforcers of existing power relationships and hierarchies for sociability, civic engagement or other domains.

Fourth, understanding mobile communication and media in a global context includes not only the examination of usage patterns and trends but also the investigation of larger structural forces such as national culture, industrial structure and technological convergence, especially the cultural and institutional contexts that shape the interactions between and among governments, corporations, and citizens (Craig 2007; Qiu and Thompson 2007). As most studies have centered on user experiences, there has been a striking lack of political, economic, infrastructural, or industrial analysis of mobile communication and media. In a similar way, limited attention has been paid to developers, entrepreneurs, artists, or activists and how they produce, design, and use mobile media and technologies (Hestres 2013).

Fifth, studies have shown that apps can increase users' spatial, social and behavioral awareness such as time use (Runyan et al. 2013) or sleep (Shirazi et al. 2013). There are also studies pointing out how mobile use has reconfigured user activities into habit (Bayer et al. 2013). The interesting tension between the growing awareness about one's self, interpersonal and built environment and the taken-for-granted habitual use offers promising venues to explore how mobile technologies reshape human cognition and identity.

References

Agar, Jon. 2003. *Constant Touch. A Global History of the Mobile Phone*. Cambridge: Icon Books.

Bailard, Catie Snow. 2009. Mobile Phone Diffusion and Corruption in Africa. *Political Communication* 26(3). 333–353.

Bayer, Joseph, Rich Ling & Scott Campbell. 2013. Mobile Automaticity: A Theory of Mobile and Social Habits. In *Proceeding of Mobile Communicatio Pre-Conference Workshop*, London School of Economics, 16–17 June.

Berlingerio, Michele, Francesco Calabrese, Giusy Di Lorenzo, Rahul Nair, Fabio Pinelli & Marco Luca Sbodio. 2013. AllAboard: a System for Exploring Urban Mobility and Optimizing Public Transport Using Cellphone Data. In Hendrik Blockeel, Kristian Kersting, Siegfried Nijssen, Filip Železný (eds.), *Machine Learning and Knowledge Discovery in Databases*, 663–666. Springer-Verlag Berlin Heidelberg.

Bertel, Troels Fibæk. 2013. 'It's Like I Trust It so Much That I Don't Really Check Where It Is I'm Going before I Leave': Informational Uses of Smartphones Among Danish Youth. *Mobile Media & Communication* 1(3). 299–313.

Boase, Jeffrey & Rich Ling. 2013. Measuring Mobile Phone Use: Self-Report Versus Log Data. *Journal of Computer Mediated Communication* 18(4). 508–519.

Brown, Katie, Scott Campbell & Rich Ling. 2011. Mobile Phones Bridging the Digital Divide for Teens in the US? *Future Internet 2011 [Special Issue Social Transformations from the Mobile Internet]* 3(2). 144–158.

Bucicovschi, Orest, Rex W. Douglass, David A. Meyer, Megha Ram, David Rideout, Dongjin Song. 2013. Analyzing Social Divisions Using Cell Phone Data. Paper presented at the Third International Conference on the Analysis of Mobile Phone Datasets, NetMob D4D Challenge. MIT, Cambridge MA, 1–3 May.

Burrell, Jenna. 2010. Evaluating Shared Access: Social Equality and the Circulation of Mobile Phones in Rural Uganda. *Journal of Computer-Mediated Communication* 15(2). 230–250.

Campbell, Scott. 2007. Cross Cultural Comparison of Perceptions and Uses of Mobile Telephony. *New Media and Society*. 9(2). 343–363.

Campbell, Scott & Tracy C. Russo. 2003. The Social Construction of Mobile Telephony: An Application of the Social Influence Model to Perceptions and Uses of Mobile Phones Within Personal Communication Networks. Communication Monograph 70(4). 317–374.

Campbell, Scott & Nojin Kwak. 2010. Mobile Communication and Civic Life: Linking Patterns of Use to Civic and Political Engagement. *Journal of Communication* 60: 536–555.

Campbell, Scott W. 2013. Mobile Media and Communication: A New Field, or Just a New Journal?. *Mobile Media & Communication* 1(1). 8–13.

Campbell, Scott W. & Michael J. Kelley. 2006. Mobile Phone Use in AA Networks: An Exploratory Study. *Journal of Applied Communication Research* 34(2). 191–208.

Campbell, Scott W. & Nojin Kwak. 2011a. Mobile Communication and Civil Society: Linking Patterns and Places of Use to Engagement with Others in Public. *Human Communication Research* 37(2). 207–222.

Campbell, Scott W. & Nojin Kwak. 2011b. Political Involvement in 'mobilized' Society: The Interactive Relationships Among Mobile Communication, Network Characteristics, and Political Participation. *Journal of Communication* 61(6). 1005–1024.

Campbell, Scott W. & Nojin Kwak. 2012. Mobile Communication and Strong Network Ties: Shrinking or Expanding Spheres of Public Discourse?. *New Media & Society* 14(2). 262–280.
Chaudhuri, Anindya. 2012. ICT for Development: Solutions Seeking Problems. *Journal of Information Technology* 27(4). 326–338.
Chen, Katherine, Yi-Ning Lo, Y. N. Ran Wei, Xiaoge Xu & Guoliang Zhang. 2012. A Comparative Study of the Relationship Between Mobile Phone Use and Social Capital Among Asian College Students. Paper presented at the International Communication Association (ICA) Pre-conference Workshop on Mobile Communication, Community and Locative Media Practices: From the Everyday to the Revolutionary, Phoenix, AZ, 23–24 May.
Chen, Wenhong & Travis Givens, T. 2013. Mobile Donation in America. *Mobile Media & Communication* 1(2). 196–212.
Chib, Arul & Vivien Hsueh-Hua Chen. 2011. Midwives with Mobiles: A Dialectical Perspective on Gender Arising from Technology Introduction in Rural Indonesia. *New Media & Society* 13(3). 486–501.
Cohen, Akiba A. & Dafna Lemish. 2005. When Bombs Go Off the Mobiles Ring: The Aftermath of Terrorist Attacks. In Kristof Nyiri (ed.), *A Sense of Place: The Global and the Local in Mobile Communication*, 117–128. Vienna: Passagen Verlag.
Collings, Simon. 2012. *Phone Charging Micro-businesses in Tanzania and Uganda*. GVEP International. http://www.gvepinternational.org/sites/default/files/phone_charging_businesses_report_with_gsma_final_for_web_0.pdf?utm_medium=referral & utm_source=pulsenews (Accessed 12 April 2014).
Craig, Robert T. 2007. Issue Forum Introduction: Mobile Media and Communication: What Are the Important Questions?. *Communication Monographs* 74(3). 386–388.
De Zúñiga, Homero Gil and Sebastian Valenzuela. 2011. The Mediating Path to a Stronger Citizenship: Online and Offline Networks, Weak Ties, and Civic Engagement. *Communication Research* 38(3). 397–421.
Dimmick, John, John C. Feaster & Gregory J. Hoplamazian. 2011. News in the Interstices: The Niches of Mobile Media in Space and Time. *New Media & Society* 13(1). 23–39.
Donner, Jonathan. 2009. Blurring Livelihoods and Lives: The Social Uses of Mobile Phones and Socioeconomic Development. *Innovations: Technology, Governance, Globalization* 4(1). 91–101.
Donner, Jonathan & Marcela X. Escobari. 2010. A Review of Evidence on Mobile Use by Micro and Small Enterprises in Developing Countries. *Journal of International Development* 22(5). 641–658.
Duggan, Maeve & Lee Rainie. 2012. Cell Phone Activities 2012. Washington, DC: Pew Internet & American Life Project. http://www.pewinternet.org/files/old-media/Files/Reports/2012/PIP_CellActivities_11.25.pdf (Accessed 11 April 2014).
Dutton, Willim & Frank Naiona. 2003. The Social Dynamics of Wireless on September 11: Reconfiguring Access. In A. Michael Noll (ed.) *Crisis Communication*, 69–82. Lanham, MD: Rowan and Littlefield.
Eagle, Nathan. 2009. Txteagle: Mobile Crowdsourcing. In *Internationalization, Design and Global Development*, 447–456. Springer. http://link.springer.com/chapter/10.1007/978-3-642-02767-3_50 (Accessed 12 April)
Eagle, Nathan, Alex S. Pentland & David Lazer. 2009. Inferring Friendship Network Structure by Using Mobile Phone Data. *Proceedings of the National Academy of Sciences of the United States of America* 106(36): 15274–15278.
García-Montes, José M., Domingo Caballero-Munoz & Marino Perez-Alvarez. 2006. Changes in the Self Resulting from the Use of Mobile Phones. *Media, Culture & Society* 28(1). 67–82.
Gergen, Kenneth J. 2003. Self and Community and the New Floating Worlds. In Kristof Nyiri (ed.), *Mobile Democracy: Essays on Society, Self and Politics*, 61–69. Vienna: Passagen Verlag.

Gergen, Kenneth J. 2008. Mobile Communication and the Transformation of Democratic Process. In James E. Katz (ed.), *The Handbook of Mobile Communication Studies*, 297–310. Cambridge (MA): MIT Press.

Goggin, Gerard. 2011. Telephone Media: An Old Story. In Park David, Nicolas Jankowski, and Steve Jones (eds.), *The Long History of New Media: Technology, Historiography, and Contextualizing Newness*, 231–252. New York: Peter Lang.

GSMA – Groupe Speciale Mobile Association. 2012. Striving and Surviving: Exploring the Lives of Women at the Base of the Pyramid. GSMA mWomenProgramme. London: GSMA.

GSMA – Groupe Speciale Mobile Association. 2013. GSMA Intelligence: Definitive Data and Analysis for the Mobile Industry. https://gsmaintelligence.com/

Gurman, Tilli A., Sara E. Rubin & Amira A. Roess. 2012. Effectiveness of mHealth Behavior Change Communication Interventions in Developing Countries: a Systematic Review of the Literature. *Journal of Health Communication* 17 (sup1). 82–104.

Habuchi, Ichiyo. 2005. Accelerating Reflexivity. In Mizuko Ito, Daishuke Okabe, and Misa Matsuda (eds.), *Personal, Portable, Pedestrian: Mobile Phones in Japanese Life*, 165–182. Cambridge, MA: MIT Press.

Haddon, Leslie. 2003. Domestication and Mobile Telephony. In James E. Katz (ed.), *Machines That Become Us*, 43–56. New Brunswick, NJ: Transaction.

Hampton, Keith N., Lauren F. Sessions & Eun Ja Her. 2011. Core Networks, Social Isolation, and New Media: Internet and Mobile Phone Use, Network Size, and Diversity. *Information, Communication & Society* 14(1). 130–155.

Hampton, Keith N., Oren Livio & Lauren Session Goulet. 2010. The Social Life of Wireless Urban Spaces: Internet Use, Social Networks, and the Public Realm. *Journal of Communication* 60(4). 701–722.

Heath, Thomas. 2010. U.S. Cellphone Users Donate $ 22 Million to Haiti Earthquake Relief via Text. *The Washington Post*, January 19. http://www.washingtonpost.com/wp-dyn/content/article/2010/01/18/AR2010011803792.html (Accessed 12 April 2014).

Hestres, Luis. 2013. App Neutrality: Apple's App Store and Freedom of Expression Online. International Journal of Communication 7. 1265–1280. http://papers.ssrn.com/sol3/papers.cfm?abstract_id=2032050 (Accessed 12 April 2014).

Hirzalla, Fadi, Liesbet Van Zoonen & Jan De Ridder, J. 2010. Internet Use and Political Participation: Reflections on the Mobilization/normalization Controversy. *The Information Society* 27(1). 1–15.

Hofheinz, Albrecht. 2011. Nextopia? Beyond Revolution 2.01. *International Journal of Communication* 5(18). 1417–1434. http://ijoc.org/index.php/ijoc/article/viewFile/1186/629 (Accessed 12 April 2104).

Horrigan, John. 2009. The Mobile Difference. Pew Internet & American Life Project. Washington, DC http://www.pewinternet.org/2009/03/25/the-mobile-difference/ (Accessed 12 April 2104).

Humphreys, Lee. 2007. Mobile Sociality and Spatial Practice: A Qualitative Field Study of New Social Networking Technologies. *Unpublished Doctoral Dissertation, University of Pennsylvania, Pennsylvania.*

Humphreys, Lee, Thilo Von Pape & Veronika Karnowski. 2013. Evolving Mobile Media: Uses and Conceptualizations of the Mobile Internet. *Journal of Computer-Mediated Communication* 18(4). 491–507.

Igarashi, Tasuku, Jiro Takai & Toshikazu Yoshida. 2005. Gender Differences in Social Network Development via Mobile Phone Text Messages: A Longitudinal Study. *Journal of Social and Personal Relationships* 22(5). 691–713.

Ishii, Kenichi. 2006. Implications of Mobility: The Uses of Personal Communication Media in Everyday Life. *Journal of Communication* 56(2). 346–365.

Ito, Mizuko. 2005. Introduction: Personal, Portable, Pedestrian. In Mizuko Ito, Daishuke Okabe, and Misa Matsuda (eds.), *Personal, Portable, Pedestrian: Mobile Phones in Japanese Life*, 1–16. Cambridge, MA: The MIT Press.

ITU – International Telecommunication Union. 2013a. Mobile Cellular, Subscribers Per 100 People. http://www.itu.int/ITU-D/ict/statistics/

ITU – International Telecommunication Union. 2013b. *The World in 2013: ICT Facts and Figures*. Geneva: International Telecommunications Union.

Jagun, Abi Richard Heeks, Jason Whalley. 2008. The Impact of Mobile Telephony on Developing Country Micro-enterprise: A Nigerian Case Study. *Information Technologies and International Development* 4(4). 47–65.

Ji, Pan & Marko M. Skoric. 2013. Gender and Social Resources: Digital Divides of Social Network Sites and Mobile Phone Use in Singapore. *Chinese Journal of Communication* 6(2). 221–239.

Jun, Liu. 2013. *Mobilized by Mobile Media. How Chinese People Use Mobile Phones to Change Politics and Democracy*. Copenhagen: Copenhagen University Dissertation.

Juul, Jesper. 2012. *A Casual Revolution: Reinventing Video Games and Their Players*. The MIT Press.

Katz, James E. 2007. Mobile Media and Communication: Some Important Questions. *Communication Monographs* 74(3). 389–394.

Katz, James E. & Mark A. Aakhus. 2002. Conclusion: Making Meaning of Mobiles a Theory of Apparatgeist. In James E. Katz and Mark Aakhus (eds.), *Perpetual Contact: Mobile Communication, Private Talk, Public Performance*, 301–320. Cambridge: Cambridge University Press.

Katz, James E. & R. E. Rice. 2002. The Telephone as an Instrument of Faith, Hope, Terror and Redemption: America, 9–11. *Prometheus* 20(3). 247–253.

Korhonen, Hannu, Hannamari Saarenpää & Jeanne Paavilainen. 2008. Pervasive Mobile Games – A New Mindset for Players and Developers. In *Fun and Games*, 21–32. Springer.

Leyshon, Michael, Sean DiGiovanna & Briavel Holcomb. 2013. Mobile Technologies and Youthful Exploration: Stimulus or Inhibitor?. *Urban Studies* 50(3). 587–605.

Licoppe, Christian. 2004. Connected Presence: The Emergence of a New Repertoire for Managing Social Relationships in a Changing Communications Technoscape. *Environment and Planning D: Society and Space* 22(1). 135–156.

Lima, Antonio, Manilo De Domenico, Veljico Pejovic & Mirco Musolesi. 2013. Exploiting Cellular Data for Disease Containment and Information Campaigns Strategies in Country-Wide Epidemics. Paper presented at the Third International Conference on the Analysis of Mobile Phone Datasets, NetMob D4D Challenge. MIT, Cambridge MA, 1–3 May.

Lin, Trisha Tsui-Chuan & Shirley Hsiao-Li Sun. 2010. Connection as a Form of Resisting Control: Mobile Phone Usage of Foreign Domestic Workers in Singapore. *Media Asia: An Asian Communication Quarterly* 37(4). 183–214.

Lindmark, Sven. 2002. Evolution of Techno-economic Systems: An Investigation of the History of Mobile Communications. Gothenberg, Sweden: Chalmers University of Technology, Department of industrial management and economics dissertation.

Ling, Rich. 2004. *The Mobile Connection: The Cell Phone's Impact on Society*. San Francisco: Morgan Kaufmann.

Ling, Rich. 2008. *New Tech, New Ties: How Mobile Communication Is Reshaping Social Cohesion*. Cambridge: MIT Press.

Ling, Rich. 2012. *Taken for Grantedness: The Embedding of Mobile Communication into Society*. Cambridge, MA: MIT Press.

Ling, Rich, Troels Bertel & Pål Roe Sundsøy. 2012. The Socio-demographics of Texting: An Analysis of Traffic Data. *New Media & Society* 14(2). 280–297.

Ling, Rich & Jonathan Donner. 2009. *Mobile Communication*. Digital and Media Society Series. London: Polity.

Ling, Rich & Heather A. Horst. 2011. Introduction, Mobile Communication in the Global South: Special Issue of New Media & Society. *New Media & Society* 13(3). 3–6.

Ling, Rich & Yttri, B. 2002. Hyper-coordination via Mobile Phones in Norway. In James E. Katz and Mark Aakhus (eds.), *Perpetual Contact: Mobile Communication, Private Talk, Public Performance*, 139–169. Cambridge: Cambridge University Press.

Lopez, Marc Hugo, Ana Gonzalez-Barrera & Elieen Patten. 2013. Closing the Digital Divide: Latinos and Technology Adoption. *Washington, DC: Pew Internet & American Life Project.* http://www.pewhispanic.org/files/2013/03/Latinos_Social_Media_and_Mobile_Tech_03-2013_final.pdf (Accessed 12 April 2014).

Maitland, Donald. 1984. The Missing Link. *Independent Commission for Worldwide Telecommunications Development Report, December*. Geneve. http://www.itu.int/osg/spu/sfo/missinglink/The_Missing_Ling_A4-E.pdf (Accessed 12 April 2104).

McEwen, Rhonda N. 2009. A World More Intimate: Exploring the Role of Mobile Phones in Maintaining and Extending Social Networks. Toronto: University of Toronto, Faculty of Information dissertation.

McEwen, Rhonda N. 2011. Tools of the Trade: Drugs, Law and Mobile Phones in Canada. *New Media & Society* 13(1). 134–150.

McLuhan, Marshall. 2004. *Understanding Me: Lectures and Interviews*. Cambridge: MIT Press.

Mills, John, Paul Egglestone, Omer Rashid & Heli Väätäjä. 2012. MoJo in Action: The Use of Mobiles in Conflict, Community, and Cross-platform Journalism. *Continuum* 26(5). 669–683.

Miritello, Giovanna, Esteban Moro, Rubén Lara, Rocío Martínez-López, John Belchamber, Sam GB Roberts & Robin IM Dunbar. 2013. Time as a Limited Resource: Communication Strategy in Mobile Phone Networks. *Social Networks* 35(1). 89–95.

Mosemghvdlishvili, Lela & Jeroen Jansz. 2013. Negotability of Technology and Its Limitations: The Politics of App Development. *Information, Communication & Society* 16(10). 1596–1618.

Mossberger, Karen, Caroline J. Tolbert & Allison Hamilton. 2012. Measuring Digital Citizenship: Mobile Access and Broadband. *International Journal of Communication* 6. 2492–2528.

Norris, Pippa & Ronald Inglehart. 2009. *Cosmopolitan Communications: Cultural Diversity in a Globalized World*. Cambridge University Press.

Onnela, J. P., J. Saramäki, G. Hyvönen, D. Szabó, K. Lazer, J. K. Kaski & A. L. Barabási. 2007. Structure and Tie Strengths in Mobile Communication Networks. *Proceedings of the National Academy of Sciences* 104(18). 7332–7336.

Oulasvirta, Antti, Tye Rattenbury, Lingyi Ma & EevaRaita. 2012. Habits Make Smartphone Use More Pervasive. *Personal and Ubiquitous Computing* 16(1). 105–114.

Peng, Tai-Quan, Lun Zhang, Zhi-Jin Zhong & Jonathan JH Zhu. 2012. Mapping the Landscape of Internet Studies: Text Mining of Social Science Journal Articles 2000–2009. *New Media & Society* 15(5). 644–664.

Piette, Jhon D., Jeremy B. Sussman, Paul N. Pfeiffer, Maria J. Silveira, Satinder Singh & Mariel S. Lavieri. 2013. Maximizing the Value of Mobile Health Monitoring by Avoiding Redundant Patient Reports: Prediction of Depression-Related Symptoms and Adherence Problems in Automated Health Assessment Services. *Journal of Medical Internet Research* 15(7). e118.

Poell, Thomas. 2013. Social Media and the Transformation of Activist Communication: Exploring the Social Media Ecology of the 2010 Toronto G20 Protests. *Information, Communication & Society* 0(0). 1–16.

Przybylski, Andrew K. & Netta Weinstein. 2013. Can You Connect with Me Now? How the Presence of Mobile Communication Technology Influences Face-to-face Conversation Quality. *Journal of Social and Personal Relationships* 30(3). 237–246.

Purcell, Kristen. 2011. Half of Adult Cell Phone Owners Have Apps on Their Phones. Washington, DC: Pew Internet & American Life Project. http://www.pewinternet.org/2011/11/02/half-of-adult-cell-phone-owners-have-apps-on-their-phones/ (Accessed 12 April 2014).

Putnam, Robert. 2000. *Bowling Alone: The Collapse and Revival of American Community*. New York: Touchstone.

Qiu, Jack Linchuan. 2009. *Working-class Network Society: Communication Technology and the Information Have-less in Urban China*. MIT Press.

Qiu, Jack Linchuan & Eric C. Thompson. 2007. Editorial: Mobile Communication and Asian Modernities. *New Media & Society* 9(6). 895–901.

Qiu, Jack Linchuan. 2008. Mobile Civil Society in Asia: A Comparative Study of People Power II and the Nosamo Movement. *Javnost-The Public* 15(5). 5–24.

Reid, Donna & Fraser Reid. 2004. Insights into the Social and Psychological Effects of SMS Text Messaging. http://www.ebusinessforum.gr/old/content/downloads/Reidetal_SocialEffectsOfTextMessaging.pdf (Accessed 12 April 2014).

Rheingold, Howard. 2002. *Smart Mobs*. Cambridge, MA: Persius.

Rice, Ronald E. & James E. Katz. 2003. Comparing Internet and Mobile Phone Usage: Digital Divides of Usage, Adoption, and Dropouts. *Telecommunications Policy* 27(8–9). 597–623.

Ring, D. H. 1947. Mobile Telephony – Wide Area Coverage. Bell Telephone Labratories. http://www.privateline.com/archive/Ringcellreport1947.pdf (Accessed 12 April 2104).

Rojas, Hernando & Eulalia Puig-i-Abril. 2009. Mobilizers Mobilized: Information, Expression, Mobilization and Participation in the Digital Age. *Journal of Computer-Mediated Communication* 14(4). 902–927.

Runyan, Jason D., Timothy A. Steenbergh, Charles Bainbridge, Douglas A. Daugherty, Lorne Oke & Brian N. Fry. 2013. A Smartphone Ecological Momentary Assessment/Intervention 'App' for Collecting Real-Time Data and Promoting Self-Awareness. *PloS One* 8(8). e71325.

Sanner, Terje Aksel, Lars Kristian Roland & Kristin Braa. 2012. From Pilot to Scale: Towards an mHealth Typology for Low-resource Contexts. *Health Policy and Technology* 1(3). 155–164.

Sawhney, Harmeet. 2007. Strategies for Increasing the Conceptual Yield of New Technologies Research. *Communication Monographs* 74(3). 395–401.

Selwyn, Neil. 2004. Reconsidering Political and Popular Understandings of the Digital Divide. *New Media & Society* 6(3). 341–362.

Shirazi, Alireza Sahami, James Clawson, Yashar Hassanpour, Mohammad J. Tourian, Albrecht Schmidt, Ed H. Chi, Marko Borazio, and Kristof Van Laerhoven. 2013. Already Up? Using Mobile Phones to Track & Share Sleep Behavior. *International Journal of Human-Computer Studies* 71(9). 878–888.

Shklovski, Irina & Adriana de Souza e Silva. 2013. An Urban Encounter: Realizing Online Connectedness through Local Urban Play. *Information, Communication & Society* 16(3). 340–361.

Silverstone, Roger, Eric Hirsch & David Morley. 1992. Information and Communication Technologies and Moral Economy of the Household. In R. Silverstone and E. Hirsch (eds.), *Consuming Technologies: Media and Information in Domestic Spaces*, 15–31. London: Rutledge.

Smith, Aaron. 2011a. Americans and Text Messaging. Washington, DC: Pew Internet & American Life Project. http://www.pewinternet.org/2011/09/19/americans-and-text-messaging/ (Accessed 12 April 2014).

Smith, Aaron. 2011b. Americans and Their Cell Phones. Washington, DC: Pew Internet & American Life Project. http://www.pewinternet.org/2011/08/15/americans-and-their-cell-phones/ (Accessed 12 April 2014).

Smith, Aaron. 2012. Real Time Charitable Giving. *Washington, DC: Pew Internet & American Life Project*. http://www.pewinternet.org/2012/01/12/real-time-charitable-giving/ (Accessed 12 April 2014).

Smith, Aaron. 2013. Smartphone Ownership 2013. Washington, DC: Pew Internet & American Life Project. http://www.pewinternet.org/2013/06/05/smartphone-ownership-2013/ (Accessed 12 April 2014).

Song, Chaoming, Zehui Qu, Nicholas Blumm, and Albert-Laszlo Barabasi. 2010. Limits of Predictability in Human Mobility. *Science* 327(5968). 1018–1021.

Sundsøy, Pål Roe, Rich Ling, Johannes Bjelland, Geoffrey Canright & Kenth Engø-Monsen. 2012. The Activation of Core Social Networks in the Wake of the 22 July Oslo Bombing. In *Proceedings of IEEE/ACM International Conference on Advances in Social Networks Analysis and Mining* (ASONAM), Istanbul, Turkey, 26–29 August.

Sutko, Daniel M & Adriana de Souza e Silva. 2011. Location-aware mobile media and urban sociability. *New Media & Society* 13, 807–23.

The mGive Foundation. 2011. Text Giving Donor Survey Report. http://www.nonprofitfederation.org/sites/default/files/mGive_Text_Giving_Donor_Survey_Report.pdf (Accessed 12 April 2014).

Thompson, Kristin, Kristen Purcell & Lee Rainie. 2013. Arts Organizations and Digital Technologies. Washington, DC: Pew Internet & American Life Project. http://www.pewinternet.org/files/old-media/Files/Reports/2013/PIP_ArtsandTechnology_PDF.pdf (Accessed 12 April 2014).

Thorson, Kjerstin, Kevin Driscoll, Brian Ekdale, Stephanie Edgerly, Liana Gamber Thompson, Andrew Schrock, Lana Swartz, Emile K. Vraga & Chris Wells. 2013. YouTube, Twitter and the Occupy Movement: Connecting Content and Circulation Practices. *Information, Communication & Society* 16(3). 421–451.

Vincent, Jane. 2006. Emotional Attachment and Mobile Phones. *Knowledge, Technology & Policy* 19(1). 39–44.

Wallis, Cara. 2011. Mobile Phones Without Guarantees: The Promises of Technology and the Contingencies of Culture. *New Media and Society* 13(3). 471–485.

Watkins, Jerry, Kathi R. Kitner & Dina Mehta. 2012. Mobile and Smartphone Use in Urban and Rural India. *Continuum* 26(5). 685–697.

Westlund, Oscar. 2011. *Cross-media News Work: Sensemaking of the Mobile Media (R)evolution.* Göteborg: University of Göthenburg.

Joanna Kulesza

17 Legal issues in a networked world

Abstract: The author discusses key legal challenges faced by the global information society. They are the consequence of significant dichotomies between individual liberties and common security, ones well known to international law and politics, yet gaining a new dimension when faced through the prism of the transnational Internet. The author starts off with a brief sketch of the evolution of international Internet governance as the background for all ongoing discussion on the appropriate legal framework for the global network. The unique multistakeholder model of its governnace requires a unique approach to resolving legal issues bound therewith. Then individual legal challenges to be faced by the international community are discussed. They include local notions of privacy confronted with the global need for cybersecurity, national perceptions of decency and public order set against online expression, exercised across territorial borders as well as the right to knowledge limited by intellectual property rights. Kulesza points to two crucial criteria in confronting all legal challenges online: the need to reinterpret the notions of jurisdiction and due diligence applied to online communications. Such reinterpretation can be made based on the rich body of work provided by international organizations and well established international practice.

Keywords: internet governance, human rights, privacy, cybersecurity, freedom of speech

1 Historical aspects: Internet as a source of legal uncertainty

1.1 Legal status of the network and its governance

In 1969 the legal status of Internet's predecessor – the Advanced Research Projects Agency Network (ARPAnet) – rose no controversies: it was a part of a research project funded by the Advanced Research Projects Agency and implemented primarily by a group of young researchers: by Jon Postel, Steve Crocker and Vinton Cerf (Mueller 2004: 74–75). The initially four-nodded network grew rapidly with interconnected computers in numerous U.S. research facilities (Hafner and Lyon 1999: 104). At that time the machines on the network belonged to universities or governmental facilities, the telecommunication infrastructure used for data transfers – to private companies, while the code the machines operated on within the project was officially supervised by ARPA, yet subject to no property rights. Quite the contrary – ARPA invested in order to make the code more popular (Mueller 2004: 84; Stiegler 2012: 22).

The nature of the ARPA defense project, aimed at protecting crucial state data from a potential, centralized, possibly a nuclear attack, deemed a strong input from a U.S. military a given (Mueller 2004: 75–76). With the strong research potential that the network held in 1983 it was agreed among the academics and the military to split the network into two separate circuits, leaving only a part of it as a continuation of the initial military project, renaming it a "military network" (MILnet) (Abbate 2000: 185). With the first international connections between the U.S., Norway and the UK being made in 1973, a decade later MILnet was detached from the rest of the interconnected computers, which enabled to offer academics worldwide the "international network" – INTERnet – to exchange research results (Byung-Keun 2005: 101). When the Internet was born the physical devices used for communicating belonged to privately owned telecommunication companies (in the U.S. it was primarily AT&T) and universities running the research, while the standards used for computer code were, just as they are today, indifferent to any legal regulation. Online content, when subject to copyright, belonged to its authors (or their employers, if the parties so decided in their employment contract). ARPA (at that time already renamed to DARPA – Defense Advanced Research Projects Agency) handed over its supervision of the international project to U.S. National Science Foundation (NSF), making the Internet a purely academic exercise (Muller 2004: 85). As the academic network grew, its architects realized the need for a centralized organization of their efforts (Mueller 2004: 90–91; Byung-Keun 2005: 93). Rather then solving technical problems on their own, they called for an open forum, not an expert group, as the usual practice had it, where all interested parties could seek solutions to new technical challenges. So the Internet Engineering Task Force (IETF) came onto the Internet governance stage, with initially a couple of dozen researchers meeting in 1986 (Mueller 2004: 91). The main task of the IETF was to resolve technical problems encountered in every day operation of the network. Working through Requests for Comments (RfCs) on each individual problem, the quickly growing community sought consensus before deciding to use a particular technical standard for all the interconnected devices. The RfCs are not legal acts, they are mere technical standards, yet their actual status may easily be compared to that of the law of cyberspace. Any device that is supposed to work online, must follow the technical specifications agreed upon by the international group of volunteers working under the IETF agenda. IETF has until this day no legal personality – it cannot be subject to litigation nor own property. It operates based on membership fees and officially is a part of the Internet Society. It is open to all volunteers who wish to improve the network and creates, although not exclusively, the code that the network runs on (Alvestrand 2004).

With Tim Berner-Lee's 1989 idea of interconnecting webpages through hyperlinks the evident commercial potential of the network rapidly grew (Post 2010: 101–102). In 1991 NSF eventually gave in to the growing number of requests and allowed for commercial use of the network (Abbate 2000: 196–197). The dotcom bubble was growing, making small online entrepreneurs millionaires overnight. One easy way to get very rich very fast online was to invest in domain names. The Domain Name System (DNS) – an Internet phonebook of a kind, allowing to translate any sequence of numbers which identifies a particular website into an easily memorable sequence of words – was at the time managed by two contractors of the U.S. government: the Internet Assigned Names Authority (IANA), run by Jon Postel and a company named Network Solutions, now still well operating in the domain name business as VeriSign (Mueller et al. 2007: 238). DNS management is now considered key to running the network.

As the commercial potential of the network rapidly grew, the U.S. government was subjected to growing pressure to release its control over the web. While domestic companies were using the all-soothing "invisible hand of free market" argument to convince the Clinton/Gore administration to surrender the DNS supervision to business, the International Telecommunications Union (ITU), a UN successor of the 1865 funded International Telegraph Union, tasked to manage world telecommunications, together with the Internet Society opted for an institutionalized and internationalized approach. The White House wished to avoid the dangers of UN red tape and was equally reluctant to lend an ear to the European proposals for an international Internet treaty, setting the rules on sharing power over the network (Mueller et al. 2007: 239). Following a 1998 recommendation from the Department of Commerce, Bill Clinton created a new, non-commercial entity designed to manage the DNS in the name of public interest, keeping the arguably most significant Internet resource outside the reach of U.S. business and international politics alike. In 1998 the Internet Corporation for Assigned Names and Numbers (ICANN) was created to "administer policy for the Internet name and address system" (NTIA 1998).

1.2 Internet governance and ICANN

The ICANN does not own the Internet nor does it own the DNS, but it holds, among other significant competences, the power to create individual domains and delegate the rights to manage them. Since only within or through domains may any significant online interaction take place, ICANN is the most powerful entity managing the Internet. At the same time it was designed as a private U.S. body, subject to California laws and under contractual obligations with the U.S. Department of Commerce. Initially no international supervision was intended for any ICANN decision. Both those characteristics quickly gave rise to strong international criticism. In 2003 the UN called upon its first World Summit on the Information

Society (WSIS) putting the phrase "Internet governance" on national and international political agendas. It demanded for the identification of norms and standards upon which this universal resource that is the Internet may and ought to be managed. In order to meet that task it called upon a group of experts – members of the Working Group on Internet Governance (WGIG) – who in November of 2004 were to answer most urging questions dealing with this global phenomena that the cyberspace had become. They were asked to prepare a working definition of Internet governance, depict public policy issues essential to managing the network and propose a consensus on the respective roles of governments, international forums, private sector and civil society (WSIS 2003b). In its 2005 Report the WGIG identified "Internet governance" as "the development and application by governments, the private sector and civil society, in their respective roles, of shared principles, norms, rules, decision-making procedures, and programmes that shape the evolution and use of the Internet" and that definition was later adopted by the WSIS (2005: 1). This definition is perceived as embodying the basic principle of Internet governance – multistakeholderism or "multistakeholder principle". Because of its multifaceted character, the Internet may only be effectively governed if all groups of entities involved in its technical and social functioning cooperate. The theory of Internet layers sees the cyberspace as a set of concentric spheres (Benkler 2006: 392; Lessig 2006: 144; Abbate 2000: 151–153). The most inner one – the physical layer – is constructed of hardware, telecommunication cables and satellites. It includes all the elements that are physically necessary for the Internet to function. Those elements are usually subject to private ownership and are easily identifiable as subjected to national laws and/or international regulations. The middle layer is the layer of code, also called the logical layer. It covers all the software, including Internet standards, that allows the hardware elements to connect and enable information to one another. This layer may be well represented by the aforementioned IETF – the factual competence to introduce technical standards usually stays out of the hands of governments, international organizations or even telecommunication companies. It is often developed by technical professionals, technically inclined users or software developers. Eventually there is the most outer content layer, which includes everything that is available online and through the use of the network. This layer is created by the users and is of most interest to national authorities, wishing to limit access to information constituting national security threats or materials considered "indecent". The very architecture of the network – one which was designed to avoid any central control – makes it however impossible to regulate the content layer without reference to and cooperation with entities shaping the two other layers.

The WSIS definition of Internet governance, derived directly from political science and regime theory is perceived as the most accurate one so far, yet other stipulations of the 2005 WSIS Tunis Agenda make the distinction between the functions of individual stakeholders blurred (Mueller et al. 2007: 241). The attempted line of division between the competences of states and business was drawn between "public policy" and "technical management" or "day to day operation", respectively, while in practice it is impossible to identify individual tasks as falling exclusively into one of the two spheres (Mueller et al. 2007: 242). Therefore as much as the definition itself is a good reflection of the multifaceted character of the network, the following attempt at dividing individual tasks may be considered one of the reasons for the still present confusion on the roles of individual stakeholders in governing the network, as it is impossible to achieve effective management without cooperation of all stakeholders involved.

One of the reasons for this methodological confusion might be the tense political atmosphere that accompanied the WSIS meetings, during which a "war over ICANN" unfolded (Rogers 2006: 26–35; Mayer-Schönberger and Ziewitz 2007: 184–186; Malcolm 2008: 321–355). Numerous governments, including those of Brazil, Russia and India as well as the European Commission, representing European states' policy on the issue, demanded the U.S. to share the exclusive authority it had over ICANN. The Bush administration very strongly resented any obligation to share power of what once was an exclusively national defense project. This exclusivity policy was upheld during both WSIS meetings where no compromise on power sharing was met. It was only in 2009 that U.S. President Obama, keeping his election-time promises, included members of foreign governments, represented in the ICANN Governmental Advisory Committee, who so far served only with a non-binding advisory opinion to the ICANN Board, in the yearly ICANN review process, a competence thus far exclusive to the U.S. Department of Commerce since 1998 (Gasser et al. 2010; ICANN 2009).

The WSIS process brought one new element to the way Internet governance was perceived. As a symbolic gesture U.S. President Bush agreed in 2005 on the creation of an UN supervised Internet Governance Forum (IGF) – a loose formula of international cooperation enabling the sharing of ideas on all Internet related issues. Reflecting the multistakeholderism principle, enshrined in the WSIS definition of Internet governance, the forum is open to all stakeholder groups, wishing to discuss issues crucial to online communications. The yearly meetings of IGF, commenced in 2005, hold no deciding power or legal authority. The lengthy discussions in the talk-shop formula are perceived as a continuation of the WSIS process, which had limited success (Mueller et al. 2007: 242). Crucial norms and principles for online governance have still not been identified and an international consensus on Internet governance perceived as an identified set of issues enshrined in a binding political compromise, seems distant. Yet the WSIS process and accompanying diplomatic, social and scholarly efforts allow to identify key challenges for legal regulation of cyberspace and their intensity allows to depict emerging trends in cyberspace law.

2 Key legal challenges for the information society

2.1 Introductory remarks

One of the tasks of the WGIG was to identify key challenges laying before the international community when it comes to governing the Internet. A list of such issues was included in the 2005 WGIG report and regarding the steady pace of the consensus seeking processes remains accurate today (WGIG 2005: 5). The broad approach to the problem, adopted by the WGIG members, resulted in a list of four groups of issues: those relating to infrastructure and the management of critical Internet resources; ones relating to the use of the Internet; issues relevant to the Internet yet impacting other media and spheres of life, such as intellectual property rights (IPRs) or international trade and eventually i ssues relating to the further development of the network and its governance, with particular reference to developing countries. A detailed list of those issues included: 1. administration of the root zone files and system; 2. interconnection costs; 3. Internet stability, security and cybercrime; 4. spam; 5. meaningful participation in global policy development; 6. capacity-building; 7. allocation of domain names; 8. IP addressing; 9. intellectual property rights; 10. freedom of expression; 11. data protection and privacy rights; 12. consumer rights and 13. multilingualism (WGIG 2005: 5–8).

As time showed, some of the issues gained more significance and political impact than others and in 2015 seem critical for further functioning of the network. Due to the limited volume of this chapter only most pressing legal and policy issues are discussed below. The analysis provided is based on the identification of conflicting interests of states, business and individuals confronted in the cyberrealm.

2.2 National notions of privacy and state security in transboundary cyberspace

Current challenges in Internet governance reflect a controversy long present in international relations – the interrelationship between the wellbeing of an individual and that of a collective. The answer to the question on whose interests should supersede depends strongly on culture, history and religion shared among those giving it. While European and American policies focus on the protection of an individual, the e.g. Asian perspective seems focused on the wellbeing of a community as a whole. Those irresolvable differences gained more significance during the universal discussion on the catalogue of human rights, most intense in mid-20th century. The divergence in national policies seems well reflected in the number and execution of various human rights treaties – numerous in Europe or Americas, few in Asia (Davis and Galligan 2001: 1–25).

The present trend in international politics, focused on the war with terror, including combating international cyberterrorism, emphasized the weakness of the existing human rights compromise in general, and on privacy in particular, when faced with issues of national security. U.S. seems to be leading the way in the war on terror and cybersurveillance. The 2001 U.S. Patriot Act together with the 2008 Foreign Intelligence Surveillance Amendments Act (FISAA), the legal foundations of the infamous PRISM program, allowing for universal, extensive cybersurveillance of all users of telecommunication services serve as primary examples of national policies fruitlessly seeking a compromise on the two values discussed. Even though both: European states and the U.S. root their domestic policies in human rights guarantees, the 2013 PRISM controversy shook up the fragile transatlantic compromise on individual privacy protection. Europe sees individual privacy as a human right, enshrined in both: the European Convention on Human Rights and EU law, primarily through its personal data protection regulations. As such, individual privacy may only be invaded in a case of "necessity", based on an act of law and subject to a thorough court assessment. According to the Convention for the Protection of Human Rights and Fundamental Freedoms (also named the European Convention on Human Rights, ECHR) from 1953 "everyone has the right to respect for his private and family life, his home and his correspondence". Its Article 8 disallows any "interference by a public authority with the exercise of this right", yet foresees for few exceptions, when such an interference is allowed. Such exceptional situations include invasions that are "in accordance with the law" and "necessary in a democratic society in the interests of national security, public safety or the economic well-being of the country, for the prevention of disorder or crime, for the protection of health or morals, or for the protection of the rights and freedoms of others". Article 8 ECHR, cited above, is accompanied by rich jurisprudence of the European Court of Human Rights, detailing individual rights and obligations vis a vis the right to privacy.

This perception of privacy as a human right is also reflected in the existing international human rights guarantees, in particular within Article 12 of the Universal Declaration of Human Rights (UDHR) and Article 17 of the International Covenant on Civil and Political Rights (ICCPR). Article 12 UDHR allows no one to be subjected to arbitrary interference with their privacy, while its Article 29 states that exercising that right, as well as any other enshrined in the Declaration, is "subject only to such limitations as are determined by law solely for the purpose of securing due recognition and respect for the rights and freedoms of others and of meeting the just requirements of morality, public order and the general welfare in a democratic society" (UDHR 1948). The 1973 ICCPR – a treaty based on the non-binding UDHR – confirms that principle by stating in Article 17 that no one may be subjected to arbitrary or unlawful interference with their privacy. This general declaration was put into more detail in the 1988 General Comment No. 16 on Article 17 provided by the UN Human Rights Committee (HRC 1988). According to this document states are required to adopt measures to ensure that the prohibition against privacy interferences and attacks is effective. It also includes a positive obligation of states to actively protect individual privacy against any interference. The HRC stated that "effective measures have to be taken by States to ensure that information concerning a person's private life does not reach the hands of persons who are not authorized by law to receive, process and use it" (HRC 1988: 21). This obligation is of particular importance at the age of cyberspace when individual privacy may be infringed not only by national authorities but those of a foreign state, as was the case with the PRISM surveillance programme. The current understanding of the right to privacy puts therefore a distinct positive obligation on the state to actively undertake effective privacy protection measures, even if the possible intrusion originates from entities other than its own. What seems particularly important in an age of uberveillance: "an omnipresent electronic surveillance facilitated by technology" (Butler 2009: 1094), the HRC clearly stated that any "surveillance, whether electronic or otherwise, interceptions of telephonic, telegraphic and other forms of communication, wire-tapping and recording of conversations should be prohibited." (HRC 1988).

The European concept of personal data protection, which builds upon the notion of personal data as a tool for safeguarding individual privacy and originates from the 1997 personal data protection Directive (European Parliament 1995), even though adopted by 1980 OECD guidelines (OECD 1980) and incorporated in numerous non-European states, is foreign to the United States of America and only partially enforceable within other American or Asian states. While Europe sees privacy and personal data protection as a human right, the U.S. perceives information on an individual rather as a commodity then an individual prerogative. This discrepancy gave rise to the late 1990s Safe Harbor Agreements controversy – a rough consensus reached by the European Commission and the U.S. Department of Commerce allowing for exporting personal data of EU residents to U.S. business. The controversy around the agreement focused on the limited scope of protection granted in the U.S. and low effectiveness of the agreement itself – U.S. companies dealing in European data were to take up an obligation to respect EU data protection laws, yet in practice there was little to no scrutiny over them actually meeting their obligations (Farrell 2002: 105–124). The PRISM controversy opened a new era in the transatlantic argument over personal data protection and made the scale of the problem well visible – territorially perceived human rights protection is far insufficient at a time of the transboundary cyberspace. What is needed is an effective, international compromise, possibly one based on the existing human rights treaties. Without one no effective privacy protection for individual users may be present in the online environment.

2.3 Internet filtering, free speech and the rights of others

Just as the existing human rights compromise is insufficient to effectively protect privacy online, so do the numerous UN documents on free speech hold no effective guarantee for Internet users expressing their ideas online. Just as privacy, the right to free speech, comprising of the right to hold, receive and impart information and ideas is not an absolute one. Article 29 par. 2 UDHR and Article 19 par. 3 ICCPR both provide for free speech limitations, however emphasize that they may only be introduced in case-specific circumstances, done solely for the grounds of guaranteeing the respect of the rights of others or for the protection of national security, public order, public health or morals and based on an act of law. Furthermore any infraction must be proportionate – should free expression by limited, such a limitation may only follow a case-specific assessment of the degree of interference confronted with the importance of the purpose of the restriction (Deibert 2008: 81; Yutaka and Yutaka 2002: 186–189). While the interpretation of various limitative clauses, such as protection of morality or public order surely leads to controversy, there is no doubt that any blanket, key-words based Internet filtering, affected country-wide, may not be considered intact with international law, as it infringes the complimentary freedoms to seek and receive information of all individuals within the territorial jurisdiction of the filtering state. Their right to obtain and share information is limited without a particular court decision dealing with heir specific case. The political and legal reality is however far distant from that envisaged in the tangible international compromise of the UDHR and the ICCPR, as the scale of state-imposed or state-legitimated online censorship is growing (Deibert 2008: 9).

Globally over 620 million people were unable to access online content for reasons of state imposed censorship in 2012 (ONI 2012). Next to China the Open Network Initiative (ONI) names also Belarus, Cuba, Russia, Kazakhstan, Malaysia and Vietnam among the filtering countries (Deibert 2008: 346, 179, 396, 384, 268, 326 respectively). While the ONI analysis is focused on censorship enacted for political reasons, states declaring their devotion to the human rights ideology use the very same techniques to safeguard national notions of morality with e.g. defamation laws or to protect monetary interests of companies or business sectors, just to name the controversies surrounding services like Napster, Kaaza or most recently The Pirate Bay and Megaupload. Also filtering done by libraries or schools for the protection of morality or children rights is an example of such key-word based, blanket filtering. It disables access to certain categories of content, usually based on an act of national law, allowing for such practice, even though it is contrary to international human rights guarantees, as not being case specific or based on a court decision.

A different issue are the so-called notice-and-takedown procedures, obliging ISPs, informed of the potentially infringing character of particular content they host, to disable access thereto. Such ISP imposed censorship, although aimed at particular content and enacted for the protection of individual rights, usually lacks court supervision, making it equally infringing to individual right to seek and access information. What is more, the ISPs – primarily private companies – act under the threat of taking over legal liability for content they fail to promptly block access to, which brings about an undesired chilling effect – ISPs would remove any content that brings a threat of a lawsuit rather than risk legal liability (Strowel 2009).

As already discussed setting the limits of free speech depends strongly on local culture, values, history and religion. This characteristic makes it most difficult to achieve a universal yet practical compromise on the limits of free speech online, but looking at the human rights law and recent achievements of international criminal law a basic common ground on the most widely condemned practices can be found. Child pornography and human trade or trafficking are penalized in criminal codes of all civilized countries and they might serve as a safe starting point for the discussion on the issue of limits of free speech online. It is equally clear that a compromise on hate speech, terrorism prevention or promoting genocide online belong to ones most difficult issues in contemporary international relations. Setting an online standard for those grounds of limiting free speech should be left as one of the last points to be discussed when seeking a practical compromise. The urgent need for identifying one however may serve as one of the crucial reasons for initiating a serious debate on limits of state power over Internet content. Key to regulating the issues where no compromise can be found might be the clear "delimitation" of cyberspace by a set of guidelines on the sequence of applying the international jurisdictional principles (Kohl 2007; Malcolm 2008; Weber 2010; Kulesza 2012).

Regarding the contemporary state practice a world without any online censorship seems difficult to imagine. Assuming that states would agree on a universal standard for online censorship covering both: technical implementation and legal boundaries, a different issue would be that of a body competent and transparent enough to execute such a difficult compromise. In many European states, such as Sweden or Germany, "black lists" of harmful content are set together by enforcement agencies cooperating with civil society, represented by e.g. NGOs. This multistakeholder model, including e.g. Interpol and/or Amnesty International filing for national court orders to limit access to certain URLs might be considered a worldwide solution for protecting Internet users from undesirable, harmful content, such as information on human trafficking or child pornography. An even more challenging issue is that of politically motivated censorship which, according to the authorities exercising it, occurs in order to protect the internal security of a given state. Since political censorship is very difficult to assess, this area of consensus seeking must be left as the last one to be discussed and compromised on, accompanied by a clear set of jurisdictional rules. The very wording of a consensus on free speech online might be similar to that of Article 19 UDHR or article 10 of ECHR. Should the HRC and the European Court of Human Rights (further herein: ECtHR) guidelines be followed by all actors in good faith a functional compromise for free speech online may be found. It seems therefore as if the current gridlock on online free speech is more about differing policies then legal regulations.

2.4 Copyright intellectual property rights and the right to access knowledge

Another long standing controversy that was highlighted by the specifics of online communication was the conflict between the individual right to property, intellectual property in particular, and the right to access knowledge. Services like Napster, Kaaza or The Pirate Bay tested the practical application of those two rights confronted. While the right of an author to enjoy the benefits of the use of their work or the right of an entrepreneur to enjoy their well deserved reputation remain undisputed, the cyberspace and its users had put those well-established legal terms to a difficult test.

The key legal term to be considered when discussing the copyright controversy is fair use. While numerous international treaties grant authors the right to enjoy the fruits of their intellectual labour, they also acknowledge individuals' right to access their works, quote them or search inspiration. While the line between an inspiration and an adaptation of other's work has always been difficult to trace, the right to learn from and enjoy the works of others, described with the term "fair use" seemed relatively well defined, until what once was a few school friends borrowing a handbook from one another changed into an online crowd of thousands Facebook "friends" sharing films or freshly released music albums across borders. The core of the copyright dispute lies between the right to access knowledge, as expressed in Article 27 UDHR which grants everyone the "right to freely participate in the cultural life of the community, to enjoy the arts and to share in scientific advancement and its benefits" and the protection of property as described within Article 17 UDHR. The notion of fair use lies at the root of promoting human creativity and development, expressed in Article 27 UDHR. It aims to allow authors deriving from the work of their predecessors and contemporaries and so to push further the evolution of mankind. The legal prerequisites for exercising fair use seem similar in most countries having copyright regulation, regardless of the fact that European and U.S. concepts of copyright differ entirely (Senftleben 2004: 112–113).

The conflict of interests between users and the multimedia industry originates from the interpretation of what fair use is and where its limits lie – the so-called three step test foreseen in Article 9(2) of the 1886 Berne Convention for the Protection of Literary and Artistic Works and repeated in numerous copyright treaties afterwards (Senftleben 2004: 1–2). Article 9(2) of the Berne Convention allows state parties to the convention to introduce national regulations permitting the reproduction of copyrighted works "in certain special cases, provided that such reproduction does not conflict with a normal exploitation of the work and does not unreasonably prejudice the legitimate interests of the author". While the media industry opts for the narrowest possible interpretation of the three step clause, including quotation rights, but disallowing individual, private use of copyrighted works claiming they "unreasonably prejudice the legitimate interests of the author", the users and entrepreneurs wish to use the full capabilities of what the network has to offer – the instantaneous, universal, high-quality exchange of digitized works. National government differ in assessing the situation – while U.S. follows the arguments of the industry, many E.U. states attempt to stand for users' rights. The controversy is well demonstrated by the 2002 Kaaza case, when the Netherlands, consequentially nicknamed "the pirate's paradise" granted Kaaza, a peer-network operator established on a Pacific island of Vanatua, immunity from its national laws as a subject outside its territorial jurisdiction. This decision was considered highly controversial, as it was made shortly after the very same service using the very same manner of operation was considered a "contributory infringer" of U.S. protected copyright by a California court. The latest clash of opinion was visible on the streets of European cities in 2012, when thousands of Y-generation Internet users demonstrated against the U.S. promoted Anti-Counterfeiting Trade Agreement (ACTA) – a copyright treaty considered by protesters a threat to online freedom of information sharing. Although their fears were by far belated – ACTA foresaw for little change to what was already made international and national law by the 1994 WTO Agreement on Trade Related Aspects of Intellectual Property Rights (TRIPS) – the demonstrations were a good representation of the intellectual property conflict present in 21st century politics.

The solution of this problem seems relatively easy to find. As the WTO Director General stated before the ACTA demonstrations stormed Europe, the application of copyright in its present form must be altered or copyright laws risk becoming obsolete (WIPO Director General 2012). There are numerous interesting alternatives, like the Creative Commons "licenses". In the face of limited practical capability to effectively exercise their extended understanding of copyright protection granted by numerous international law treaties the multimedia industry needs to re-examine sources of possible income and seek new ways of making profit (Benkler 2006: 59–133). The solution to the problem would be exploring new business models, which proved successful for open source communities, such as e.g. Red Hat Linux or crowd sourcing initiatives, like Wikimedia, rather than insisting on incriminating forever younger generations of "digital natives" (Lessig 2009, xvii, speaking on the irrationality of raising a "generation of criminals"). The open access philosophy, well represented by the Creative Commons licensing models, offers a well-balanced equilibrium of simultaneously protecting rights of authors and enabling non-commercial use of copyrighted, non-derivative works by default. Giving the authority to decide upon the fate and use of their works back into the hands of authors seems a legitimate solution. Again, it will be the political attitude that will be decisive when it comes to any significant change.

Another aspect of the copyright debate is the one on trademarks and domain names. Unlike copyright, based on a universal compromise enshrined in the Berne Convention and following international treaties, granting a minimum, universal level of copyright protection worldwide, trademark rights are granted territorially – every state grants each trademark owner protection within its territory for a limited time, yet only if they file for such protection and pay the appropriate fee. There is no universal trademark protection – obtaining exclusivity to use a certain mark requires re-applying in each country separately (or asking an attorney or a company, to do that for oneself). This seems particularly troublesome in the universal cyberspace, governed along the DNS standards, where each domain name may appear only once in a certain domain. The contractually granted "right" to a domain name which follows from a private law agreement between a registrar and an applicant, has little to do with trademark law. As a rule it is not required to present any trademark related exclusive right in order to register a domain name, and rightfully so, since such a requirement would deem domain name application and management very difficult, if not practically impossible. Therefore as much as there may be numerous rights to a famous mark, such as Budweiser, "owned" by two different entities: one in U.S. and one in the Czech Republic, there may only be one "budweiser" domain name in the ".com" domain. This characteristic of the cyberspace creates the biggest challenge for contemporary trademark law and with the enlargement of top level domains (TLDs) catalogue which began in 2013, that problem is bound to grow. Domain name litigation is difficult and cost-consuming due to the complexity of the trademark laws jigsaw. ICANN together with the World Intellectual Property Organization (WIPO) have introduced a settlement procedure for domain names: the Uniform Dispute Resolution Policy (UDRP). Despite its numerous faults, including above all particular unpredictability, it is regarded as the best solution in place when it comes to online arbitration. Offering out-of-court dispute settlement seems the most pragmatic way to go about the jurisdictional challenge of the Internet. Rather than applying complex and conflicting national regulations, one universal set of rules applicable to online disputes of a particular character, deployed by specialized and authorized entities may prove the best solution.

3 Future legal challenges of the information society

3.1 General remarks

The scope of legal challenges which influence international telecommunication policies reaches far beyond the limited scope of this chapter. Among those not discussed in detail, yet which hold significance in international information policies, one should name: 1. new media and the question of applicability of traditional press laws, relating to journalistic secrecy or professional ethics to new forms of mass media, such as weblogs or social journalism; 2. question on the applicability of traditional labour laws to teleworking contracts, in particular the applicability of workers' protection guarantees and insurance; 3. electronic trade, with particular emphasis on international consumer protection, including the validity of click-agreements; online advertising and gambling; 4. developmental issues – questions regarding the costs of building Internet infrastructure in developing countries. Yet when attempting to discuss the challenges that are yet to gain significance at least **two** problem areas must be named: jurisdiction and the question of cybersecurity, with particular emphasis on the role of states in protecting it.

3.2 Transboundary jurisdiction

As already explained, Internet is a transnational medium with no physical nor electronic borders in place, running primarily on computer code. Laws are territorial, established and enforced within physically delimited borders, where the competence of states to establish and enforce "their" laws is called jurisdiction. Regarding this dissimilarity laws are intrinsically ill-suited to govern the Internet. This simple observation inspired early Internet researchers (Johnson and Post 1996), yet has never attracted the attention of policy makers. The existing matrix of norms on conflicting state jurisdictions is complex and difficult to apply (Kulesza 2012: 1–30 and literature quoted therein). Without a revision of jurisdictional principles for their applicability to online relations no effective compromise for Internet governance may be found. Every national law affecting the cyberspace – be it through taking content offline or blocking access to certain information – holds an inherent transboundary effect. The principle of effective jurisdiction, allowing state to exercise its laws over actions initiated outside its borders, yet causing effects within, is the biggest contemporary threat to the rule of law. No Internet user may be safe from litigation or criminal proceedings in a distant jurisdiction, should his actions cause effects contrary to national laws, while even the principle of double penality does not protect him from responsibility (Teulfel 2012: 403–405 who covers the Thai Joe Gordon case). This uncertainty is also dangerous for Internet companies, offering their services worldwide, at the same time attempting to protect themselves from a worldwide liability. They attempt to do so with self-imposed "rules of service", where they call upon the jurisdiction of the state of their incorporation, requiring all customers to comply. The effectiveness of such contractual rules may be easily questioned, not only as an effect of applying international private law, but also for the simple fact that such a contact is finalized with a click of the mouse and that might be regarded as not providing proper customer protection, as per e.g. EU law.

As depicted above, no comprehensive international accord on limits of state jurisdiction online is in place. At the same time a discussion on state responsibility for transboundary harm online is being initiated. Jurisdiction is one of the keys to assessing liability for harm inflicted by online actions.

3.3 Due diligence and cybersecurity

Due diligence is a term common to national trade law practice, but also to international law on state responsibility. While individual states hold no responsibility for actions other than their own, that is those of their representatives or individuals they authorise to act on their behalf, they may be held responsible for their omissions, including ones that lead to significant transboundary harm (García-Amador 1957: 121; ILC 2001: 38–39). This means that if a state is obliged to take action aimed at preventing certain events from happening, while its failure to do so may give rise to its responsibility. One of the obligations of due conduct is an obligation of prevention – states are obliged to take all reasonable measures to prevent transboundary harm originating from their territory or individuals under their control. This obligations is a direct consequence of the principle of sovereignty – since all states are considered equally independent, they are under an equal obligation to refrain from interfering in affaires of others.

These international law principles may easily be applied to cyberspace. Respecting others' sovereignty requires states to act with due diligence in preventing transboundary harm inflicted by individuals under its jurisdiction or control. In the era of cyberwar and cyberespionage detailing the due diligence obligation for cyberspace is of crucial importance. The existing due diligence principle in international law is compiled of nine elements. First is its close combination with the principle of good faith, requiring all international law subjects to fulfil their obligations with an actual intent of meeting the aim of an obligation taken. It is especially significant with regard to the so-called obligations of conduct (as opposed to the obligations of result) – ones where the actual manner in which the states act when fulfilling the obligation is the main subject thereof. One of the obligations of conduct is an obligation to prevent significant transboundary harm. Due diligence is also closely interrelated with the principle of good neighbourliness, fundamental to international environmental law. It requires states to refrain from causing harm in the territories or to the interests of other states. Whether a state has met its due diligence obligation will be assessed territorially – the assessment will be made primarily, yet not exclusively, with regard to activities performed within state borders. An interpretative guideline for applying the due diligence principle is the principle of sustainable development, which, just as due diligence, needs to be considered in all policy making decisions. The due diligence obligation also requires states to take "all appropriate measures" when preventing transboundary harm, including introducing administrative and other procedures. When assessing the fulfilment of due diligence a theoretical model of "good government" shall be applied. This theoretical model will be combined with an assessment of the state of the art in particular area of practice. A "good government" ought to be aware of state of the art technologies or procedures and have applied them within its individual technological and financial capabilities. The required standard of care is relative also to the degree of potential risks – the greater the risk, the higher the due diligence standard required in a particular case. In order to learn the contemporary state of art states need to engage in international cooperation – the seventh prerequisite for meeting the due diligence obligation. This obligation has a continuing character – a single impact assessment of a particular dangerous activity is not considered a fulfilment of the due diligence requirement. Finally, when exercising due diligence states must refrain from any discrimination – the same norms and regulations must be applied to all entities subject to state laws, jurisdiction or control (Kulesza 2013: 299–300).

The principle of due diligence is best visible in international environmental law, yet it has been recognized by international courts and tribunals since early 19th century. It has been set into most detail in the rich body of international environmental law treaties. While environmental law has been quite influential on Internet governance research, the analogy may also be considered when assessing state obligations online. According to international law any risk-originating activity, that is activity causing the danger of significant transboundary harm, be it operating a nuclear power plant or managing a fleet of oil tankers, is subject to state authorisation. With the growing significance of computer-based infrastructure, when mass catastrophes could be caused by the malfunction of a single machine, as was the case with the ones at uranium enrichment facilities in Iran compromised in 2010 by the Stuxnet virus, operating computer networks might be considered a risk originating activity. The term "critical infrastructure" covers ought to be understood as covering also computer networks significant to the protection of states and their communities. At the same time operating a computer network – within a bank, a water utility station or in mass transportation systems – rarely requires a specific state authorisation. With the dispute on state responsibility for individual cyberattacks increasing the question of due diligence in cyberspace is bound to come into the debate. When seeking the answers, the environmental law model is a given. Whether however, as is the case with oil transportation or nuclear power production, Internet service providers should be considered risk-activity operators and subjected to particular due diligence obligations, including participation in an obligatory liability fund, remains an open question. Introducing such security obligations however, while arguably increasing online security, would change the face of Internet services as we know them.

4 Summary

Legal aspects of the information society are one of the leading issues when it comes to Internet governance. The possible future directions of Internet governance may be reduced to a simple alternative: maintaining the global resource of cyberspace, presumed a compromise on basic values and their practical applications may be found or abandoning the initial concept of a global network for one comprised of smaller, national networks, loosely interconnected. The concept of cyberbalkanisation – a splitting of the net into smaller, national networks organised according to national laws or local standards may seem the safe choice for securing the network, yet it will also mean the end of 21st century's greatest information resource. Given that the global society wants to maintain this groundbreaking invention for the benefits of entire human kind, the only rational way forward is to commence the search for an international compromise on a set of basic principles for a legal order applicable online – to identify and knowledge the basic principles of international Internet law (Kulesza 2012: 125–156).

References

Abbate, Janet. 2000. *Inventing the Internet.* Massachussets: MIT Press.

Alvestrand, Harald. 2004. *Request for Comments: 3935, A Mission Statement for the IETF.* http://www.ietf.org/rfc/rfc3935.txt (accessed 11 April 2014).

Benkler, Yoachi. 2006. *The Wealth of Networks: How Social Production Transforms Markets and Freedom.* Yale: Yale University Press.

Butler, Sue (ed.). 2009. "Dartmoor". *Macquarie Dictionary.* 5th ed. Sydney: Macquarie Dictionary Publishers Pty Ltd.

Byung-Keun, Kim. 2005. *Internationalizing the Internet: The Co-evolution of Influence and Technology.* Cheltenham: Edward Elgar Publishing.

Davis, Thmas W. D. & Brian Galligan. 2011. *Human Rights in Asia.* Cheltenham: Edward Elgar Publishing.

Deibert, Ronald, John Palfrey, Rafal Rohozinski, Jonathan Zittrain & Janis Gross Stein (eds.). 2008. *Access Denied: The Practice and Policy of Global Internet Filtering.* Cambridge, MA: MIT Press.

Farrell, Henry. 2002. Negotiating Privacy Across Arenas: The EU-US "Safe Harbor". In Adrienne Windhoff-Héritier (ed.), *Common Goods: Reinventing European And International* Governance, 105–124. Lanham, Maryland: Rowman & Littlefield.

García-Amador, F. V. 1957. *Second Report of the Special Reporteur.* U.N. Doc. A/CN.4/106

Gasser, Urs, Herbert Burkert, John Palfrey & Jonathan Zittrain. 2010. *Accountability and Transparency at ICANN: An Independent Review*, Harvard: Harvard University. http://dash.harvard.edu/bitstream/handle/1/4555854/atrt-review-berkman-final-report-20oct10-en.pdf?sequence=1 (Accessed 24 April 2014).

Hafner Kathie & Matthew Lyon. 1999. *Where Wizards Stay Up Late: The Origins Of The Internet.* New York: Simon and Schuster.

HRC – Human Rights Committee. 1988. *General Comment No. 16 on Article 17.*

ICANN – Internet Corporation for Assigned Names and Numbers. 2009. *Affirmation of Commitments by the United States Department of Commerce and ICANN.* http://www.icann.org/en/about/agreements/aoc/affirmation-of-commitments-30sep09-en.htm (Accessed 11 April 2014).

ILC – International Law Commission. 2001. *Report.* U.N. Doc. A/56/10.

Kohl, Uta. 2007. *Jurisdiction And The Internet: Regulatory Competence over Online Activity.* Cambridge, MA: Cambridge University Press.
Kulesza, Joanna. 2012. *International Internet Law.* Oxon: Routledge.
Kulesza, Joanna. 2013. *Należyta staranność w prawie międzynarodowym* [due diligence in international law]. Poznan/Lodz: Ars boni et aequi.
Johnson, David R. & David G. Post. 1996. Law And Borders: The Rise of Law in Cyberspace. *Stanford Law Review* 48. 1367–1379.
Lessig, Lawrence. 2006. *Code: version 2.0.* New York: Basic Books.
Lessig, Lawrence. 2009. *Remix – Making Art and Commerce Thrive in the Hybrid Economy.* New York: A & C Black.
Malcolm, Jeremy. 2008. *Multi-Stakeholder Governance and the Internet Governance Forum.* Perth: Terminus Press.
Mayer-Schönberger Viktor & Malte Ziewitz. 2007. Jefferson Rebuffed: The United States And The Future Of Internet Governance. *Columbia Science & Technology Law Review* 8(188). 184–205.
Mueller, Milton. 2010. *Networks and States; The Global Politics of Internet Governance.* Massachusetts: MIT Press.
Muller, Milton. 2004. *Ruling the Root: Internet Governance and the Taming of Cyberspace.* Massachusetts: MIT Press.
NTIA – National Telecommunications and Information Administration, United States Department of Commerce. 1998. *Management of Internet Names and Addresses ('the White Paper'), Docket Number: 980212036-8146.* http://www.ntia.doc.gov/ntiahome/domainname/6_5_98dns.htm (Accessed 11 April 2014).
OECD – Organization for Economic Co-operation and Development. 1980. *Guidelines on the Protection of Privacy and Trans-border Flow of Personal Data.* http://www.oecd.org/internet/ieconomy/oecdguidelinesontheprotectionofprivacyandtransborderflowsofpersonaldata.htm (Accessed 11 April 2014).
ONI – Open Network Analysis Team. 2012. *Global Internet filtering in 2012 at a glance* (April 3rd, 2012). https://opennet.net/blog/2012/04/global-internet-filtering-2012-glance (Accessed 11 April 2014).
Post, David. G. 2010. *In Search of Jefferson's Moose: Notes on the State of Cyberspace.* New York: Oxford University Press.
Rogers, Kevin M. 2006. Internet Governance: The United States won the battle, but will the Internet win the war? *Hertfordshire Law Journal* 4(1). 26–35.
Senftleben, Martin. 2004. *Copyright, Limitations, and the Three-step Test: An Analysis of the Three-step Test in International and EC Copyright Law.* London/New York: Kluwer Law International.
Stiegler, Zack. 2012. *Regulating the Web: Network Neutrality and the Fate of the Open Internet.* Lanham, Maryland: Lexington Books.
Strowel, Alain. 2009. *Peer-to-Peer File Sharing and Secondary Liability in Copyright Law.* Cheltenham: Edward Elgar Publishing.
Teulfel, Mark. 2012. *Thailand Jahrbuch Band 2 nach Ratchaprasong.* Berlin: epubli.
UN – United Nations. 1948. The Universal Declaration of Human Rights (UDHR).
Weber, Rolf H. 2010. *Shaping Internet Governance, Regulatory Challenges.* Berlin/Heidelberg/New York: Springer.
WGIG – Working Group on Internet Governance. 2005. Report of the Working Group on Internet Governance. Château de Bossey. www.wgig.org/docs/WGIGREPORT.pdf (Accessed 11 April 2014).
WIPO – World Intellectual Property Organization. Director General Addresses the Future of Copyright. 2011. Geneve, February 24th. http://www.wipo.int/pressroom/en/articles/2011/article_0005.html (Accessed 11 April 2014).

WSIS – World Summit on the Information Society. 2003a. *Declaration of Principles; Building the Information Society: a global challenge in the new Millennium.* http://www.itu.int/wsis/docs/geneva/official/dop.html (Accessed 11 April 2014).
WSIS – World Summit on the Information Society. 2003b. *Plan of Action.* http://www.itu.int/wsis/docs/geneva/official/poa.html (Accessed 11 April 2014).
WSIS – World Summit on the Information Society. 2005. *Tunis Agenda For The Information Society.* http://www.itu.int/wsis/docs2/tunis/off/6rev1.html (Accessed 11 April 2014).
Yutaka, Arai & Yutaka Arai-Takahashi. 2002. *The Margin of Appreciation Doctrine and the Principle of Proportionality in the Jurisprudence of the ECHR.* Oxford, UK: Intersentia nv.

For further readings

Affirmation of Commitments by the United States Department of Commerce and the Internet Corporation for Assigned Names and Numbers. 2009. https://www.icann.org/en/about/agreements/aoc/affirmation-of-commitments-30sep09-en.htm (Accessed 24 April 2014).
Barendt, Eric. M. 2005. *Freedom of speech.* New York: Oxford University Press.
Baumer, David L., Julia B. Earp, J. C. Poindexter. 2004. Internet privacy law: a comparison between the United States and the European Union. *Computers & Security* 5(23). 400–412.
Boyle, Alan E. 1990. State responsibility and international liability for injurious consequences of acts not prohibited by international law: A necessary distinction? *International and Comparative Law Quarterly* 39(1). 21–24.
Brownlie, Ian. 2008. *Principles of public international law.* Oxford, UK: Oxford University Press.
Crawford, James. 2002. *The International Law Commission's Articles on State Responsibility: Introduction, Text and Commentaries.* Cambridge, UK: Cambridge University Press.
Delibasis, Dimitrios. 2007. *The Right to National Self-defense: In Information Warfare Operations.* Bury St. Edmonds: Arena books.
Drake William J. (ed.). 2005. *Reforming Internet governance: perspectives from the Working Group on Internet Governance (WGIG).* New York: United Nations Publications.
European Parliament. 1995. Directive 95/46/EC of the European Parliament and of the Council of 24 October 1995 on the protection of individuals with regard to the processing of personal data and on the free movement of such data. *Official Journal L* 281. 31–50.
Gutwirth, Serge, Yves Poullet & Paul de Hert (eds). 2010. *Data Protection in a Profiled World.* Berlin/Heidelberg/New York: Springer.
Komaitis, Kostantinos. 2010. *The current state of domain name regulation: domain names as second class citizens in a mark-dominated world.* Oxon: Routledge.
Kramer, Fraklin D., Stuart H. Starr & Larry Wentz. 2009. *Cyberpower and National Security.* Dulles: Potomac Books Inc.
Kurbalija, Jovan. 2010. *An Introduction to Internet Governance.* Msida: Diplo Foundation.
Lessig, Lawrence. 2002. *The future of ideas: the fate of the commons in a connected world.* New York: Vintage Books.
Mayer-Schönberger, Viktor. 2009. *Delete – The Virtue of Forgetting in the Digital Age.* Princeton: Pinceton University Press.
Petković, Milan & Willen Jonker (eds.). 2007. *Security, privacy and trust in modern data management.* Berlin/Heidelberg/New York: Springer.
Reding, Vivien. 2009. *The Future of Internet Governance: Towards an Accountable ICANN: Commissioner Reding's Weekly Videomessage Theme* (May 4th, 2009). http://pl.scribd.com/doc/14969947/The-Future-of-Internet-Governance-Towards-an-Accountable-ICANN-by-Viviane-Reding (Accessed 14 April 2014).

Segura-Serrano, Antonio. 2006. Internet Regulation and the Role of International Law. In Armin von Bogdandy and Rüdiger Wolfrum (eds.), *10 Max Planck Yearbook of United National Law*, 191–272. Leiden: Koninklijke Brill.
Segura-Serrano, Antonio. 2006. Internet Regulation: A Hard-Law Proposal. *Jean Monnet Working Paper*, 10/06. http://www.jeanmonnetprogram.org/papers/06/061001.pdf (Accessed 24 April 2014).
Van Schewick, Barbara. 2010. *Internet Architecture and Innovation*. Massachusetts: MIT Press.
Wang, Faye Fangfei. 2010. *Internet Jurisdiction and Choice of Law: Legal Practices in the EU, US and China*. Cambridge: Cambridge University Press.
Weber, Rolf H. & Romana Weber. 2010. *Internet of Things: Legal Perspectives*. Berlin/Heidelberg/New York: Springer.
Winn, Jane K. 2006. *Consumer protection in the age of the 'information economy*. Aldershot/Burlington: Ashgate Publishing, Ltd.
Xue, Hong. 2010. Diversity: Achieving an Internet that is Really for All. In William J. Drake (ed.), *Internet Governance: Creating Opportunities for All*, 25–33. New York: United Nations.
Zittrain, Jonathan. L. 2008. *The future of the Internet and how to stop it*. Cambridge: Yale University Press.

Adriano Fabris

18 Ethical issues in Internet communication

Abstract: My intention is to discuss in this chapter the numerous and multifarious Ethical Issues in Internet Communication. Once I have defined the term "ethics", its issues and its various forms, I will address the specific ethical problems which emerge in connection with the world of the Internet and examine the different strategies generally used for tackling them. To this end, I will differentiate and single out two ways in which we can successfully relate to the Internet. On the one hand it is necessary to individuate the issues regarding the Internet's impact on society: it is worth reflecting preliminarily on these issues in order to understand the changes in human behaviour caused by the emerging technologies. On the other hand, it is possible to discern behaviour problems relating to how we use the web: a behaviour that can be defined as good or bad, fair or unfair, starting form criteria and principles that have to be established and justified. I will term the former aspect "Ethics of the Internet" and the latter "Ethics in the Internet". Both will be developed in my paper.

Keywords: ethics, communication, internet, social network, responsibility

1 Introduction

Over the past years Information and Communication Technologies (ICTs) have profoundly changed our relationship systems and the way in which human beings relate to the world and are connected with each other. The field of these technologies comprises Internet Communication as well. Although its history is fairly recent, its impact on our lives is profound and rapidly growing.

The focus upon the way technologies affect human actions has led, in recent years, to the emergence of ethical reflections on the topic. These reflections concern both the general field of the ICTs and the specific Internet one. In a way which parallels the elaboration of an Information and Computer Ethics (ICE) (some of the most recent publications on the topic are: Drushel and German 2011; Floridi 2010; Floridi 2013; Johnson 2009; Migga Kizza 2013; Ploug 2009; Quinn 2012; Reynolds 2013; Sanders and Brinkman 2013; Stamatellos 2007; Tavani 2012) an Internet Ethics, concerning especially Internet Communication, has been developed within the same field. This is the aspect dealt with by the current chapter.

My intention is to discuss in a precise and articulate way the numerous and multifarious Ethical Issues in Internet Communication. These issues have been addressed in a thorough – albeit mainly disordered – way in a number of works. The aim of this paper is to provide, instead, a general survey, a mapping of the various problems that concern the performance of human agency by means of the Internet. I will try to provide a systematic discussion of these problems and of the way in which they can be examined from an ethical perspective.

I will proceed as follows. Once I have defined the term "ethics", its issues and its various forms, I will address the specific ethical problems which emerge in connection with the world of the Internet and examine the different strategies generally used for tackling them. It is not my intention, however, to address all these issues by adopting a bottom-up method. That is to say, I do not wish to start with specific case-studies in order to identify a number of ethical criteria enabling us to make definite choices, which can be justified and be subject to argumentation (Barger 2008). My aim, as I have already stated, is to provide a general framework, a comprehensive ethical horizon, within which these multifarious problems can be incorporated and systematically discussed. This horizon coincides with an ethics of responsibility.

To this end, I will differentiate and single out two ways in which we can successfully relate to the Internet, two directions along which both a reflection and an ethically motivated behaviour towards the gradual development of Internet Communication can be developed. As a matter of fact, it is necessary to identify, *on the one hand*, the issues regarding the Internet's impact on society, on the world we live in, on our very way of thinking and living: it is worth reflecting preliminarily on these issues in order to understand the changes in human behaviour caused by the emerging technologies. *On the other hand*, it is possible to discern behaviour problems relating to how we use the web, to the actions performed on the web and by means of the web: a behaviour that can be defined as good or bad, fair or unfair, starting from criteria and principles that have to be established and justified. I will term the former aspect *Ethics of the Internet* (EoI) and the latter *Ethics in the Internet* (EiI).

As far as the EoI is concerned, I will address some of the ways in which ICTs, and in particular Internet Communication, broaden human potential, affect human behaviour, induce the human being to make new choices. This is mainly caused by the fact that a powerfully attractive virtual reality exists, and develops, together with the reality of everyday experience. I will examine these topics both from a theoretical and a meta-ethical perspective.

As far as the EiI is concerned, I will discuss some of the more concrete issues that emerge in every day uses of the Internet – considered in its various forms (up to Web 2.0) – and the ways in which these issues have been tackled so far from an ethical and deontological point of view. I will show the need to introduce and justify a number of general rules which may, for instance, guide people in the writing of emails, in their Social Network participation or in the use of a search engine.

The way in which one usually attempts to deal with these issues for achieving this result is bound to the elaboration of specific behavioural rules. These rules establish a series of duties and regulations that are generally listed in specific codes, and are mainly intended to restrict the activities of Internet professionals. But, bearing in mind Social Networks and user-generated contents, these codes should be applied to whoever uses the web. I will point out the advantages and disadvantages of an approach such as this one and, above all, show that the adoption of a code has to be ethically motivated. This is the reason why deontology, professional deontology in the first place, relates to ethics. In other words, it is crucial that deontology be founded upon ethics and, more specifically, that an ethics of responsibility – able to affect all netsurfers – be elaborated.

Hence, by making reference to these problems and these distinctions I will attempt to draw my general map of the ethical issues in Internet Communication. On this map it will be possible to place, and discuss, problems which have already been examined by leading experts in the area, and which define the scope of a correct, good, virtuous use of the Web. In addition, it will be possible to include, in this frame, emerging issues and issues that, due to the many directions of Internet's growth (suffice it to mention Web 3.0), are going to powerfully emerge in the future.

2 Ethics and ICE

Ethics is a discipline that studies the criteria and the principles of conduct, which helps distinguishing the standards by which principles can be judged to be right or wrong, which provides reasons able to guide an agent – who is free to make choices –, which motivates the choice of what is held to be good rather than what is believed to be bad. There are quite a few philosophical approaches which serve this purpose, not only in the history of Ethics, but also in contemporary debate. Still, some ethical questions are constantly iterated throughout this history and this debate. They are, above all: What am I doing? What should I do? What is the sense of my actions? Nowadays, the doctrines which attempt to answer these fundamental questions refer mainly to two strategies. On the one hand, ethics is conceptualized as a *theoretical reflection* on human actions and on the criteria for judging those actions, on the concept of "good" and on how we use our moral vocabulary in a competent and grounded way. In this case it is preferable to speak about meta-ethics. On the other hand, one of the major concerns is how to move from the knowledge and the enunciation of ethical principles to their actual realization. The issues that emerge here regard the application of moral criteria to specific cases; they concern the education and the formation of character and aim at the realization of a meaningful conduct (for a general overview: Singer 2006).

The major trends in contemporary ethics reappear, albeit slightly changed, in the field of that specific kind of activity consisting in actions performed by means of the new technologies: in our case, by means of the ICTs. But a very important change occurs in this context. Ethics is not expected to investigate the conduct of human beings only, or human conduct enhanced by technologies. Since the very same technologies are able to promote specific and autonomous actions (Dusek 2006; Fabris 2012), ethics also has to investigate the criteria which guide technological agency and the interaction between human beings and the new artificial agents. Consider, for instance, the development of Roboethics (Wallach and Allen 2009).

It is necessary, therefore, to apply to this situation of constant technological development the research elaborated by contemporary ethics. In today's debate this is accomplished by privileging mainly two aspects. On the one hand, new reflections on the values of reference in technological agency are vigorously promoted. On the other hand, an attempt is made to individuate a system of fundamental norms able to guide the human being in his/her decision making process. The ultimate challenge is thus to elaborate an ethics which is truly adequate to the behavioural changes brought about by the ICT.

As I have already mentioned, the ethical values of reference are not only the ones pertaining to human actions, but also the ones inherent in the agency of the new technologies. Technologies, as a matter of fact, are not neutral. They embody a number of values, which are reflected in the actions they perform. There are, for instance, values which are "embedded" in computer systems, independently of the particular ways of using them (Flanagan et al. 2008). From this point of view technological artifacts are capable of either promoting or arming the realization of values when they are used. For instance, as far as the field of interpersonal relations is concerned, they can promote the broadening of democratic systems.

Analogously, the normative approach characterizing modern and contemporary ethics reappears in the new technological context as well. And yet, it also undergoes an important transformation. ICTs alter our old habits, change the way we act, and thus it is very hard to regulate them on the basis of a single and absolute moral principle, whose application is seen as dutiful. One should rather start from the intrinsic plurality of values which characterizes our era and take into account the difficulty in immediately applying general normative criteria to concrete situations. A new conception is worthy of being elaborated, a conception which is "*empirically informed, realistic and practical*, so as to provide guidance and direction in cases where information technology is actually used" (van den Hoven in Floridi 2010: 61).

As it can be seen, ICTs not only require to be managed from an ethical point of view, but they have even succeeded in widening the meaning attributed by philosophical tradition to a discipline such as ethics. In order to address these changes Luciano Floridi has elaborated an original ICE approach. Floridi takes into account the fact that we live today in a new reality, termed "infosphere", in which, consequently, the role and the function of the moral agent undergo significant changes. Floridi formulates four fundamental laws so that the moral agent be able to orient him/herself in this situation. Essentially, these concepts help avoiding the entropy of the system within which s/he operates (Floridi 2010: 92–93; 2013).

3 Ethics and the Internet

The specific discussion on Internet Ethics (IE) can thus be incorporated both within this general reflection on ethics, which addresses the fundamental questions tackled by the discipline and the way in which these questions are dealt with in the contemporary debate, and within the frame of the new ICTs development. I have already mentioned that there are mainly two ways in which we generally relate to the Internet. So, in relation to this matter, there are two directions along which an IE can be elaborated

The EoI focuses on the impact the Internet has on society, on the world we live in, on our way of thinking, irrespective of whether we relate to the web professionally or not. The EoI deals thus with the structure of the web and with the consequences of the various aspects of the Internet on our life (Cavalier 2005; Bouchon 2005). Even without being programmers, webmasters, Internet professionals or just mere web users (albeit this is really atypical as far as the new generations are concerned), we still nevertheless are part of a world that is being changed by the Internet.

In particular, we have to examine three characteristics of Internet's structure and of the consequences it entails. The Internet allows us to enter a virtual reality, causes inequities in accessing the opportunities offered by it, but at the same time it nonetheless transforms data, which can be accessed by a growing amount of people, into something that cannot be fully controlled and managed by the human subject. Our behaviour and the way we regard and respond to this behaviour undergo a substantial change in this context.

Nowadays we are able to live and to act in a number of worlds at the same time, whether they be real (as in the case of offline experience) or virtual; we are offered plenty of opportunities for accessing information and can easily interact with other human beings (but only if we live in specific parts of the world and have an adequate education); we are able to collect an overabundant amount of information (which makes it quite difficult to orient oneself), and to get easily connected (a fact that might compromise privacy and the personal property of what is put on the web). Hence, we must reflect on these specific issues and ethically evaluate them.

The EiI investigates, instead, a set of attitudes, held to be good and right, performed by whoever uses the Internet. This involves three categories of people or three different activities which might relate to the same person.

First of all, there are Internet professionals who work as programmers, professional developers or webmasters, and for whom the Internet is a most essential tool. Their occupation might lead them to "digital dilemmas" (Berkman and Shumway 2003), which involve the making of very specific choices. There are, furthermore, people who also use the Internet for professional reasons and who carry out their work by means of the Internet, albeit they do not see the web as the most essential object of their work. Consider for instance the occupations in which the Internet is used for real-time communication, as a solution for exchanging news, or for data elaboration. Consider also the people who use the Internet for their research (McKee and Porter 2009). There are, finally, people who do not inhabit the world of the Internet for professional reasons, but who rather see it as a way of establishing relations or getting information and entertainment. As a matter of fact this is quite a widespread experience in the world of Web 2.0.

The issue which emerges in all these cases is how to regulate the activities made possible by the Internet. The individual who conducts activities on the Internet, is seen as a free individual, inasmuch as s/he is free to choose whether to connect up or not; once connected to the Internet, the individual may behave well or badly. Of course, the Internet is a new environment which we have to interact with and which we can evaluate the implications of. But the way in which we really interact with it and make specific choices within it, depends on us, and brings into play the responsibility, which is ultimately our own only.

But what are the points of reference able to guide us in our decision making? The Internet does not give any indications, it does not prescribe anything. This might be the reason why the Internet is perceived as a place of freedom: because it leaves the choice to the user's free will. Nevertheless, a frame of reference able to guide our decision making, a set of criteria allowing us to know right from wrong, appears in any case to be of utmost importance.

In the field of the EiI this is mostly provided by a specific tool: the behavioural codes. They establish the criteria of a proper conduct in using the Internet and in availing oneself of the opportunities supplied by it. These criteria concern, albeit in different ways, both the people working on the development of the Internet and, more generally, on the development of the new technological applications, and Internet professionals or people who use the Internet for spending their free time online.

But it is right here that some of the problems emerge. If ethics is identified with a set of codes, there is the risk that ethical issues be confused with legal ones. There is a substantial difference between these two fields: law has to do with justice and with fostering a righteous society, whereas ethics has to do, above all, with what is good for the human being. Furthermore, even if our preference went to normative ethics, it still wouldn't be correct to assume that ethics consists in a mere list of obligations, because otherwise ethics would be merely transformed in "etiquette". Or better, in our case, in a "netiquette". I will focus on this subject in a short while.

4 Some EoI issues

Let's go back, first of all, to the EoI. As I have already stated, the EoI deals with the way we evaluate and ethically judge the various aspects that belong to the Internet's structure, no matter if we are actively involved in the field or not. I would like to elaborate on three of these aspects which can be examined from other points of view as well, but which have specific ethical implications. This involves various consequences of virtuality, the inequities in Internet access, the proliferation and transparency of information but also, in parallel, the disrespectful way in which information is sometimes handled.

4.1 Ethics of virtuality

The first aspect of the EoI concerns the changes which, from an ethical point of view, are brought about by the presence of a virtual reality next to the reality which characterises offline experience. Virtual reality changes the way we understand and experience space, time, the relationship with other human beings (Crowston et al. 2007). It is therefore imperative to elaborate proper ways of relating this reality to the reality of everyday life. This is the main ethical task we have to face in regard to this issue.

The word "virtual" refers in fact to the "being" which is endowed with *virtus*, that is with the ability to fulfil one's potential, since it is empowered to do so, and to fulfil all the possibilities that characterize it. Or, better still, since the capacity to fulfil its potential is always inherent in the virtual being, so that its possibilities attain ontological depth and achieve autonomy, it is possible to argue that the virtual being possesses its own specific reality. This is the reason why it is neither a contradiction nor an oxymoron to talk about "virtual reality" (Fabris 2007).

We are dealing here with the fact that we tend to consider the reality of offline experience as equivalent to virtual reality and see no differences between them. Better yet, we are driven, in our everyday experience, towards the adoption of a mentality which endorses this indiscriminateness and considers it to be a value. Nowadays virtual reality not only tends to obtain more and more consistency, that is, it gradually comes to resemble the reality of everyday life, but, in parallel, the very offline reality tends to lose depth, becomes insubstantial, because it is manipulated in a more and more profound way.

All this has to be considered and judged from an ethical perspective, that is to say, in a way of dealing with our behaviour and our life, which is based on universally shareable principles. As a matter of fact, everyday reality is not a place where role-playing games are played. Notwithstanding the convergence of some typically everyday processes, according to which the virtual is realized and the real is virtualized, what ethics ought to aim at is the individuation of a number of criteria that enable a fair relationship between what is offline and what is online.

The major prerequisite of the EoI consists, indeed, in discerning the various senses of "reality" in which human beings can live. This distinction allows the individual to express oneself and one's abilities within a variety of fields. A process such as this one contrasts all attempts at reducing to a single dimension the articulated sphere of activities and thought which characterizes everyone of us: an attempt that has to be rejected because its realization would deny the plurality of our nature and its openness to new possibilities. And still, the process of reducing the human being and its sphere of action to a single dimension not only is a temptation which systematically recurs in history, but it also comes into view in relation to virtual reality. It seems that the commandment with which we ought to comply is: "Stay always connected!". That is: "Let yourself be absorbed by what is virtual!". Faced with this issue, ethics teaches us to salvage other aspects of our life as well. To take them back, if necessary. Even if this implies turning the computer off.

4.2 Fairness and digital divide

Another aspect to be examined in relation to the Internet's structure and to its consequences on our life concerns, from a geo-political perspective, the so-called "digital divide" and, from a perspective taking into consideration the internal structuring of a society, the so-called "cultural divide". It is true that the Internet enormously expands our possibilities for information access and knowledge exchange. But it is also true that these opportunities are not ensured to everybody in the same way. Better still, Internet's geography is characterized by a global disequilibrium between places which are connected to the Internet and places which are not, or are inadequately so. Similarly, there is an analogous disequilibrium between areas directly traversed by telecommunications infrastructures and areas which have to refer to the infrastructures of others. Usually this is referred to as "digital divide". It does not only concern inequalities of a technological kind. Indeed, the digital divide affects all the possibilities of innovation, of knowledge development and of economical growth, which depend on the actual access to the Internet.

An analogous divide can be discerned within the same society when not all the groups or people partaking in it have the cultural tools, the open-mindedness and the education for availing themselves of the opportunities provided by the

Internet. We are talking, namely, of the so-called "cultural divide". The problem, in this case, is related to society's structure and concerns the ability of its members to adapt to the constantly changing and ever-evolving technologies. The issues that come into play here concern, above all, educational forms, generational differences, the fact that the members of a society are expected to stay up-to-date; even though it is quite evident that the global inequities brought forth by the digital divide also play a role in the emergence of cultural inequities within a society (Walsham 2001).

Both concepts contribute to drawing the attention to a real situation of inequality and to a scarce respect for the equal opportunities which should be ensured to every human being. This is the reason why both the digital and the cultural divide are of utmost importance in ethics.

First and foremost, it is a question of justice. In this frame, justice is identified with fairness, that is to say, with guaranteeing everybody an equal access to the same opportunities (Rawls 1999). Hence, the role of ethics is not only to individuate the criteria enabling that access, but also to account for their legitimacy and to explicate their endorsement. From this perspective, every intervention in the social sphere and every political initiative, both national and international, ought to be based on the way in which ethics develops these topics.

4.3 Uses and abuses of transparency

A third aspect of the EoI concerns another consequence of the Internet's development. It is related, more specifically, both to availability and accessibility issues, be they global or pertaining to individuals, and to the accuracy of information, to issues concerning intellectual property, privacy and the correct administration of online data. All this can be considered from two different points of view: both as a consequence of the Internet's very structure, that is to say of the way the Internet is built (this is what I will focus on now); and as a specific problem to be addressed in the field of the EiI, that is to say as a number of ethical issues related to the specific conduct of the human being within this structure (I will examine this topic in the next paragraph). Therefore, the reference to this third aspect will allow me to move the discussion from the EoI towards the EiI.

If we address these issues it becomes clear that, due to its very structure, the Internet puts us in an ambiguous situation. On the one hand, we are subjected to an information overload – which makes it quite difficult to orient ourselves – and to an easiness of connections which broadens our relational competencies. On the other hand, this entails a transparency in the relationships, which, while undeniably considered to be a value, runs the risk, at the same time, of dissolving individual privacy and cancelling the possibility to safeguard the intellectual property right over the creative works published on the Internet.

At a closer look, we first ought to notice that before being a consequence of an improper Internet use, this situation is an outcome of the Internet's very structure. As we know, the Internet is structured as a "hypertext" (Nelson 1981). The hypertext is an open text, made of other texts, branching off indefinitely. Virtually, every part is connected to the other and can be integrated and built in an unlimited way. This is the reason why the Internet is able to provide us with an enormous amount of data, which can be easily uploaded and shared, joined and mingled as happens in Web 2.0. Consequently, we run the risk of incurring into a communicative overdose, in relation to which the Internet does not offer any guidance criteria (a different thesis is to be found in Enderlin 2003). These criteria have to be introduced from the outside, starting from the interests and the abilities of the people acting by means of the Internet. Another outcome, furthermore, is the pulling down of every barrier hindering the free flow of information, the availability of data to anyone and data intermingling. If we must comply with the imperative, which I have previously referred to, to stay connected by all means, transparency has to be global.

The most evident and mostly discussed ways in which this transparency is realized concern the dissolution of individual privacy and the annulling, in the net, of the possibility to safeguard intellectual property. Everything that is put online falls, virtually, into the public domain, and as such it can be shared and disseminated. It doesn't matter if this involves sensitive data or information which – if it falls into the wrong hands – might cause harmful consequences. It doesn't matter if this data is then used for commercial reasons. Everything has to be shared and taken part in: this is what the Internet makes possible. And – as is theorized by some – if it is possible then it is good.

The same applies to intellectual property, that is to say the property over the products of the human intellect. As a matter of fact, the development of the ICTs offers new ways of creating, presenting and disseminating intellectual creations. It is very difficult, therefore, to reestablish the conditions of an adequate, let's say economical, acknowledgment of the work which has produced these creations. The fact that they are online, and that, most likely, everyone is capable of altering them, leads one to think that they are everybody's property. This is, as I have already stated, due to the Internet's very structure: an open, boundlessly expanding structure, which pulls down every barrier and enables the sharing of everything which is put online. Internet users can either accept these consequences or disregard them, they can choose to avail themselves of the possibility, for instance, to download copyrighted material without paying it. They can even theorize the end of these requirements.

As a matter of fact, some people – the advocates, to cite an instance, of the so-called "hacker ethics" (Himanen 2010) – uphold the view that the elimination of every barrier is legitimate and necessary, whereas others attempt to contain the indiscriminate use of the Internet with the aim of regulating it in new ways. They do this by rooting their reasoning in philosophical argumentation and by attempting, once again, to justify the respect of individual privacy and the safeguarding of intellectual property (Migga Kizza 2013). A utilitarian perspective is usually referred to in this case, even though this leads to the contraposition between what is useful for some Internet users and the usefulness and opportunities characterising the system as such.

In any case, new forms of regulation have to be created and truly shared so that these argumentations be efficient and really of use. In fact, the problems at stake here are concrete ones. As far as the respect of privacy is concerned, what comes into play, for instance, is the possibility of using the information obtained by a search engine or by a Social Network for purposes which might be far removed from the ones a netsurfer normally has when looking for information or using specific services. As far as the safeguarding of intellectual property is concerned, what comes into play is the permission to freely use software or to download audio and video files which are not in the public domain. The tools which are normally used for reaching this goal are legal: the regulation codes are a case in point. But by introducing this topic we are already led towards the examination of the EiI.

5 EiI issues

The EiI intends to regulate the conduct of the people who act by means of the Internet. It does not deal, therefore, with general issues, concerning the Internet's structure or issues where a moral judgment is required, as is the case of the EoI. The EiI concerns, rather, concrete problems which arise when we act on the Internet. These problems can be examined from *two main perspectives*.

The first one refers to the various categories of people who use the Internet. We have already examined this: these categories involve professionals for whom the Internet is a most basic tool (programmers, software creators etc.), people using the Internet as a support for their professional activity (albeit this activity is not directly involved with the web), people using the Internet for their research and, finally, people using the Internet for reasons which are not bound to any professional activity. In all these cases, apart from a number of general and shared norms, there are some specific behavioural criteria, which these various categories are expected to comply with.

The second perspective refers to ethical issues and to the challenges posed by the management and by the actual use of online communication tools. This is the case, in particular, of the platforms and environments enabling specific interactions: such as email, Social Networks (like Facebook, LinkedIn, Twitter, and other messaging tools), search engines (like Google), and other online research resources (Drushel and German 2011). The use of these online tools allows us to perform specific actions which need to be regulated. In addition, a further problem comes into view, not only in relation to the individual conduct of the person using these tools, but also to the way the platforms and the networks are built, organized and developed. It is thus necessary to regulate the way their services are provided and the way in which data is made use of once the societies which supply these services enter into possession of it.

In both perspectives, however, the ethical issues that most frequently come into view are addressed in an analogous way: reference is made to a series of prescriptions which are indicated in regulation codes. This mainly happens when it is not possible to rely on commonsense and on the criterion of immediate accordance between individual and collective utility, since the former, the individual utility, runs the risk to prevail over the latter.

These codes are multifarious. As far as professional Internet users are concerned (people who work on the Internet and for the Internet, and not only by means of the Internet), there are mainly binding codes that refer to their specific responsibility (Bynum and Rogerson 2004). As far as the wider group of Internet users is concerned, the one comprising people who use the Internet both as a support for their professional activity and in their free time, there are general behavioural codes. More specifically, these codes aim at indicating the way in which specific multimedia texts should or shouldn't be written. These rules facilitate an appropriate and respectful behaviour towards the other users when writing, for instance, an email or a blog, when partaking in a chat or in a Social Network (for a proper email administration see Rooksby 2002).

In addition to these codes, which intend to regulate the activities conducted on the Internet by appealing chiefly to the individual interests and commitments of the Internet users, there are other codes, which are more attentive towards the interests of the people involved in the communication processes: the recipient of a message or the people who are the subject matter of the message. This is the case, for instance, of the codes concerning the Public Administration's use of the Internet. This is the case also, and above all, of the codes that take into consideration the consequences of networking practices among underage users: minors are regarded both as netsurfers (and thus exposed to the viewing of sites which might perturb them), and as objects of improper activities conducted on the web (online solicitation, exchange of pedo-pornographic files and so on). In these cases, in addition to the formulation of rules on fair Internet use, concrete initiatives of control and monitoring are promoted, which prohibit thus the indiscriminate freedom of navigation.

In addition to the just mentioned codes referring to the various typologies of Internet users and to the ways in which these people are involved in online communication, there are codes which involve the computing professionals and the companies providing the services. Operators and companies use these codes for stating their policy and the way in which they intend to regulate the web community they offer the services to.

As a matter of fact, if we wish to access these services, we usually find a series of statements and principles listed in minute detail and aimed at defining the appropriateness of our conduct, even though we rarely pay attention to, or read carefully, the many pages which state what joining, let's say, the community, entails. So we immediately click the icon that indicates we accept the terms. Let's think, for instance, about the norms for using Facebook. The general impression, overall, is that these codes are meant to safeguard the company's interests rather than the interests of those people who avail themselves of its services.

Finally, there is a very last typology of codes, of a more general and programmatic kind. They are elaborated by the various associations engaged in computer development and, in general, in the development of new technologies; at the same time they aspire to provide criteria for a global regulation of the cyberspace. This category comprises, for instance, ACM's code (Association for Computing Machinery) and analogous codes promoted by the IEEE (Institute for Electronic and Electrical Engineers), by the DPMA (Data Processing Management Association) and by the ICCP (Institute for Certification of Computing Professionals). These codes are aimed towards guaranteeing the dignity of all the people involved in ICT processes; the personal integrity and honesty of professionals; the responsibility in carrying on one's profession; the confidentiality of information; the public safety, health, and welfare, both in the online and in the off-line world; the participation in professional societies to improve standards of the professions; the notion that knowledge and access to technology is equivalent to social power, and that this social power is something which has to be accessible to everybody.

And yet, the managing of ethical issues in Internet Communication by means of the elaboration of concrete behavioural codes, as is often the case with the EiI, allows some perplexities to come into view. It is true that codes have the merit of clearly individuating what is allowed and what is prohibited while using the tools which constitute the environment of the Internet. But in order to function and be efficacious codes can refer only to mechanisms able to punish an ascertained transgression. Here is where two different types of problems emerge.

On the one hand, this situation might, in effect, necessitate that an institution supervise the activities performed within a specific environment: an institution acknowledged by everybody and able to really sanction the transgressions. But it is rather difficult to constitute such an institution and, in any case, its activity cannot be globally extended to the Internet as a whole. It might undeniably concern the specific use of a platform or the access to a particular service, but only within clearly established limits. In this case, if a transgression is noticed, the heaviest sanction would consist in being excluded from the enjoyment of these very same services: that is, of being excluded from the community. But since the Internet is not governed and supervised by a single authority, it is not possible, at the present, to refer to a single institution exerting control on a global level and able to impose sanctions to penalise every transgression. But maybe it is better this way, since there always is the risk that this control institution be changed into something similar to the "Big Brother" depicted by George Orwell (2008).

On the other hand, the reference to the codes seems to solve a series of ethical problems with the help of an approach which belongs, rather, to jurisprudence. But ethics is radically different from jurisprudence in scope and in subject matter. I have already touched on this topic: while jurisprudence establishes laws of social behaviour, and hence deals with justice, ethics applies to all human beings and deals with what is morally good. This means that the principles ethics examines and validates are not presented in the form of norms that are applicable to a social context, but concern, rather, the human being as such, considered in its habitual way of acting and in its life choices. In brief, it is not for avoiding a probable penalty that one holds an ethical behaviour, but in observance and commitment to what is good.

Therefore, codes are not sufficient for regulating the behaviour on the Internet. The EiI cannot merely be based on a modus operandi which provides a punishment for every prospective transgression. What is needed is an ethical supplement; an appeal to the sense of responsibility of whoever performs actions on the Internet. This means that, similarly to the EoI, the EiI calls for the elaboration of general criteria able to clarifying what a good action is and to prompting towards the performance of good actions.

But how can this appeal be realized? We can concretely see it by analysing one of the previously mentioned issues. That is to say, the alternative, on the Internet, between the respect and safeguarding of everybody's freedom of speech and the need to control this very same freedom of expression, chiefly with social security purposes. I will employ this analysis for concluding the chapter.

6 The Internet between freedom of expression and will to control

The Internet is, in its multifarious development, a place of freedom. This has already been pointed out. As such, it encourages new forms of agency. But these new forms of agency raise the old questions which make every communicative interrelation possible. It is necessary that the communication on the Internet be based on the trust established between the various interlocutors. This trust, in turn, is based on respect. Respect is based on the presumption that what is being said by each one interlocutor is true. This is the ethical ground which enables any communication.

In order to guarantee a communicative relationship it is, therefore, not possible to say whatever one wishes to. Freedom of expression has to be regulated. The problem is the way in which this regulation is put into practice. The problem is who enacts this regulation and according to which criteria. In addition, and within this regulated space, the problem is how to use data which is shared, data which has to be gathered and mastered, also for being regulated.

Media periodically disseminate news regarding mechanisms of control and surveillance of online information. Without any doubt, surveillance and control seem to hinder the freedom of expression and frequently bear a resemblance to censorship. Nevertheless, this activity is usually justified for safety reasons. The surveillance and control of the flow of information on the Internet is, for instance, of strategic importance in preventing terrorist attacks. From this perspective the dissemination of confidential information constitutes an attack against national security. Safeguarding the confidentiality of one's communication activities and gaining unauthorized access to controlled information frequently implies the development of sophisticated systems and plans for informational attacks. It is not unusual for world powers to rely on hackers. And again, as was formerly the case with the exercise of the freedom of expression by means of media such as newspapers and television, a situation takes shape in which the only choice seems to be the one between freedom and security.

On the other hand, the Internet is widely used and acknowledged as being the place in which someday every barrier will be torn down. We have examined this topic with regard to how hard the application of Internet codes of conduct is. Copyleft has emerged as a challenge to copyright. Guy Fawkes mask – the white mask which is dressed up by many demonstrationists all over the world – is seen as a tool for running against the respect for privacy and for the ways of managing our online identity via the creation of multiple profiles we might decide to set up on the Internet. The counter-information published by Wikileaks stands out against Echelon's surveillance.

How are thus these issues to be ethically managed? The answer that I will attempt to give to this question represents a case in point, which can be applied to other ethically problematic situations, analogous to the ones previously discussed. It must be said that, in this case as well, ethics does not provide clear-cut formulas; it does not suggest straightforward procedures that can be mechanically followed. It rather provides a general orientating framework within which every individual is free to make his/her own choices.

In this case the orientating framework is provided by the sense of responsibility intrinsic to each free agent, even to the one conducting activities on the Internet. "Responsibility" (Raffoul 2010) means to take into account the consequences of one's actions: inasmuch as this characteristic distinguishes human actions – which are rational ones – from the actions of other beings. But the awareness of the consequences of one's actions must be something which is taken on in a conscious way. In other words, one must profoundly feel this principle of responsibility, be pulled towards it. This is where an additional meaning of being responsible comes into view; the one which induces us to endorse criteria and principles which, being acknowledged by us as binding, guide our actions.

The main criteria to be acknowledged in communicative acting are the ones I have stated before. They involve mutual trust between and respect for the various interlocutors. In fact, if these criteria are not endorsed and preserved, communication comes to a standstill sooner or later. This is the reason why, starting from this orientating framework based on the two meanings of "responsibility" previously mentioned, we can reject – because considered to be irresponsible – both the idea of a limitless freedom of expression (although the Internet makes it possible) and the need for a rigid censorship. More exactly, it is necessary to find, instead, an ethically motivated equilibrium point in between these two extremes.

This equilibrium point cannot affect the right to freedom of expression from the outside: this would be an additional example of censorship. It must, instead, be the result of a process of a transparent self-regulation common to the various Internet users. As such it may vary in accordance to the growth or the decrease of security instances, due to specific historical situations, and to the need to realize the transparent and widespread information made possible by the Internet.

The criterion for identifying this equilibrium point could be termed a criterion of *responsible transparency*. On the basis of this criterion, the growing transparency made possible by the Internet's structure is certainly adhered to: but only within the limits established in observance of the trust without which communication would be impossible; and in compliance with the respect shown to individuals engaged in a communicative process, without which it would come to a standstill. It is we who establish these limits, in a responsible way.

This criterion can be used for addressing and tackling not only the issues examined so far, but also a number of other issues which the Internet raises and will continue to raise for whoever intends to conduct responsible and motivated actions on it. It will, furthermore, be possible to orientate oneself with reference to this principle in the numerous and complex issues discussed so far and pertaining both to the EoI and to the EiI. But above all, it will be possible to act ethically while using the Internet.

References

Barger, Robert N. 2008. *Computer Ethics: A Case-based Approach.* Cambridge: Cambridge U.P.

Berkman, Robert I. & Christopher A. Shumway. 2003. *Digital Dilemmas: Ethical Issues for Online Media Professionals.* Ames: Wiley-Blackwell/Iowa State Press.

Bouchon, Catherine. 2005. *Macht das Internet einsam?.* München: Grin.

Bynum, Terrell Ward & Simon Rogerson (eds.). 2004. *Computer Ethics and Professional Responsibility*, Oxford: Blackwell.

Cavalier, Robert. J (ed.). 2005. *Impact of Internet on Our Moral Lives.* Albany: State University of New York Press.

Crowston, Kevin, Sandra Sieber & Eleanor Wynn (eds.). 2007. *Virtuality and Virtualization.* New York: Springer.

Drushel, Bruce & Kathleen German (eds.). 2011. *Ethics of Emerging Media.* New York: Continuum.

Dusek, Val (ed.). 2006. *Philosophy of Technology: An Introduction*. Oxford: Wiley-Blackwell.
Enderlin, Renate. 2003. *Das Internet als sinnstiftendes Medium*. München: Grin.
Fabris, Adriano. 2007. *Etica del virtuale*. Milano: Vita & Pensiero.
Fabris, Adriano (ed.). 2012. *Etica delle nuove tecnologie*. Brescia: La Scuola.
Flanagan, Mary & Daniel Howe & Helen Nissenbaum. 2008. Embodying Values in Technology: Theory and Practice. In Jeroen van den Hoven and John Weckert (eds.), *Information Technology and Moral Philosophy*, 322–353. Cambridge: Cambridge U.P.
Floridi, Luciano (ed.). 2010. *The Cambridge Handbook of Information and Computer Ethics*, Cambridge: Cambridge U.P.
Floridi, Luciano. 2013. *The Ethics of Information*, Oxford: Oxford U.P.
Himanen, Pekka. 2010. *The Hacker Ethic: A Radical Approach to the Philosophy of Business*. New York: Random House.
Van den Hoven, Jeroen, and Weckert, John (eds.). 2008. *Information Technology and Moral Philosophy*, Cambridge: Cambridge U.P.
Van den Hoven, Jeroen. 2010. The Use of Normative Theories in Computer Ethics. In Luciano Floridi (ed.), *Information and Computer Ethics*, 59–76. Cambridge: Cambridge U.P.
Johnson, Deborah G. 2009. *Computer Ethics*. London–New York: Pearson.
McKee, Heidi A. & James E. Porter. 2009. *The Ethics of Internet Research. A Rhetorical, Case-Based Process*. Frankfurt a. M.: Lang.
Migga Kizza, Joseph. 2013. *Ethical and Social Issues in the Information Age*. London/Heidelberg/New York/Dordrecht: Springer.
Nelson, Theodor Holm. 1981. *Literary Machine. The Report concerning Word Processing, Electronic Publishing, Hypertext, Thinkertoys, Tomorrow's Intellectual Revolution, and certain other Topics including Knowledge, Education and Freedom*. Sausalito: Mindful.
Orwell, George. 2008. *1984*. London/New York: Pearson.
Ploug, Thomas. 2009. *Ethics in Cyberspace*. London/Heidelberg/New York/Dordrecht: Springer.
Quinn, Mike. 2012. *Ethics for the Information Age*. London/New York: Pearson.
Raffoul, François. 2010. *The Origins of Responsibiliy*. Bloomington/Indianapolis: Indiana U.P.
Rawls, John. 1999. *A Theory of Justice*. Harvard: Harvard U.P.
Reynolds, George W. 2013. *Ethics in Information Technology*. Nashville: Southwestern Publishing Group.
Rooksby, Emma. 2002. *Email and Ethics*. London–New York: Routledge.
Sanders, Alton F. & William Brinkman. 2013. *Ethics in a Computing Culture*. London: Delmar/Cengage.
Singer, Peter (ed.). 2006. *A Companion to Ethics*. Oxford: Wiley-Blackwell.
Stamatellos, Giannis. 2007. *Computer Ethics: A Global Perspective*. Sudbury: Jones and Bartlett.
Tavani, Herman T. 2012. *Ethics and Technology: Controversies, Questions and Strategies for Ethical Computing*. Hoboken: Wiley.
Wallach, Wendell & Colin Allen. 2009. *Moral Machines: Teaching Robots Right from Wrong*. Oxford: Oxford U.P.
Walsham, Geoffrey. 2001. *Making a World of Difference: IT in a Global Context*. Hoboken: Wiley.

III. Communication technologies and new practices of communication in the information and communication society

Rolf T. Wigand

19 Commerce

Abstract: In an economy based on the division of labor, commerce takes on the task to balance the spatiotemporal, qualitative and quantitative friction between production and consumption. More generally, commerce is any exchange of goods and services. This typically occurs along a market hierarchy starting with the manufacturer, on to the wholesaler, then retailer and, eventually the product or service is purchased by the consumer. Market dynamics and competition are examined based on such concepts as the division of labor, exchange and competition. One theory is highlighted: Transaction Cost Theory, especially as it allows us to focus on transaction and coordination costs. Particular attention is paid to markets and electronic markets. The efficiency of electronic markets, especially their gain in efficiency over traditional markets is recognized. It is emphasized that the very essence of an electronic market is the human being, i.e. man as actor. Lastly, the author presents a model of how value added comes about in commerce: Value comes from business processes with proper alignment of information and communication technologies, business strategy/goals and business processes.

Keywords: commerce, competition, disintermediation, division of labor, electronic market, exchange, market dynamics, market hierarchy, Transaction Cost Theory, value and value added

1 Introduction and overview

In the broadest possible view, commerce is any exchange of goods or services. Initially hunters and gatherers but also people with agricultural cultivations or herders and nomads exchanged based on the value and the need for the good. Early physical goods that were traded included shells, obsidian and solid copper. Trading emerged out of the need or desire for certain goods trader A did not have, but were available by trader B, C, etc. Thus, eventually, markets were created. Goods that had to be moved considerable distances such as salt or gold were perceived to be of much higher value than locally produced or found goods. In time specialized routes developed to move these goods in high demand and of high value. Special means of transportation developed such as the caravans using camels, a designated road network such as the Silk Road or the Amber Road, trade routes using ships emerged for the spice trade or powerful trading organizations such as the Hanseatic League (Hansa) and the trading power of Venice where only

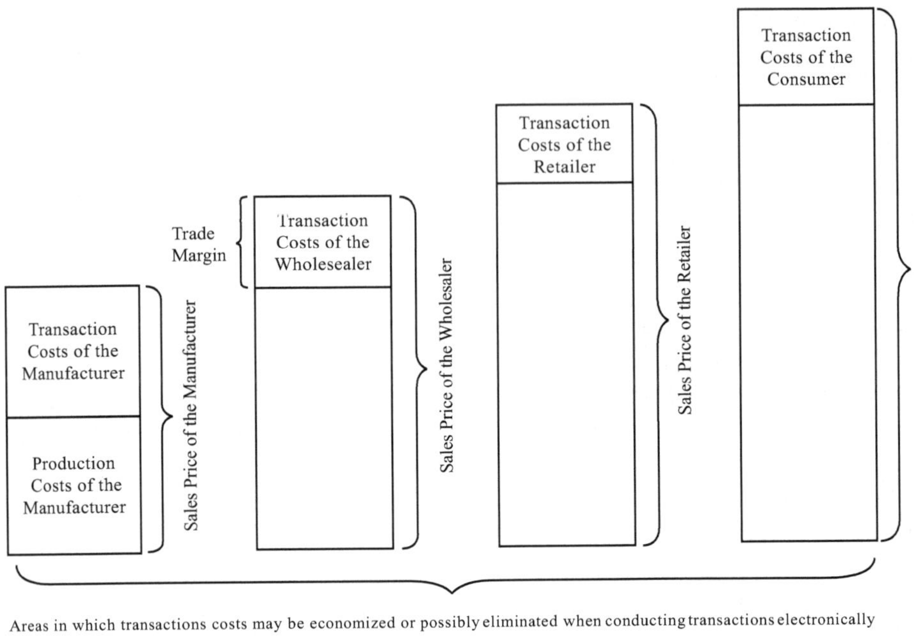

Fig. 1: Market hierarchy and transaction costs in a stepwise fashion (Wigand 1995d).

noble people were entitled for long-distance and overseas trade. The Hansa, Venice and other powerful trade organizations and cities (sometimes city-states) established and advanced techniques of trade, organizational forms of coordination and methods of finance, but also means of business development, the concepts of insurance, risk as well as standards, patent and trade protection. Moreover they thrived with many technical, organizational and managerial innovations, including the means of controlling financial and money-markets. One result of such efforts was the establishment of eventually famous centers of trade for many goods (e.g., spices, gold, silk, precious stones). Today the goods traded over long distances, often referred to as trading lanes connecting continents, have changed and they may include car parts, precious metals, wine, cotton and electronic goods. This even continues today with many cities and regions focusing on trade including Amsterdam, Hong Kong, New York, Shanghai, Silicon Valley and Singapore. Then but also today businesses of all kinds still engage in traditional commerce among each other or with consumers.

Specialized trading firms (e.g., traders, dealers) evolve when they can recognize and realize a certain monetary value added that they can handle and bridge better and quicker, as well as more cost-effectively than this could be accomplished directly by producers or consumers. Typically they procure physical goods, alter them at best minimally, and link these with services and/or rights and trade these eventually to other businesses or consumers. This may happen in a wholesale or retail fashion. Early on in the development of commerce we can recognize distribution chains or levels (hierarchies) such as goods being manufactured by a producer who then distributes these goods via a wholesaler who in turn passes them on to retailers. Eventually consumers go directly to these retailers to purchase their goods. From an economics perspective commerce and trade take on the role of a mediator between production and consumer in the form of coordinating the passing along and facilitating the transfer of goods. Moreover, the consumer gets informed about the supply offered by producers. Today with the advancement of newer information and communication technologies (ICT) and the Internet, it is possible to by-pass or leap frog over one or more of these distribution levels, i.e. a wholesaler and a retailer may be eliminated (disintermediated) in the traditional distribution chain (Benjamin and Wigand 1995). These distribution levels are depicted in Figure 1 above.

2 Market dynamics and competition

The *satisfaction of human needs* is the fundamental intention of economic activity. Human needs are unlimited, while the resources for their respective satisfaction are scarce. Such *scarcity of goods* vis-à-vis to endless needs has led to the formation of economic institutions which cannot eliminate, but at best can minimize this scarcity. This predicament sets the stage for various economic phenomena, some abstract such as exchange, division of labor and competition and some more plastic or concrete such as markets and corporations.

The effective pursuit of scarcity reduction requires the allocation of production factors in a manner maximizing need satisfaction. Economizing implies making decisions in regards to scarce resources. One way to diminish the scarcity problem is through the emergence of the concept of the division of labor, the realization of advantages earned through specialization and the use of production detours. Exchange activities between economic subjects are invariably tied to these forms of scarcity reduction. Thus, exchange is a fundamental phenomenon of economic activity.

The *division of labor* is the organizational breakdown of a work task into several subtasks and their respective assignment to an individual to execute this subtask as part of a production process. Division of labor and specialization are mutually dependent concepts. A person's limited temporal and cognitive ability to deal with comprehensive assignments on his/her own provide the motivation for these concepts. This leads to a systematic breaking down of complex assignments into single tasks until the completion of these partial tasks lies within the scope of an individual's capacity. In most societies a high degree of division of labor results from this process. Concentrating on single task components allows for the development of specific knowledge, abilities and processes with which these tasks can be completed in an efficient manner. The principle of creating and using specialized skills, known as far back as Aristotle (384–322 B.C.) and more recently by Adam Smith (1776), leads to considerable productivity increases in the management of partial tasks. In *An Inquiry into the Nature and Causes of the Wealth of Nations* (1776), Adam Smith foresaw the essence of industrialism by recognizing that division of labor makes possible a qualitative increase in productivity. This he illustrated via the manufacture of pins. In contrast to Plato (ca. 427–347 B.C.), Smith famously argued that the difference between a street porter and a philosopher was as much a consequence of the division of labor as its cause. Therefore, while for Plato the level of specialization determined by the division of labor was externally determined, for Smith it was the dynamic engine of economic progress itself. However, in a further chapter of the same book Smith criticizes the division of labor saying it leads to a 'mental mutilation' in workers; they become ignorant and insular as their working lives are confined to a single repetitive task. The contradiction has led to some debate over Smith's opinion of the division of labor.

Thus a more productive economizing of given resources suggests and implies the possibility of satisfying a larger number of needs. This mechanism can be observed at all levels of economic activity, beginning with personal, intrafirm and interfirm specialization and the division of labor all the way to sectoral, regional, national and international specialization and the division of labor.

Another way to increase a resource's potential for need satisfaction can be found in the creation of *production detours*. This term was coined by Carl Menger (1871), the Austrian national economist. Menger classifies goods according to their proximity to end consumption. Consumer goods are goods of first order. They are produced from pre-products using explicit production means. These production factors represent goods of second order, which, once again, stem from pre-products and production means of a higher order. The expression 'production tour' describes a good's replacement in a higher order, bringing its productive use further away from end consumption (e.g., the use of grain as seed instead of food; the use of materials and labor to produce tools). The production detour implies sacrificing present consumption for the sake of future productivity increases and, in turn, an increase in future consumption options. Production detours lead to an increase in the potential of need satisfaction on the basis of given resources.

It behooves us to examine this setting theoretically as this effort will guide us in our explorations. There are numerous theoretical approaches we could choose for this purpose, especially from overall organization theory, including Principal Agent Theory, Contract Theory, Property Rights Theory, the Division of Labor, but also marketing, diffusion, information retrieval and strategic networking. For the present purposes we chose the *Transaction Cost Theory* (TCT) approach by Ronald Coase (1937) and Oliver Williamson (1975, 1985) as it offers great utility and functionality. Moreover, TCT may be viewed as probably the overarching all-purpose theory for this perspective and our present purpose.

The coordination of economic activities demands the deployment of resources and this, in turn, generates costs which are defined as coordination or transaction costs. Transaction costs are the "production" costs of coordination. Generally, they comprise the partial tasks of initiation, negotiation, completion, control, and adaptation of a transaction relationship (Picot et al. 2008; Wigand et al. 1997). They include those information and communication costs necessary for the preparation, implementation and supervision of division of labor and exchange. The level and amount of transaction costs is determined primarily by the diversity and characteristics of the transaction costs. Thus, the challenge in coordination lies in finding fitting coordination patterns for partial tasks given the circumstances and allowing for the frictionless completion of task-related exchanges among the participants with the goal to minimize transaction costs.

Economists have classified transactions among and within organizations as those that a) support coordination between buyers and sellers, i.e. market transactions, and those b) supporting coordination within the firm. Figure 1 above depicts a typical market hierarchy progressing from 'manufacturer' to 'wholesaler', 'retailer,' and 'consumer'. The associated respective of transaction costs is shown as well at each hierarchical level. Williamson (1981) points out that the choice of transaction depends on a number of factors, including asset specificity, the parties' interests in the transaction, and ambiguity and uncertainty in describing the transaction. Transactions may then be broken down into production and coordination costs (e.g., Wigand et al. 1997; Benjamin and Wigand 1995; Picot 1982; Malone et al. 1987; Williamson 1985; Coase 1937). In this context, coordination costs include the transaction (governance) costs of the information processing necessary to coordinate the work of people and machines performing primary processes (Malone et al. 1987 p. 485). Transaction costs may be viewed as the economic equivalent of friction in a physical system, i.e. if friction is too great, no or at least impeded movement will occur suggesting that if transaction costs are high, no or little economic activity is likely to occur. These costs are comprised of the following four types:

- Search costs, the cost of searching for products, sellers, and buyers
- Contracting costs, the cost of setting up and carrying out the contract
- Monitoring costs, the cost ensuring that the terms of the contract have been met
- Adaptation costs, the cost incurred in making changes during the life of the contract.

Firms will choose those transactions that economize on coordination costs. As information technology continues its rapid cost performance improvement, the unit cost of coordination transactions will approach zero, thus enabling the design of innovative coordination transactions to fit new business needs (Benjamin and Wigand 1995). The ever-increasing and innovative use of the World Wide Web

(WWW) to conduct business and WWW-related forms of electronic commerce are clear examples of firms' desires to economize on transaction costs. Figure 1 above suggests that transaction cost savings may be achieved through the use of information technology within the entire market hierarchy and resulting market or industry value chain. Benjamin and Wigand (1995) present an example of the purchase of a high-quality shirt with empirical cost figures clearly demonstrating actual savings in transaction costs, resulting in lower purchase costs for the consumer. Moreover, this example demonstrates nicely how the potential elimination of entire levels within the market hierarchy (e.g., wholesaler, retailer) may occur. The latter is clearly observable in several markets and industries, where disintermediation has occurred and reintermediation may be necessary, as, otherwise, no commerce or trade can happen, i.e. sellers and buyers have no means to meet or find each other (Wigand 1997). One may argue that with cheap coordinative transactions, interconnected networks and their strategic deployment, and easily accessible databases, there would be a proportional shift of economic activity to cheaper electronic communications channels to conduct a firm's business.

3 Definition of market and electronic market

Commerce is conducted within markets. The term market is indeed a multidimensional concept. A market is conceived of all goal-seeking firms, government agencies or individuals producing some commodity, as well as all firms, government agencies or individuals purchasing the commodity. Within this market, the exchange of goods and services takes place. It should be noted though that an electronic market is not a physical market, i.e. it is not a place to go to as an individual such as a market place in a town square. In a competitive situation, a competitive market is characterized by:

1. Many buyers and sellers;
2. Homogeneous products; and
3. Easy entrance to and departure from the market.

The first characteristic suggests also that no individual producer or consumer is sufficiently large enough to influence the price of the commodity. Consequently, each market participant accepts the price of the commodity and behaves as if this price is beyond his/her control. Information is an essential ingredient for the functioning of any market and is exchanged frequently between buyer and seller such as when price information is exchanged (e.g., Hirshleifer 1980). The second characteristic denotes, in addition to what is already stated in (2) above, that consumers do not have strong preferences which producers they buy from and that they will seek out sellers with the lowest price. The third characteristic suggests that supply and demand adjustments can and do come about easily and occur quickly.

The concept of electronic markets has been discussed since the 1970s, and more formal discussion is reflected in the literature since at least 1987 and several authors have reviewed this topic (e.g., Schmid 1993). The term electronic market denotes a market place in which business transaction within a given market are carried out or supported, at least partially, electronically utilizing ICT networks. An electronic market can also be characterized as one in which a prevailing market intensity requires a certain volume of activities that, in turn, becomes attractive to be carried out electronically. Consequently, electronic markets emerge through the mediation of market transactions.

An electronic market may not be viewed in an absolute sense, i.e. not necessarily all activities will be carried out electronically in a given market. In essence one may view an electronic market as a very wide or a very narrow concept and one should realize that there are steps in between: one may envision hierarchies that can function electronically, but also as an electronic market, as well as hybrid electronic forms in between hierarchies and market are conceivable such as the modular firm and the strategic network. This has been explored in detail by Picot, Reichwald and Wigand (2008, p. 165 and chapter 2, "Market Dynamics and Competition: The Fundamental Role of Information"). Moreover, such an electronic market makes available the electronic inquiry by firms about products and services offered by other firms within a particular industry branch and where such inquiry makes possible the comparisons of offers, as well as the eventual electronic purchase of such products and services. The often idealized, optimum electronic market that is fully automatized, operating around the clock without imperfect human interference and still be understood as a market is not possible. It is important to distinguish here, as emphasized by Schmid (1993), Wigand (1995c) and others, that such an electronic setup through networks is not envisioned as a market, as this would constitute merely an automated processing system of some sort. A market in this sense, and an electronic market as well, necessitates that (at least minimally) human actors are engaged, trigger an action, make a decision or choice, express a preferred behavior or maybe avoid something, but it is essentially this human involvement that characterizes a market rather than a fully automated process (Wigand 1997). Often past experiences, typically *joint* past experiences, play an important role, but also such essential features as trust, faith, confidence and credibility characterize an actor's behavior in a market (Schmid 1993; Wigand 1995c, 1997). All of these attributes would be most difficult to capture by an automated processing system in order to make crucial decisions (e.g., Crowston and Wigand, 1999). An electronic market then is not any different.

There are other limitations or forms of boundaries within electronic markets. People will *meet* in an electronic market with certain expectations. One expects that the price rises; another expects that the price decreases. Both, however, agree

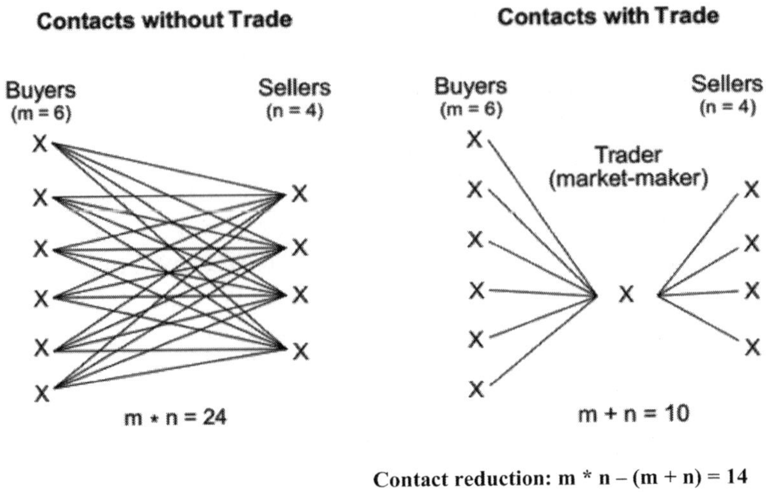

Fig. 2: The reduction of necessary contacts through the market-maker/electronic market, resulting in considerable market efficiency by intermediation.

that they can talk and do business. It may be that only certain people are attracted to electronic markets or would consider doing business there. It is clear, however, that ICT can support and help in this process, but, in the final analysis, the very essence of the electronic market will always be the human being, i.e. man as actor.

Malone, Yates and Benjamin (1987, 1989), Benjamin and Yates (1991), Benjamin and Wigand (1995), Wigand (1995b), Wigand and Benjamin (1995) as well as Wigand (2011) argue that organizations increasingly coordinate their activities electronically and that this coordination takes the form of single source electronic sales channels (one supplier and many buyers coordinated through hierarchical transactions) or electronic markets. Moreover, they argue that electronic markets are a more efficient form of coordination for certain classes of product transactions, especially those where asset specificity is low and where products are easily describable. An example is the issuance of electronic requests for proposals (RFPs) by firms that can be retrieved or inspected electronically. Similarly, it may be possible to submit a full-fledged proposal or offer electronically.

Electronic markets can potentially introduce considerable amounts of efficiency as depicted in Figure 2 (Benjamin and Wigand, 1995). This is sometimes also referred to as the Baligh-Richartz Effect (Baligh and Richartz, 1967).

This example demonstrates how in this market with 24 firms the market-maker makes possible a reduction of 10 contacts by reducing the necessary contacts in this market to get complete information from 24 to 10, resulting in an efficiency gain of 14. Thus one can argue that with cheap coordination costs or transaction costs, interconnected networks and easily accessible data bases, a proportional shift of economic activity from single source sales channels to markets is likely to occur.

4 Commerce enabled by value added through information and communication technology

Looking abstractly at organizational processes one realizes quickly that a large number of activities transform a specific input into a specific output. In order to describe a process, activities are arranged such that they depict the flow of activities. From a more formal perspective one may utilize here the notations of function and process modeling. For our present purposes, a much broader and general, i.e. non-technical, perspective is pursued. In accordance with Davenport's (1993) deliberations on the necessity of *process innovation*, i.e. the fusion of ICT with the human factor, the term process here is understood in a broader sense than the mere chaining of elementary transformations. The concept of process innovation is assisted by new ICT, but an essential factor making the process successful in the first place depends upon the motivation of the employee and the engagement of the upper management with regard to strategic vision. ICT is the cornerstone of this paradigm and presents itself in the role of the enabler (or, conversely the impeder).

The actual goal in the design of commerce, especially in electronic commerce, organization structures as well as entrepreneurial processes and their subsequent respective synchronization with information and communication technologies (ICT) is the generation or increase of *value* (Wigand, 1988; Wigand 1995a, 1995d). Such value, e.g., may take on many forms:

- Enable us to leverage information
- Create value for key stakeholders
- Transform our business
- Create potentially new markets, maybe niches
- Create new channels and applications
- Create competitive advantage
- Enable potentially strategic transformations
- Other

If such value cannot be generated, one must seriously question the entire effort and undertaking in the first place. Thus the mere infusion of new ICT does not necessarily deliver the desired result, the above mentioned value. The underlying relationships are much more complex.

In the past many firms and managers assumed that the desired added value was brought about directly through the introduction or the actual application of ICT (Wigand 1995d, 1985). Reflecting on many projects and efforts that broadly speaking resemble nothing more than the sheer automation of already existing bureaucratic processes, the anticipated high productivity improvements did not come about through such efforts. This discrepancy is explained by the false assumption that ICT directly cause added value and that, consequently, a number

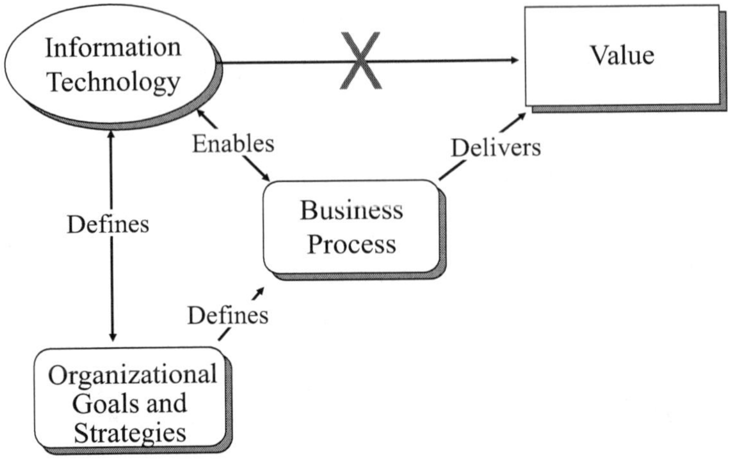

Fig. 3: Value comes from business processes: Adding value within the organizational fit of information and communication technology, business processes and goals/strategies (Wigand 1995a, 1995d).

of results are brought about on the side of efficiency (e.g., return on investment), effectiveness (e.g., quality of management information), but also with regard to transformations within the firm (e.g., innovations, competitive advantage) (Wigand 1995a, 1995d). This is depicted in Figure 3 in the top line by the direct relationship between 'Information Technology' and 'Value'. This falsely suggests a direct causal relationship, analogous to a hypodermic needle model in which medicine directly and immediately or in time cures a medical problem when injected. This would imply that the deployment of an ICT directly and immediately 'triggers' the desired value. The assumption that the deployment of ICT and value added function within such a trivial relationship is, as we know today, fundamentally incorrect.

Of greater importance here is to pay attention to the 'Business Strategy' or the firm's goals. These demands placed on 'information technology' have to be seen in conjunction with demands resulting from the concrete organization of 'Business Processes'. Adequate means in this context the relationship is reciprocal: on the one hand business strategy, on the other hand information technology is the enabler for new strategies and processes. The challenge is to identify the optimal point for the deployment of information and communication technologies, i.e. the optimal organizational fit and alignment (Wigand 1995a, 1995d).

In order to achieve the appropriate organizational and strategic fit several iterations of implementation efforts may be necessary, each followed by improvements and adaptation processes and analogous to the self-regulatory mechanism of a cybernetic system. Only when the optimal relationships among information technology, business strategies and goals, and processes have been developed, all mistakes and misalignments have been corrected and the organization has adapted to the changes, it is then and only then that one may expect the anticipated improvements in efficiency and effectiveness, i.e. the *adding of value* occurs (Wigand 1995d). The relationships are depicted in Figure 3 above.

Striving toward innovative information technology solutions on the one hand and the appropriate organizational fit on the other hand are the key for overcoming performance boundaries of the organization. Not only are the traditional boundaries of the firm affected with regard to geographic limitations, distances, speed and working time, but also awareness limitations of the employees in the organization. Through the continual balancing of information technology, business strategies, goals and entrepreneurial processes on the one hand and the awareness about the expansion of entrepreneurial boundaries and limits on the other hand, it becomes possible to realize the vision of the horizontally integrated firm. This is supported and encouraged by the steadily growing potential of ICT.

References

Baligh, Helmy H. & Leon E. Richartz. 1967. *Vertical market structures*. Boston: Allyn and Bacon.

Benjamin, Robert I. & Rolf T. Wigand. 1995. Electronic Markets and Virtual Value Chains on the Information Highway. *Sloan Management Review, Winter*. 62–72.

Benjamin, Robert I. & JoAnne Yates. 1991. The Past and Present as a Window on the Future. In Scott Morton, M. S. (ed.), *The Corporation of the 1990s,* 61–92. New York: Oxford University Press.

Coase, Ronald. 1937. The nature of the firm. *Economica* 14(16). 386–405.

Crowston, Kevin & Rolf T. Wigand. 1999. Real-estate war in cyberspace: an emerging electronic market? *Electronic Markets* 9(1/2). 1–8.

Davenport, Thomas H. 1993. *Process Innovation – Reengineering Work through Information Technology*. Boston, MA: Harvard Business School.

Hirshleifer, Jack 1980. *Price, Theory and Application*. Englewood Cliffs, NJ: Prentice-Hall.

Malone, Thomas, JoAnne Yates & Robert I. Benjamin. 1987. Electronic Markets and Electronic Hierarchies. *Communications of the ACM* 30(6). 484–497.

Malone, Thomas, JoAnne Yates & Robert I. Benjamin. 1989. The Logic of Electronic Markets. *Harvard Business Review* 3. 166–172.

Menger, Carl. 1871. *Grundsätze der Volkswirtschaftslehre*. Vienna/Leipzig: Braumüller.

Picot, Arnold. 1982. Transaktionskostenansatz in der Organisationstheorie: Stand der Diskussion und Aussagewert. *Die Betriebswirtschaft* 2. 267–284.

Picot, Arnold, Ralf Reichwald & Rolf T. Wigand. 2008. *Information, Organization and Management*. Berlin, Heidelberg: Springer-Verlag.

Schmid, Beat. 1993. Elektronische Märkte. *Wirtschaftsinformatik* 5. 465–480.

Smith, Adam. 1776. *An Inquiry into the Nature and Causes of the Wealth of Nations*. London: W. Strahan and T. Cadell Publishers.

Wigand, Rolf T. 2011. 20 Years of Research in Electronic Markets and Networked Business: An Interview with Thomas Malone. *Electronic Markets* 19/20. 5–17.

Wigand, Rolf T. 1995a. Doing business on the Information Superhighway: are we adding value? *Business Journal* 9(24). 3–15.

Wigand, Rolf T. 1995b. Electronic commerce and reduced transaction costs: Firms' migration into highly interconnected electronic markets. *Electronic Markets* 16/17. 1–5.

Wigand, Rolf T. 1995c. Information technology and payoff: The productivity paradox revisited. Paper Presented to the Annual Conference of the International Communication Association, Albuquerque, NM, May 25–29.

Wigand, Rolf T. 1995d. The Information Superhighway and Electronic Commerce: Effects of Electronic Markets. Paper presented to the Annual Conference of the International Communication Association, Albuquerque, NM, 25–29 May.

Wigand, Rolf T. 1985. Integrated Communications and Work Efficiency: Impacts on Organizational Structure and Power. *Information Services and Use* 5. 241–258.

Wigand, Rolf T. 1997. Electronic commerce: definition, theory and context. *The Information Society* 13, 1–16.

Wigand, Rolf T. 1988. Fünf Grundsätze für die erfolgreiche Einführung des Informations-Managements. *Information Management* 3(2). 24–30.

Wigand, Rolf T. & Benjamin, Robert I. 1995. Electronic Commerce: Effects on Electronic Markets. *Journal of Computer-Mediated Communication*, 3, December, URL: onlinelibrary.wiley.com/enhanced/doi/10.1111/j.1083-6101.1995.tb00166.x/ (accessed on July 23, 2014).

Wigand, Rolf T., Arnold Picot, & Ralf Reichwald. 1997. *Information, Organization and Management: Expanding Markets and Corporate Boundaries*. Chichester, England: Wiley.

Williamson, Oliver E. 1985. *The Economic Institutions of Capitalism, Firms, Markets, Relational Contracting*, 11th ed. New York: Free Press.

Williamson, Oliver E. 1975. *Markets and Hierarchies: Analysis and Antitrust Implications*. New York: Free Press.

Williamson, Oliver E. 1981. The Modern Corporation: Origin, Evolution, Attributes. *Journal of Economic Literature* 19. 1537–1568.

Kevin B. Wright

20 Workplace relationships: Telework, work-life balance, social support, negative features, and individual/organizational outcomes

Abstract: This chapter examines several prominent interpersonal relationship issues in the workplace that have been affected by the introduction of new technologies. Toward that end, it explores several theoretical frameworks that have been applied to the study of new communication technologies and workplace relationships, telecommuting, work life balance, and negative behaviors in the workplace associated with new communication technologies, such as cyberbullying and cyber-surveillance in the workplace. Finally, it explores the strengths and limitations of this research as well as areas that would benefit from additional attention by researchers in the future.

Keywords: computer-mediated communication, social support, work life balance, cyberSurveillance, distributed teams

The advent of computer-mediated communication (CMC) technologies have led to profound changes in the workplace, including telecommuting, distributed work groups, and electronic outsourcing (Boswell and Olson-Buchanan 2007; Gajendran and Harrison 2007). New communication technologies can allow for new forms of work by providing the opportunity for people with common interests to connect, overcoming limitations of space and time. Moreover, new communication technologies can alter traditional organizational structures, allowing for greater information exchange among more people and enhancing socialization. However, while new communication technologies certainly can have a positive effect on relationships within the workplace, other aspects of the technology (such as reduced cues and contextual information) can lead to increased misunderstandings, depersonalization, organizational conflict, and more recent issues like cyberbullying in the workplace (D'Urso 2006; Priviteria and Campbell 2009; Turnage 2007). Thus, several limitations of current communication technologies can have a negative impact on workplace relationships as well as important individual outcomes for works, such as stress, burnout, and job satisfaction.

In particular, advances in new communication technologies have transformed the traditional workplace, providing opportunities for employees to telecommute and accomplish work tasks from home (Thatcher and Zhu 2006). However, telecommuting appears to have some negative relational implications within the workplace, such as employees who predominately work from home perceiving that they are isolated from others within the organizations or the feeling that not being able to have face-to-face time with supervisors may limit opportunities for advancement (Gajendran and Harrison 2007). In addition, telecommuting has led to a number of issues for employees in terms of work life balance.

Moreover, new communication technologies can sometimes introduce new stressors into the workplace, including the need to continually master new communication technologies and applications in an effort to be economically viable within the organization (Holt and Brockett 2012; Scott and Timmerman 2005). Such stressors can sometimes exacerbate interpersonal conflicts within the workplace or damage relationships among workers as workers experience added strain and burnout. At the same time, new communication technologies provide opportunities for employees to connect with other workers within and outside of traditional workplace hours through new forms of cyber-support, including on-line support groups and work-related on-line communities with people can share important resources. Finally, new communication technologies have brought new issues to the workplace, including the advent of relatively new phenomena such as cyber-bullying and cyber-surveillance of employees.

The purpose of this chapter is to examine several prominent interpersonal relationship issues in the workplace that have been affected by the introduction of new technologies. Toward that end, it explores several theoretical frameworks that have been applied to the study of new communication technologies and workplace relationships, telecommuting, work life balance, and negative behaviors in the workplace associated with new communication technologies, such as cyberbullying and cyber-surveillance in the workplace. The following sections explore each of these areas in detail, including attention to areas that would benefit from additional attention by researchers in the future.

1 Theoretical perspectives on new technologies in the workplace

Early theoretical approaches to the use of new communication technologies in the work place, such as Daft and Lengel's (1984) Media Richness Theory and the "cues filtered out" perspective (Culnan and Marcus 1987) posited that CMC is less socially-oriented and less personal than face-to-face communication. These scholars theorized that e-mail is a lean medium because it lacks the capability for immediate feedback, uses only a single communication channel (textual message) that limits multiple nonverbal cues (such as paralinguistic cues or body language). Daft and Lengel (1984) argue that email and other text-based forms CMC should be used for less equivocal tasks whereas tasks that may involve a higher degree of ambiguity should use richer channels (such as face-to-face communication).

However, later studies (Rice and Love 1987; Parks and Floyd 1996) found that individuals spend a great deal of time interpreting lean media (such as email), and they adapt to the limitations of text-based media by altering their written messages to convey information that might not be explicit (such as emotions, etc.). Walther's (1992) Social Information Processing Theory posits that while there are reduced cues in CMC, individuals are driven to socially interact with other, and they will use individual, cognitive, knowledge-generating strategies to test their assumptions of relational partners in on-line situations. While this process typically takes longer than in face-to-face interaction, these strategies allow individuals to experience rich communication via a lean medium (i.e., email) over time.

Other researchers (Fulk 1993; Schmitz and Fulk 1991), drawing upon Social Influence Theory have suggested that media attitudes and choices are largely shaped by structural characteristics in the organization and organizational culture, including expectations about communication requirements and culturally-specific interpretations of how a medium meets these requirements. Finally, several researchers have applied a uses-and-gratifications perspective to new communication technology use within work organizations (Dobos 1992; Flanagin and Metzger 2001; Hovick et al. 2003; Stephens 2007). From this perspective, certain media tend to fulfill or gratify various needs, such as information-seeking, social integration, and convenience. This body of research has examined interpersonal motivations for using new communication technologies, including the development and maintenance of workplace relationships.

Each of these theories and perspectives serve an important function in terms of providing a framework for understanding communication behaviors and processes related to new communication technology use in work place settings. As new relational and computer-mediated communication phenomena continue to surface in the workplace, it is important to seek larger theoretical explains for behaviors and practices. In addition, to these more established theories and perspectives that have been applied to the study of workplace relationships, it is important for scholars to continue work to develop theory in this area in addition to empirical studies.

2 New communication technologies and telecommuting

Paralleling advances in new communication technology, the practice of telecommuting, which is also known as telework or remote work, has become a widespread practice within organizations in the United States and worldwide (Baker et al. 2006; Thatcher and Zhu 2006). Telecommuting can be defined as an alternative work arrangement in which employees perform tasks elsewhere that are normally done in a primary or central workplace, for at least some portion of their work schedule, using electronic media to interact with others inside and outside the organization (Bailey and Kurland 2002). Research on the benefits associated with telecommuting is mixed. Some studies have reported a number of benefits of telecommuting for the individual worker, including improved work–life balance, heightened morale, and increased productivity (Baker et al. 2006; Brownson 2004). However, other studies have found negative relational consequences of telecommuting for individuals, such as increased social isolation from the organization, career stagnation, and family conflict (Gajendran and Harrison 2007).

Specifically, some scholars have found that telecommuting enhances perceived autonomy by providing employees with choice over the location, scheduling, and means of work. Other researchers have examined the effects of telecommuting on work life balance (see section below), with some studies finding that telecommuting leads to better integration of work and family roles (Baker et al. 2006; Thatcher and Zhu 2006), while others have found that telecommuting may intensify work life conflict by increasing the permeability of work and family boundaries, role ambiguity, or lead to reduced identification with the organization (as a result of fewer face-to-face interactions and richness of communication with other employees) (Baker et al. 2006; Daft and Lengel 1984; Gajendran and Harrison 2007). In addition, some studies have provided evidence that telecommuting may weaken interpersonal bonds with coworkers and supervisors over time (Gajendran and Harrison 2007; Golden 2006; Mackie-Lewis 1998).

An important moderating variable is *telecommuting intensity*, or the extent or amount of scheduled time that employees spend working on tasks away from a central work location. High-intensity telecommuters may spend the majority (or all) of their workdays away from a central location while low-intensity telecommuters often spend the majority of their workdays at a conventional (physical) work location, only telecommuting 1 or 2 days a week. Gajendran and Harrison (2007) found that telecommuting intensity influenced perceptions of worker autonomy, work–family conflict, and workplace relationship quality. Employees who spend the majority of their working time away from office often have fewer opportunities for rich face-to-face interactions with supervisors and coworkers. In addition, a reduced physical presence in the organization may lead to fewer opportunities for spontaneous informal interactions (Mackie-Lewis 1998). This can lead to diminished perceptions of intimacy and immediacy, which are crucial to effective interpersonal communication within the workplace.

3 New communication technologies and work–life balance

In recent years, scholars are beginning to investigate the ways in which technology may contribute to increasing levels of workplace stress and work life imbalance

(Ayyagari et al. 2011; Boswell and Olson-Buchanan 2007; Gajendran and Harrison 2007). New communication technologies, such as smart phones, I-Pads, portable computers, and an ever-changing array of mobile communication technologies push the boundaries of the traditional office space, a new type of employer-employee relationship appears to be emerging as employer expectations grow to incorporate expected use of technology outside the workplace (Wright et al. 2014). However, the freedom of perpetual connectedness may also irreversibly blur the lines that once spatially separated work life and family life for some workers (Davis 2010). Organizations are increasingly under pressure to implement work practices intended to facilitate employees' efforts to fulfill both their employment-related and their personal responsibilities (Richman et al. 2008). However, many organizations are just beginning to understand the impact of new communication technologies on work life balance, and it appears that relatively few organizations have formal policies for helping workers cope with the stress of being almost perpetually connected to the workplace. Email, Skype, and texting have blurred the boundaries between work and home life. Although new communication technologies allow for greater connectivity to and flexibility in managing work demands, they may give employees little time to disengage from the workplace, especially if the stress of work-related issues interfere with a workers ability to enjoy time with family and friends (Boswell and Olson-Buchanan 2007).

Competing demands between work and home have become more relevant for employees in recent years due to demographic changes in the workplace, such as rising numbers of women in the labor force, an aging population, longer working hours, and more sophisticated communication technologies that enable nearly constant contact with the workplace. Work life imbalance typically takes two different forms: 1. work obligations interfere with family responsibilities; and 2. family responsibilities interfere with work obligations (Kossek and Ozeki 1998). Both forms of work life imbalance have been linked to a variety of negative outcomes for employees, including increased stress and burnout, higher absenteeism, psychiatric and substance abuse disorders, physical health problems, reduced job satisfaction, and employee turnover intentions (Burke and Greenglass 1999; Goff et al. 1990; Frone 2000; Martins et al. 2002; Netemeyer et al. 1996).

However, there is some evidence that certain characteristics of technology are predictors of workplace stress (Ayyagar et al. 2011). These researchers identified work overload and role ambiguity as technology-driven stressors resulting from the excessive use of ICTs. Ayyagari et al. (2011) state "constant connectivity provided by information and communication technologies increases the workload by enhancing the speed of work flow and expectations of productivity" (Ayyagar et al. 2011: 848). These authors call for more research on what is termed "technostress," or characteristics of technology that can lead to increased levels of stress and related outcomes such as job burnout. Other research supports this conceptualization of "technostress," finding that technological stressors affected both job satisfaction and emotional exhaustion in a population of teachers (Beam et al. 2003).

Yet, other research suggests that for some employees, new communication technologies may actually improve work/life balance, particularly when workers perceive them positively and believe these technologies will allow them additional to spend time at home with children and telecommute via computer to accomplish work duties at home (Chow and Keng-Howe 2006; Gajendran and Harrison 2007). Positive and negative perceptions of new communication technologies have also been linked to their use outside of the work place, with people who view them positively reporting greater usage than people who view them negatively (Fenner and Renn 2010; Gajendran and Harrison 2007).

4 Social support and well-being in the workplace

Workplace social support can be defined as the degree to which individuals perceive that their well-being is valued by workplace sources, such as coworkers and supervisors and the perception that these sources provide help to support this well-being (Dormann and Zapf 1999; House 1981). Moreover, workplace social support also refers positive workplace interactions that lead to the provision of resources to reinforce a particular type of role demand within the organization (Eisenberger et al. 2002). Employees who have greater access to workplace social support tend to obtain job psychological resources over individuals who have less workplace support, and these resources often provide a buffer to workplace stress. In addition, workers who have strong workplace support networks tend to have great emotional and psychological resources for coping with daily stressors that occur on the job (Dormann and Zapf 1999; House 1981). New communication technologies appear to bring new stressors to the traditional workplace. For example, new communication technologies have increased the amount of time workers spend on work outside of traditional office hours, workloads are often distributed on 24–7 operating systems, work often occurs at a faster pace due to new communication technologies than by traditional means of communication and information exchange, and employees may feel increased pressure to become proficient with a rising number of new communication technologies and applications in order to feel secure in their jobs (Holt and Brockett 2012; Scott and Timmerman 2005). According to Wiesenfeld, Raghuram and Garud (2001), virtual workers who perceive that they are socially integrated with other organization members are likely to assume that others view them as organization members, and thus may be more likely to identify themselves as organization members. Moreover, the social support they perceive from others online may also function to reassure teleworkers that they are part of the (physical) organization despite being geographically separated from it. Future research on workplace social support theory and interventions should focus on the conditions under which new communication technologies contribute to workplace stress as well as how both face-to-face social support and computer-mediated support in the workplace may help to buffer stress or help employees to avoid stress. Such research should also take into consideration key stress-related outcomes in organizational settings, including job satisfaction, employee health outcomes, well-being, burnout, and employee turnover.

4.1 E-mentoring

One specific type of supportive relationship that is common in most work organizations is the mentor-protégé relationship. In the era of computer-mediated communication, telecommuting has given rise to increased e-mentoring, which has been defined as "use of email or computer conferencing systems to support a mentoring relationship when a face-to-face relationship would be impractical" (O'Neill et al. 1996: 39). Mentor–protégé relationships within the workplace have long been recognized as key workplace interpersonal relationships where a protégé can benefit from support, resources, and the organizational experience of another employee who has obtained success within the organization or one's discipline. E-ementoring can also be conceived as a computer-mediated relationship between a senior individual who is the mentor, for a lesser skilled protégé with the goal of developing the protégé in a way that helps him or her succeed (Bierema and Merriam 2002). In most cases, mentors have had longer careers and/or have obtained a higher status within the organization than protégés. However, social support within these types of relationships can work both ways, with more seasoned employees learning new information or skills from the individuals they choose to mentor. In such cases, e-mentoring is a mutually beneficial relationship that is highly versatile and can be adapted to work in a variety of settings.

In many workplaces, supervisors create structured mentoring programs with the organization. Yet, e-mentors need not be part of worker's immediate organization due to the advent of computer-mediated communication. E-mentoring may occur between peers, one-on-one, and it can include relationships between current employees and retirees from the organization or discipline. In many cases, e-mentors work in complimentary or competing organizations, and new communication technologies allow mentors and protégés to communicate with each other and transcend geographical and temporal constraints (Ensher et al. 2003; Thompson et al. 2010). As a result, many employees seek out individuals from other organizations via the Internet and develop relationships that often serve to benefit them within their particular organization. E-mentoring has been relatively understudied within the communication discipline, and future research is needed to understand the benefits and limitations of mentor-protégé relationships with new communication technology contexts as well as distinct communication patterns that may occur in these relationships (Ensher et al. 2003; Thompson et al. 2010).

5 Work place bullying, cyber-surveillance, and new communication technologies

A dark side of new communication technologies and relationships in the workplace is the rise of behaviors that are potentially damaging to organizations, including cyberbullying and cyber-surveillance in the workplace. Both cyber-bullying and cyber-surveillance appear to increase stress in the workplace. However, relatively few communication researchers have examined the impact of these new negative forms of interpersonal communication in the workplace on important organizational outcomes, such as employee burnout and turnover. The following sections explore these phenomena in greater detail, and it is hoped that researchers will devote more attention to these issues in the future.

5.1 Cyberbulling in the workplace

While cyberbullying has mostly been examined among adolescents, there is evidence that suggests it is becoming a more common practice also in workplaces. Workplace bullying has been defined as repeated behavior that offends, humiliates, sabotages, intimidates, or negatively affects someone's work when there is an imbalance of power. Cyberbulling is a new form of workplace bullying in which the behaviors described above occur in computer-mediated form (Bartlett and Bartlett 2011; McKenna 2007; Priviteria and Campbell 2009), such as within texts, emails, and social media platforms (e.g., Facebook and Twitter). Cyberbullying behaviors include using communication technology to send derogatory or threatening messages directly to the victim or indirectly to others, to forward personal and confidential communication or images of the victim for others to see and to publicly post denigrating messages. Larger organizations with a higher ratio of male to female workers tend to exhibit greater prevalence of workplace bullying, including cyberbullying (Bartlett and Bartlett 2011).

In addition, male workers are more likely to engage in workplace bullying behaviors than female workers (Priviteria and Campbell 2009). Cyberbullying, like other types of workplace bullying, appears to have a negative effect on several important individual and organizational outcomes, including psychological well-being, low staff morale, reduced job satisfaction, increased absenteeism, and loss of productivity (Bartlett and Bartlett 2011). Cyberbullying can also contribute to negative perceptions of the organization and may affect future recruitment of employees. Priviteria and Campbell (2009) found that nearly one third of their sample of manufacturing workers indicated that they had experienced cyberbullying in the workplace, including the spreading of gossip and excessive monitoring of one's work. Moreover, these authors contend that the experience of online workplace bullying can have negative consequences on the victim's physical health and emotional well-being. In addition, online workplace bullying can also extend into the victim's social and family relationships, it can affect future career advancements such as job promotion (Bartlett and Bartlett 2011). Furthermore, it can lead to low staff morale, reduced organizational commitment, reduced job satisfaction, and the breakdown of work relationships and teams (Bartlett and Bartlett 2011). Clearly, this is an area of new communication technology use and relationships in the workplace that deserves attention by communication scholars in the future.

5.2 Cyber-surveillance in the workplace

Workplace surveillance has become more prevalent in recent years in a variety of work environments (D'Urso 2006; Firoz et al. 2006; Riedy and Wen 2010), with over 80 % of U.S. organizations using some form of employee surveillance (Riedy and Wen 2010). Technological advances in recent years have led to the emergence of new surveillance possibilities such as global position monitoring to determine the exact location of employees, devices that track key strokes taken per minute or time spent away from the computer used to determine worker productivity, and software (i.e. desktop monitors) that enables supervisors to watch an employee's screen as he or she works. In addition, organizations often access emails and other files stored or deleted by employees in an effort to identify troublesome messages or behaviors that could potentially damage the reputation of a company or lead to litigation (e.g., sexual harassment cases, etc.).

Managers often defend the use of cyber-surveillance as a means of protection against employee misconduct or behaviors that can damage the company financially, particularly in terms of litigation costs, information that could undermine investor confidence in the business, security breaches, and loss of productivity. In addition, it is well known that employees often use company time to indulge in cyber-slacking activities like sending personal emails, podcasting, blogging, chatting on social networking sites such as Facebook. Workers often view cyber-surveillance measures as an invasion of privacy, and surveillance that is viewed as inappropriate or excessive can lower employee morale and lead to increased job stress (D'Urso 2006). However, the effects of cyber-surveillance on employee morale may be related to perceptions about how the cyber-surveillance is being used by supervisors. For example, Urabaczewski and Jessup (2002) found that satisfaction with electronic monitoring was lower when it was used for control purposes (e.g., monitoring Internet use) than when it was used for electronic performance monitoring.

Future research is needed to study the effects of cyber-surveillance on key organizational outcomes such as employee stress, job satisfaction, and organizational commitment. Such individual outcomes may be related to interpersonal relationship issues in the workplace. For example, Tabak and Smith (2005) contend that cyber-surveillance undermines interpersonal trust between employees and management, and this is a promising area of study for interpersonal communication scholars who are interested in the impact of cyber-surveillance in the workplace.

6 Conclusion

As new communication technologies continue to be adopted by workplace organizations, many of the innovations and challenges outlined in this chapter will continue to affect workplace relationships and important organizational outcomes. These technologies enable teleworkers to achieve a greater degree of work-life balance by allowing for people to work from home. New communication technologies allow organizational members to give and receive social support that may help to improve organizational identification and reduce stress. E-mentoring relationships allow seasoned workers to interact with protégés in a more convenient manner than in the face-to-face world. However, new communication technologies in the workplace also bring negative aspects, such as cyberbullying (both between coworkers and between supervisors and workers) as well as increased cyber-surveillance of new communication technology use in the workplace. These phenomena appear to influence work relationships in a variety of ways, and both have been linked to decreases in important organizational variables like job satisfaction, stress, and turnover. As new innovations are developed, communication researchers should continue to examine the effects of these technologies on interpersonal relationships within the workplace.

References

Ayyagari, Ramakrishna, Varun Grover & Russell Purvis. 2011. Technostress: Technological antecedents and implications. *MIS Quarterly* 35(4). 831–858.

Bailey, Diane E. & Nancy B. Kurland. 2002. A review of telework research: Findings, new directions and lessons for the study of modern work. *Journal of Organizational Behavior* 23. 383–400.

Baker, Paul M. A., Nathan W. Moon & Andrew Ward. 2006. Virtual exclusion and telework: Barriers and opportunities of technocentric workplace accommodation policy. *Work: A Journal of Prevention, Assessment and Rehabilitation* 27. 421–430.

Bartlett, James E. Michelle E. & Bartlett. 2011. Workplace bullying: An integrative literature review. *Advances in Developing Human Resources* 13. 69–84.

Beam, Randal, Eunseong Kim & Paul S. Voakes. 2003. Technology-induced stressors, job satisfaction and workplace exhaustion among journalism and mass communication faculty. *Journalism and Mass Communication Educator* 57(4). 335–51.

Bierema, Laura L. & Sahran B. Merriam. 2002. E-mentoring: Using computer-mediated communication to enhance the mentoring process. *Innovative Higher Learning* 26(3). 211–227.

Boswell, Wendy R. & Julie B. Olson-Buchanan. 2007. The use of communication technologies after hours: The role of work attitudes and work-life conflict. *Journal of Management* 33(4). 592–610.

Brownson, Kenneth. 2004. The benefits of a work-at-home program. *Health Care Manager* 23(2). 141–144.

Burke, R. J. & Esther R. Greenglass. 1999. Work-life congruence and work-life concerns among nursing staff. *Canadian Journal of Nursing Leadership* 12(2). 21–29.

Chow, Irene Hau-siu & Irene Keng-Howe. 2006. The effect of alternative work schedules on employee performance. *International Journal of Employment Studies* 14(1). 105–130.

Culnan, Mary & M. Lynne Markus. 1987. Information technologies: Electronic media and intraorganizational communication. In Frederic Jablin, Linda Putnam, Karlene Roberts and Lyman Porter (eds.), *Handbook of Organizational Communication*, 420–443. Beverly Hills, CA: Sage.

Daft, Richard & Robert Lengel. 1984. Information richness: A new approach to managerial behavioral and organizational design. *Research in Organizational Behavior* 6. 191–233.

Davis, Jennifer Deering. 2010. *The modern, mobile me: an exploration of smartphones, being always on, and our relations with work in the United States*. University of Texas, Austin, Doctoral dissertation. Available electronically from http://hdl.handle.net/2152/ETD-UT-2010-05-760 (Accessed 13 April 2014).

Dobos, Jean. 1992. Gratification models of satisfaction and choice of communication channels in organization. *Communication Research* 19. 29–51.

Dormann, Christian & Dieter Zapf. 1999. Social support, social stressors at work, and depressive symptoms: testing for main and moderating effects with structural equations in a three-wave longitudinal study. *Journal of Applied Psychology* 84(6). 874–884.

D'Urso, Scott C. 2006. Who's watching us at work? Toward a structural-perceptual model of electronic monitoring and surveillance in organizations. *Communication Theory* 16(3). 281–303.

Eisenberger, Robert, Florence Stinglhamber, Christian Vandenberghe, Ivan L. Sucharski & Linda Rhoades. 2002. Perceived supervisor support: Contributions to perceived organizational support and employee retention. *Journal of Applied Psychology* 87(3). 565–573.

Ensher, Ellen A., Christian Heun & Anita Blanchard. 2003. Online mentoring and computer-mediated communication: New directions and research. *Journal of Vocational Behavior* 63(2003). 264–288.

Firoz, Nadeem, Ramin Taghi & Jitka Souckova. 2006. Emails in the workplace: The electronic equivalent of DNA evidence. *Journal of American Academy of Business* 8(2). 71–78.

Flanagin, Andrew J. & Miriam J. Metzger. 2001. Internet use in the contemporary media environment. *Human Communication Research* 27(1). 153–181.

Fenner, Grant H., & Robert W. Renn. 2010. Technology-assisted supplemental work and work-to-family conflict: The role of instrumentality beliefs, organizational expectations and time management. *Human Relations* 63(1). 63–82.

Frone, Michael R. 2000. Work-family conflict and employee psychiatric disorders: The national comorbidity survey. *Journal of Applied Psychology* 85(6). 888–895.

Fulk, Janet. 1993. Social construction of communication technology. *Academy of Management Journal* 36(5). 921–950.

Gajendran, Ravi S. & David A. Harrison. 2007. The good, the bad, and the unknown about telecommuting: Meta-analysis of psychological mediators and individual consequences. *Journal of Applied Psychology*, 92. 1524–1541.

Goff, Stephen J., Michael K. Mount & Rosemary L. Jamison. 1990. Employee supported child care, work/family conflict, and absenteeism: A field study. *Personnel Psychology* 43(4). 793–809.

Golden, Timothy D. 2006. The role of relationships in understanding telecommuter satisfaction. *Journal of Organizational Behavior* 27(3). 319–340.

Holt, Lila & Ralph G. Brockett. 2012. Self direction and factors influencing technology use: Examining the relationships for the 21st century workplace. *Computers in Human Behavior*, 28(6). 2075–2082.

House, James S. 1981. *Work stress and social support.* Reading, MA: Addison-Wesley.

Hovick, Shelly R. A., Renée A. Meyers & C. Eric Timmerman. 2003. E-mail communication in workplace romantic relationships. *Communication Studies* 54(4). 468–482.

Kossek, Ellen Ernst & Cynthia Ozeki. 1998. Work-family conflict, policies, and the job-life satisfaction relationship: A review and directions for organizational behavior-human resources research. *Journal of Applied Psychology* 8(2). 139–149.

Mackie-Lewis, Stephanie Anne. 1998. *Changing place and the time of work: The impact of telecommuting on employees' personal networks and psycho- logical well-being.* University of Michigan, Ann Arbor, Unpublished doctoral dissertation.

Martins, Luis L., Jimberly A. Eddleston & John F. Veiga. 2002. Moderators of the relationship between work-family conflict and career satisfaction. *Academy of Management Journal* 45(2). 399–409.

McKenna, Phil. 2007. The rise of cyberbullying. *New Scientist* 195. 26–27.

Netemeyer, Richard G., James S. Boles & Robert McMurrian. 1996. Development and validation of work-family conflict and family-work conflict scales. *Journal of Applied Psychology* 81(4). 400–410.

O'Neill, D. Kevin, Rory Wagner & Louis M. Gomez. 1996. Online mentors: Experimenting in science class. *Educational Leadership* 54(3). 39–42.

Parks, Malcom R. & Kory F. Floyd. 1996. Making friends in cyberspace. *Journal of Communication* 46(1). 80–94.

Priviteria, Carmel & Marylin Anne Campbell. 2009. Cyberbullying: The new face of workplace bullying? *CyberPsychology and Behavior* 12(4). 395–400.

Rice, Ronald E. & Gail Love. 1987. Electronic emotion: A content and network analysis of a computer-mediated communication network. *Communication Research* 14(1). 85–108.

Richman, Amy L., Janet T. Civian, Laurie L. Shannon, E. Jeffrey Hill & Robert T. Brennan. 2008. The relationship of perceived flexibility, supportive work-life policies, and use of formal flexible arrangements and occasional flexibility to employee engagement and expected retention. *Community, Work, & Family* 11(2). 183–197.

Riedy, M. K. & Joseph H. Wen. 2010. Electronic surveillance of Internet access in the American workplace: Implications for management. *Information & Communication Technology Law* 19(1) 87–89.

Schmitz, Joseph & Janet Fulk. 1991. Organizational colleagues, media richness, and electronic mail – A test of the social influence model of technology use. *CommunicationResearch* 18(4). 487–523.

Scott, Craig R. & C. Eric Timmerman. 2005. Relating computer, communication, and computer-mediated communication apprehensions to new communication technology use in the workplace. *Communication Research* 32(6). 683–725.

Stephens, Keri K. 2007. The successive use of information and communication technologies at work. *Communication Theory* 17. 486–507.

Tabak, Filiz & William P. Smith. 2005. Privacy and electronic monitoring in the workplace: A model of managerial cognition and relational trust development. *Employee Responsibilities and Rights Journal* 17(3). 173–189.

Thatcher, Sherry M. B. & Xiumei Zhu 2006. Changing identities in a changing workplace: Identification, identity enactment, self-verification, and telecommuting. *Academy of Management Journal* 31(4). 1076–1088.
Thompson, Lynne, Mike Jeffries & Keith Topping. 2010. E-mentoring for e-learning development. *Innovations in Education and Teaching International* 47(3). 305–315.
Turnage, Anna K. 2007. Email flaming behaviors and organizational conflict. *Journal of Computer-Mediated Communication* 13. 43–59.
Urbaczewski, Andrew & Leonard M. Jessup. 2002. Does electronic monitoring of employee Internet usage work? *Communications in the ACM* 45(1). 80–83.
Walther, Joseph B. 1992. Interpersonal effects in computer-mediated interaction: A relational perspective. *Communication Research* 19(1). 52–90.
Wiesenfeld, Batia M., Sumita Raghuram & Raghu Garud. 2001. Organizational identification among virtual workers: The role of need for affiliation and perceived work-based social support. *Journal of Management* 27(2). 213–229.
Wright, Kevin B., Abendschein, Bryan, Wombacher, Kevin, O'Connor, Michaela, Hoffman, Megan, Dempsey, Molly, Krull, Christopher, Dewes, Audrey, & Shelton, Audrey. (2014). Work-related communication technology outside of regular work hours and work life conflict: The influence of communication technologies on perceived work life conflict, burnout, job satisfaction, and turnover intentions. *Management Communication Quarterly*, Available: http://mcq.sagepub.com/content/early/2014/05/13/0893318914533332

Anne Linke

21 Marketing and public relations

Abstract: The digital evolution from the Web 1.0 to the Web 2.0 has created many challenges for enterprises and their corporate communication. Those who were not afraid of the new communication environment and who did not want to miss the trend started experimenting with these new platforms quickly. This became necessary as more and more of their stakeholders used the new opportunities for publishing and exchanging user generated content actively However, the shortcomings of such an actionism appeared and the search for adequate communication management processes and structures started. This chapter addresses the associated process of professionalization as well as the necessary rules and resources.

Keywords: strategic communication, communication management, social media, structuration, governance, professionalization

1 Strategic corporate communication in social media

Social media – new, interactive, and easy to operate communication technologies – as well as new practices of using them have led to a major change for individual and organizational communication. Motivated by strong demand and usage by their stakeholders, more and more companies are using social media platforms for communication and business purposes (e.g. Barnes 2010; Duhé 2012a: 1–2; Fink et al. 2011; Wright and Hinson 2009, Wright and Hinson 2010, Wright and Hinson 2011; Young et al. 2008). But, integrating them into the existing media mix of print or broadcasting activities presents many challenges, mainly caused by the dynamics and the complexity of the numerous new channels. This means in addition to operative issues, more general questions regarding the *management of communication* need to be asked. Instead of jumping on the bandwagon of the social media hype or only thinking about the Facebook page design, new strategies and approaches are needed. The new online environment brings together people with different (sociodemographic) characteristics, interests, demands, and usage patterns. For them, it offers diverse sources of information and platforms for public debate as well as for opinion-forming. Consequently, being present here or knowing about these developments has become an essential issue for organization of all kinds. Because of the special characteristics of the new online communication, the boundaries of communication units within the organizations are dissolving so that many authors nowadays talk about *corporate communication* instead of singular disciplines like Marketing or Public Relations (PR) (e.g. Christensen and Cornelissen 2011, 2013; Pleil and Zerfaß 2014; Zerfaß 2010). It can be defined as follows: "[...] corporate communication embraces all communication processes which contribute to the definition of tasks, and to said tasks' realization, within profit-oriented economic organizations, and which, in particular, contribute to the internal and external coordination of actions and to the clarification of interests defining the relation between companies and their specific stakeholders." (Zerfaß 2008: 66; see also Austin and Pinkleton 2006; Holtzhausen and Zerfaß 2013). These communication activities aim to inform relevant stakeholders and to construct meaning via bidirectional exchange. With the help of these processes, stakeholders shall be influenced strategically in order to support business processes and create profitability on the one side and legitimacy ("licence to operate") on the other (Zerfaß 2008: 68). Depending on which stakeholders to focus on primarily, different sub disciplines can be distinguished: internal communication e. g. concentrates mainly on employees and the collaborative production processes, marketing on selling products to customers and buying others from suppliers, and PR on cultivating social and political relationships (Zerfaß 2008: 80). However, with complex modern flows of communication, the boundaries between these spheres become more blurred, which is mainly caused by the extensive exchange between networked online stakeholders and their emancipation (Castells 2010). By way of illustration, marketing campaigns are also received by journalists and employees act as ambassadors for an enterprise in public discussions. Many stakeholders do not wait for communication that is pushed into their direction but demand information, actively pull it towards them and become publishers themselves. The web 2.0 enables this because of quick, easy, and global publication possibilities. On the whole, this creates a new communication environment for organizations, which have to adapt accordingly and change their corporate communication to adequately deal with the challenges and chances.

This chapter explains these circumstances in the new communication sphere with reference to relevant literature and empirical studies from communication and management science. Moreover, it goes beyond the often chosen focus of mere social media activities and refers to their structural prerequisites in the form of necessary rules and resources. With the help of the structuration theory by Anthony Giddens (1997), it shall be explained how such structures shape the communication actions in the social web and how the latter reproduces or changes the structures. Building on that, a conclusion is drawn regarding the overall professionalization of online communication from the first phase of experiments until the strategic integration into adequate management structures and processes.

2 A new communication environment

Like any other innovation, it at first spreads among specialized groups: social media were, at first, primarily used by technology elites or so called "early adopters" (Rogers 2003). But, after overcoming a tipping point they have reached mass exploitation in the meanwhile. The increased use of social media by many different stakeholders results in a *changed communication environment* that enterprises have to deal with. For example, the specific requirements in social media with regard to participation and dialogue must be responded to adequately, possibly with responses that are quite different from the traditional approaches. The specificity of the newly arisen communication space is mainly linked to the fact that in today's society former linguistic, spatial or temporal boundaries of the public sphere are overcome (Bentele and Nothhaft 2010: 112; Graham 1999: 86–88; Miller 2011: 61). The technical development of the internet has also led to public and private spheres no longer being isolated to the same degree. In addition, individual and mass communication now mix more intensively. Enterprises are increasingly operating publically, so communication measures in the public sphere are "no longer a goal but also [...] corrective entrepreneurial activities" (Karmazin 2007: 72; see also van Dijk 2012: 182–183). Furthermore, participation in public discussion by a number of different actors increases due to simpler publication opportunities by the audience itself. On the internet, thus, a new public sphere has emerged (Graham 1999: 21–23; Miller 2011: 113–115). This does not mean that old power structures, such as the power of mass media, completely disappear. The online pages of influential newspapers and TV stations are still very well received and consequently influence public opinion. The new communication spaces differ in their structures, themes, processes and rules. New reference groups and opinion leaders shape public communication, as the influence and interactivity change (Downey and Fenton 2003; Fraas et al. 2012: 10–11, 30–33; Miller 2011: 14–15; van Dijk 2012: 189–191; Zerfaß 2010: 420). Summing it up, Zerfaß (2010: 418) points out that the emergence of the internet has changed the public communication processes, due to the global availability of communication channels to a variety of different participants.

This development should not be understood too one-sidedly as the abolition of the existing order, but should be viewed as its extension with specific characteristics. Hence, the challenge for professional communicators lies in *integrating the new platforms into the media mix* and into the overall strategic approach of corporate communication.

Pleil (2010: 14) refers to the sum of the new communication spaces in demarcation of the traditional media public as *"premedial space"* and compares it with an overpopulated marketplace on which acquaintances who met just recently are brought together because of a specific issue. In such a situation, the topic of a small group jumps to other small groups – just like in social media. Here, according to the author, the three levels of the public – interpersonal, assembly or theme and media publicity – can no longer be separated clearly. This could influence individuals to communicate at all levels and makes them no longer dependent on journalists as mediators (Pleil 2012: 20–21). On the one hand side, the breaking away of old hierarchies enables new freedom for individuals. On the other hand side, this does not necessarily mean equality or democratization as the technical possibilities also open up a new power struggle: "Whenever we find ourselves in human beings a new environment there is a struggle for power and control. The fact that this environment is virtual rather than physical makes very little difference" (Brown 2009: 83; see also Schelske 2006: 1291–31). For corporate communication this means a *shift in the rules and resources*. Apart from the technical change, a shift from general groups to communities and the emphasis on authenticity can be attested. In addition, new modes of persuasion – specific information retrieval by the recipient rather than information dissemination via the mass media (broadcasting) – as well as the change of authoritative competence and communication rules need to be considered (Zerfaß 2007: 41; see also Hauptmann 2012: 100–103). Many scholars herein see an opportunity for emancipation and democratization, because of the move away from hierarchical structures towards more particpatory interaction patterns (Brown 2009: 2; Coombs 1998: 289; Sweetser 2010). Optimistic approaches assume that "new communication technologies can save democracy by restoring dialogic and participatory communication in the public sphere" (Hiebert 2005: 1). Democracies could only exist, according to those scholars, when conflicting interests occupy the public sphere, which in turn was improved by all unprecedented communication innovations (writing, book publishing, radio and TV). Hiebert argues that all these innovations have led to fights between the ever same antagonists: those who sought democratic dialogues, and those who used them to control. He compares the democratization tendencies of social media with the liberation of prehistoric oral societies from witchcraft healers through scripture (Hiebert 2005: 2).

But, some critical authors doubt that this democratization actually takes place (e.g. Kielholz 2008: 25; McChesney 2013; Moore 1995: xiv; Mosco 2004). Mosco (2004: 24) describes the democratization claim as "seductive tales containing promises unfulfilled or even unfulfillable". One example is that although large populations have access to the web, only in the highly industrialized countries can they really use it and if so mainly under conditions of commercialization and power differences (Ebersbach et al. 2010: 26). Moreover, the new techniques can also be used for authoritative top-down communication, as well as for monitoring purposes. Computers in general can act as intelligent agents in order to predict interests and reactions or to manipulate (Hiebert 2005: 6; Kielholz 2008: 25; Miller 2011: 116–118).

3 Corporate communication 2.0

Many scholars believe that fundamental changes of corporate communication are caused by the digital evolution so, analogous to the term "Web 2.0" they talk about "Marketing 2.0", "PR 2.0", or "Corporate Communication 2.0" (Breakenridge 2008; Hazelton et al. 2007; Meckel 2008; Solis and Breakenridge 2009). Macnamara (2010: 31) emphasizes that the new forms of communication offer the opportunity to improve corporate communications. Phillips and Young (2009: 1) even claim that "for public relations the unavoidable conclusion is that nothing will ever be the same again". While the original World Wide Web as an "enabler of existing methodologies" (Pavlik 2007: 9) primarily presented new ways of communicating information, the web 2.0 implies a fundamental change. This goes beyond technology and includes the *need for new communication approaches* on a strategic, conceptual and operative level (Duhé 2012a: 1–2; Duhé 2012b: 64–65; Macnamara 2010: 324; Meckel 2008: 490). Also empirically, many communication professionals agree that social media have changed the PR practice. In a European wide study among communication managers, the results indicate that:

> The majority of communication professionals in Europe think that social media influence the perceptions of the organisation and the construction of reality by the use of this type of media. Almost three quarters (72.5 per cent) of the professionals think that social media content changes the perceptions of external stakeholders about the organisation and 57 per cent perceive this to be true for employees (Zerfass et al. 2013: 25).

In different studies however, the attested *degree of changes* ranges from the mere integration of new communication channels to the beforehand cited extreme of a completely new start (e.g. Fink and Zerfaß 2010: 21; Fink et al. 2011; Rossmann 2010; Wright and Hinson 2010: 12). Pradel (2001: 230) summarizes the most important changes in three classes:

1. Complexity and turbulence contradict established patterns or processes.
2. Solidification, standardization and innovation weakness jeopardize brand building.
3. The progressive depersonalization of the digital world demands atmosphere, images and relationship building.

In an adequate modern understanding of corporate communication, communication departments do not "construct" the world by persuasively communicating and trying to achieve certain rational effects on their audience. Instead, the communication pulse also comes from the stakeholders that understand themselves as "investigative multipliers" rather than passive recipients. Solis (2012) sees the change in the behavior of the stakeholders as fundamental and therefore calls his book *The end of business as usual*. He refers to the emancipatory phenomenon as "social consumerism" and describes that consumers connect in a comprehensive

Tab. 1: Comparison of the three types of online PR (*Source*: modified from Pleil 2007: 18).

Characteristics	Types of Online PR		
	Digital PR	**Internet PR**	**Cluetrain PR**
Environment	Internet Galaxy		Google World
Communication	monological	monological (with indirect feedback channel)	dialogical, network oriented
Userś Role	recipients	recipients with limited action options	communication partners organized in a network
Demands	PR competences of the analogous world and understanding of hypermediality	Content Management, social research	increased strategic and interpersonal skills; special position of trust; advice to other members of the organization
Expenditure	relatively small, often an additional service by the existing PR department	high technical effort Need for their own site (s) for online PR	very time-consuming; continuous task (“always on”)
Role of Online PR	executive	channeling	manufacture of openness, empowerment of stakeholders
Main Aim	information	persuasion	understanding, argumentation

and efficient information network to create their own content (Solis 2012: xiii). Networked stakeholders (“audience of audiences”) share good and bad experiences online and have great power over companies (Solis 2012: 60–62). The “digital darwinism” – the economic struggle for survival in the digital age – can according to the author only be won by the best adjusted companies: “Businesses that recognize disruption and develop a culture and process for innovation now and over time will survive.” (Solis 2012: xiv). Traditional, organization-centered approaches are therefore misleading (Argenti and Barnes 2009; Gossieaux and Moran, 2010: 81–83; Luoma-aho & Vos 2010: 2; Sheldrake 2011).

Pleil (2007: 18) explains the details of these changes by comparing *three types of Online PR*: “Digital PR”, “Internet PR”, and a third form he calls “Cluetrain PR” based on 95 theses describing the changes of the internet that is called “Cluetrain Manifesto” (see table 1). Whilst the first two ones are mostly applicable in the traditional “Internet Galaxy” of the www, the latter, more dialogical and network oriented form applies for the “Google World” of social media and the web 2.0. Here, stakeholders need to be understood as communication partners that are organized in a network where they trust themselves more than any corporate source and where intensive exchange takes place. Consequently, they (as emancipated actors) can only be influenced by argumentation if understanding is reached. This is by far more time-consuming and costly than information dissemination in the traditional online sphere. Although a general shift to Cluetrain PR approach appears advisable, such an extensively interactive approach is cost prohibitive and, thus, a mixture of the online PR approaches are required to maximize efficient communication between companies and the stakeholders.

Changes in the *design of corporate communications* are required. For example, the need to "[...] minimise the extent to which the symbolic, interpretive paradigm of public relations affects their thinking and institutionalise public relations as a strategic management, behavioural paradigm." (Grunig 2009: 20) In their book *Naked Conversations* Scoble and Israel (2006) suggest an individualized, direct style of public communication as well as a shift away from the controlled, unidirectional communications towards decentralized, interactive models, "it's about stripping out all the crap that gets in the way of understanding and trust" (Scoble and Israel 2006: 1). In the "Conversational Era", the social media age, blogs allow the direct access to journalistic stakeholders without intermediate steps. For this, companies must listen to them and understand them seriously as a partner (Scoble and Israel 2006: 227–230).

However, the change should *not be overstated*, because not everything that was valid before is automatically obsolete. In their publication *Digital Strategies for Powerful Corporate Commu-nications*, Argenti and Barnes (2009: 19) sum this up as follows "„[B]esides acknowledging the rapidly changing business environment, companies must adapt without changing what they stand for or compromising their principle." In addition, looking back on the history of corporate communication, Pavlik (2007: 4) mentions the general tendency that its strategies and tactics have always adapted to the media used.

4 Necessary structures: Rules and resources

New concepts of corporate communication need to be supported by appropriate structures and processes, as can be explained by the *structuration theory* of Giddens (1997). According to the social scientist structures and actions interact interdependently so that actions of individuals only succeed if everyone can resort to structures in the sense of a common stock of rules and resources. In the context of modern online communication, this calls for a comprehensive understanding of so called "social media governance" as a necessary structural prerequisite. Social Media Guidelines alone are not enough to guide corporate social media communication. Instead, a comprehensive framework is required that allows on the one hand communicative action on social media platforms, and also restricts them on the other hand. Because of the duality of structures and actions, the regulatory

Tab. 2: Categories of Social Media Governance (*Source*: adapted from Linke 2014: 173).

Categories of Social Media Governance		
Control System	**Support System**	**Feedback System**
Goals/Strategies Social Media Guidelines Juridical Knowledge Social Media Trainings Key Performance Indicators	Management Support Freedom Personnel Responsibilities and Cooperation Forms Technological Access Budgets Social Media Trainings	Monitoring Instruments Key Performance Indicators

framework is reproduced and changed through communication over time. Orlikowski (1992: 410) in this regard compares four different forms of influence: technology as a product of human actions, technologies as a medium of human actions, institutional conditions of interaction with technology and institutional consequences of interaction with technology. The emergence of governance accordingly takes place in a complex process with different effect levels.

Social Media Governance as a requirement for strategic social media communication can be defined as any formal or informal conditions for the actions of the members of an organization on the social web, consisting of rules (modes of understanding and argumentation) and resources (allocative and authoritative). Despite the theoretically proofed relevance of governance, this comprehensive concept has only recently been studied (Linke 2014; Monks and Minow 2011; Rice and Gattiker 2001: 572; Sinclair and Vogus 2011). Some empirical data can be found for Europe (e.g. Fink and Zerfaß 2010; Fink et al. 2011; Linke 2014; Linke and Zerfass 2012; Oliveira et al. 2014), Australia and Asia (Macnamara and Zerfass 2012), and the USA (Owyang 2011). For Germany, the authors of the social media governance studies attest that 70.6 % of the surveyed organizations have established only a few aspects of governance, 22.3 % an average number and 7.1 % many. The latter had increased by only four percentage points over the previous year (Fink et al. 2011: 57). In these studies, the most comprehensive governance understanding is present, which is also based in a theoretical context. Linke (2014: 173) compares different studies on this topics and presents a list of relevant structural aspects which she divides (based on a factor analysis) into three categories: control, support and feedback system (see table 2). In German enterprises, the most commonly established governance items are "technical ways to access social media during working hours" (76.9 %). The "endorsement by the management" is present in 71.7 % of respondents, while 62.4 % have the "necessary freedom to decide independently", and 53.7 % have "defined responsibilities and cooperation processes". Room for improvement show aspects of "legal protection" (27.6 %) as well as "key performance indicators" (KPIs) (31.3 %) (Linke: 2014: 172). The dimensions of governance can also be divided into "rules" and "resources" in accordance with the understanding of Giddens. On average, the companies create more rules than resources (*M* = 3.06 and 2.74, *SD* = 1.80 and 1.74) (Linke: 2014: 172).

In Europe as a whole, less than a third of the organizations have created the necessary conditions (Zerfass et al. 2010: 88). However, the study is limited to guidelines, monitoring tools, overviews of the social media activities, training and KPIs (Zerfass et al. 2010: 79). In Australia and Asia only 35 % of surveyed organizations have guidelines, only 33 % trainings and only 20 % a strategy (Macnamara and Zerfass 2012: 297; for more theoretical information see Boudreaux 2011). The Altimeter Group analyzed in their non-academic but comprehensive study similar structural framework conditions under the umbrella concept of "Social Business Readiness" (Owyang 2011). The authors list four points: objectives and policies, editorial and response processes, training, and a central interdisciplinary "Center of Excellence" (Owyang 2011: 6–8). Among the tested advanced enterprises, 72 % have a permanent social media training program, compared with 34 % in the total sample (Owyang 2011: 11–12). Overall, only 26 % of the companies have established a "baseline process to reinforce social media policy and education." Other non-academic studies support the observation that investments in the general conditions for social media corporate communication are rising (e.g. Burson-Marsteller 2010; Young et al. 2008). From the empirical insights, it can be concluded that the importance of comprehensive governance structures for social media is increasingly recognized in practice and theory. Companies now invest in the necessary framework for strategic social media communication, but still have a lot of catching up to do. In order to be able to deal with the variety of different platforms and their complexity, such a framework is needed for effective strategic communication. However, it has to be established with an *appropriate long-term change process*.

5 Professionalization of online communication management

With the digital evolution and the rise of social media, a new communication sphere was created that imposes many challenges to corporate communication. At first, many organizations did not want to miss the trend and started *experimenting* with these new platforms as more and more of their stakeholders used them actively searching for recognition and information. Soon, the shortcomings of this quite unstrategic actionism appeared. Adequate social media communication requires quick response rates, a certain etiquette and complexity management. This needs advanced process management and structures in the sense of the above mentioned social media governance (e.g. Argenti and Barnes 2009; Macnamara and Zerfass 2012; Owyang 2011; Sheldrake 2011; Solis 2012). Empirical studies show that more and more organizations invest in this *consolidation* and have therefore started a new era of professionalized cluetrain communication. Only with a strategically managed change process this can be accomplished. Relevant factors to consider are environmental as well as specific organizational conditions. For instance, it can also be advisable to reduce social media activities to a minimum if the relevant stakeholders are not present there or if it does not help the organizational aims (Linke 2014: 208).

Different *paths for the professionalization* of social media communication are possible and detectable in empirical data (Linke 2014: 197–198). Firstly, organizations that acknowledge the altered communication situation can experiment with the new technologies. Such a more "impulsive strategy" begins with social media communication and then creates social media governance structures, which in turn alters the process of communication management. Referring to Giddens' (1997) structuration theory, the communicative actions of social media communication require changes in the structures like e.g. a strategy or enough personnel to handle the tasks. With increasing experience and more intensive communication activities this becomes clear and helps to focus. However, also the opposite way appears logical. In a "planning-oriented approach", initially, an adequate communication management is established. Specific rules and resources for social media communication are created, like social media guidelines or trainings. These structures then enable extensive or advanced social media activities. The first impulsive strategy needs a positive way of thinking, because instead of having concerns regarding the new communication space it is being experimented with it. Only in the course of the experimental phase do the organizations notice that they also need the appropriate conditions. In contrast, the other approach seems cautious or risk-avoiding. Both approaches need to be understood as the outer poles of a scale without one of them per se being the better option. Both involve advantages and disadvantages and need to consider the specifics of the company and its environment (Linke 2014: 198). On the whole, they show corporate communication can specifically adapt to the communication technologies and the altered online environment to the mutual benefit of both the stakeholders and the organisations.

References

Argenti, Paul A. & Courtny M. Barnes. 2009. *Digital Strategies for Powerful Corporate Communications*. New York: McGraw-Hill Education – Europe.

Austin, Erica W. & Bruce E. Pinkleton. 2006. *Strategic public relations management: Planning and managing effective communication programs*. 2nd Ed. Mahwah, NJ: Lawrence Erlbaum Associates.

Barnes, Nora Ganim. 2010. *The 2010 Inc. 500 Update: Most Blog, Friend And Tweet But Some Industries Still Shun Social Media*. Center for Marketing Research. Dartmouth: University of Massachusetts, Retrieved from www.umassd.edu/media/umass dartmouth/cmr/studiesandresearch/2010inc500.pdf [23. 06. 2014].

Bentele, Günter & Howard Nothhaft. 2010. Strategic Communication and the public sphere from a European perspective. *International Journal of Strategic Communication* 4(2). 93–116.

Boudreaux, Chris. 2011. Social Media Policies. In Robert Wollan, Nick Smith, Catherine Zhou (eds.), *The social media management handbook: Everything you need to know to get social media working in your business*, 274–285. Hoboken, NJ: Wiley.

Breakenridge, Deirdre. 2008. *PR 2.0: New media, new tools, new audiences*. Upper Saddle River, NJ: FT Press.

Brown, Rob. 2009. *Public relations and the social web: How to use social media and Web 2.0 in communications*. London: Kogan Page.

Burson-Marsteller. 2010. *The Global Social Media Check-up: Insight from the Burson-Marsteller Evidence-Based Communications Group*. Retrieved from http://www.burson-marsteller.com/Innovation_and_insights/blogs_and_podcasts/BM_Blog/Documents/Burson-Marsteller%202010 %20Global%20Social%20Media%20Check-up%20white%20paper.pdf [23. 06. 2014].

Castells, Manuel. 2010. *The rise of the network society* (2. edn.). *The information age:* Vol. 1. Malden, Mass: Wiley-Blackwell.

Christensen, Lars T. & Joep Cornelissen. 2011. Corporate and organizational communication in conversation. *Management Communication Quarterly* 25(3). 383–414.

Christensen, Lars T. & Joep Cornelissen. 2013. Bridging Corporate and Organizational Communication: Review, Development and a Look to the Future. In Ansgar Zerfaß, Lars Rademacher, & Stefan Wehmeier (eds.), *Organisationskommunikation und Public Relations: Forschungsparadigmen und neue Perspektiven*, 43–72. Wiesbaden: VS Verlag für Sozialwissenschaften.

Coombs, W. Timothy. 1998. The internet as potential equalizer: New leverage for confronting social irresponsibility. *Public Relations Review* 24(3). 289–303.

Downey, John & Natalie Fenton. 2003. New Media, Counter Publicity and the Public Sphere. *New Media & Society* 5(2). 185–202.

Duhé, Sandra. 2012a. A Thematic Analysis of 30 Years of Public Relations Literature Addressing the Potential and Pitfalls of New Media. In Sandra Duhé (ed.), *New Media and Public Relations*, 1–2. 2nd edn. New York: Peter Lang.

Duhé, Sandra. 2012b. Public Relations and Complexity Thinking in the Age of Transparency. In Sandra Duhé (ed.), *New Media and Public Relations*. 2nd. ed. 57–76. New York: Peter Lang.

Ebersbach, Anja, Markus Glaser & Richard Heigl. 2010. *Social Web*. 2nd rev. Ed. Konstanz: UVK.

Fink, Stephan & Ansgar Zerfaß. 2010. *Social Media Governance 2010. Results of a survey among communications professionals from companies, governmental institutions and non-profit organizations in Germany*. Leipzig/Wiesbaden: University of Leipzig/FFPR.

Fink, Stephan, Ansgar Zerfaß & Anne Linke. 2011. *Social Media Governance 2011 – Expertise Levels, Structures and Strategies of Companies, Governmental Institutions and Non-Profit Organizations communicating on the Social Web. Results of an empirical study of communications professionals*. Leipzig/Wiesbaden: University of Leipzig/FFPR.

Fraas, Claudia, Stefan Meier & Christian Pentzold. 2012. *Online-Kommunikation: Grundlagen, Praxisfelder und Methoden [Online Communication: Basics, Practical Fields and Methods]*. München: Oldenbourg.

Giddens, Anthony. 1997. *The Constitution of Society: Outline of the Theory of Structuration*. Cambridge: Polity.

Gossieaux, Francois, & Ed Moran. 2010. *The hyper-social organization: Eclipse your competition by leveraging social media*. Maidenhead: McGraw-Hill Professional.

Graham, Gordon. 1999. *The Internet: A Philosophical Inquiry*. New York: Taylor & Francis.

Grunig, James E. 2009. Paradigms of global public relations in an age of digitalisation. *PRism, 6*(2). Retrieved from http://www.prismjournal.org/fileadmin/Praxis/Files/globalPR/GRUNIG.pdf [23. 06. 2014].

Hauptmann, Stefan. 2012. *Social Media in Organisationen: Strukturation und computervermittelte Kommunikation [Social Media in Organisations: Strukturation and computer mediated Communication].* Wiesbaden: Springer Gabler.

Hazelton, Vincent, Jill Harrison-Rexrode & William Kennan. 2007. New technologies in the formation of personal and public relations: social capital and social media. In Sandra C. Duhé (ed.), *New media and public relations,* 91–105. New York: Peter Lang.

Hiebert, Ray E. 2005. Commentary: new technologies, public relations, and democracy. *Public Relations Review* 31(1). 1–9.

Holtzhausen, Derina & Ansgar Zerfaß. 2013. Strategic Communication – Pillars and Perspectives of an Alternative Paradigm. In Ansgar Zerfaß, Lars Rademacher & Stefan Wehmeier (eds.), *Organisationskommunikation und Public Relations: Forschungspradigmen und neue Perspektiven,* 73–94. Wiesbaden: VS Verlag für Sozialwissenschaften.

Karmasin, Matthias. 2007. Stakeholder-Management als Grundlage der Unternehmens-kommunikation [Stakeholder Management as a basis for Corporate Communication]. In M. Piwinger & Ansgar Zerfaß (eds.), *Handbuch Unternehmenskommunikation,* 71–87. Wiesbaden: Gabler.

Kielholz, Annette. 2008. *Online-Kommunikation: Die Psychologie der neuen Medien für die Berufspraxis [Online Communication: The psychology of new media].* Heidelberg: Springer.

Linke, Anne & Ansgar Zerfass. 2012. Future trends of social media use in strategic communication: Results of a Delphi study. *Public Communication Review* 2(2). 17–29.

Linke, Anne. 2014. *Management der Online Kommunikation von Unternehmen: Steuerungskonzepte, Multi-Loop-Prozesse und Governance [Management of the Online Communication of Enterprises: regulation concepts, multi loop processes and governance].* Dissertation eingereicht an der Universität Leipzig.

Luoma-aho, Vilma & Marita Vos. 2010. Towards a more dynamic stakeholder model: acknowledging multiple issue arenas. *Corporate Communications: An International Journal* 15(3). 315–331.

Macnamara, Jim. 2010. *The 21st century media (r)evolution: Emergent communication practices.* New York: Lang.

Macnamara, Jim & Ansgar Zerfass. 2012. Social media communication in organizations: The challenges of balancing openness, strategy and ma-nagement. *International Journal of Strategic Communication* 6(4). 287–308.

McChesney, Robert W. 2013. *Digital disconnect: How capitalism is turning the Internet against democracy.* New York: New Press.

Meckel, Miriam. 2008. Unternehmenskommunikation 2.0 [Corporate Communication 2.0]. In Miriam Meckel & Beat F. Schmid (eds.), *Unternehmenskommunikation: Kommunikations-management aus Sicht der Unternehmensführung.* 2nd ed. 471–492. Wiesbaden: Gabler | GWV.

Miller, Vincent. 2011. *Understanding digital culture.* Los Angeles: SAGE.

Monks, Robert A. G. & Nell Minow. 2011. *Corporate governance.* 5th Ed. Hoboken, NJ: John Wiley & Sons.

Moore, Dinty W. 1995. *The emperor's virtual clothes: The naked truth about Internet culture.* Chapel Hill, N.C: Algonquin Books.

Mosco, Vincent. 2004. *The digital sublime: Myth, power, and cyberspace.* Cambridge, Mass: MIT Press.

Oliveira, Evandro, Anne Linke, Margarida Kunsch & Ansgar Zerfass. 2014. *How global is the web 2.0 hype? An international comparison of the status quo of social media communication in Brazil, Germany and Portugal.* Paper to be presented (already accepted) at Euprera Annual Conference 2014, Brussels, Belgium.

Orlikowski, Wanda J. 1992. The Duality of Technology: Rethinking the Concept of Technology in Organizations. *Organization Science* 3(3). 398–427.

Owyang, Jeremiah K. 2011. *Social Business Readiness: How Advanced Companies Prepare Internally*. Retrieved from http://www.altimetergroup.com [23. 06. 2014].

Pavlik, John V. 2007. *Mapping the Consequences of Technology on Public Relations*. Retrieved from http://www.instituteforpr.org [23. 06. 2014].

Phillips, David & Philp Young. 2009. *Online Public Relations: A Practical Guide to Developing an Online Strategy in the World of Social Media*. 2nd Ed. *PR in Practice Series*. London: Kogan Page.

Pleil, Thomas. 2007. *Online-PR im Web 2.0: Fallbeispiele aus Wirtschaft und Politik [Online PR in the Web 2.0: Case studies from economics and politics]*. Konstanz: UVK.

Pleil, Thomas. 2010. Web-Monitoring: Kommunizieren setzt Zuhören voraus [Web Monitoring: Communication requires Listening]. In Patrick Brauckmann (ed.), *Kommunikationswissenschaft: Web-Monitoring. Gewinnung und Analyse von Daten über das Kommunikationsverhalten im Internet,* 11–18. Konstanz: UVK.

Pleil, Thomas. 2012. Kommunikation in der digitalen Welt [Communication in the digital World]. In Ansgar Zerfaß & Thomas Pleil (eds.), *Handbuch Online PR,* 17–38. Konstanz: UVK.

Pleil, Thomas & Ansgar Zerfaß. 2014. Internet und Social Media in der Unternehmenskommunikation [Internet and Social Media for Corporate Communication]. In Ansgar Zerfaß & Thomas Pleil (eds.), *Handbuch Online PR,* 731–753. Konstanz: UVK.

Pradel, Marcus. 2001. *Dynamisches Kommunikationsmanagement: Optimierung der Marketingkommunikation als Lernprozess [Dynamic Communication Management: Optimization of Marketing Communication as a Learning Process]*. Wiesbaden: Gabler.

Rice, Ronald E. & Urs E. Gattiker. 2001. New Media and Organizational Structuring. In Frederic M. Jablin & Linda L. Putnam (eds.), *The new handbook of organizational communication: Advances in theory, research, and methods,* 544–584. Thousand Oaks, CA: SAGE.

Rogers, Everett M. 2003. *Diffusion of innovations*. 5th Ed. New York: Free Press.

Rossmann, Alexander. 2010. *Next Corporate Communication: Qualitative Marktforschung. Perspektiven von Social Media für Marketing und Unternehmenskommunikation [Next Corporate Communication: Qualitative Market Research. Perspectives of Social Media for Marketing and Corporate Communication]*. Next Corporate Communication Konferenz, St. Gallen. Retrieved from http://de.slideshare.net/UNISGresearch [23. 06. 2014].

Schelske, Andreas. 2006. *Soziologie vernetzter Medien: Grundlagen computervermittelter Vergesellschaftung. Interaktive Medien [Sociology of interconnected media: basics of computer enabled socialization. interactive media]*. München: Oldenbourg.

Scoble, Robert & Shel Israel. 2006. *Naked Conversations: How Blogs are Changing the Way Businesses Talk with Customers*. Hoboken, NJ: Wiley.

Sheldrake, Philip. 2011. *The business of influence: Reframing marketing and PR for the digital age*. Chichester: Wiley.

Sinclair, Jollean K. & Clinton E. Vogus. 2011. Adoption of social networking sites: an exploratory adaptive structuration perspective for global organizations. *Information Technology and Management* 12(4). 293–314.

Solis, Brian. 2012. *The end of business as usual: Rewire the way you work to succeed in the consumer revolution*. Hoboken, NJ: John Wiley.

Solis, Brian, & Breckenridge, Deirdre. 2009. *Putting the public back in public relations: How social media is reinventing the aging business of PR*. Upper Saddle River, NJ: FT Press.

Sweetser, Kaye D. 2010. A Losing Strategy: The Impact of Nondisclosure in Social Media on Relationships. *Journal of Public Relations Research* 22(3). 288–312.

van Dijk, Jan A. 2012. *The network society*. 3rd Ed. London: SAGE.

Wright, Donald K. & Michelle Hinson. 2009. Examining how public relations practitioners actually are using social media. *Public Relations Journal* 3(3). Retrieved from http://www.prsa.org/intelligence/prjournal/documents/2012wrighthinson.pdf [23. 06. 2014].

Wright, Donald K. & Michelle Hinson. 2010. An analysis of new communications media use in public relations: Results of a five-year trend study. *Public Relations Journal* 4(2). Retrieved from http://www.prsa.org/Intelligence/PRJournal/Documents/2010tWrightHinson.pdf [23. 06. 2014].

Young, G. Oliver, Eric G. Brown, TJ Keitt, Jeremiah K. Owyang, Rob Koplowitz, & Heidi Shey. 2008. *Global Enterprise Web 2.0 Market Forecast: 2007 To 2013*. Retrieved from http:// www.forrester.com [23. 06. 2014].

Zerfaß, Ansgar. 2007. Von der Einkanal-Kommunikation zum Dialog – wenn Empfänger zu Akteuren werden [From one channel communication to dialogue – wenn recipients become actors]. In Thomas Ellerbeck & Klaus Siebenhaar (eds.), *Vernetzte Welt: Veränderungen der Kommunikation durch neue Medien und Mobilfunk*, 31–48. Berlin: B & S Siebenhaar.

Zerfass, Ansgar. 2008. Corporate communication revisited: integrating business strategy and strategic communication. In: Ansgar Zerfass, Betteke van Ruler & Krishnamurthy Sriramesh (eds.), *Public relations research: European and international perspectives and innovations*, 65–96. Wiesbaden: VS Verlag für Sozialwissenschaften.

Zerfaß, Ansgar. 2010. *Unternehmensführung und Öffentlichkeitsarbeit: Grundlegung einer Theorie der Unternehmenskommunikation und Public Relations* [*Corporate Leadership and Public Relations: basics of a theory of Corporate Communication and Public Relations*. 3rd Ed. Wiesbaden: Springer.

Zerfass, Ansgar, Ralph Tench, Piet Verhoeven, Dejan Vercic & Angeles Moreno. 2010. *European Communication Monitor 2010: Status Quo and Challenges for Public Relations in Europe: Results of an Empirical Survey in 46 Countries*. Brussels: EACD/EUPRERA, Helios Media.

Zerfass, Ansgar, & Angeles Moreno, Ralph Tench, Dejan Vercic & Piet Verhoeven. 2013. *European Communication Monitor 2013*. A Changing Landscape-Managing Crises, Digital Communication and CEO Positioning in Europe. Results of a Survey in 43 Countries, Brussels.

Tomasz Janowski

22 From electronic governance to policy-driven electronic governance – evolution of technology use in government

Abstract: Policy-Driven Electronic Governance (EGOV) captures an idea that through the appropriate use of Information and Communication Technologies (ICT) we can transform the working of government organizations and how they interact with citizens, businesses and other arms of government in order to achieve a variety of high-value public policy goals. Such goals generally include delivering better public services, reforming public institutions, engaging citizens in government decision-making, and increasing the performance of public policies in health, education, security, employment and other sectors. The aim of this chapter is to outline the evolution of government use of technology, and to introduce Policy-Driven EGOV as the latest phase in this evolution by proposing: conceptual model for Policy-Driven EGOV, evidence of emerging presence of Policy-Driven EGOV among national EGOV strategies, challenges in progressing from EGOV to Policy-Driven EGOV, and a research and policy agenda to build a foundation for this new phase.

Keywords: information and communication technology, government, public policy, electronic government, electronic governance, policy-driven electronic governance

While the provision of public services and infrastructure, formulation and implementation of public policies, and maintenance of social order and security remain the core functions of any government, such functions must be increasingly performed in both physical and digital worlds. With 86 % of the world's population using mobile phones and 33 % using the Internet by end of 2011, with fixed broadband prices falling by 75 % and the price of ICT services by 30 % between 2008 and 2011, with mobile-cellular services becoming the prime revenue source in developing countries, and with broadband becoming part of national infrastructures in countries around the world (ITU 2012), no government can afford leaving this space unattended or ungoverned. In order to establish their presence and authority in the digital world and to improve their internal operations, government organizations are increasingly adopting the latest in mobile, cloud, social, virtual and other technologies following their private sector counterparts (Accenture 2013), and transforming themselves in the process.

The concept of Electronic Government captures an idea that the appropriate use of ICT and the resulting transformation can positively change the internal working of government and lead towards: more efficient operations, better public service delivery, and more effective public administration reform including the reform of public finances, leadership, human resource management, transparency, procurement, regulation and many others (OECD 2011). This understanding is reflected by the early but still relevant definition of Electronic Government as "the use of information and communication technologies, and particularly the Internet, as a tool to achieve better government" (OECD 2003). However, the impact of ICT extends beyond government boundaries, and includes how government interacts with various outside entities. We refer to this broader concept as Electronic Governance (EGOV) "the application of technology by government to transform itself and its interactions with customers, in order to create impact on the society" (Estevez and Janowski 2013). The goals of such transformation include engaging citizens in government decision-making, building Knowledge Society for all (UNESCO 2009) and increasing the performance of public policies in health, education, economy, security, and other sectors, in addition to Electronic Government-specific goals. The concepts of Electronic Government and Electronic Governance are contrasted in (Grönlund and Horan 2004).

In this chapter we make an argument that the difference between Electronic Government and Electronic Governance is not just conceptual but part of a larger evolution of the use of technology by governments. The chapter tracks this evolution by identifying four distinctive development phases with corresponding sets of goals, challenges and limitations: 1. Technology in Government – deploying technology in government, 2. Electronic Government – transforming government organizations through technology, 3. Electronic Governance – transforming relationships between government and citizens, businesses and other non-state actors through technology, and 4. Policy-Driven Electronic Governance – supporting policy and development goals in different locations and sectors through technology and how it transforms government and its relationships with citizens and other non-state actors. Policy-Driven Electronic Governance, the latest and still largely undeveloped phase, is the subject of the remainder of this chapter where we present: a conceptual model for Policy-Driven EGOV, offer some evidence that various elements of Policy-Driven EGOV already appear among national EGOV strategies, outline some challenges in progressing from EGOV to Policy-Driven EGOV, and propose a research and policy agenda to build a foundation for this new phase.

The rest of this chapter is structured as follows. Section 2 outlines the evolution of the use of technology by governments and identifies four distinctive phases in this evolution. Section 3 proposes a conceptual model to explain the latest phase in this evolution – Policy-Driven EGOV. Section 4 provides some evidence that Policy-Driven EGOV is starting to take roots among national EGOV strategies by most advanced countries in this area. Section 5 outlines a research and policy agenda for Policy-Driven EGOV. The final Section 6 provides some conclusions.

1 Evolving use of technology by government

Over the past two decades, the goals of government investment in technology has changed significantly, from increasing the quality and efficiency of internal government operations, through delivering better public services and making them available across both traditional and electronic channels, to facilitating administrative and institutional reform in government, engaging citizens in policy- and decision-making processes, and (more recently) directly supporting policy and development goals in health, education, economy, security and other sectors. With evolving goals, the type of initiatives undertaken by governments to fulfill them evolved as well. The aim of this section is to identify four phases in this evolution: 1. technological – Technology in Government, 2. organizational – Electronic Government, 3. socio-economic – Electronic Governance, and 4. context-specific – Policy-Driven Electronic Governance. These phases are depicted in Table 1 and described as follows.

Tab. 1: Evolving Use of Technology by Government

Phases		Context	Goals
Phase 1	Technology in Government	Technological	Increasing the quality and efficiency of internal government operations
			Delivering public services across traditional and electronic channels
Phase 2	Electronic Government	Organizational	Facilitating administrative and institutional reform in government
Phase 3	Electronic Governance	Socio-economic	Engaging citizens and other non-state actors in decision-making processes
Phase 4	Policy-Driven Electronic Governance	Context-specific	Supporting policy and development goals in specific locations and sectors

1.1 Phase 1 – technology in government

Phase 1 “Technology in Government” focuses essentially on deploying technological environment in government (Ojo et al. 2011) in order to digitize government information, equip government officials with the appropriate technological tools, automate government processes and work practices and others. The goals pursued in this phase include: establishing government portals to publish government information and services online (Rorissa and Demissie 2010), empowering government officials with various technological tools (Scholl et al. 2007) and introducing information systems to manage existing business processes (Weerakkody et al. 2006). The challenges to fulfill such goals include: connecting agencies, citizens and businesses to the Internet (ITU 2012); ensuring that the systems run by different government organizations are interoperable (Scholl and Klischewski 2007); and connecting legacy systems to other systems and to the Internet (Lam 2005). However, as Phase 1 focuses on technological development alone, it is unable to help improve existing business processes and work practices, to ensure interagency cooperation or real-time responsiveness by government organizations to make available seamless and transactional services, or to demonstrate that government investment in technology can create sustainable benefits to the public.

1.2 Phase 2 – electronic government

Learning from the limitations of the Phase 1, Phase 2 "Electronic Government" is not focused on technology alone but on utilizing technology to improve organizational environment within government, on transforming the internal working of government through technology. The goals pursued at this phase include: aligning technology and business processes within a government organization (Kim et al. 2007), training government officials in new technological skills (Edelmann et al. 2013), enabling cross-agency collaboration (Ramon Gil-Garcia et al. 2007), promoting whole-of-government (Christensen and Lægreid 2007), developing seamless services (Estevez 2009) and building one-stop portals and offering seamless services according to the needs of citizens and not the internal structure of government (Wimmer 2002). The challenges to fulfill such goals include: hierarchical organization and command-and-control operation, inward looking culture and lack of collaboration in government, lack of compatible technological infrastructure and common standards between agencies, and lack of customer orientation and resistance to change (Ke and Wei 2004). However, as Phase 2 initiatives focus on organizational improvement in government alone, they are unable to increase the usage of electronic channels by the public (Carter and Bélanger 2005) to match increased supply of government information and services electronically, to carry out public consultation on design and content of different delivery channels (Anthopoulos et al. 2007), to build capacity among different groups of citizens in using electronic channels (van Deursen and van Dijk 2009), or to demonstrate that improvements in internal government organization can create sustainable benefits to the public.

1.3 Phase 3 – electronic governance

Learning from the limitations of Phase 2, Phase 3 "Electronic Governance" is not focused on organizational improvement in government alone but on improving relationships between government on the one side, and citizens, businesses and other non-state actors on the another (Reddick 2005). The goals pursued at this phase include: making government portals accessible to all groups in the society (West 2005), utilizing social media to engage citizens in government decision-making (Bertot et al. 2010), making government data available online for businesses and non-profits to build useful services for citizens (Hendler et al. 2012), promoting the use of electronic procurement for businesses to be able to sell goods and services to government (Padhi and Mohapatra 2010), integrating public, private and non-profit services into one service space (Chan et al. 2008), and co-creating of public services (Linders 2012). In general, government is expected to create and maintain a platform for all relevant actors to create public value through collaboration and innovation (Janssen and Estevez 2013), and this role requires a range of legal, institutional, cultural and other transformations. The challenges to fulfill such goals include: large differences among different groups in the society between access to technology (e.g. digital skills) and ability to benefit from access to technology (e.g. information society education); lack of trust in government and particularly its willingness and ability to address societal needs; and increased engagement of non-state actors in delivering public services, executing public programs and spending public funds, raising the risk of corruption and lack of accountability (Langford and Roy 2009). Unlike Phase 1 and 2, Phase 3 can directly create public value, for instance by ensuring the actual usage of electronic public services by addressing both supply (Electronic Government) and demand (Digital Divide) sides. However, while improving relationships between government and its constituencies as part of Phase 3 is important, it does not automatically lead to improved conditions of these constituencies.

1.4 Phase 4 – policy-driven electronic governance

Learning from the limitation of Phase 3, Phase 4 "Policy-Driven Electronic Governance" is not just focused on improving the relationships between government and its constituencies but on improving the conditions of these constituencies through better organization within government and improved relationships with government due to transformative use of technology. However, in order to contribute to improving the conditions of citizens, businesses and other non-state actors, Policy-Driven EGOV cannot restrict itself to working on the national level or focus on addressing cross-sectoral issues alone. It must focus on a specific application environment (both location and sector) and tailor its response to the needs and circumstances of this environment in terms of the choice of locally-relevant and/or sector-specific goals, locally-acceptable and sectorally-feasible ways of pursuing such goals, and managing the impact of meeting such goals on the local environment and the sector involved. The locations can vary across all levels of government – national, provincial and local. The sectors can cover health, education, economy, environment, governance, security and other sectors. Policy-Driven EGOV is expected to directly support policy and development goals in different locations and sectors, while taking into account location- and sector-specific goals and circumstances. The concept is elaborated in subsequent sections.

2 Policy-driven electronic governance – model

This section proposes a conceptual model for Policy-Driven Electronic Governance. Given the focus on improving the conditions of citizens, businesses and other non-state actors through technology-enabled improvements within government and in relationships between government and its constituencies, Policy-Driven EGOV naturally connects to the issues of governance and development. The model relies upon well-established link between both, with governance playing a key role in any development effort and good governance considered a pre-condition to achieving any form of development.

The concept of Policy-Driven Electronic Governance can be applied to a range of public policies. Here we apply this concept to Sustainable Development (SD) "the development that meets the needs of the present generation without compromising the ability of the future generations to meet their own needs" (World Commission on Environment and Development 1987), and propose a model for SD-Driven Electronic Governance (EGOV4SD) to capture the notion that EGOV supports simultaneous pursuit of social, economic, environmental dimensions of sustainability (Estevez and Janowski 2013). Each dimension includes a variety of concerns: social dimension includes among others poverty and inequality, hunger and malnutrition, gender imbalance, illiteracy, maternal and infant mortality, and access to communication; environmental dimension includes among others climate change, biodiversity loss, deforestation, water scarcity, land degradation and fish stock depletion; and economic dimension includes among others transportation, energy consumption, logistics and growth. In addition, transitional dimension captures ongoing pursuit of sustainable practices and lifestyles and includes among others environmentally-friendly technology and practice, integration of environmental dimensions into policies and plans, energy from renewable sources, green accounting and access of under-privileged groups.

In order to balance the pursuit of such goals, SD faces specific governance issues including: 1. engaging the whole society and economy – citizens, businesses and other non-state actors in sustainability transition, 2. integrating public policies across different levels and sectors to ensure coherency of government decisions, and 3. adopting a long-term perspective in public policy development to address inter-generational concerns. Technology in turn can play an important role in addressing such SD-specific governance issues, for instance by facilitating feedback from citizens (including the youth) to public policy drafts through electronic

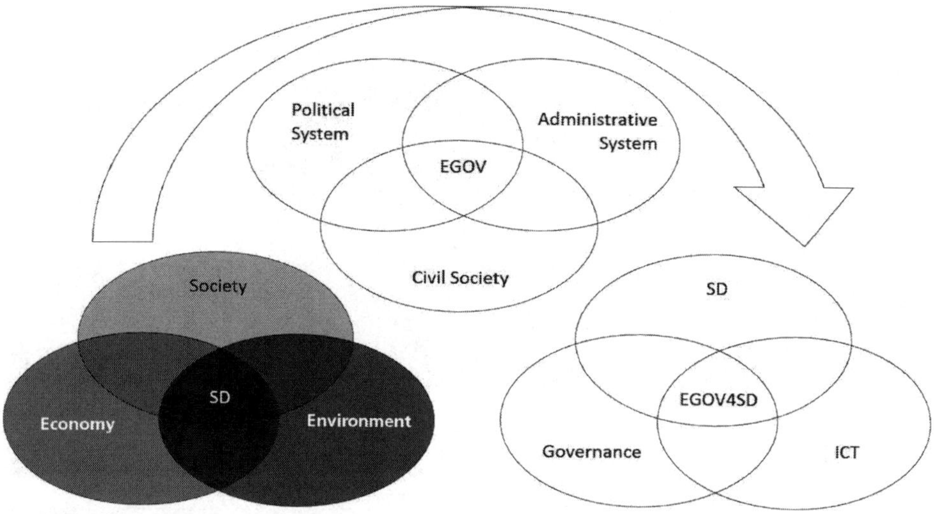

Fig. 1: Conceptual model for EGOV4SD.

participation and deliberation channels, crowdsourcing and promoting ideas how to utilize technology and innovation for sustainability, consolidating whole-of-government responses to public policy challenges and making openly available policy- and development-related data for better research, models and decisions.

Figure 1 above depicts possible conceptual model for EGOV4SD. In this model, SD is depicted at the intersection of economic, social and environmental sustainability representing simultaneous pursuit of the goals related to these dimension; EGOV is depicted at the intersection of administrative system, political system and civil society, representing how technology can be used to transform the working of administrative and political systems and their relationships with the society; EGOV4SD is depicted at the intersection of technology, governance and SD, representing technology-enabled governance of SD.

3 Policy-driven electronic governance – evidence

This section presents some evidence that EGOV development, as observed at the national, regional and global levels, is increasingly contributing to various dimensions of sustainability. In the process, we validate the proposed model for EGOV4SD. The evidence was drawn from three national EGOV strategies from Singapore (Government of Singapore 2011), Korea (Ministry of Public Administration and Security 2011), and Estonia (Government of Estonia 2006), regional strategy pursued by the European Union (European Commission 2010), and two global EGOV ranking instruments published by the United Nations (UNDESA 2012) and Waseda University (Waseda University 2012), and is depicted in Table 2.

Tab. 2: SD dimensions among EGOV strategies.

SINGAPORE 2015	KOREA 2015
Next generation infocomm infrastructure	Joint growth among e-government stakeholders
Innovation centers and entrepreneurship	Increased transparency in government operations
Infocomm competency framework	Low carbon through green IT systems
Electronic health records	User oriented service integration

EUROPEAN UNION 2015	ESTONIA 2013
Improve services to cater for different needs	One service space - public, private and third sectors
Invite third parties in EGOV development	Paperless document management
Involve stakeholders in public policy processes	Traceability of the use of one's own data
Reduce carbon footprint	Internet in rural areas

UNITED NATIONS 2010	WASEDA 2012
Whole-of-governmant approach	Cloud computing and resource utilization
Multi-channel service delivery	Disaster management and business continuity
Towards economic sustainability	Digital inclusion and ICT for ageing society
Promoting awareness of environmental issues	Mobile government

Society	
Economy	
Environment	

According to this sample, economic sustainability is pursued through e.g. innovation centers and entrepreneurship, joint growth among e-government stakeholders, third party involvement in EGOV development, and integrating public, private and third sector services into one service space; social sustainability is pursued through e.g. establishment of electronic health records, improvements to services that cater to different needs, involving stakeholders in public policy processes, traceability of the use of one's own data, Internet in rural areas, multi-channel service delivery and digital inclusion and ICT for ageing society; environmental sustainability is pursued through e.g. reducing carbon footprint through green IT systems, promoting awareness of environmental issues and cloud computing and resource utilization. In addition, increased transparency in government operations and disaster management and business continuity contribute to transitional sustainability through building trust in government and maintaining a reliable technological environment.

4 Policy-driven electronic governance – agenda

This section explains the challenges in performing the transition from EGOV to Policy-Driven EGOV including: 1. shortage of research and understanding about developing EGOV in specific locations and sectors, 2. absence of EGOV policy and development instruments adapted to requirements and conditions in particular locations and sectors, 3. shortage of human capacity within such locations and sectors to be able to build and utilize such instruments, 4. lack of models and experience of engaging universities and other non-private actors in EGOV initiatives, and 5. fundamentally different nature of the transition from EGOV to Policy-Driven EGOV – narrowing the application context, compared to earlier transitions – widening the application context, making incremental development difficult. We also propose an action agenda to overcome such challenges.

4.1 Location- and sector-specific EGOV research

While there is a growing experience and body of research on how to plan, develop and sustain EGOV initiatives in general, systematic studies and cases of how EGOV initiatives are being deployed in different locations and sectors are scarce. Even scarcer is research into theories, methods and tools that facilitate location- and sector-focused EGOV development. Some of the probing questions are: In what aspects is the choice of a particular location or sector affecting EGOV development? Which development stages – planning, design, implementation, operation, sustainability – are affected and how? How to adapt location- and sector-independent EGOV development instruments to particular locations and sectors to optimize their performance? How to transfer adaptation experience between locations and sectors? A focused research effort is required to develop a better understanding of location- and sector-aware EGOV and to explore and answer these and other relevant questions.

4.2 Location- and sector-specific EGOV instruments

As each location and sector comes with its own set of conditions, goals and acceptable ways of pursuing such goals given the conditions, Policy-Driven EGOV must rely on location- and sector-specific policies and development instruments. For example, applying generic one-size-fits-all EGOV maturity stages like e.g. information, interaction, transactions and data-sharing to track progress in EGOV development (the higher maturity, the better) may be appropriate for some countries but not for others. Likewise, measuring the performance of Policy-Driven EGOV should rely on the indicators that reflect locally-defined policy goals, not on the one-size-fits-all generic benchmark instruments. For example, context-aware benchmarking would allow locations or sectors to learn from their peers – locations and sectors in similar development conditions, or leaders – locations and sectors most successful in pursuing the relevant public policy goals. In these and other cases, a focused research, development and policy efforts are required to build, apply and institutionalize the use of such instruments.

4.3 Location- and sector-specific EGOV capacity

While the knowledge of the local or sectoral conditions, ownership of the local or sectoral development goals and awareness of locally- or sectorally-acceptable ways of pursing such goals are critical for successful planning and implementation of Policy-Driven EGOV, the capacity to engage in such planning and implementation is increasingly scarce for lower levels of government and possibly within different sectors. Instead of focusing the resources and capacity at the central level for visibility and prestige, a focused effort is required to build human and institutional capacity at the local levels, choosing the right level to balance effectiveness and efficiency of the response, and promoting collaboration between levels. A similar effort is required to refocus Policy-Driven EGOV initiatives from cross-sectoral issues referring to public administration, public services, public policy, etc. to sectoral issues to address the needs of health, education, security, economy, environment and other sectors. In both cases, location- and sector-specific EGOV education programs are also required to enable a new generation of government leaders, managers and experts to emerge.

4.4 Location- and sector-specific EGOV networks

The outcome of Policy-Driven EGOV depends on government being able to engage citizens, businesses, academia, non-profits and other non-state actors in various network forms aimed at formulating and pursuing location- and sector-specific development goals through EGOV initiatives. However, such networks are still scarce particularly among local governments, so is the experience with establishing and operating them. Within multi-stakeholder Policy-Driven EGOV networks, academia could contribute to planning and design of EGOV initiatives; the private sector could contribute to development and implementation; while the non-profit sector would ensure the delivery of benefits from EGOV initiatives to the target group of stakeholders, thus contributing to their sustainability. As part of such networks, local universities and the universities located within specific sectors have a key role to play in formulating location- and sector-specific policies, in constructing development instruments to support such policies, and in building local capacity to apply such instruments. In addition, such networks could also facilitate the transfer of local-level and sector-specific EGOV innovations within and between countries, thus contributing to accelerating local development.

4.5 From EGOV to policy-driven EGOV

Concerning the evolving use of technology by government, the nature of the transition from phase 3 (EGOV) to 4 (Policy-Driven EGOV) is to narrow (localize or specialize) the application context, whereas all transitions from phase 1 (Technology in Government) to phase 2 (Electronic Government) and from phase 2 (Electronic Government) to phase 3 (Electronic Governance) is to gradually expand the application context. Given this fundamental difference, it is difficult to carry out the transition from EGOV to Policy-Driven EGOV by building incrementally upon earlier phases. The transition requires investment into: 1. research and innovation including location- and sector-specific EGOV research – Section 4.1, 2. policy support including development of location- and sector-specific EGOV policies and instruments – Section 4.2, 3. capacity building including building location- and sector-specific EGOV capacity at both individual and institutional levels – Section 4.3, and 4. network development including the development of multi-stakeholder location- and sector-specific EGOV networks – Section 4.4. Lacking location- and sector-specific EGOV implementation evidence, the transition also requires the ability to run controlled experiments in applying EGOV to various location- and sector-specific policy and development goals, and to develop and validate theories while learning from such experiments. Table 3 presents an integrated agenda for transition towards Policy-Driven EGOV.

Tab. 3: From EGOV to policy-driven EGOV – integrated agenda.

		Research and Innovation	**Policy Support**	**Capacity Building**	**Network Development**
1	Location- and sector-specific EGOV research	x			
2	Location- and sector-specific EGOV instruments	x	x		
3	Location- and sector-specific EGOV capacity			x	
4	Location- and sector-specific EGOV networks				x
5	Location- and sector-specific EGOV agenda	x	x	x	x

5 Conclusions

This chapter focused on the use of Information and Communication Technologies in government and how this usage has been evolving over the past two decades through the progression of various development goals: 1. improvements in internal government operations, 2. delivery of better public services, 3. institutional and administrative reform in government, 4. engagement of citizens and other non-stage actors in government decision-making processes, and 5. direct contributions to policy and development goals in health, education, economy, environment and other sectors. The chapter identified four phases in this evolution, characterized

Tab. 4: Characterizing phases in technology use by government.

Phases		Characterization		
		Transformation of government?	Includes non-state actors?	Location- and sector-specific?
1	Technology in Government	no	no	no
2	Electronic Government	yes	no	no
3	Electronic Governance	yes	yes	no
4	Policy-Driven Electronic Governance	yes	yes	yes

by the presence of transformation in government to accompany technological development, the nature of this transformation being internal to government or also involving relationships between the government and its various stakeholders, and whether this transformation is location- and sector-independent or it responds to the needs and conditions of different locations or sectors. The resulting characterization is depicted in Table 4.

The chapter then focused on the latest "Policy-Driven Electronic Governance" phase, providing a conceptual model for this phase based on a Sustainable Development framework, some evidence of the presence of Policy-Driven Electronic Governance among existing EGOV strategies at the national, regional and global levels, and an action agenda to facilitate the transition from EGOV to Policy-Driven EGOV.

Acknowledgements

The author would like to thank Elsa Estevez for collaboration, discussions and very useful feedback received on earlier versions of this chapter.

References

Accenture. 2013. *Accenture Technology Vision 2013 – Every Business is Digital Business.* http://www.accenture.com/SiteCollectionDocuments/PDF/Accenture-Technology-Vision-2013-Executive-Summary.pdf (Accessed 12 April 2014).

Anthopoulos, Leo G., Panagiotis Siozos & Ioannis A. Tsoukalas. 2007. Applying participatory design and collaboration in digital public services for discovering and re-designing e-Government services. *Government Information Quarterly* 24(2). 353–376. http://dde.teilar.gr/publications/141/141fulltext.pdf (Accessed 12 April 2014).

Bertot, John C., Paul T. Jaeger & Justin M. Grimes. 2010. Using ICTs to create a culture of transparency: E-government and social media as openness and anti-corruption tools for societies. *Government Information Quarterly* 27(3). 264–271. http://www.sciencedirect.com/science/article/pii/S0740624X10000201 (Accessed 12 April 2014).

Carter, Lemuria & France Bélanger. 2005. The utilization of e-government services: citizen trust, innovation and acceptance factors. *Information Systems Journal* 15(1). 5–25.

Chan, Calvin M. L., Yimeng Lau & Shan L. Pan. 2008. E-government implementation: A macro analysis of Singapore's e-government initiatives. *Government Information Quarterly* 25(2). 239–255. http://www.sciencedirect.com/science/article/pii/S0740624X06001651 (Accessed 12 April 2014).

Christensen, Tom & Per Lægreid. 2007. The Whole-of-Government Approach to Public Sector Reform. *Public Administration Review* 67(6). 1059–1066.

Edelmann, Noella, Peter Parycek & Judith Schossbock. 2013. Exit the E-government Ivory Tower: A Training Strategy – A Case Study from Austria. In *2013 46th Hawai'i International Conference on System Sciences*, 1674–1683. IEEE Computer Society Washington, DC, USA.

Estevez, Elsa 2009. *Programmable Messaging for Electronic Government.* National University of the South PhD Thesis.

Estevez, Elsa & Tomasz Janowski. 2013. Electronic Governance for Sustainable Development – Conceptual framework and state of research. *Government Information Quarterly* 30. 94–109.

European Commission. 2010. *The European eGovernment Action Plan 2011–2015.* Brussels, 15 December 2010.

Government of Estonia. 2006. Estonian Information Society Strategy 2013. http://www.epractice.eu/files/media/media_186.pdf (Accessed 12 April 2014).

Government of Singapore. 2011. *e-Government Master Plan 2011–2015, Connecting People, Enriching Lives.* http://www.egov.gov.sg/egov-masterplans/egov-2015/vision-strategic-thrusts (Accessed 12 April 2014).

Grönlund, Åke & Thomas A. Horan. 2004. Introducing e-Gov: History, Definitions, and Issues. *Communications of the Association for Information Systems* 15(1). 713–729.

Hendler, James, Jeanne Holm, C. Musialek & G. Thomas. 2012. US Government Linked Open Data: Semantic.data.gov. *IEEE Intelligent Systems* 27(3). 25–31.

ITU – International Telecommunication Union. 2012. *Measuring the Information Society.* http://www.itu.int/ITU-D/ict/publications/idi/ (Accessed 12 April 2014).

Janssen, Marijin & Elsa Estevez. 2013. Lean government and platform-based governance – Doing more with less. *Government Information Quarterly* 30(1). S1–S8.

Ke, Weiling & Kwok Kee Wei. 2004. Successful e-government in Singapore. *Communications of the ACM* 47(6). 95–99.

Kim, Hyun Jeong, Gary Pan & Shan Ling Pan. 2007. Managing IT-enabled transformation in the public sector: A case study on e-government in South Korea. *Government Information Quarterly* 24(2). 338–352.

Lam, Wing. 2005. Barriers to e-government integration. *Journal of Enterprise Information Management* 18(5). 511–530.

Langford, John & Jeffrey Roy. 2009. Building shared accountability into service transformation partnerships. *International Journal of Public Policy* 4(3/4). 232.

Linders, Dennis. 2012. From e-government to we-government: Defining a typology for citizen coproduction in the age of social media. *Government Information Quarterly* 29(4). 446–454.

Ministry of Public Administration and Security. 2011. *Smart Government Implementation Plan 2011–2015.* In "Digital Society Development in Korea". In "Digital Society Development of Korea", http://egovexport.or.kr/site/egovexport/upload/book/digital_society_development_of_korea.pdf

OECD – Organisation for Economic Co-operation and Development. 2003. *The e-Government Imperative.* OECD Publishing.

OECD – Organisation for Economic Co-operation and Development. 2011. *Government at a Glance 2011*. OECD Publishing.
Ojo, Adegboyega, Tomasz Janowski & Esla Estevez. 2011. Whole-of-government approach to information technology strategy management: Building a sustainable collaborative technology environment in government. *Information Polity* 16(3). 243–260.
Padhi, Sidhartha S. & Pratap K. J. Mohapatra. 2010. Adoption of e-procurement in the government departments. *Electronic Government, an International Journal* 7(1). 41.
Ramon Gil-Garcia J., InduShobha Chengalur-Smith & Peter Duchessi. 2007. Collaborative e-Government: impediments and benefits of information-sharing projects in the public sector. *European Journal of Information Systems* 16(2). 121–133.
Reddick, Christopher G. 2005. Citizen interaction with e-government: From the streets to servers? *Government Information Quarterly* 22(1). 38–57.
Rorissa, Abebe & Dawit Demissie. 2010. An analysis of African e-Government service websites. *Government Information Quarterly* 27(2). 161–169.
Scholl, Hans Jochen, Raya Fidel, Shuhua Monica Liua, Michael Paulsmeyer & Kris Unsworth. 2007. E-government field force automation: Promises, challenges, and stakeholders. In *Lecture Notes in Computer Science (including subseries Lecture Notes in Artificial Intelligence and Lecture Notes in Bioinformatics)*, Vol. 4656, 127–142. http://www.scopus.com/inward/record.url?eid=2-s2.0-38049054789 & partnerID=tZOtx3y1 (Accessed 12 April).
Scholl, Hans Jochen & Ralf Klischewski. 2007. E-Government Integration and Interoperability: Framing the Research Agenda. *International Journal of Public Administration* 30(8–9). 889–920.
UNDESA – UN Depertment of Economic and Social Affairs. 2012. *UN E-Government Survey 2012 – E-Government for the People*. United Nations, New York, 2012.
UNESCO – UN Educational Scientific and Cultural Organization. 2009. *National Information Society Policy: A Template*. UNESCO, Paris, November 2009.
Van Deursen, A. J. A. M. & J. A. G. M van Dijk. 2009. Improving digital skills for the use of online public information and services. *Government Information Quarterly* 26(2). 333–340.
Waseda University. 2012. *Waseda University International e-Government Ranking 2012*. http://www.waseda.jp/eng/news11/120224_egov.html (accessed 12 April 2014).
Weerakkody, Vishanth, Simon Baire, Jyoti Choudrie. 2006. E-Government: The Need for Effective Process Management in the Public Sector. In *Proceedings of the 39th Annual Hawai'i International Conference on System Sciences (HICSS'06)*. Vol. 4, 74b–74b. IEEE.
West, Darrell M. 2005. Equity and Accessibility in E-Government. *Journal of E-Government* 1(2). 31–43.
Wimmer, Maria A. 2002. A European perspective towards online one-stop government: the eGOV project. *Electronic Commerce Research and Applications* 1(1). 92–103.
World Commission on Environment and Development. 1987. *Our Common Future*. Oxford: Oxford University Press.

Aziz Douai

23 Technology and terrorism: Media symbiosis and the "dark side" of the web

Abstract: This chapter discusses the complex relationship between technology and the deployment of terrorism as a political weapon in contemporary societies. To propagate mass panic and thus influence policies, terrorists exploit the mass media's ability and willingness to publicize indiscriminate violence, creating "media spectacles of terror" (Kellner, 2003). In addition to censuring media's coverage of terrorist attacks, some governments have in turn exploited the threat of terrorism for their own ideological agendas by abetting "moral panics." Moreover, new technologies, as the new discourse surrounding cyber-terrorism indicates, have followed the same path. Technologies' arcane and double-edged usage nurtures human fears and anxieties about terrorism threats. The Internet's own architecture has allowed terrorists to use it to evade detection, communicate, recruit and organize. Yet, the weak evidence for cyber-terrorist attacks has not prevented an alarmist official discourse to hype up the threat, conflating "hacktivism" with "cyber-terrorism." Such discourse has the potential to create new "moral panics" about the Internet as well as justify intrusive surveillance of regular citizens' online activities as the 2013 Edward Snowden's leaks of National Security Agency eavesdropping details reveal.

Keywords: cyberterrorism, terrorism, moral panics, propaganda, media events, social shaping of technology

The relationship between technology and terrorism is both contentious and complex, as recent terrorism incidents demonstrate. The attacks of September 11, 2001, on the US have entered the history books as the most spectacular terrorist attack of the 21st century not only because of the huge loss of lives, or the perpetrators' ruthlessness and sophisticated level of coordination, but also because they exhibited how civilian technologies can be wielded as an incredible tool of destruction outside traditional battle fronts. In this case, the perpetrators turned passenger planes into huge bombs that caused the collapse of New York's World Trade Center towers (Kellner 2003). The immeasurable harm that the attack inflicted created a climate of fear among populations, which is exactly the main purpose of terrorism (Laqueur 1987; Tuman 2010). Moreover, "fear of terrorism," like fear of crime, affects both populations and political leaders in many ways (Altheide 2004). Because fear of terrorism can lead to public and "moral panics," regular citizens' feeling of being "under siege," or feeling vulnerable to unpredictable terrorist attacks, can affect the daily behaviours and the social life of entire populations. Politicians have used this "fear of terrorism" to enact policies and justify actions that restrict civil liberties, or armed interventions abroad (Altheide 2006; Cole 2003). As research suggests (Landau et al. 2004), angry and fearful people are likely to support their governments in attacking and retaliating against harming terrorist groups.

The other effect of those terrorist attacks lies in making prominent the issue of how technology has turned into a new battlefront between governments and terrorist networks. Terrorist networks are acutely cognizant that communication technologies provide novel and effective ways to communicate with their cells, evade detection, and organize new crippling attacks (Denning 2007). By the same token, governments and security agencies are aware that possessing advanced technologies offer them an edge over terrorists in terms of detection, surveillance, and countering terrorist activities. Yet, these governments and agencies painfully recognize that technological infrastructures, such as critical telecommunication networks, are vulnerable to exploitation and sudden attacks by these terrorist networks. For these reasons, recent versions of the US National Security Strategy have devoted great attention to how terrorism threatens the United States' and the world's critical infrastructures. In its 2013 National Security Strategy for instance, the Obama administration plainly re-emphasizes its commitment to cyber security: "We must recognize and address our cyber security vulnerabilities. Our reliance on cyber networks for maintaining critical infrastructure means that we must prepare to provide robust cyber security" (Obama 2013: 5).

This chapter teases out these interrelated issues about terrorism and communication technologies to address a central question: What is the nature of the relationship between technology and terrorism? Communication scholarship has long investigated this relationship, concluding that a "mutual shaping" and "symbiosis" occurs between technology and terrorism (Wilkinson 1987). This is based on the notion that the "social shaping" of technologies encompasses terrorism, whether at the level of how modern mass media have traditionally covered terrorist violence, how technologies are used by terrorist organizations to further their political goals, or how governments have used the same technologies in their counter-terrorism efforts and to promote their ideological agendas. In this respect, the chapter begins with a brief explication of terrorism to revisit and clarify the terms of the debate. The second major section of the chapter discusses the influential and contentious role the modern mass media (as traditional mass communication technologies) have played in the coverage of terrorism. This section presents the concepts of terrorism as a "global media event" (Weimann 1987) and "media spectacles of terror" (Kellner 2003), and discusses the politics of fear terrorism generates (Altheide 2004). The final section analyzes how "new" communication technologies influence public perceptions of the threat of terrorism, specifically discussing cyber-terrorism, cyber-jihad and the issue of "moral panics."

1 An old battleground: Explicating terrorism

Technology's relationship with warfare and terrorism is as old as human civilization itself. Fires and beacons were one of the first known communication/technology systems to be used as a means of military communication. They provided an early warning system against enemy attacks, helped in troop deployment, and the conquest of new territories. The Greeks perfected such a system as highlighted in *Agamemnnon*'s Aschylus question: "Yet who so swift could speed the message here?" referring to news of the fall of Troy and how it was conveyed "beacon to beacon" across large distances (Palmer 2007). Moreover, whether it was the compass, the printing press, the telegraph, the phone, radio, TV or the Internet, all other communication technologies have had a dual use in war and peace by state and non-state actors alike. State-actors, such as countries and their militaries, have used these technologies to increase the lethality of their forces (Sloan 1988). Terrorists and other non-state actors have used the same communication technologies to resist and attack their opponents. There is also the issue of "state terrorism" in which some states are involved in terrorism through direct or indirect mobilization of military force to terrorize populations (Sloan 1988). While both non-state actors and state actors tend to make similar use of communication technologies to advance their political agendas, this chapter focuses only on "terrorism" perpetrated by non-state actors.

Prior to discussing their intricate relationship, a series of legitimate questions have to be addressed: What is terrorism? Are there different "brands" of terrorism? Who are these terrorists? What are their goals? While academics have studied terrorism for many decades, these interrelated questions continue to be problematic because of their inherent moral and conceptual ambiguities. On the moral side, the term "terrorism" is very "loaded" and lacks objectivity because as soon as it is uttered, a "moral" judgment is passed. For many reasons, terrorism connotes criminal and reprehensible activity and branding a person as a "terrorist" is not only pejorative but it can actually land that person in jail. The moral ambiguity leads into a conceptual quandary that stems from the difficulty of classifying attacks as acts of terrorism because of the subjective judgments endowed on the concept. The apt summation of this moral and conceptual quandary has long been embodied in the equivocal contention that "'one man's terrorist is another man's freedom fighter'" (Laqueur 1987: 7).

Despite such difficulty, frequent attempts to define the "term" have created even more confusion spawning more than a hundred definition of terrorism. Linguists concur, however, that the term derives from "the Latin word *terrere,* meaning 'to tremble,' and [w]hen combined with the French suffix *isme,* referencing 'to practice,' it becomes more like 'to practice the trembling,' or 'to cause or create the trembling'" (Tuman 2010: 4). Most definitions have consistently drawn from Laqueur's classic definition of terrorism in his book, *The Age of Terrorism,* as:

> The use or the threat of the use of violence, a method of combat, or a strategy to achieve certain targets. ... [I]t aims to induce a state of fear in the victim, that is ruthless and does not conform with humanitarian rules. ... [P]ublicity is an essential factor in the terrorist strategy (Laqueur 1987: 143).

Scholars and terrorism analysts mostly concur that terrorist organizations, insurgents, and other armed non-state actors resort to spectacular attacks to instill fear in populations to influence governments' policies and decisions (Kellner 2003). Terrorism is the "poor man's weapon" (Teichman 1989), the argument goes, because this type of warfare is asymmetrical, that is, a lopsided balance exists in the sheer power of adversaries.

Lacquer's definition also points out to an important aspect of terrorism: publicity. The more widely disseminated and publicized a terrorist act is, the greater the fear it will generate in the population, and the more likely it will be that governments bend to their pressure. In other terms, terrorism's primary goal is influencing public opinion and communication technologies may have played a "facilitating" role. In covering terrorist acts, the modern mass media have provided effective vehicles that massively disseminate terrorists' "exploits" and demands to wide audiences. The media grant so much visibility that without mass media coverage terrorist acts will possibly not be as influential on public opinion. Former British Prime Minister Margaret Thatcher called the media the "oxygen" of terrorists and suggested that reporters should ignore terrorist acts to discourage terrorism (Camphuijsen and Vissers 2012). Therefore, it is not shocking to hear government officials asking the media to refrain from covering, or at least restrain their coverage of, terrorism. So, what is the nature of this relationship?

2 Open symbiosis: Mass media and terrorism

Modern mass media's complex relationship with terrorism has been described as "symbiotic" to emphasize the intricate relationship between the two. As Wilkinson explains:

> In sociology the term symbiosis is taken to mean relations of mutual dependence between different groups within a community when the groups are unlike each other and their relations are complementary. It would be foolish to deny that modern media technology, communications satellites' and the rapid spread of television have had a marked effect in increasing the publicity potential of terrorism. A dramatic illustration was the seizure and massacre of Israeli athletes by Black September terrorists at the 1972 Munich Olympics. It is estimated that these events were relayed to worldwide TV audience of over 500 million (Wilkinson 1997: 52).

While highlighting the centrality of contemporary media in society, this "symbiotic" relationship emphasizes that the psychological impact of "terror" cannot be fully realized unless its threat is widely publicized via mass communication technologies. Rather than being *culpable*, the mass media remain *vulnerable* to the manipulation of terrorist organizations. Open and democratic societies expect the news media to shed light on social and political issues and threats facing these societies, and covering terrorist threats and attacks are no different. Cognizant of this "symbiotic" relationship, terrorist organizations have resorted to spectacular attacks that create what Weimann (1987) labels as global "media events" or Kellner (2003) describes as "media spectacles of terror." The "symbiosis" persists because, as Kellner explains, terrorism "works in part through spectacle, using dramatic images and montage to catch attention, hoping thereby to catalyze unanticipated events that will spread further terror through domestic populations" (Kellner 2003: 87). Within our media-saturated contemporary societies, news organizations cannot resist covering such spectacular attacks as those perpetrated on USS Cole in Yemen, US Embassies in Africa, or the attacks of September 11, 2001.

In this respect, modern mass media provide a powerful example of how communication technologies have become a battlefront between these adversaries as they seek to take advantage of technology. Terrorists usually court the limelight of the media to influence the public and government policies. For terrorists, spectacular attacks are more likely to generate wall-to-wall coverage from television, newspapers and other mass media. Terrorist organizations have been savvy enough to "use" this media coverage to broadcast their messages and grievances to international publics during the hostage taking by Palestinians at the 1972 Munich Olympics, the hijacking of TWA flight 847 in 1985, the 9/11 attacks on the US, the bombings of Bali and Mumbai, and the Subway attacks in London and Madrid.

That modern mass media constitute an effective propaganda tool, and thus a central part of the communication strategies of terrorist organizations, is recognized and openly acknowledged by these organizations themselves. According to Al-Qaeda leader Ayman al-Zawahiri, "We are in a battle, and more than half of this battle is taking place in the battlefield of the media. We are in a media battle for the hearts and minds of our *umma* [community of Muslims]" (Rogan 2007: 89). Lynch (2006) explains Al-Qaeda's media strategy's savvy exploitation of the new media environment:

> The centrality of the Arab mass media to Al-Qaeda's political strategy has long been evident. From spectacular terror attacks designed for maximal media exposure, to carefully timed videos from Osama bin Laden and his lieutenant, Ayman al-Zawahiri, to the burgeoning realm of jihadi Internet forums, Al-Qaeda the organization has increasingly become indistinguishable from Al-Qaeda the media phenomenon (Lynch 2006: 50).

Satellite television and the Internet have been central to this Al-Qaeda "media phenomenon" because they are transcending the traditional boundaries of the state and they can easily evade government's censorship. The media war is as important as military "hard power" because, in the words of French analyst, Gilles

Kepel, "rhetoric and satellite propaganda can be on equal footing with unmanned bombers and cruise-missiles" (Lynch 2006: 50).

In the age of globalization, media organizations and professionals have grappled with this contentious position because their primary responsibility is to inform the public of current events including disasters, war and conflict. Some analysts have accused the media of opportunism, stating that media organizations' coverage of terrorism is driven by ratings. At other times, the public accuses the media of sensationalizing its coverage of terrorists. After the *Rolling Stone* published a cover photo of accused Boston Marathon bomber Dzhokhar Tsarnaev, a public backlash objected to the magazine's "glamorization" of the bomber. Ed Kelly, president of the Professional Fire Fighters of Massachusetts, called it "insulting." He told CNN that "using Tsarnaev's booking photo might have been one thing, but a photo that shows 'the innocence of youth' gives the wrong message" (Cannold, Cuevas and Sterling 2013: n.p.). Matt Taibbi, *Rolling Stone*'s contributing editor, responded to this criticism by emphasizing the investigative journalistic spirit of the magazine, and argued that "[t]errorists are a fact of our modern lives and we need to understand them, because understanding is the key to stopping them" (Taibbi 2013: n.p.).

In the aftermath of 9/11, a similar backlash met the Qatar-based Al Jazeera Television's airing of Osama Bin Laden's and other Al-Qaeda's videos and statements. Accusing it of being the official "mouthpiece of Al-Qaeda," US officials and some Western journalists raised questions about how Al-Jazeera Television obtained its Bin Laden videos and statements (Miles 2005). The channel's coverage of the US invasion of Iraq and the insurgents' attacks drew the ire of the US administration, and US President George Bush and Defense Secretary Donald Rumsfeld publicly upbraided the channel's bias and "collusion" with terrorists. Not only did its reporting fuel anti-American sentiments among Arabs, but al Jazeera had "a pattern of not making judgments about the accuracy of the propaganda," Rumsfeld claimed (Pew Research Center Staff 2006). Al Jazeera's reporters and editors argued, however, for the inherent "news value" of these terrorists organizations' tapes and that any other Western organization would jump at such an opportunity. In an interview with the Associated Press, Al Jazeera's Editor-in-Chief, Ahmed al-Sheikh reiterated this argument: "This is news ... Any other station would jump at this. Why label us?" (Associated Press 2006).

3 Technology and the politics of terrorism: From propaganda to moral panics

Communication technologies have had a clear impact on how terrorism has been waged and countered, most clearly evident in the "politics" surrounding it. Specifically, terrorism and the ensuing propaganda exploit human fears to spread "moral panics" among the populace (Cohen 1972; Altheide 2006), and thus influence domestic policies and international relations. Moreover, terrorism's "symbiotic" relationship with media technologies implies that media coverage offers free publicity and propaganda for terrorists and their causes, and augments fear among the population, as explained above. Yet, some communication scholars have also observed that governments have engaged in the same "politics of fear" after terrorist attacks (Altheide 2006). Analyzing Bush's official discourse, or "Bushspeak", Kellner (2003) argues that, not unlike the terrorists, the Bush administration had "also deployed terror spectacle to promote their geo-political ends" by neglecting to emphasize democratic values in their war on terror. Kellner (2003) draws attention to discursive similarities between Al-Qaeda and "Bushspeak":

> Interestingly, Bush Administration discourses, like those of bin Laden and radical Islamists, are fundamentally Manichean, positing a binary opposition between Good and Evil, Us and Them, civilization and barbarism. Bush's Manichean dualism replicates as well the Friend/Enemy opposition of Carl Schmidt upon which Nazi politics were based. Osama bin Laden and al Qaeda provided the face of an enemy to replace the "evil Empire" of Soviet Communism, which was the face of the Other in the Cold War. The terrorist Other, however, does not reside in a specific country with particular military targets and forces, but is part of an invisible empire supported by a multiplicity of groups and states. This amorphous terrorist Enemy, then, allows the crusader for Good to attack any country or group that is supporting terrorism, thus promoting a foundation for a new doctrine of preemptive strikes and perennial war (Kellner 2003: 92).

Thus, Bush's "war on terror" itself had been rooted in a "politics of fear" that exploited terrorist spectacles to further a hawkish ideological and geopolitical agenda of "regime change" without recourse to international institutions and due process. The same "politics of fear" enabled the Bush Administration to justify the invasion of Iraq by purposefully conflating the Iraqi regime with terrorism, as many studies have demonstrated (Landau et al. 2004).

Media sociologists have argued that terrorism's "politics of fear" work more effectively because they engender "moral panics" (Cohen 1972). As an analytical concept, a "moral panic" refers to how the media's distorted coverage of crime, such as the frequency of covering violent crime, heightens the public's "fear of crime" and their perception of the likelihood of being victimized. Through creating these "moral panics," the media influence the criminal justice agenda outside the political process because alarmed and "panicked" publics pressure the political and justice systems to implement more authoritarian law and order policies to keep them safe. While "moral panics" describe exaggerated public perceptions of social problems, they have real-world consequences on policy making. Cohen suggests a five-step process through which "moral panics" evolve. The first stage is called "labeling" in which a social phenomenon or a social group is "defined" as "deviant" or a "threat" to the majority's values or interests. In the second stage, called "amplification," the media depict this "threat" in a stereotypical fashion that makes the "threat" easily identifiable by the public. As media coverage of the problem increases, public concern rapidly builds up and its demands for decisive action grow, which is the third stage of a "moral panic." Authorities and political leaders respond to the public's demands for action to contain the threat in the fourth stage. Finally, the fifth stage of a "moral panic" refers to how the social and moral orders are "preserved" when the threat is eliminated and the moral panic recedes.

While exaggerated media coverage whips up fear of terrorism, the public's alarm and the manufactured "moral panic" dictate officials' response be immediate and severe to ensure the public's safety. In a study that looked at the 9/11 attacks on the US, critical criminologists, Rothe and Muzzatti (2004) focus on the stage with the most serious consequence of a "moral panic," the fourth stage, i.e. the "disproportionate" reactions of the Bush Administration and other government officials. In addition to Congress' War Declaration authorizing Bush to use unchecked military power in the "war on terror," crowned by the Iraq War, they explained that:

> Other disproportionate reactions by the polity included the creation of a large bureaucracy to ensure domestic security: the Office of Homeland Security. The Bush Administration also instituted an alert system to keep the public informed as to what level or how much fear "we" should be experiencing: The Homeland Security Advisory System (Rothe and Muzzatti 2004: 341).

President Bush signed several Executive Orders that authorized military tribunals and other extra-judicial measures to apprehend and incarcerate terrorism suspects. With the Homeland Security bureaucracy, the passage of the Uniting and Strengthening America by Providing Appropriate Tools Required to Intercept and Obstruct Terrorism Act (USA Patriot Act) in 2001 was clearly designed to demonstrate to the American public that their political leaders were taking whatever measures necessary to protect the "homeland."

Through media coverage, "moral panics" and the "politics of fear" intersect because they both promote fear and danger as part of the experience of daily life, and clamp on political dissent regarding national security issues. As Altheide (2006: 434) suggests, these "politics of fear" have become "a dominant motif for news and popular culture" and thus pertinent to the experience of everyday life. Within this mindset, politicians have resorted to the systematic "deviling" of terrorists, portraying them as "folk devils," and the construction of Manichean narratives in the service of their ideological agendas. Mass media technologies remain a centerpiece in conveying and mediating these ideological agendas.

4 Navigating the “dark side”: New technologies, new terrorism?

The bulk of the above discussion referred to “older” communication technologies of traditional mass media, but how have new communication technologies, such as the Internet, affected public perception of terrorism? What is the “Dark Side” of the web anyway and how does it inform technology’s relationship with terrorism? In *The Net Delusion: The Dark Side of Internet Freedom*, Evgeny Morozov posits a contrarian position vis-à-vis what he calls the “Google Doctrine” (Morozov 2012: xiii). The “Google Doctrine” refers to a relatively nascent US worldview that heralds the power of new communication technologies to spread democracy and freedom. According to this view, once citizens living under authoritarian regimes use Twitter or other Web 2.0 tools, they would challenge their oppressors and rise up in revolt. Morozov argues that these new communication tools could have opposite effects whereby authoritarian regimes deploy these tools to spy on their democratic opponents, jail dissenters and thereby suffocate any chance for democracy. As Kalathil and Boas (2003) also argue, closed regimes’ ability to thwart the “democratic potential” of the web is well documented. From China and Vietnam to Singapore and the United Arab Emirates, authoritarians have controlled citizens’ access to the Internet or went on to establish their national “Intranet” to curtail dissent and free speech as in the case of Cuba (Kalathil and Boas 2003).

The “Dark Side” also encompasses non-state actors’ newly discovered capacity to exploit the openness of the web to wreak new havocs on businesses and governments. In this respect, Sageman (2008) has described the Internet as a “virtual training camp” for a new generation of “wannabe Jihadis,” where recruits are radicalized and ready to commit terrorist acts without ever having had face-to-face interactions with Al-Qaeda leaders and operatives (p. 123). Hackers’ power to steal individuals’ identity, cripple media organizations, multinational corporations, and governments’ agencies exacerbate the perception of the Web’s dangers (Schwartau 2000; Yourdon 2002). Based on these complex threats and risks, this “Dark Side” of the Net has to be fully considered in any discussion of the Internet and terrorism. In other terms, the Web has indeed empowered citizens and compressed space, but it has equally been used by both terrorist organizations and security organizations to sow the seeds of war in the “global village” (McLuhan and Fiore 1968). To put this relationship into a communication perspective, the rest of this chapter will analyze cyber-terrorism as a poignant illustration of how new communication technologies shape, and are in turn shaped by, society and its fears.

Simply put, cyber-terrorism refers to “the convergence of terrorism and cyberspace” and how terrorists may deploy the Internet to inflict destruction on critical infrastructure and to instill fear in populations and thus achieve their political goals. Mark Pollitt, a former FBI Special Agent, suggested one of the first working definitions: “Cyberterrorism is the premeditated, politically motivated attack against information, computer systems, computer programs, and data which result in violence against noncombatant targets by sub national groups or clandestine agents" (Pollitt 1998: 9). A well-known cyber-terrorism expert, Denning (2000) provides a rather precise definition:

> Cyber terrorism is the convergence of terrorism and cyberspace. It is generally understood to mean unlawful attacks and threats of attack against computers, networks, and the information stored therein when done to intimidate or coerce a government or its people in furtherance of political or social objectives. Further, to qualify as cyber terrorism, an attack should result in violence against persons or property, or at least cause enough harm to generate fear. Attacks that lead to death or bodily injury, explosions, plane crashes, water contamination, or severe economic loss would be examples. Serious attacks against critical infrastructures could be acts of cyber terrorism, depending on their impact. Attacks that disrupt nonessential services or that are mainly a costly nuisance would not (Denning 2000: n.p.).

Unlike Pollitt's, Denning's definition emphasizes the primacy of intimidation or coercion of governments to achieve political ends. The above definition further outlines how cyber-terrorists can wield the power of information technology to wreak havoc and terrorize populations. Cyber-terrorism thus possesses all the classic attributes of traditional terrorism as discussed in the previous section of the chapter. It does transfer the battle into cyberspace, making the Internet a lethal weapon of terror at the hands of these non-state actors.

5 Cyber-terrorism's rhetoric and reality

Al-Qaeda's 9/11 attacks have increased the salience and urgency of the issue of "cyber-terrorism" among academic scholars, policy makers and the popular press. However, the number of attacks that can be described as "cyber-terrorism" remains strikingly small. According to many experts (Denning 2007; Stohl 2007; Singer 2012), Al-Qaeda has never mounted a full-fledged cyber attack at the scale identified both in the definition and the rhetoric surrounding cyber-terrorism. The closest to a cyber-terrorism incident that could qualify under this scope occurred in Australia where Vitek Boden, a disgruntled municipal candidate, manipulated the software that controls a sewage treatment and released untreated sewage into local waterways. The attack had environmental consequences in Queensland, Australia, but it did not cause any loss of life. A 2007 cyber-attack on Estonia's government websites did not cause the mass panic, popular terror, or indiscriminate violence associated with terrorism's modus operandi (Singer 2012).

Singer presents the stark discrepancy between the rhetoric and the reality of cyber-terrorism this way:

> *About 31,300*. That is roughly the number of magazine and journal articles written so far that discuss the phenomenon of cyber terrorism.

> *Zero*. That is the number of people that who been hurt or killed by cyber terrorism at the time this went to press (Singer 2012: n.p.) [emphasis mine].

It is not clear whether the above count includes the amount of news coverage of cyber-terrorism in newspapers, online media, and radio and television newscasts. But the hyped-up fear of cyber-terrorism obscures the fact that cyber-terrorists are incapable of waging attacks that would meaningfully change people's daily life (Stohl 2007; Singer 2012; Denning 2001). According to the U.S. Naval Academy's George R. Lucas Jr., inflicting massive attacks via cyberspace "simply outstrips the intellectual, organizational and personnel capacities of even the most well-funded and well-organized terrorist organization, as well as those of even the most sophisticated international criminal enterprises" (Singer 2012: n.p.).

From a strictly communication perspective, however, the threat of cyber-terrorism has been exponentially exaggerated because it merges two deep-seated human fears: 1. fear of terrorism's unpredictable victimization; and 2. fear of technology. Fear of technology has been embedded in human's beings' evolution and the social imaginary because technology is both "conducive" and "subversive," a constructive and destructive force at the same time. As a "conducive" and "constructive" force, technology offers society the capacity to tame their natural environment, eliminate disease, and transcend the constraints of physical space. Through medical technologies, human life has become longer, healthier and more tolerable. Communication and transportation technologies have made borders irrelevant. At the same time, the "subversive" dimension of technologies brings with them the fear and possibility of indiscriminate destruction and mayhem as nuclear energy exemplifies. The capacity for technology to run amok coexists in nuclear technology from Hiroshima and Chernobyl catastrophes to nuclear powered submarines and power generation. Exploiting these perennial fears, the official and popular press's discourse about cyber-terrorism has resulted in constructing this threat as a "digital Armageddon" or a "digital Pearl Harbour," creating the conditions for a new "moral panic" (Stohl 2007).

Since "moral panics" and the "politics of fear" draw some legitimacy from "reality" (Altheide 2004), cyber-terrorism discourse is still based on evidence of terrorists using the Internet to further their goals. Publicly accessible evidence confirms that Al-Qaeda members have used the Internet as an operational and tactical platform to circulate encrypted email and messages, establishing password-protected chat rooms and websites, and circulating propaganda videos among their followers and the world wide web's audiences. One of Al-Qaeda's websites, Ansar-al-Islam became notorious upon posting a video of the beheading of American hostage, Nick Berg, by Al-Qaeda's Iraqi branch. In addition to encrypted communications, Al-Qaeda affiliated websites host terrorist training manuals and bomb-making instructions, and the organization's sympathisers have engaged in "denial of service" and web-defacement attacks (Denning 2007; Singer 2012). While their overt purpose includes organization, communication and recruitment, these activities do not conform to the definition of cyber-terrorism as "terrorizing populations" to achieve social and political goals. In fact, these cyber-Jihadists may be engaging in computer-mediated activities that can best be described as "hacktivism" (Denning 2001; Stohl 2007).

The official rhetoric about real and imagined risks of cyber-terrorism has similarly been used to justify indiscriminate "digital surveillance" of the general population. Irrefutable evidence of this mass surveillance has come to light after Edward Snowden, a former National Security Agency (NSA) contractor, exposed in his 2013 leaks the unprecedented collection and mining of meta-data and Internet communications by US and British intelligence services. NSA's covert efforts of widespread Internet surveillance involved compelling telecommunications companies to give it "backdoor" access into their systems as well as using top secret programs like "XKeyscore" to collect "nearly everything a user does on the internet" (Greenwald 2013: n.p.). As Snowden's NSA leaks revealed, aggregating all Internet communications has exponentially moved beyond fighting terrorist networks to establishing "Dragnet mass surveillance" that watches regular citizens and foreign leaders alike (Goodman and González 2013). While counter-terrorism efforts have been approved by US Congressional legislations, like the Cyber Security Enhancement Act of 2002 which was passed to stiffen penalties for cybercrime and cyber-terrorism, most covert surveillance activities remain out of the public domain and thus escape judicial and congressional oversights, creating legitimate fears about a "surveillance state" running amok and infringing upon citizens' privacy and democratic rights (Harris 2010).

Moreover, cyber-terrorism has become a compelling narrative about new forms of terrorism due to the nature of the Internet and the blending of technology and terrorism. In fact, the convergence of terrorism and cyberspace is an inevitable outcome of the new "information society" or what Castells (2005) describes as the "network society," that is:

> A social structure based on networks operated by information and communication technologies based in microelectronics and digital computer networks that generate, process, and distribute information on the basis of the knowledge accumulated in the nodes of the networks (Castells 2005: 7).

Cyber-terrorism discourse capitalizes on the fear that terrorists might exploit the "nodes of the networks" and thereby disrupt economic and social life. Cyber-terrorists may well hope to take advantage of the "networked" and "open" nature of contemporary information society and how these digital networks constitute new vulnerabilities and "soft targets." Beyond creating a "moral panic" about cyberspace, cyber-terrorism discourse lacks compelling and empirical evidence to back up its assertions about the capabilities of terrorist organizations to deploy the web as a terrorist weapon of choice.

6 Conclusion

In this chapter, I have argued that the "social shaping" of technology implies that technologies have shaped how terrorism has been deployed as a political weapon in contemporary societies. To propagate mass panic and thus influence policies, terrorists exploit the mass media's ability and willingness to publicize indiscriminate violence, and create "media spectacles of terror" (Kellner 2003). The news media continue to cover terrorist attacks due to professional and institutional considerations, providing an indispensable "oxygen" to terrorists. To describe this media and terrorism special relationship as "symbiotic" underscores the mutual dependence that exists between terrorism and mass media. In addition to censuring media's coverage of terrorist attacks, some governments have in turn exploited the threat of terrorism for their own ideological agendas. By virtue of a "politics of fear" and abetting "moral panics," they have used terrorist violence and threats of terrorism to gain public support for domestic policies and foreign interventions.

The "new" discourse surrounding cyber-terrorism demonstrates that new technologies have not veered off this well-trodden path. Technologies' arcane and double-edged usage nurtures human fears and anxieties about terrorism threats. The Internet's native architecture has allowed terrorists to use it to evade detection, communicate, recruit and organize. The weak evidence for cyber-terrorist attacks has not prevented an alarmist official discourse to hype up the threat, which conflates "hacktivism" with "cyber-terrorism." Alarmist discourse has the potential to create new "moral panics" about the Internet to justify intrusive surveillance of regular citizens' online activities as the 2013 Edward Snowden's leaks of NSA eavesdropping details demonstrate.

References

Altheide, David L. 2004. Consuming terrorism. *Symbolic Interaction* 27(3). 289–308.

Altheide, David L. 2006. Terrorism and the politics of fear. *Cultural Studies ↔ Critical Methodologies* 6(4). 415–439.

Associated Press. 2006. Al Qaida tapes often come through Al Jazeera: Network defends airing news; critics say it gives terrorists a podium. NBC, January 20, 2006. http://www.nbcnews.com/id/10948626/ns/world_news-terrorism/t/al-qaida-tapes-often-come-through-al-jazeera/# (Accessed 13 April 2014).

Camphuijsen, Marjolein and Esther Vissers. 2012. Terrorism and the mass media: A symbiotic relationship? *Social Cosmos* 3(1) 14–24.

Cannold, Juile, Mayra Cuevas and Joe Sterling. 2013. *Rolling Stone* cover of bombing suspect called 'slap' to Boston. *CNN*, July 18, 2013. http://www.cnn.com/2013/07/17/studentnews/tsarnaev-rolling-stone-cover/index.html (Accessed 13 April 2014).

Castells, Manuel. 2005. The network society: From knowledge to policy. In Manuel Castells & Gustavo Cardoso (eds.), *The network society: From knowledge to policy*, 3–23. Center for Transatlantic Relations, Paul H. Nitze School of Advanced International Studies, Johns Hopkins University.

Cohen, Stanley. 1972. *Folk devils and moral panics: The creation of the mods and rockers*. Oxford: Basil.

Cole, David. 2003. *Enemy aliens: Double standards and constitutional freedoms in the war on terrorism*. New York: New Press.

Denning, Dorothy E. 2000, May 23. Cyberterrorism. Testimony before the Special Oversight Panel on Terrorism Committee on Armed Services U.S. House of Representatives. *The Terrorism Research Center*. http://www.stealth-iss.com/documents/pdf/CYBERTERRORISM.pdf (Accessed 7 May 2014).

Denning, Dorothy E. 2001. Activism, hacktivism, and cyberterrorism: the Internet as a tool for influencing foreign policy. In John Arquilla & David Ronfeldt (ed.) *Networks and netwars: The future of terror, crime, and militancy*, 239–288. Arlington, VA: Rand Corporation.

Denning, Dorothy E. 2007. A view of cyberterrorism 5 years later. In Kennet Einar Himma (ed.), *Internet security: Hacking, counterhacking, and society*, 123–140. Sudbury, MA: Jones & Bartlett Learning.

Goodman, Amy and Juan González. 2013. Edward Snowden speaks out against NSA "Dragnet Mass Surveillance." Democracy Now, October 14, 2013. http://www.democracynow.org/2013/10/14/edward_snowden_speaks_out_against_nsa (Accessed May 10, 2014).

Greenwald, Glen. 2013. XKeyscore: NSA tool collects 'nearly everything a user does on the internet'. *The Guardian*, Wednesday 31 July 2013. http://www.theguardian.com/world/2013/jul/31/nsa-top-secret-program-online-data (Accessed May 10, 2014).

Harris, Shane. 2010. *The watchers: The rise of America's surveillance state*. New York: The Penguin Press.

Kalathil, Shanthi and Taylor C. Boas. 2003. *Open networks, closed regimes : The impact of the internet on authoritarian rule*. Washington DC: Carnegie Endowment for International Peace.

Kellner, Douglas. 2003. September 11, spectacles of terror, and media manipulation: a critique of Jihadist and Bush media politics. *Logos* 2(1). 86–102.

Landau, Mark J., Sheldon Solomon, Jeff Greenberg, Florette Cohen, Tom Pyszczynski, Jamie Arndt, Claude H. Miller, Daniel M. Ogilvie & Alison Cook. 2004. Deliver us from evil: The effects of mortality salience and reminders of 9/11 on support for President George W. Bush. *Personality and Social Psychology Bulletin* 30(9). 1136–1150.

Laqueur, Walter. 1987. *Age of terrorism*. 2nd edn. Boston: Little, Brown.

Lynch, Marc. 2006. Al-Qaeda's media strategies. *The National Interest* 83(1). 50–56. http://www.marclynch.com/wp-content/uploads/2012/03/out.pdf (Accessed 13 April 2014).

McLuhan, Marshall & Quentin Fiore. 1968. *War and peace in the global village*. New York: Bantam books.

Miles, Hugh. 2005. *Al Jazeera: The inside story of the Arab news channel that is challenging the West*. New York: Grove Press.

Morozov, Evgeny. 2012. *The net delusion: The dark side of Internet freedom*. New York: Public Affairs Store.

Obama, Barack. 2013. *The national security strategy of the United States of America*. http://www.utexas.edu/lbj/sites/default/files/file/news/National%20Security%20Strategy%202013%20(Final%20Draft).pdf (Accessed 13 April 2014).

Palmer, Allen. 2007. Following the historical paths of global communication. In Yahya R. Kamalipour (ed.), *Global Communication*. 2nd edn. 1–20. Belmont, CA: Wadsworth.

Pew Research Center Staff. 2006. Al Jazeera's global gamble: Al Jazeera timeline. Pew Research Journalism Project, August 22, 2006. http://www.journalism.org/2006/08/22/al-jazeera-timeline (Accessed 13 April 2014).

Pollitt, Mark M. 1998. Cyberterrorism – fact or fancy? *Computer Fraud & Security* 1998(2). 8–10.

Rogan, Hanna. 2007. Abu Reuter and the E-Jihad: Virtual battlefronts from Iraq to the Horn of Africa. *Georgetown Journal of International Affairs* 8(2). 89–96.

Rothe, Dawn & Stephen Muzzatti, S. L. 2004. Enemies everywhere: Terrorism, moral panic, and US civil society. *Critical Criminology* 12(3). 327–350.

Sageman, Marc. 2008. *Leaderless Jihad: Terror networks in the Twenty-First Century*. Philadelphia: University of Pennsylvania Press.

Schwartau, Winn. 2000. *Cybershock: Surviving hackers, phreakers, identity thieves, Internet terrorists, and weapons of mass disruption*. New York: Thunder's Mouth Press.

Singer, Peter W. 2012. The cyber terror bogeyman. Armed Forces Journal, The Brookings Institution. *Brookings*. http://www.brookings.edu/research/articles/2012/11/cyber-terror-singer (Accessed 14 April 2104).

Sloan, Stephen. 1991. Technology and terrorism: Privatizing public violence. *Technology and Society Magazine, IEEE* 10(2). 8–14.

Stohl, Michael. 2007. Cyber terrorism: A clear and present danger, the sum of all fears, breaking point or patriot games? *Crime, law and social change* 46(4–5). 223–238.

Taibbi, Matt. 2013. Explaining the *Rolling Stone* Cover, by a Boston Native. *Rolling Stone*, July 19, 2013. http://www.rollingstone.com/politics/blogs/taibblog/explaining-the-rolling-stone-cover-by-a-boston-native-20130719#ixzz2iPIM3JxE (Accessed 13 April 2104).

Tuman, Joseph S. 2010. *Communicating terror: The rhetorical dimensions of terrorism*. SAGE Publication, Inc.

Teichman, Jenny. 1989. How to define terrorism. *Philosophy* 64(250). 505–517.

Wilkinson, Paul. 1997. The media and terrorism: A reassessment. *Terrorism and Political Violence* 9(2). 51–64.

Weimann, Gabriel. 1987. Media events: The case of international terrorism. *Journal of Broadcasting & Electronic Media* 31(1). 21–39.

Yourdon, Edward. 2002. *Byte wars: The impact of September 11 on information technology*. NJ: Prentice Hall.

Pauline Hope Cheong and Daniel Arasa

24 Religion

Abstract: This chapter provides a discussion of religion's dynamic developments alongside contemporary communication technology connections and appropriations, focusing on the debate and issues surrounding the emergent paradoxes and challenges in the areas of religious authority, community and identity. The chapter spotlights key insights from prior reviews, and examines the dialectics, i.e. forces of interaction and relational tensions, in the relationships between religion and the Internet, involving the latest digital and social media applications. Illustrations are drawn from various religions, particularly the Catholic Church, the largest religious institution in the world.

Keywords: religion, authority, community, identity, digital and social media

> Ever since the eighteenth-century Enlightenment, which proclaimed the inevitable "secularization" of society, it has generally been assumed that the first of these tendencies would historically supercede the second, that the advance of scientific technology, with its rational rigors grounded in practical experience and material knowledge, signaled the demise of religious authority and enthusiasm based upon blind faith and superstition. Religion, presumably, belonged to the primitive past, secular science and technology to the mature future. *Yet today we are seeing the simultaneous flourishing of both, not only side by side but hand in hand.* While religious leaders promote their revival of spirit through an avid and accomplished use of the latest technological advances, scientists and technologists increasingly attest publicly to the value of their work in the pursuit of divine knowledge (Noble 1997: 4) [emphasis mine].

The inclusion of the theme of religion in this volume on Communication Technology may seem somewhat oxymoronic as sacred beliefs and traditions are paired with the most popular technological applications of our time. Yet as many historians and social commentators like David Noble remind us, religion and technology are not incongruous developments but are cognate fields, with profound convergences and challenges.

Arguably, as individualized and institutionalized religious practices and movements embrace the Web to improve their reach, religious communication on digital and social media platforms take an increasingly significant role together with more traditional venues for such discourse. It may be, however, that religious enchantment associated with digital media use problematizes established faith rites and the conception of religious identity and community. Changes in forms of spiritual organizing in "the church" and "the temple" can also be conceived as intertwined with a range of other forms of social and political developments, such that digital media facilitates challenges to the authority of clergy and elites, as well as promotes innovative reconfigurations of church norms and expectations. Furthermore, as "the Internet" is a moving target, there may be past traditions that are enduring as well as new ideas such as church branding or celebrity that are being envisioned and enacted in contemporary social and religious life.

This chapter provides a discussion of religion's dynamic developments alongside contemporary digital and social media connections and appropriations, focusing on the debate and issues surrounding the emergent tensions and challenges in religious authority, community and identity. We are cognizant that prior comprehensive surveys on the topic "Religion and the Internet" exists; indeed, Cheong (in press) provides an overview of the topic Religion and the Internet, in an encyclopedia on *Religion and American Cultures,* and Cheong and Ess (2012) review the relational and hybridizing pathways in religion, social media and culture in the international anthology *Digital Religion, Social Media and Culture: Perspectives, Practices, Futures.* See also thematic contributions in Campbell (ed.) *Digital Religion: Understanding Religious Practice in New Media Worlds* (2013), and Arasa, Cantoni and Ruiz (eds). *Religious Internet Communication. Facts, Trends and Experiences in the Catholic Church* (2010). In addition, noteworthy are three special issues on religion in the Journal of Media and Religion (2002), the Journal of Computer Mediated Communication (2007), and Information, Communication and Society (2012).

Against the background of this growing scholarship, this chapter will spotlight key insights from prior reviews, while extending our understanding on this topic by examining the dialectics, i.e. forces of interaction and relational tensions (Cheong et al. 2012) in the relationships between religion and the Internet, involving the latest digital and social media applications. Given our length limitations, insights here will be grounded in examples from the Catholic Church, the largest religious institution in the world with more than 1.2 billion followers (Vatican Information Service, 2013a), but connections to other religions may be made using the dialectical approach in this topical review.

1 Research in religion and communication technologies: Waves and paradoxes

Broadly, scholarship in the field of religion and the internet in the past two decades has proceeded in tandem with the conceptual and methodological developments in new media and internet studies. Scholars are generally in agreement that research on religion and the internet can be characterized by different waves or phases, with different logics underpinning the assumptions and data collection on computer mediated religion (Campbell 2013; Cheong 2013). For instance, it is contended that the earliest phase of research was marked by an emphasis on understanding the then new phenomenon of "cyberfaith", "cyber-religion" and "virtual church", with investigations focused on understanding pioneering paths of access to spiritual information, and an expanding repertoire of religiously inspired text-based and interactive rituals online.

Research following the initial wave of cyber enthusiasm largely echoed concerns voiced in the larger body of internet studies for the need to incorporate the role of history and philosophy to contextualize transformations in wider society and culture. Accordingly, the conjoining of the online and the offline dimensions of religion has surfaced as an important theme in recent years. Against this background, is a reinvigorated emphasis on "networks", including the "rise of the network society" (Castells 2000), and more recently the proposition that social life is being "networked" in light of the trifold "social network, Internet and mobile revolutions" (Rainie and Wellman 2012). Likewise, it has been suggested that digital religion is also being integrated into everyday life rhythms and conduct, weaved in hybridized multimodal connections in contemporary "convergence culture" (Jenkins 2006).

Furthermore, a small but growing corpus of recent research studies observe how mediated connections entail multiple opportunities, uneven gains and conflicting tensions in religious information sharing, community building and authority practices. A dialectical perspective to digital media and culture offers unique opportunities to recognize the simultaneous presence of two relational forces of interaction, and accepts as ordinary the interdependence and complementary aspects of seeming opposites. Thinking dialectically about religion and communication technologies helps identify key tensions and paradoxes, which are in need of further exploration, in order to understand the diversity and complexity of contemporary social and religious life (Cheong and Ess 2012; Spadaro 2012).

1.1 Religious authority: Erosion-augmentation

A key paradox in religious authority relations is its enervation and centralization of control. Increased access to religious information and interpretations online provide followers with alternative resources and discourse to challenge traditional teachings and disrupt hierarchies. Yet these same online resources may be appropriated by clergy to serve as a source of education for themselves and their laity, while their elite legitimacy is simultaneously enhanced when they move beyond top-down instruction to personal and face-to-face sharing and mentorship practices that influence the interpretation of religious texts, including the publication and dissemination of their own religious materials (Cheong et al. 2011).

Social media make possible more egalitarian forms of communication between laity and clergy, thereby prompting a renegotiation of power and authority. At the same time, religious authority may use the latest suite of communication technologies to facilitate the co-creation of information and expertise under conditions where laity cooperation is elicited by retaining discretionary power among the leadership to determine informational and interpersonal outcomes such that they do not undermine the organization (Cheong et al. 2011). New instantiations of religious authority are evolving. Likewise, the rise of religious branding and celebrification across various media illustrates how religious authority is at once more accessible and banal, and yet more mystical and inscrutable.

For instance, consider the papal authority and adoption of the internet by the Catholic church. On the one hand, there appears to be a more conservative approach to digital and social media amidst various concerns, yet appropriation of novel technological applications by the church and its regional and local leadership evince how authorial relations are changing and may be strengthened by digital media engagement.

In recent times, the Catholic Church's hierarchy has integrated the Internet into the Church's internal and external communications at a practical and a theoretical level. Previous Popes have issued several statements about the Internet, acknowledging and permitting its use (John Paul II 2001, 2002, 2005; Benedict XVI 2009, 2010, 2011, 2013). Even Church documents affirm that "hanging back timidly (from using it [Internet]) for fear of technology or for some other reason is not acceptable" for Catholics (PCSC 2002a: 10). Yet at the same time, there is ideological resistance to endorse the internet as the main or exclusive channel for human relations (for e.g. John Paul II 2002; Benedict XVI 2011; 2013). Thus, the virtual world is seen as complement to, but not a substitute for essential elements of the Christian faith. Accordingly, authorial and community relations may also be virtual but have necessarily a face-to-face component. Pope Francis's recent statement about indulgences (Vatican Information Service, 2013b) illustrate this – his tweet reminded followers about the need of devotion and committed practice connected to the church and theology to obtain them, but nothing was said about obtaining virtual indulgences via Twitter. Similarly, the use of the Confession app was also acknowledged by Pope Benedict XVI but users were reminded that the absolution of sins still required the presence of a real priest in the local church (Cheong and Ess 2012).

Furthermore, it is noteworthy that the Vatican has exercised considerable effort to expand its communication repertoire at the institutional or corporate level (e.g. Vatican radio, website, newspaper L'Osservatore Romano, and television production center known as Centro Televisivo Vaticano) and be present on social media (e.g. @Pontifex twitter handle and Vatican YouTube channel), alongside mediations of significant events (like the Year of the Faith through www.annusfidei.va). Nonetheless, its vast use of the internet is primarily for broadcast (through websites especially), with a more conservative approach regarding social media outreach. The dialectic of authority erosion-augmentation is observed for example in the control vs. promotion of the use of Vatican channel on YouTube, which now broadcasts news coverage of the main activities of Pope Francis in multiple languages on a daily basis. The comments section was disabled (Campbell 2012) but comments in some cases are now allowed and appear to be submitted to moderation and thus restricting the type of interactions that take place online. Although practical aspects are involved in that decision (lack of personnel to handle users' interaction), there appears to be apprehension or fear of losing over the representation of the Vatican online.

Moreover, it is important to distinguish between the individual use of digital media by Church leaders and the corporative communication actions of Church institutions or para-church organizations (e.g. dioceses or bishop's conferences, nonprofit faith-based organizations like hospitals or charities, educational institutions like universities or colleges, etc.). Even if all of them deliver the Church's teachings, they may constitute and represent different authority relations as seen in, for instance, the numerous blogs run by Church leaders like bishops and cardinals versus the official stories and updates by the Papacy. As such, the contestation and re-interpretation of Church statements is probably one of the biggest challenges that Church institutions have to cope with as Church teachings and decisions of religious elites are now more firmly embedded in the "society of conversation" (Verdû Macía 2005), constantly subject to analysis, criticism, denial and debate in the digital realm.

A specific example within the Catholic Church was the public opinion crisis created with the remission of the *excommunio* (excommunication) to the Lefevbrian bishops by Pope Benedict XVI in January 2009 (Congregation for Bishops 2009). Lefevbrians are the followers of Marcel Lefevbre (1905–1991), French Roman Catholic archbishop who opposed to the changes within the Church associated with the Second Vatican Council. In 1970, Lefebvre founded the Society of St. Pius X and, in 1988, against the express prohibition of Pope John Paul II, he consecrated four bishops to continue his work. The Holy See immediately declared that he and the other bishops who had participated in the ceremony had incurred automatic excommunication under Catholic canon law, so becoming schismatic and separated from it. That act of indulgence that the Pope desired to have with those considered schismatic for the Catholic Church was portrayed by many as an irresponsible decision, especially when mass media chose to publish an interview with Richard Williamson, one of the four Lefevbrian bishops, highlighting his denial of the Holocaust and other scandalous opinions. A few weeks later, the Pope wrote a sorrowful letter to express his concern over the public reaction and even the aggressiveness within the Catholic Church towards his decision (Benedict XVI 2009b). In that letter, among other propositions, Benedict XVI mentioned the need for the Church's hierarchy to pay more attention to the internet before making or announcing decisions that interested public opinion, setting a lesson for the governance of Church organizations: Church leaders need to be responsive to digital media because of their impact on public opinion and, indirectly, within the Church's internal debate. Yet there are certainly limits on the role of communication technologies, since church doctrine and practices are arguably more influential forces in shaping the episteme in the Catholic tradition, not the popular opinion of the masses. The Pope and bishops are leaders of the community (universal and local), but at the service of the other members. Hence, decisions and pronouncements are made through processes of discernment and consultation under the governance of the synod. Technologies play an essential role at this level of consultation, but canons and laws cannot – according to church's doctrine – be simply decided by majority vote.

Finally, social media use by the Pope allows the church to communicate its authority. For instance, while mass media coverage offers us a look at the Pope's life, as seen in the coverage of the World Youth day at Rio de Janeiro 2013, where Pope Francis was described as being mobbed by enthusiastic crowds, mainstream media's selective news frames and coverage focused more on his comment on the gay lifestyle, than on his statements about poverty and socio-economic injustices in a globalizing age, which may stem from their slanted and biased interpretation of religion. In these situations, social media facilitates the representation of the church in the way it wants it to be portrayed, potentially bypassing the agenda of the mass media. Even though it is not the Pope himself who tweets, but his collaborators (the Pope limits himself to provide some kind of approval to the messages), his spiritual influence and authority through Twitter is made present or invoked and "presentified" by other followers and on various media (Brummans et al. 2013).

1.2 Community: Global-local

Tensions in the area of religious community often entail if and how new forms of social and religious interaction constitute a community. In light of mediated networks, social ties may now be more loosely coupled and shared by those beyond one's immediate geographic location and kinship lines. The emergence of new socialities on virtual platforms and immersive digital worlds raise the possibilities for the imagining and operationalization of transnational religious communities. Yet one paradox in the expansion or globalization of mediated religious community is that it often necessarily entails the fostering of local ties. These ties may simultaneously encompass communal rituals and face to face meetings, even while churches and temples provide more opportunities for individualized practices of accessing spiritual resources online as they establish and grow their digital presence. Therefore, dialectics in the logic of the networked and place-based community may co-exist.

Consider for example how communication technology provides opportunities to foster a sense of transnational community among faith believers. The recent World Youth Day – WYD 2013, was a massive encounter that Pope Francis had in Brazil on July 2013 with 3.8 million youths from all over the world. This mediated event was an opportunity for the Catholic Church to showcase the Pope in face to face meetings, visiting slums, and leading rituals including corporate worship of hundreds of thousands on Copacabana Beach. The events were followed live online via web streaming or watched asynchronously by millions, illustrating the use of digital media as an instrument to unite local communities to a global event.

Yet although the Catholic community may be united by temporal, physical and mediated events or what may be understood as "media spectacles" (Kellner 2003) like the World Youth Day, Catholic believers are also instructed that their spiritual ties extend beyond physical events or digital links (what theology denotes as the "communion of saints") which is historically rooted in a specific deity (Jesus), born in a specific period (2000 years ago), in a particular place (Palestine). In this sense, the internet affects local Catholic communities in their regular practices (government, internal communication, diffusion of teaching, etc.) but it does not shake the sense of global community already existing within the faith (it is pertinent to note here that *catholic* means universal). However, it is important to recognize that the main aim of the Church is the salvation of souls through the preaching of the Gospel and the distribution of the Sacraments, and both necessarily happen through personal contact and community engagement. According to Pope John Paul II, the internet is "a unique supplement and support in both preparing for the encounter with Christ in community, and sustaining the new believer in the journey of faith which then begins." (John Paul II 2002: 3), but "electronically mediated relationships can never take the place of the direct human contact required for genuine evangelization" (PCSC 2002a: 5). Correspondingly, religious community is necessarily embodied by local residents, even while technology affords global affinities and fealty.

1.3 Identity: Stable, modernist-ephemeral, multiplex

It is often said that we are witnessing a shift in the sense of identity to more relational selves, facilitated through the multiple communicative possibilities of new media (Ess 2011; Gergen 2011). Digital and social media platforms provide new avenues for individuals and collectives to self-represent their identity, with heightened visibility and oftentimes accompanying expectations of candid intimacy through the creation and sharing of personal profiles, updates and Facebook timelines. Yet, digital media also facilitate online interactants' capacity and autonomy to track, tailor, continuously update, privatize and even monetize and sell their personal profiles. The hybrid notion of "networked individualism" (Wellman 2001) for instance, highlights modernist notions of the individual self-affiliated with literary and print as media technologies, and postmodern notions of multiple, fluid and ephemeral selves affiliated with dynamic electronic media connections.

Consequently, a paradox of religious identity in our networked age concerns how one's religious identity represented through communication technologies becomes simultaneously more authentic and accessible, as well as more edited and polished than in face to face settings. And this dialectic has several implications for religion. Many declare themselves as followers of a religion and are explicit about their faith beliefs in their online personal profiles. This heightened publicity of spiritual piety and devotion in their bios and social media updates may raise the relevance and significance of the church or temple to their friends. Yet some believers also do not hesitate to discuss, act and speak contrarily to the moral standards and doctrinal principles of their confession, which may, in turn, mar and distort the organizational identity of their church or temple.

Challenges in light of the stable, modernist-ephemeral, multiplex identity dialectic may be illustrated by the perennial issue of determining the "catholicity" of a website (or any other digital initiative) and, therefore, the Catholic identity of its contents. In the case of the possible confusion of Catholic identity, advisory boards of the Holy See have proposed that people responsible for websites make a clear distinction between Catholic doctrine and personal opinions as well as participate in the voluntary requirement of certification by Catholic officials regarding sites with doctrinal and catechetical material (PCSC 2002: 8). Yet on top of the practical problems and lack of resources to comply with the advice by the Holy See, catholicity is not easily definable since online content is highly dynamic and ever-changing. Therefore, sites that could be considered within the orthodoxy at the initial point of approval may offer content which does not fit with official guidelines at a later time. Consequently, Church teachings are now necessarily generic in their approach, appealing to the freedom and responsibility of Catholic individuals and organizations to represent what they deem to be a scripturally inspired or authentic Catholic identity in the digital arena.

Furthermore, religious identity construction and enactment may encompass more rationally oriented decisions as well as more emotive aspects as online users potentially engage a more complete human sensorium – fostered by textual, oral, visual and kinesthetic capabilities of digital and social media. This mediation of identity performances, however, may be limited by multiple meso and micro level digital divides, in terms of personal choice and preferences, as well as structural barriers constraining internet access based on one's identity in terms of age, gender and class.

2 Conclusion

We began this chapter with Noble (1997)'s sage observation and now we want to close the chapter with a related observation of our own regarding the simultaneous flourishing of science and religion. It is interesting to note how "ubiquity" was originally a theological position maintaining the omnipresence of Christ yet it is remarkable how we increasingly refer to the ubiquity of the internet and communication technologies in our wired environments today. Yet even as technical innovations advance the portability and (in)visibility of our communication devices, it is important to focus not just on media-centric developments but also consider the communicative dialectics even as both technology and religion thrive in our contemporary age. For while this encyclopedia is titled communication and technology, our discussion makes clear that the relationship between religious communication and technology is a profound and intertwined one, co-mingled with relational tensions and conflicting visions of authority, community and identity, attesting to its passel of future research possibilities.

References

Arasa, Daniel. 2009. Il Magistero della Chiesa sulla comunicazione. In Jean Marie La Porte (ed.), *Introduzione alla comunicazione istituzionale della Chiesa*, 11–40. Edusc: Rome.

Arasa, Daniel, Lorenzo Cantoni & Lucio A. Ruiz (eds.). 2010. *Religious Internet Communication. Facts, Trends and Experiences in the Catholic Church*. Edusc: Rome.

Benedict XVI. 2009a. New Technologies, New Relationships. Promoting a Culture of Respect, Dialogue and Friendship. *Message for the 43rd World Communications Day*. Vatican, May 24, 2009. http://www.vatican.va/holy_father/benedict_xvi/messages/communications/documents/hf_ben-xvi_mes_20090124_43rd-world-communications-day_en.html (Accessed 13 April 2014).

Benedict XVI. 2009b. *Letter to the Bishops of the Catholic Church concerning the remission of the excommunication of the four Bishops consecrated by Archbishop Lefebvre*. Vatican, March 10, 2009. http://www.vatican.va/holy_father/benedict_xvi/letters/2009/documents/hf_ben-xvi_let_20090310_remissione-scomunica_en.html (Accessed 13 April 2014).

Benedict XVI. 2010. The Priest and Pastoral Ministry in a Digital World: New Media at the Service of the Word. *Message for the 44th World Communications Day*. Vatican, May 16, 2010. http://www.vatican.va/holy_father/benedict_xvi/messages/communications/documents/hf_ben-xvi_mes_20100124_44th-world-communications-day_en.html (Accessed 13 April 2014).

Benedict XVI. 2011. Truth, Proclamation and Authenticity of Life in the Digital Age. *Message for the 45th World Communications Day*. Vatican, June 5, 2011. http://www.vatican.va/holy_father/benedict_xvi/messages/communications/documents/hf_ben-xvi_mes_20110124_45th-world-communications-day_en.html (Accessed 13 April 2014).

Benedict XVI. 2013. Social Networks: portals of truth and faith; new spaces for evangelization. *Message for the 47th World Communications Day*. Vatican, May 12, 2013. http://www.vatican.va/holy_father/benedict_xvi/messages/communications/documents/hf_ben-xvi_mes_20130124_47th-world-communications-day_en.html (Accessed 13 April 2014).

Brummans, Boris H. J. M., Jennie M. Hwang & Pauline Hope Cheong. 2013. Mindful Authoring through Invocation: Leaders' Constitution of a Spiritual Organization. *Management Communication Quarterly* 27(3). 346–372.

Campbell, Heidi (ed.). 2013. *Digital Religion: Understanding Religious Practice in New Media Worlds*. New York: Routledge.

Campbell, Heidi. 2013. Community. In Heidi Campbell (ed.), *Digital Religion: Understanding Religious Practice in New Media Worlds*. 57–71. New York: Routledge.

Campbell, Heidi. 2012. How Religious Communities Negotiate New Media Religiously. In Pauline Hope Cheong, Peter Fischer-Nielsen, Stefan Gelfgren & Charles Ess (eds.), *Digital Religion, Social Media and Culture: Perspectives, Practices, Futures*, 81–96. New York: Peter Lang.

Castells, Manuel. 2000. *The Rise of the Network Society, The Information Age: Economy, Society and Culture* Vol. I. Cambridge, MA; Oxford, UK: Blackwell.

Cheong, Pauline Hope. 2013. Authority. In Heidi Campbell (ed.), *Digital Religion: Understanding Religious Practice in New Media Worlds*, 72–87. NY: Routledge.

Cheong, Pauline Hope. (in press). Religion and the Internet. In Gary Laderman & Luis Leon (eds.), *Religion and American Cultures. An Encyclopedia of Traditions, Diversity and Popular Expressions.* Santa Barbara, California: ABC-CLIO.

Cheong, Pauline Hope & Charles Ess. 2012. Religion 2.0? Relational and hybridizing pathways in religion, social media and culture. In Pauline Hope Cheong, Peter Fischer-Nielsen, Stefan Gelfgren & Charles Ess (eds.), *Digital Religion, Social Media and Culture: Perspectives, Practices, Futures*, 1–24. New York: Peter Lang.

Cheong, Pauline Hope, Judith N. Martin & Leah Macfadyen. 2012. Mediated intercultural communication matters: Understanding new media, dialectics and social change. In Pauline Hope Cheong, Judith N. Martin & Leah Macfadyen (eds.), *New Media and Intercultural Communication: Identity, Community and Politics*, 1–20. New York: Peter Lang.

Cheong, Pauline Hope, Shirlena H. Hwang & Jessie P. H. Poon. 2011. Religious Communication and Epistemic Authority of Leaders in Wired Faith Organizations. *Journal of Communication* 61(5). 938–958.

Congregation for Bishops. 2009. *Decree remitting the excommunication "latae sententiae" of the bishops of the Society of St. Pius X.* Rome, January 15, 2009. http://www.vatican.va/roman_curia/congregations/cbishops/documents/rc_con_cbishops_doc_20090121_remissione-scomunica_en.html (Accessed 13 April 2014).

Ess, Charles. 2011. Self, Community, and Ethics in Digital Mediatized Worlds. In Charles Ess and May Thorseth (eds.), *Trust and Virtual Worlds: Contemporary Perspectives*, 3–30. Oxford: Peter Lang.

Gergen, Kenneth. 2011. *Relational Being: Beyond Self and Community.* Oxford: Oxford University Press.

Jenkins, Henry. 2006. *Convergence culture: Where old and new media collide.* New York: New York University Press.

John Paul II. 2001. Preach from the Housetops: The Gospel in the Age of Global Communications. *Message for the 35th World Communications Day.* Vatican, January 24, 2001. http://www.vatican.va/holy_father/john_paul_ii/messages/communications/documents/hf_jp-ii_mes_20010124_world-communications-day_en.html (Accessed 13 April 2014).

John Paul II. 2002. Internet: A New Forum for Proclaiming the Gospel. *Message for the 36th World Communications Day.* Vatican, May 12, 2002. http://www.vatican.va/holy_father/john_paul_ii/messages/communications/documents/hf_jp-ii_mes_20020122_world-communications-day_en.html (Accessed 13 April 2014).

John Paul II. 2005. The Rapid Development. *Acta Apostolicae Sedis* (AAS) 97. 265–274. Vatican, January 24, 2005.

Kellner, Douglas. 2003. Media Culture and the Triumph of the Spectacle. London: Routledge.

Noble, David. F. 1997. The Religion of Technology, the Divinity of Man and the Spirit of Invention. New York: Alfred A. Knopf.

PCSC – Pontifical Council for Social Communications. 2002a. *The Church and Internet.* LEV: Vatican City.

PCSC – Pontifical Council for Social Communications. 2002b. *Ethics in Internet.* LEV: Vatican City 2002b.

Spadaro, Antonio. 2012. *Cyberteologia. Pensare il cristianesimo al tempo della rete.* Vita e Pensiero: Milan.

Verdù Macía, Vincent. 2005. *Yo y tú, objetos de lujo.* Editorial Debate: Madrid.

VIS – Vatican Information Service. 2013a. *Statistical Yearbook of the Catholic Church.* LEV: Vatican City.

VIS – Vatican Information Service. 2013b. Papal Indulgences for World Youth Day. *Vatican Information Service*, July 9, 2013. www.news.va/en/news/papal-indulgences-for-world-youth-day (Accessed 14 April 2014).

Wellman, Barry. 2001. Physical Place and Cyber Place: The Rise of Networked Individualism. *International Journal of Urban and Regional Research* 25(2). 227–52.

Thomas C. Reeves and Patricia M. Reeves

25 Learning

Abstract: Learning is fundamental to humans as a species and to each person as an individual. This chapter examines the nature of learning in the information and communication age by addressing key questions related to both formal learning in schools and informal learning through experience. The theoretical perspectives on learning presented in this chapter include behaviorist, cognitivist, humanist, constructivist, constructionist, social, and connectivist orientations. Practical instructional design principles derived from each theoretical orientation are included. The chapter concludes with speculation about how learning may be changed by communications and technology over the next thirty years and a call for enhanced educational design research to guarantee that this near-term future will be as positive and socially responsible as possible.

Keywords: instructional design principles, learning, learning sciences, learning theories

Learning is fundamental to humans as a species and to each of us as individuals. Both formal learning in schools and informal learning through experience are influenced by the times in which we live. This chapter examines the nature of learning in the information and communication age by addressing several key questions. For example, how do the processes and outcomes of learning today differ from how learning was accomplished in earlier ages, e.g., in the time before personal computers and cellular communications became commonplace? How is the concept of learning evolving as information and communication technologies become ever more ubiquitous in the lives of humans of all ages, from the infant learning about shapes and colors with a smart phone app to the senior citizen learning to perform post-surgery rehabilitation exercises via personal videoconferencing and digital biotelemetry? What will it mean to learn in the not-too-distant future when information and communication technologies actually become part of our bodies, e.g., when today's Google Glass becomes tomorrow's Google Implant (Trenholm 2010)?

Answering these kinds of questions is a formidable task, especially in a format as brief as a chapter. Entire volumes are regularly written to address learning (Bransford et al. 1999; Spector et al. 2014), and thus setting some limitations is necessary. Accordingly, the perspectives on learning presented in this chapter will begin at the end of World War II in 1945 when behaviorism was the dominant learning theory and programmed instruction was just beginning to emerge as an innovative learning technology. In addition, the prognostications summarized in this chapter will encompass predictions made by learning technologists and other experts up until 2045, a mere 30 years from publication of this *Handbook of Communication and Technology*. Covering this hundred-year timespan is still a daunting challenge, and further limitations are necessary. Although it would be valuable to take a more global perspective on learning, this chapter will be restricted to the research and development emerging primarily from Europe and North America. Despite these and other unspecified limitations, it is hoped that this chapter makes a useful contribution to this *Handbook*.

1 The recent history of learning

Our understanding of learning has evolved dramatically over the past 70 years. At the end of World War II, educational psychology and the emerging field of educational technology were dominated by the behavioral learning theories of Watson (1913), Skinner (1935), and other behaviorists that had emerged over the previous thirty-odd years (Benjamin 2007). Behaviorists conceptualized learning as a change in observable behavior and largely dismissed as irrelevant speculations about what might be going on in the minds of learners. From the behavioral perspective, learning emerges when desirable behaviors (e.g., table manners) are solicited and reinforced by environmental conditions (e.g., the smiles and compliments of parents).

Behaviorism reached its peak during the decade from 1955 to 1965 with the emergence and proliferation of analog teaching machines and printed programmed instruction (Lumsdaine and Glaser 1960). The programmed instruction movement was spearheaded by B. F. Skinner in a 1954 paper titled "The Science of Learning and the Art of Teaching" (Reiser 2001), in which Skinner argued that the contingencies of reinforcement and other operant conditioning principles that he and other behaviorists had applied so successfully in training rats and pigeons could be used to teach humans from primary school through military training. North American readers of this chapter who were schooled during that time period may recall the small programmed instruction booklets that were a widely-used learning technology in classrooms at that time. These materials required the learner to read a small amount of text, answer a multiple-choice question closely linked to the information in the text, and, depending on the nature of the response, "branch" to another brief paragraph of content if the response was correct or to a remedial text if an incorrect response was given.

Instructional design principles derived from behaviorist learning theory include:

- Active learner participation in responding to instructional stimuli is much more effective than passive observation of instruction.
- Behaviors that are rewarded are more likely to be repeated than behaviors that are ignored.

- Immediate reinforcement is more effective in most cases than delayed reinforcement.
- Intermittent reinforcement is more effective than constant reinforcement.
- Extinction, i.e., withholding reinforcement of a behavior or ignoring it can result in the decrease of that behavior.
- Extensive practice with positive reinforcement is a powerful method for developing fine motor skills such as playing a piano or injecting a vaccine.
- Information should be presented in small amounts so that responses to questions about the information can be reinforced frequently.

Although later eclipsed by cognitive learning theories, behaviorism remains strong in many quarters. As reviewed by Burton, Moore and Magliaro (1996), behavioral strategies of stimulus-response conditioning are at the core of many contemporary learning technologies such as direct instruction (Englemann 1980) and computer-based tutoring systems (Halff 1988). Another product of behaviorism, the specification of behavioral learning objectives intended to be objectively measured, remains inherent in virtually all instructional systems design models (Gustafson and Branch 2002).

Cognitive learning theories, a subset of the field of cognitive science, emerged in response to what psychologists saw as a failure of behaviorism to explain more complex forms of learning than the simple rote memorization outcomes accomplished through programmed instruction (Schunk 2012). Cognitive learning focuses on the largely invisible, but powerfully evident, functions of the brain (e.g., imagination and creativity) and the relationships among them. Cognitive psychologists have also dealt directly with the role of different types of motivation (e.g., intrinsic and extrinsic) in learning whereas behaviorists dismissed the role of motivation beyond the most basic animal drives for safety and pleasure (Schunk et al. 2013).

In addition, many cognitive theorists have focused their research on memory functions (e.g., short-term, long-term and working) (Anderson 2000). Some cognitive psychologists have likened human learning to the information processing functions of computers in which different types of memory are also critical. Cognitive theories that include information processing are especially popular among cognitive and computer scientists involved in developing intelligent tutoring systems and other applications of artificial intelligence to promote learning (VanLehn 2011).

Altogether, cognitivists place much more emphasis on internal mental states than behavioral psychologists and have even developed hierarchies of the cognitive functions of the brain ranging from factoids to mental models. For example, Kyllonen and Shute (1989) developed a taxonomy representing the spectrum of internal mental states with which cognitive psychologists are concerned, beginning with simple propositions (e.g., stating that China sells more manufactured products than any other nation), proceeding through schema, rules, general rules, skills, general skills, automatic skills, and finally, mental models (e.g., interpreting the trade relations between China and the United States as a form of economic symbiosis). The "higher" forms of knowledge are particularly important because these mental models are the basis for generalizable problem-solving abilities and creativity (Halford 1993; Sternberg 1999).

Cognitive scientists recognize that a wide variety of learning strategies may have to be employed in any given instructional setting depending upon the type of knowledge to be constructed. Learning strategies include memorization, direct instruction, drill-and-practice, deduction, and induction (Schank and Cleary 1995). Various forms of computer-based instruction or e-learning differ in their capacity to implement these different learning strategies. Whereas a straightforward computer-based tutoring system may provide adequate opportunities for direct instruction and drill-and-practice, some type of sophisticated interactive learning simulation (Landriscina 2013) or a serious game (Blumberg et al. 2013) may be required to support deductive and inductive learning strategies.

Instructional design principles derived from cognitivist learning theory include:

- Working (short-term) memory has inherent limitations that place a cognitive load on learning processes that must be carefully managed during instruction. Instruction should be designed to minimize the cognitive load of extraneous factors (e.g., the user interface of a computer program) and maximize the cognitive capacity to process information needed for learning.
- There are two separate channels that humans use for processing information during instruction, the auditory and visual, and thus, instruction should provide verbal and graphical information in ways that clearly guide the learners to identify and process relevant words and images.
- Learning occurs when humans identify, filter, select, link, organize, and integrate information under the guidance of a tutor, which may be human or computer-based.
- Content to be learned should be organized from simple to complex.

Whereas behaviorist and cognitive learning theories have obvious connections with communications and technology, humanist learning theory (Rogers 1969) may be perceived as having fewer evident links. Still, humanism as a learning theory is an important development that traces its roots back to the child-centered curriculum developed in Italy by Maria Montessori (1964) and the self-actualization theory of the American psychologist Abraham Maslow (1962). Humanism as a learning theory emphasizes the importance of free will, an innate desire for growth and competence, and the need for individuals to fulfill their highest potential. Humanistic psychology informs learning and practice in fields such as social work (Payne 2011) and education (Veugelers 2011). Humanism also has strong ties to the concepts of andragogy, self-directed learning, and transformative learning found in adult learning theory (Merriam and Bierema 2014). In addition, humanists often promote experiential learning which Kolb (1984) defined as "the process whereby knowledge is created through the transformation of experience" (Kolb 1984: 41).

Instructional design principles derived from humanist learning theory include:

- Teachers cannot teach learners directly. They can only facilitate their learning.
- The role of learners is to be active explorers of learning opportunities. Their intellectual curiosity should be nurtured by positive role models.
- Ideally, learners' motivation to learn should be intrinsic rather than extrinsic.
- Creativity should be strongly encouraged when learners are asked to complete real-world tasks and solve ill-structured problems.

2 Contemporary learning theories

Contemporary perspectives of learning include constructivist learning principles (Duffy and Jonassen 1992), constructionism (Papert 1990), social learning theory (Bandura 1977; Zimmerman and Schunk 2013), connectivism (Siemens 2005), and various combinations of these such as social constructivist theory (Woo and Reeves 2007). Whereas behaviorist, cognitive, and humanist learning theories arose fifty years ago or more, these more recent developments have been especially influenced by researchers and theorists working in the field now known as the "learning sciences" (Sawyer 2006) that emerged in the 1980s at leading universities in Europe (e.g., the Munich Center of the Learning Sciences at the Ludwig-Maximilians-Universität München in Germany and the Learning Sciences Research Institute at the University of Nottingham in the United Kingdom), Australasia (e.g., the Centre for Research on Computer Supported Learning and Cognition at the University of Sydney in Australia and the Learning Sciences Lab at the National Institute of Education at Nanyang Technological University in Singapore), the USA (e.g., the Learning Sciences Research Institute at the University of Illinois at Chicago and the Learning in Informal and Formal Environments (LIFE) Science of Learning Center at the University of Washington), and elsewhere.

The learning sciences is an interdisciplinary field of study focused on understanding and enhancing learning as it emerges from the experiences of people in both informal (e.g., home) and formal (e.g., school) settings. With respect to theoretical underpinnings, the learning sciences draw on principles represented in a range of cognitive, constructivist, social-constructivist, and socio-cultural theories of learning (Sawyer 2006). The research studies conducted by learning scientists often employ what is known as design-based research in the USA (Barab and Squire 2004) and educational design research in Europe (McKenney and Reeves 2012).

Constructivism is a theory of how people learn that has both fervent adherents and contentious critics (Fosnot 2013; Tobias and Duffy 2009). Constructivism posits that humans bring to any learning context a host of previously constructed concepts and mental models upon which any new knowledge or skill must be built. In other words, constructivists argue that humans are not "blank slates" upon which learning can be scrawled by teachers or instructional technologies. Constructivism is often presented in contrast to behaviorism. Learning for constructivists is anything but a step-by-step linear process. Instead, it is highly individualized, very complex, and difficult to measure or control. Often, learning for constructivists involves unlearning previous mental models and resolving conflicts among pre-existing beliefs when new experiences happen. Radical constructivists also maintain that "learning to learn" and problem-solving skills are more important outcomes for students than other kinds of outcomes such as specific content knowledge (Von Glasersfeld 1995).

Instructional design principles derived from constructivist learning theory include:

- Teachers cannot force or map their own interpretations of the world onto learners because they (teacher and learners) do not share a set of common experiences and interpretations.
- Learners should be the most active people in a course or other type of formal learning environment because they must actively engage in completing tasks or solving problems, make mistakes, reflect, and then try again for meaningful learning to occur.
- Learning is derived from trying to solve complex, challenging problems rather than passively attending to messages transmitted by a teacher.
- Knowledge and skills should be learned in contexts as much like the real-world situations in which the knowledge and skills will eventually be applied if the learning is not to be inert.
- Transfer of learning from one context to another is very difficult and, therefore, learners must be given ample opportunities to apply their knowledge and skills in multiple contexts and domains.

As a derivative of constructivist learning theory, constructionism (Papert 1990) was influenced greatly by earlier work by Vygotsky (1978), Dewey (1938), Bruner (1977), and most directly, Piaget (1932). Papert (1993), credited as the father of constructionism, warrants that for meaningful and robust learning to take place, learners must construct an external artifact that "represents" their new knowledge. Numerous studies have shown that the construction of artifacts that represent original knowledge interpretations, constructive feedback, reflective discourse, and refinement of the artifacts are effective foundations for the construction of new knowledge (Kafai and Resnick 1996). As reviewed by Jonassen and Reeves (1996), using computer programs as cognitive tools is an especially effective way of creating these external representations of new knowledge. Papert and his colleagues at the Massachusetts Institute of Technology developed Logo, a computer programming language specifically intended to foster constructionist learning (Papert 1980), and later integrated it with the popular Lego building blocks for children (Ocko et al. 1988). Today, constructionist learning projects often involve students creating interactive digital stories (Di Blas et al. 2009), video games (Kafai 2006), and robotics (Barak and Zadok 2009).

Instructional design principles derived from constructionist learning theory include:

- The primary role of a constructionist teacher is to "scaffold" learners as they construct new knowledge. Scaffolding involves providing learners with just enough help that may be needed in a new context (akin to training wheels on a bike) and then gradually removing the support as the learners make progress.
- Assessment of learning should focus on the creation of original artifacts that represent learners' knowledge and skills rather than multiple-choice tests that may just measure recall.
- Constructionist learning projects are best done in small groups that better reflect the way design is done in the real world.

Social learning theory (Bandura 1977; Zimmerman and Schunk 2013) maintains that people learn from observations of others, and thus emphasizes the importance of modeling as a teaching strategy. Social learning theory is based upon the belief that most learning takes place in social contexts, and therefore, people can and should learn in groups. Emphasizing the role of modeling in learning, Bandura (1977) wrote:

> Learning would be exceedingly laborious, not to mention hazardous, if people had to rely solely on the effects of their own actions to inform them what to do. Fortunately, most human behavior is learned observationally through modeling: from observing others one forms an idea of how new behaviors are performed, and on later occasions this coded information serves as a guide for action (Bandura 1977: 22).

Working in Russia in the 1920s and early 1930s, Vygotsky (1978) contributed two key principles that are important to comprehending how social learning theory can be applied to support learning in social contexts. First, there needs to be a "More Knowledgeable Other" (MKO) – a person who knows more or has more advanced skills than the learners. Second, the MKO needs to understand the learners' "Zone of Proximal Development" (ZPD), which is the space between the things that learners can do on their own and the things that they cannot yet do, even with assistance. While working in the ZPD and with the support of a MKO, people learn effectively. If people are trying to learn outside the ZPD, they may become frustrated or bored. The MKO can expand the ZPD by guiding learners through ever more challenging experiences so they can develop more robust expertise.

Instructional design principles derived from social learning theory include:

- Throughout life, people learn much by simply observing other people.
- Teachers should model appropriate behaviors and avoid modeling any inappropriate behaviors.
- Teachers should encourage collaborative learning, since much of learning happens within various social contexts.
- Collaborative learning activities help people to develop important interpersonal and cultural skills.
- To enable learners to work in the "Zone of Proximal Development" (ZPD), the teacher as a "More Knowledgeable Other" (MKO) should design learning experiences that are neither too easy nor too difficult for the particular learners.

The most recent of the learning theories to be considered here, connectivism (Downes 2012; Siemens 2005) maintains that "knowledge is distributed across a network of connections, and therefore that learning consists of the ability to construct and traverse those networks" (Downes 2012: 9). Connectivism incorporates chaos theory, which posits that unambiguous knowledge structures are relatively rare and that much of what passes for knowledge is so complex as to defy prediction and control (Gleick 1987). In light of this degree of chaos, connectivism views learning as a process of grappling with complexity by building networks of evolving knowledge that transcend the personal, the social, and the wider environment. Connectivism was the defining learning theory for the earliest Massive Open Online Courses (MOOCs), a type of online learning environment that has garnered much attention in recent years (Daniel 2012; Siemens 2012). Bell (2010) concluded "connectivism alone is insufficient to inform learning and its support by technology in an internetworked world" (Bell 2010: 98) and recommended that it be viewed as one theoretical stance among others that should be considered in designing a learning environment.

Instructional design principles derived from connectivist learning theory include:

- Knowledge is not a thing that can be isolated and transmitted as though it were some type of simple communication. In other words, teaching by telling rarely works.
- Because learning involves connecting widely different information sources, teachers should consider a diversity of opinions in formal presentations.
- Learning often requires access to information on the Internet and therefore students should be allowed to bring information access and communications appliances such as smart phones into formal and informal learning environments.
- Helping learners develop the capacity to know more through expanding their personal, social, and environmental environments is more important than acquiring what is currently known within any given subject area.

- Learners must be given opportunities to enhance their ability to see connections between fields, ideas, and concepts as well as to improve their decision-making capacities with respect to choosing which knowledge is most useful within a particular context.

3 Contentious perspectives on learning

In 2004, Phillips and Soltis wrote: "Theorists do not always agree about what learning is and how it occurs. Psychologists, anthropologists, linguists, neurophysiologists, philosophers, and others are still trying to understand how the mind works and how people learn" (Phillips and Soltis 2004: 3). Although progress is being made, disagreements still exist and from time to time, they boil over into what passes for vociferous arguments within the scholarly literature.

How one thinks of learning and how one conducts research on learning are obviously influenced by the mental model one has constructed of learning. Those who emphasize the preeminence of memory as a cognitive function appear to hold a mental model of learning as the process of filling and organizing some sort of organic filing cabinet or a digital storage device. For example, Kirschner, Sweller, and Clark (2006) defined learning as "a change in long-term memory" (Kirschner et al. 2006: 75) and maintained that what they perceived as the ineffectiveness of instructional models such as constructivism, discovery learning, problem-based learning, experiential learning, and inquiry-based learning primarily resulted from the failure of these alternative pedagogical methods to account for "human cognitive architecture." They argued that "The aim of all instruction is to alter long-term memory. If nothing has changed in long-term memory, nothing has been learned" (Kirschner et al. 2006: 77). In addition, they claimed that direct instruction based on cognitive learning theory is the most effective method of teaching because of its capacity to manage the "cognitive load" that instruction places on learners (Sweller 1994).

Other researchers, while recognizing the importance of memory functions, appear to hold a mental model of learning that is akin to a complex network in which the solitary human mind is just one node. For example, in response to Kirschner et al. (2006), Jonassen (2009) argued that "direct instruction" is based on a limited view of learning as the processing, storage, and retrieval of information, a perspective that doomed it to failure in addressing other interpretations of learning such as: 1. biochemical activity in the brain, 2. a relatively permanent change in behavior or behavioral dispositions, 3. conceptual change, 4. problem solving, 5. social negotiation, and 6. creativity, among others.

One's mental model of learning also has major implications for how teaching and learning should occur. Spiro and DeSchryver (2009) supported Jonassen's defense of alternative pedagogies by providing substantial evidence that "the success of direct instruction in well-structured domains (WSDs) cannot extend to ill-structured domains (ISDs), in principle, because of the very nature of those domains" (Spiro and DeSchryver 2009: 106). Proponents of direct instruction such as Kirschner et al. (2006) all too often ignore learner intention and motivation, a failure that is compounded when direct instruction is focused on ill-structured domains "characterized by being indeterminate, inexact, noncodifiable, nonalgorithmic, nonroutinizable, imperfectly predictable, nondecompositional into additive elements, and in various ways, disorderly" (Spiro and DeSchryver 2009: 107).

The debate between cognitivists and constructivists is unlikely to be resolved any time soon because the goals of learning emphasized by each camp are so divergent. Other debates about the nature of learning exist such as those who maintain that adult learners are fundamentally different from younger learners and those who maintain that learning is learning regardless of age. Scholars and practitioners who hold different mental models of learning are also likely to disagree about how communications technology should best be used to support learning. For example, a teacher implementing direct instruction would be unlikely to encourage students to use smart phones during a class whereas such devices would be viewed as integral to the design and implementation of a learning environment grounded in connectivist learning theory.

Traditional instruction in the form of an instructor lecturing in front of a classroom of students is still ubiquitous, but new and emerging information and communications technologies are playing important roles in both traditional and innovative learning environments. For example, few higher education courses operate today without instructors and students sharing access to a learning management system (LMS) that supports functions related to scheduling classes, distributing assignments, hosting online discussions, tracking assessments, reporting grades, and other activities relevant to the instructional context. In the world of corporate training where standardization of content and instructional delivery are paramount, learning content management systems (LCMS) are popular because of their capacity to document accurately who was trained to what degree of effectiveness. As described below, online and blended learning technologies are being increasingly applied across all sectors of human activity and the very meaning of learning is shifting as humans rely upon smart mobile technologies to a larger and larger extent (Vander Ark 2011).

4 Learning and emerging communication technologies

Bonk's (2009) book, *The World Is Open*, provided a comprehensive overview of how the World Wide Web and other technologies are changing the goals and processes of learning around the globe. Bonk regularly updates the ideas presented in his book in keynotes and other presentations that he gives around the world, and he provides open access to these ideas on his website at http://worldisopen.com/. Bonk has highlighted developments such as:

- Tablet computing
- Smart phones
- Digital textbooks
- E-books designed by learners
- Socially-designed learning resources
- Simulations and serious games
- Videoconferencing
- Virtual mentoring
- Online education
- Online assessment

One of the primary types of communications technologies being significantly changed is the printed textbook. Since 2012, the U.S. Secretary of Education, Arne Duncan, has been promoting the adoption of e-textbooks and smart tablets across all public schools in the nation, and many U.S. states are allocating millions of dollars to funding pilot projects (Lederman 2012). Amazon allows students to rent e-books for their courses through their Kindle reader software (Copeland 2014) and Apple's iBook Textbooks are now available in 51 countries (Grant 2014).

Video, an educational technology with a fifty plus history of "no significant differences" in learning (Schramm 1962), is making a strong comeback in education, especially through the provision of open educational resources in digital video formats. Attracting the most attention are videos provided by the Khan Academy that allow teachers from primary schools to universities to "flip" their classrooms (Khan 2013). In the flipped classroom, lectures are replaced by videos that students view at home to prepare for hands-on problem-solving activities during class time (Herreid and Schiller 2013).

Learning assessment is another component of formal education undergoing radical change through communications technologies. Student learning, traditionally measured through multiple-choice exams and essays, is now being assessed via media such as podcasts, digital productions, and high fidelity simulations (Clouder et al. 2013). Shute and Ventura (2013) described how video games are being used for "stealth assessments" wherein learners' knowledge and skills are being assessed without learners feeling like they are being tested.

It is clear that some of the most fundamental components of education and training (e.g., textbooks, lectures, and assessments) are undergoing radical transformations as a direct result of the incorporation of new communications technologies in these environments. In the last section of this chapter, the future relationships among communications technology and learning are considered.

5 The not-too-distant future of learning

> 'The Matrix has its roots in primitive arcade games,' said the voice-over, 'in early graphics programs and military experimentation with cranial jacks.' ... 'Cyberspace. A consensual hallucination experienced daily by billions of legitimate operators, in every nation, by children being taught mathematical concepts' ... 'A graphic representation of data abstracted from the banks of every computer in the human system. Unthinkable complexity' (Gibson 1984: 51).

Science fiction writers such as William Gibson, Bruce Sterling, Neal Stephenson, and others have envisioned postmodern dystopias where technology triumphs over the human spirit with the exception of a few "hackers" who struggle to fight back against the tyranny of digital meta-systems wherein all thoughts, words, and actions are captured, stored, and used to fill the manufactured appetites of the masses. Some of their cyberpunk novels have specifically focused on what it will mean to learn in the future. For example, in Stephenson's (1995) *The Diamond Age: Or, A Young Lady's Illustrated Primer*, the primary characters are three young girls who each receive an electronic book called a "Primer" that strives to guide the children toward a more intellectually stimulating and fulfilling life. Interestingly, one major goal of the more "interesting life" promoted by the Primer involves exhibiting a subversive attitude towards the status quo.

Stephenson's (1995) disruptive Primer can be contrasted with the behaviorist teaching strategies portrayed in Aldus Huxley's (1932) book *Brave New World*:

> A small boy asleep on his right side, the right arm stuck out, the right hand hanging limp over the edge of the bed. Through a round grating in the side of a box a voice speaks softly. "The Nile is the longest river in Africa and the second in length of all the rivers of the globe. Although falling short of the length of the Mississippi-Missouri, the Nile is at the head of all rivers as regards the length of its basin, which extends through 35 degrees of latitude ..."
> At breakfast the next morning, "Tommy," some one says, "do you know which is the longest river in Africa?" A shaking of the head. "But don't you remember something that begins: The Nile is the ..."
> "The – Nile – is – the – longest – river – in – Africa – and – the – second – in – length – of – all – the – rivers – of – the – globe ..." The words come rushing out. "Although – falling – short – of ..."
> "Well now, which is the longest river in Africa?"
> The eyes are blank. "I don't know."
> "But the Nile, Tommy."
> "The – Nile – is – the – longest – river – in – Africa – and – second ..."
> "Then which river is the longest, Tommy?"
> Tommy burst into tears. "I don't know," he howls.
> (Huxley 1932: 25)

Questions must be asked about whether communications and technology researchers focused on learning are ultimately leading humanity toward dystopia, utopia, or some form of society between the two extremes. Consider the massive open online courses (MOOCs) and other forms of online learning that have garnered so much attention and funding over the past two years (Baggaley 2013; Reeves and Hedberg 2014). Some argue that these "innovations" cater to the agendas of corporations that use them to amass big data to shape advertising and product development while others worry about the privacy of students who enroll in "free" MOOCs (Head 2013). It is clear that many online learning programs require people to share enormous amounts of information about themselves. How can a morally defensible balance be struck between collecting information for the purposes of affording better learning opportunities to individuals and having that same information repurposed for marketing and control? For example, it has been suggested that the financial viability of MOOCs partially rests in identifying exceptionally talented people and selling this information to corporate headhunters. What will prohibit the same information from being used to identify others as undesirable employees?

Future learning innovations may include genetic engineering or pharmaceutical-aided learning, but these probably remain beyond the temporal scope of this chapter. In any case, what will occur over the next thirty years will be shaped by the research and development underway today. An environment scan on the future of learning suggests that learning opportunities will be increasingly ubiquitous, mobile, and tailored to "just for me, just for now." On the face of it, this seems positive. But when witnessing a family of four at dinner in a restaurant all individually "interacting" with their own smart phones and interactive pads without any evident conversation among them, one has to wonder what is being learned.

One way to change the future of learning through communications technology is to engage in changing it for the better. To do so, communications and technology researchers, as well as those in related fields such as education and psychology, must collaborate in more socially responsible forms of research. Educational design research (McKenney and Reeves 2012) represents a viable option for this kind of initiative. Educational communications and technology researchers have focused primarily on the question "What works?" with respect to teaching and learning with technology with predictably dismal outcomes (Reeves 2011). Too often our research agendas have focused on things (e.g., iPads) rather than problems (e.g., the under-representation of women and minorities in fields such as science and engineering). Educational design research fundamentally changes the focus of research from "what works?" questions to "what is the problem and how can we solve it?" intentions.

Accordingly, it is time for communications and technology researchers interested in learning to focus their studies on problems rather than things. Hattie's (2009) extensive review of meta-analyses of the effectiveness of 135 variables in education clearly demonstrates that it is instructional design processes (e.g., feedback to students and mastery learning) rather than specific technologies (e.g., computer-based instruction or games) that have the most direct impact on learning.

What are some of the near-term problems that should be tackled by communications and educational researchers? One involves the digital divide and its relationship to the growing gap between the haves and have-nots in our societies (Bauerlein 2012; Stiglitz 2013). Another is the challenge of preparing young people to learn in a world where many of them as adults will never have a career or job in the traditional sense (Ali et al. 2013). Another relevant problem concerns the increasing evidence of rampant narcissism among youth today promoted by social media (Twenge and Campbell 2009). These and other societal problems should be the starting point for a bold new research agenda at the intersection of learning and communications technology. It is hoped that this chapter will provide some guidance for initiating this socially responsible line of research.

References

Ali, Saba R., Kevin Fall & Tina Hoffman. 2013. Life without work: Understanding social class changes and unemployment through theoretical integration. *Journal of Career Assessment* 21(1). 111–126.

Anderson, John R. 2000. *Learning and memory*. 4th Ed. New York: John Wiley.

Baggaley, Jon. 2013. MOOC rampant. *Distance Education* 34(3). 368–378.

Bandura, Albert. 1977. *Social learning theory*. Englewood Cliffs, NJ: Prentice Hall.

Barab, Sasha & Kurt Squire. 2004. Design-based research: Putting a stake in the ground. *The Journal of the Learning Sciences* 13(1). 1–14.

Barak, Moshe & Yair Zadok. 2009. Robotics projects and learning concepts in science, technology and problem solving. *International Journal of Technology and Design Education* 19(3). 289–307.

Bauerlein, Mark. 2012. *The digital divide: Arguments for and against Facebook, Google, texting, and the age of social networking*. New York: Penguin Group.

Bell, Frances. 2010. Connectivism: Its place in theory-informed research and innovation in technology-enabled learning. *The International Review of Research in Open and Distance Learning* 12(3). 98–118.

Benjamin, Ludy T. 2007. *A brief history of modern psychology*. Oxford, UK: Blackwell Publishing.

Blumberg, Fran C., Debby E. Almonte, Jared S. Anthony & Naoko Hashimoto. 2013. Serious games: What are they? What do they do? Why should we play them? In Karen Dill (ed.), *Oxford Handbook of Media Psychology*, 334–351. New York: Oxford University Press.

Bonk, Curtis J. 2009. *The world is open: How web technology is revolutionizing education*. San Francisco: Jossey-Bass.

Bransford, John D., Ann L. Brown & Rodney R. Cocking (eds.). 1999. *How people learn: Brain, mind, and experience*. Washington, DC: National Academy Press.

Bruner, Jerome S. 1977. *The process of education*. Cambridge, MA: Harvard University Press.

Burton, John K., David M. Moore & Susan G. Magliaro. 1996. Behaviorism and instructional technology. In David H. Jonassen (ed.), *Handbook of research on educational communications and technology*, 46–73. New York: Macmillan.

Clouder, Lynn, Christine Broughan, Steve Jewell & Graham Steventon (eds.). 2013. *Improving student engagement and development through assessment: Theory and practice in higher education*. New York: Routledge.

Copeland, Joe. 2014. Amazon's latest target: University Book Store. *Crosscut*, January 8. http://crosscut.com/2014/01/08/business/118229/amazons-latest-target-university-book-store/ (Accessed 14 April 2014).

Daniel, John. 2012. Making sense of MOOCs: Musings in a maze of myth, paradox and possibility. *Journal of Interactive Media in Education 3*. http://jime.open.ac.uk/jime/article/view/2012–18 (Accessed 14 April 2014).

Dewey, John. 1938. *Experience and education*. New York: Collier MacMillan.

Di Blas, Nicoletta, Franca Garzotto, Paolo Paolini & Amalia G. Sabiescu. 2009. Digital storytelling as a whole-class learning activity: Lessons from a three-years project. In Ido A. Iurgel, Nelson Zagalo, & Paolo Petta (eds.), *Interactive storytelling*, 14–25. Berlin: Springer.

Downes, Stephen. 2012. *Connectivism and connective knowledge: Essays on meaning and learning networks*. http://www.downes.ca/me/mybooks.htm (Accesses 14 April 2014).

Duffy, Thomas M. & David H. Jonassen. 1992. *Constructivism and the technology of instruction: A conversation*. Hillsdale, NJ: Lawrence Erlbaum Associates.

Englemann, Siegfried. 1980. *Direct instruction*. Englewood Cliffs, NJ: Educational Technology Publications.

Fosnot, Catherine Twomey. 2013. *Constructivism: Theory, perspectives, and practice*. 2nd Ed. New York: Teachers College Press.

Gibson, William. 1984. *Neuromancer*. New York: Ace Books.

Gleick, James. 1987. *Chaos: The making of a new science*. New York: Penguin Books.

Grant, Rebecca. 2014. iBook Textbooks now available in 51 countries worldwide as Apple pushes for interactive learning. *VentureBeat*. http://venturebeat.com/2014/01/21/ibook-textbooks-now-available-in-52-countries-worldwide-as-apple-pushes-for-interactive-learning/ (Accessed 13 April 2014).

Gustafson, Kent L. & Robert Maribe Branch. 2002. *Survey of instructional development models*. 4th Ed. Syracuse, NY: Eric Clearinghouse on Information.

Halff, Henry M. 1988. Curriculum and instruction in automated tutors. In Martha Campbell Polson & J. Jeffrey Richardson (eds.), *The foundations of intelligent tutoring systems*, 79–108. Hillsdale, NJ: Lawrence Erlbaum Associates.

Halford, Graeme S. 1993. *Children's understanding: The development of mental models*. Hillsdale, NJ: Lawrence Erlbaum Associates.

Hattie, Jhon A. C. 2009. *Visible learning: A synthesis of over 800 meta-analyses related to achievement*. New York: Routledge.

Head, Karen. 2013. Lessons learned from a freshman-composition MOOC. *The Chronicle of Higher Education*. http://chronicle.com/blogs/wiredcampus/lessons-learned-from-a-freshman-composition-mooc/46337 (Accessed 14 April 2014).

Herreid, Clyde Freeman & Nancy A. Schiller. 2013. Case studies and the flipped classroom. *Journal of College Science Teaching* 42(5). 62–66.

Huxley, Aldous. 1932. *Brave new world*. London: Chatto & Windus.

Jonassen, David H. & Thomas C. Reeves. 1996. Learning with technology: Using Computers as cognitive tools. In David H. Jonassen (ed.), *Handbook of Research on Educational Communications and Technology*, 693–719. New York: Macmillan.

Kafai, Yasmin. 2006. Playing and making games for learning instructionist and constructionist perspectives for game studies. *Games and culture* 1(1). 36–40.

Kafai, Yasmin & Mitchel Resnick. 1996. *Constructionism in practice*. Hillsdale, NJ: Laurence Erlbaum Associates.

Khan, Salman. 2013. What college could be like. *Communications of the ACM* 56(1). 41–43.

Kirschner, Paul A., John Sweller & Richard E. Clark. 2006. Why minimal guidance during instruction does not work: An analysis of the failure of constructivist, discovery, problem-based, experiential, and inquiry based teaching. *Educational Psychologist* 41(2). 75–86.

Kolb, David A. 1984. *Experiential learning*. Englewood Cliffs, NJ: Prentice Hall.

Kyllonen, Patrick C. & Valerie J. Shute. 1989. A taxonomy of learning skills. In Philip Laaurence Ackerman, Robert H. Sternberg, and Robert Glaser (eds.), *Learning and individual differences: Advances in theory and research*, 117–163. New York: W. H. Freeman.

Landriscina, Franco. 2013. *Simulation and learning: A model-centered approach*. New York: Springer.
Lederman, Josh. 2012. Arne Duncan calls for textbooks to become obsolete in favor of digital. *The Huffington Post*. http://www.huffingtonpost.com/2012/10/02/education-chief-wants-tex_0_n_1933469.html (Accessed 14 April 2014).
Lumsdaine, Arthur A. & Robert Glaser (eds.). 1960. *Teaching machines and programmed learning: A source book*. Washington, DC: National Education Association.
Maslow, Abraham. 1962. *Toward a psychology of being*. New York: Van Nostrand.
McKenney, Susan E. & Thomas C. Reeves. 2012. *Conducting educational design research*. London: Routledge.
Merriam, Sharan B. & Laura L. Bierema. 2014. *Adult learning: Linking theory and practice*. San Francisco: Jossey-Bass.
Montessori, Maria. 1964. *The Montessori method*. New York: Random House.
Ocko, Stephen, Seymour Papert & Mitchel Resnick 1988. Lego, logo, and science. *Technology and Learning* 2(1). 1–8.
Papert, Seymour. 1980. *Mindstorms: Children, computers, and powerful ideas*. Basic Books.
Papert, Seymour. 1990. *Constructionist learning*. Cambridge, MA: MIT Media Laboratory.
Papert, Seymour. 1993. *The children's machine: Rethinking school in the age of the computer*. New York: Basic Books.
Payne, Malcom. 2011. *Humanistic social work: Core principles in practice*. Chicago: Palgrave Macmillan.
Phillips, Dennis Charles & Jonas F. Soltis. 2004. *Perspectives on learning*. New York: Teachers College Press.
Piaget, Jean. 1932. *Moral judgment of the child*. London: Kegan Paul, Trench, Trubner and Co.
Reeves, Thomas C. 2011. Can educational research be both rigorous and relevant? *Educational Designer: Journal of the International Society for Design and Development in Education* 1(4). http://www.educationaldesigner.org/ed/volume1/issue4/ (Accessed 13 April 2013).
Reeves, Thomas C. & John G. Hedberg. 2014. MOOCs: Let's get REAL. *Educational Technology* 54(1). 67–75.
Reiser, Robert A. 2001. A history of instructional design and technology: Part II: A history of instructional design. *Educational Technology Research and Development* 49(2). 57–67.
Rogers, Carl. 1969. *Freedom to learn: A view of what education might become*. Columbus, OH: Charles E. Merrill Publishing Company.
Sawyer, R. Keith. 2006. *The Cambridge handbook of the learning sciences* 2(5). New York: Cambridge University Press.
Schank, Roger C. & Chip Cleary. 1995. *Engines for education*. New York: Routledge.
Schramm, Wilbur. 1962. Learning from instructional television. *Review of Educational Research* 32(2). 156–167.
Schunk, Dale H. 2012. *Learning theories: An educational perspective*. 6th Ed. Boston: Pearson Education.
Schunk, Dale H., Judith R Meece, Paul R. Pintrich. 2013. *Motivation in education: Theory, research, and applications* (4th ed.). Boston: Pearson Education.
Shute, Valerie J. & Matthew Ventura. 2013. *Stealth assessment: Measuring and supporting learning in video games*. Cambridge, MA: The MIT Press.
Siemens, George. 2005. Connectivism: A learning theory for the digital age. *International Journal of Instructional Technology and Distance Learning* 2(1). 3–10.
Siemens, George. 2012. MOOCs are really a platform. *Elearning space*. http://www.elearnspace.org/blog/2012/07/25/moocs-are-really-a-platform/ (Accessed 13 April 2014).
Skinner, B. F. 1935. Two types of conditioned reflex and a pseudo type. *Journal of General Psychology* 12. 66–77.

Skinner, B. F. 1954. The science of learning and the art of teaching. *Harvard Educational Review* 24(86). 99–113.

Spector, Jean Michael, M. David Merrill, Jan Elen & M. J. Bishop (eds.). 2014. *Handbook of research on educational communications and technology*. 4th Ed. New York: Springer.

Spiro, Rand J. & Michael DeSchryver. 2009. Constructivism: When it's the wrong idea and when it's the only idea. In Thomas M. Duffy & Sigmund Tobias (eds.), *Constructivist instruction: Success or failure?* 106–123. New York: Routledge.

Sternberg, Robert J. 1999. *Handbook of creativity*. New York: Cambridge University Press.

Stephenson, Neal. 1995. *The diamond age: Or, a young lady's illustrated primer*. New York: Bantam Spectra.

Stiglitz, Joseph E. 2013. *The price of inequality: How today's divided society endangers our future*. New York: W. W. Norton & Company.

Sweller, John. 1994. Cognitive load theory, learning difficulty, and instructional design. *Learning and Instruction* 4(4). 295–312.

Tobias, Sigmund & Thomas M. Duffy. 2009. *Constructivist theory applied to instruction: Success or failure?* New York: Routledge.

Trenholm, Rich. 2010. Google boss predicts Google implant will put the Web in your head by 2020. *CNET*. http://crave.cnet.co.uk/software/google-boss-predicts-google-implant-will-put-the-web-in-your-head-by-2020-50001515/ (Accessed 13 April 2014).

Twenge, Jean M. & W. Keith Campbell. 2009. *The narcissism epidemic: Living in the age of entitlement*. New York: Free Press.

Vander Ark, Tom. 2011. *Getting smart: How digital learning is changing the world*. San Francisco, CA: Jossey-Bass.

VanLehn, Kurt. 2011. The relative effectiveness of human tutoring, intelligent tutoring systems, and other tutoring systems. *Educational Psychologist* 46(4). 197–221.

Veugelers, Wiel. 2011. *Education and humanism: Linking autonomy and humanity*. Rotterdam, NL: Sense Publishers.

Von Glasersfeld, Ernst. 1995. *Radical constructivism: A way of knowing and learning*. London: Falmer Press.

Vygotsky, Lev S. 1978. *Mind in society: Development of higher psychological processes*. Cambridge, MA: Harvard University Press.

Watson, John B. 1913. Psychology as the behaviorist views it. *Psychological Review* 20. 158–177.

Woo, Younghee & Reeves, Thomas C. 2007. Meaningful interaction in web-based learning: A social constructivist interpretation. *The Internet and Higher Education* 10(1). 15–25.

Zimmerman, Barry J. & Dale H. Schunk. 2013. *Self-regulated learning and academic achievement: Theoretical perspectives*. New York: Routledge.

Gary L. Kreps

26 Communication technology and health: The advent of ehealth applications

Abstract: Powerful new health information technologies (HITs) hold tremendous promise for enhancing the delivery of health care and promoting health. These health information technologies are transforming the modern health care system by supplementing and extending traditional channels for health communication. The use of new health information technologies enable broad dissemination of relevant health information that can be personalized to the unique information needs of individuals. These ehealth communication channels can provide health care consumers and providers with the relevant health information they need exactly when they need the information and where they need the information. This new area of health communication has become known as ehealth communication. With the use of artificial intelligence, interactive health information systems can be designed to answer complex health questions, as well as provide needed support to health care providers and consumers. This chapter describes the advent of health information technologies and their current and future applications to enhance the delivery of health care and the promotion of public health.

Keywords: health information technology (HIT), ehealth, artificial intelligence (AI), electronic health records (EHRs), interactive virtual human agents, digital feedback systems, interactive decision support systems, video games, evaluation research, community-participative research

1 The advent of Ehealth communication

We are in the midst of an information revolution within the modern health care system, fueled by the growth of powerful new health information technologies (HITs) that hold tremendous promise for enhancing the delivery of health care and promoting health (Eysenbach 2000; Kreps and Neuhauser 2010; Neuhauser and Kreps 2003). A burgeoning body of research concerning the development, implementation, and application of new HITs has helped to increase understanding of the potential uses and influences of communication technologies to support health care, health promotion, and health risk prevention/response. This area of study and application has become popularly known as "ehealth" communication (Ahern et al. 2006; Atkinson and Gold 2002; Kreps and Neuhauser 2010). The major implication from this growing body of ehealth research is that HITs have tremendous potential to enhance health care and health promotion, yet there are also significant communication challenges that must be addressed in the design and implementation of these technologies.

2 The scope of Ehealth communication

E-health communication encompasses the development, adoption, and implementation of a broad range of important new health information technology tools and applications. The most pervasive ehealth communication application has probably been the widespread introduction of many online health information websites that have the potential to provide health care consumers with access to relevant information about their specific health concerns and the health concerns of their loved ones (Beckjord et al. 2007; Gallant et al. 2007, Shepperd et al. 1999). These health websites often provide both diagnostic information (helping consumers to interpret troublesome symptoms), as well as therapeutic and prognostic information (helping consumers evaluate different health care strategies for remediating health problems). Many of these health information websites, such as websites sponsored by US federal health agencies including the Department of Health and Human Services (Healthfinder), the National Library of Medicine (Medline Plus), and the Centers for Disease Control and Prevention (Diseases and Conditions) offer the public the latest comprehensive health information. Surveillance research that tracks the publics' use of online information suggests that searching for health information is one of the primary communication activities conducted via the worldwide web today (Fox and Duggan 2012; Fox and Jones 2008; Hesse et al. 2005).

Similarly, there has been rapid growth of use of online social support networks by health care consumers and caregivers. A large body of research has shown that online support groups often provide participants with much needed health information and support (Benderd et al. 2008; Eysenbach et al. 2004; Ginossar 2008; Hesse et al. 2008; Rimer et al. 2005; Wen et al. 2011; Wright and Bell 2003). In fact, online support groups have been found to be preferred over in-person support groups due to the ability of online support group users to connect with relevant others over long distances, the convenience of being able to access social support from home (particularly for home-bound patients), and the high level of anonymity provided by online support groups (Bender et al. 2008; Wen et al. 2011).

There is growing research evidence accumulating about the potential benefits of using interactive health decision support systems and tailored interactive health education programs to promote health and influence health behaviors (Bock et al. 2008; Campbell and Quintiliani 2006; Jibaja-Weiss et al. 2003; Noar et al. 2009; Rimer and Glassman 1998). These adaptive and interactive health information systems are used by both health care providers and health care consumers to help make complex health-related decisions. Tailored interactive health information systems have been found to be particularly effective in promoting and sustaining health behavior changes by providing users with pertinent and highly motivational health information (Hawkins et al. 2011; Jibaja-Weiss et al. 2003; Whitten et al. 2005). Digital feedback systems that enable consumers to monitor their health behaviors and outcomes (often referred to as the quantified self), such as monitoring blood sugar levels, heart rate, exercise, nutrition, and the results of laboratory tests, can help enhance personal awareness and reinforce healthy behaviors (Schulz 2010; Schulz et al. 2010).

Major medical centers, clinics, hospitals, and health maintenance organizations are increasingly employing health care system web portals to communicate with their constituents (both consumers and providers) (Cho et al. 2010; Gallant et al. 2007). These portals often provide users with access to health information, appointment scheduling, prescription medication refills, medical records and test results, as well as the ability to exchange messages with health care providers (Shepperd et al. 1999).

There has been steady growth in the use of tele-health programs by health care delivery systems (sometimes referred to as telemedicine) that combine video, telecommunication, and computer technologies to connect health care consumers and providers, often over great distances. There is abundant evidence showing that these tele-health programs can effectively extend and supplement traditional in-person health care delivery systems, while also promoting care coordination, and helping to reduce health care costs (Chamberlain 1994; Darkins et al. 2008; Dávalos et al. 2009; Harris et al. 2007; Matusitz and Breen 2007; Schover et al. 2012).

In recent years there has been strong encouragement from the US government for health care delivery systems (hospitals, clinics, and other health care practices) to adopt and implement electronic health record (EHR) systems. However, adoption of EHRs has been relatively slow in the US compared to other developed nations, despite growing evidence from the research literature suggesting that electronic health records can reduce medical errors, enhance health outcomes, and improved health care system efficiency (Krist and Woolf 2011; Krist et al. 2012; Mair et al. 2012; Nambison et al. 2013; Wen et al. 2010).

Online and mobile health communication devices are being increasingly used to provide consumers with health information and support, with some notable successes in influencing relevant consumer health behaviors (including diet, smoking, alcohol use, exercise, and medication use) (Bates et al. 2001; Bennett and Glasgow 2009; Castelnuovo and Simpson 2011; Fox and Duggan 2012; Graham and Abrams 2005; Gustafson et al. 2011; Kreps 2012; Neuhauser and Kreps 2010; Neuhauser et al. 2013; Person et al. 2011). Several other promising new ehealth applications include the use of virtual reality simulations for health education, video games for health education and promotion, as well as the use of virtual interactive human agents (typically operationalized with avatars) to provide health information, collect health information, and to provide therapeutic support (Bickmore et al. 2012; Bickmore et al. 2010; Eng 2005; Huang, Morency and Gratch 2011; Lieberman 2006; Lisetti et al. 2012; Morie et al. 2012; Pfeifer and Bickmore 2011; Rizzo et al. 2011; Vardoulakis et al. 2012). Social networking platforms, such as Facebook and Twitter, have also been used with some success as ehealth communication tools for health promotion efforts (Hawn 2009; Fox and Jones 2008; Marketing Leadership Council 2008).

3 Challenges to effective use of Ehealth communication applications

Ehealth communication applications show great potential for helping to increase consumer and provider access to relevant health information, enhance quality of health care, reduce health care errors, increase collaboration, and encourage the adoption of healthy behaviors (Kreps and Neuhauser 2010). Yet, with the growth of new and exciting HIT opportunities comes the daunting responsibility to design ehealth tools that communicate effectively with a diverse array of health care consumers, providers, and policy makers. These tools must be designed to effectively communicate the right information needed by different audiences at the right time, in the right place, and in the best ways to guide health care and health promotion. Ehealth tools need to be interactive, interoperable, easy to use, engaging, adaptable, and accessible for diverse audiences (Kreps 2013).

Concerns have been raised about the relative quality of health information acquired by consumers via the Internet and their abilities to understand complex health information, especially for those consumers with limited education and health literacy levels (Chang et al. 2004; Ginossar and Nelson 2010; *Hesse et al.* 2008; Kreps et al. 2007; Piette et al. 2012). The rapid growth of the Internet has produced numerous health-related websites that may not provide the latest and most accurate health information to the public (Atkinson and Gold 2002; Cassell et al. 1998). Some of these websites are offered by commercial concerns that are more interested in enticing consumers into buying their products than in providing useful health information (Eysenbach and Kohnler 2002). Consumers must carefully evaluate the sources of online health information, as well as the accuracy and timeliness of the information, to make sure the information provided is valid and reliable. It is important for consumers to become sophisticated users of online information, developing computer literacy, to know what online health information is most accurate and appropriate for guiding their health decisions (Bodie and Dutta 2008; Eysenbach and Kohnler 2002). Ferguson and Frydman (2004) refer to these sophisticated users of online health information as "e-patients."

Additional critical issues raised concerning ehealth include making sure that all of the people who need health information have access to health information technologies, know how to use the available technologies, and can understand the information provided via these ehealth tools (Kreps and Neuhauser 2010; Neuhauser and Kreps 2010; 2003; Chamock and Gann 1999; Whitten et al. 2005). There also must be strategies developed to preserve information privacy for consumers and protect consumers from predatory use of ehealth tools to sell unwarranted and untested health products and services (Kreps and Neuhauser 2010).

While many health care consumers and providers now understand that communication is a central social process in the provision of health care delivery and the promotion of public health, they do not always recognize that effective communication is a complex and fragile human process that demands strategic design, careful monitoring, and responsive adaptation. This is particularly true in the development and implementation of new health information technologies. Often, it appears that ehealth designers are more enamored with the technical elegance and innovation of new information technologies than with the utility of these tools for health care consumers and providers (Kreps and Neuhauser 2013). Are the technologies easy for these audiences to understand and use? Do new ehealth tools fit comfortably within the policies, practices, and technical infrastructure that are built into existing health and social systems? Are new ehealth tools affordable and accessible for all intended audiences? Are the messages delivered on ehealth programs designed so that diverse populations of users can understand and apply the health information provided? Are the information systems adaptive, interactive, and self-correcting? Do ehealth programs provide interesting, relevant, and engaging information for users? Too often, the answers to these questions are disappointing. Ehealth tools have to be designed to complement other health communication channels, be easy for health care professionals to use, and communicate effectively with diverse audiences of consumers (Kreps and Neuhauser 2013).

4 Ehealth communication research directions

There is a tremendous demand to conduct in-depth and rigorous evaluation research to assess the influences of ehealth programs on the diverse audiences that use them and on relevant health outcomes (Atkinson 2007; Kreps 2013; Rigby et al. 2009). Good evaluation research is needed to guide the development and refinement of ehealth communication tools and programs (Kreps 2013). In the rush to develop and implement new health information technologies the communication needs and expectations of the users of these systems have not always received adequate attention (Gustafson et al. 1999; Laakso and Tandy 2011). To enhance the usability of health information technologies there have been concerted efforts to promote the use of community-participative ehealth research to promote user-centered design of ehealth products and programs (Neuhauser 2001; Ginossar and Nelson 2010; Neuhaser and Kreps 2011; Neuhauser et al. 2013). Such community-participative research has been successfully used to design, implement, and sustain effective ehealth communication interventions concerning a number of health issues (Ginossar and Nelson 2010; Hawkins et al. 2011; Neuhauser et al. 2013; Van Bogaert et al. 2012). Evidence-based design of ehealth communication programs will help guide effective future applications of health information technologies for health care and health promotion (Rigby et al. 2009).

Acknowledgments

In the preparation of this essay, we are grateful for the assistance of Colin Agur, Caroline Chen, and Nancy R. John.

References

Ahern, David K., Jennifer M. Kreslake & Judith M. Phalen. 2006. What Is eHealth (6): Perspectives on the evolution of ehealth research. *Journal of Medical Internet Research* 8(1). e4. http://www.jmir.org/2006/1/e4/ (Accessed 14 April 2014).

Atkinson, Nancy L. 2007. Developing a questionnaire to measure attributes of ehealth innovations. *American Journal of Health Behavior* 31(6). 612–621.

Atkinson, Nancy L. & Robert S. Gold. 2002. The promise and challenge of eHealth interventions. *American Journal of Health Behavior* 26(6). 494–503.

Bates David W., Michael Cohen, Lucian L. Leape, J. Marc Overhage, Michael M. Shabot & Thomas Sheridan. 2001. Reducing the frequency of errors in medicine using information technology. *Journal of the American Medical Informatics Association* 8(4). 299–308.

Beckjord, Ellen Burke, Lila J. Finney Rutten, Linda Squiers, Neeraj K. Arora, Lindsey Volckmann, Richard P. Moser & Bradford W. Hesse. 2007. Use of the Internet to communicate with health care providers in the United States: Estimates from the 2003 and 2005 Health Information National Trends Surveys (HINTS). *Journal of Medical Internet Research* 9(3). e20, http://www.jmir.org/2007/3/e20/ (Accessed 14 April 2014).

Bender, Jacqueline L., Laura O'Grady & Alejandro R. Jadad. 2008. Supporting cancer patients through the continuum of care: A view from the age of social networks and computer-mediated communication. *Current Oncology* 15 (Suppl 2). 42–47.

Bennett, Gary G. & Russel E. Glasgow. 2009. The delivery of public health interventions via the Internet: Actualizing their potential. *Annual review of public health* 30. 273–292.

Bickmore, Timothy, Laila Bukhari, Laura Vardoulakis, Michael Paasche-Orlow & Christopher Shanahan. 2012. Hospital buddy: A persistent emotional support companion agent for hospital patients. In Yukiko Nakano; Michael Neff; Ana Paiva; Marilyn Walker (eds.), *Intelligent Virtual Agents* LNCS 7502. 492–495. Berlin/Heidelberg: Springer.

Bickmore, Timothy, Laura Pfeifer, Donna Byron, Shaula Forsythe, Lori E Henault, Brian W Jack, Rebecca Silliman, & Michael K Paasche-Orlow. 2010. Usability of Conversational Agents by

Patients with Inadequate Health Literacy: Evidence from Two Clinical Trials. *Journal of Health Communication* 15. 197–210.

Bock, Beth C., Amanda L. Graham, Jessica A. Whiteley & Jacqueline L. Stoddard. 2008. A review of web-assisted tobacco interventions (WATIs). *Journal of Medical Internet Research* 10(5). e39. http://www.jmir.org/2008/5/e39/ (Accessed 14 April 2104).

Bodie, Graham D. & Mohan Jyoti Dutta. 2008. Understanding Health Literacy for Strategic Health Marketing: eHealth Literacy, Health Disparities, and the Digital Divide. *Health Marketing Quarterly* 25(1/2). 175–203.

Campbell, Marci Kramish & Lisa M. Quintiliani. 2006. Tailored interventions in public health: Where does tailoring fit in interventions to reduce health disparities? *American Behavioral Scientist* 49(6). 775–793.

Cassell, Michael, Christine Jackson & Brian Cheuvront. 1998. Health communication on the internet: An effective channel for health behavior change? *Journal of Health Communication* 3. 71–79.

Castelnuovo, Gianluca & Susan Simpson. 2011. Ebesity – e-health for obesity – new technologies for the treatment of obesity in clinical psychology and medicine. *Clinical Practice and Epidemiology in Mental Health* 7. 5–7.

Chamberlain, Michael. 1994. New technologies in health communication. *American Behavioral Scientist* 38(2). 271–285.

Chang, Betty L., Suzan Bakken, S. Scott Brown, Thomas K. Houston, Gary L. Kreps, Rita Kukafka, Charles Safran & P. Zoe Stavri. 2004. Bridging the digital divide: Reaching vulnerable populations. *Journal of the American Medical Informatics Association* 11. 448–457.

Cho, Alex H., Nedal H. Arar, David H. Edelman, Patricia Hartwell, Eugene Z, Oddone & William S. Yancy Jr. 2010. Do diabetic veterans use the Internet? Self-reported usage, skills, and interest in using My HealtheVet Web portal. *Telemedicine and e-Health* 16(5). 595–602.

Darkins, Adam, Patricia Ryan, Rita Kobb, Linda Foster, Ellen Edmonson, Bonnie Wakefield & Anne E. Lancaster, A. E. 2008. Care Coordination/Home Telehealth: The Systematic Implementation of Health Informatics, Home Telehealth, and Disease Management to Support the Care of Veteran Patients with Chronic Conditions. *Telemedicine and e-Health* 14(10). 1118–1126.

Dávalos, Maria E., Michael T. French, Anne E. Burdick & Scott C. Simmons. 2009. Economic evaluation of telemedicine: review of the literature and research guidelines for benefit – cost analysis. *Telemedicine and e-Health* 15(10). 933–948.

Eng, Thomas R. 2005. Emerging technologies for cancer prevention and other population health challenges. *Journal of Medical Internet Research*. 7(3). e30. http://www.jmir.org/2005/3/e30/ (Accessed 14 Aprl 2104).

Eysenbach, Gunther. 2000. Recent advances: Consumer health informatics. *British Medical Journal* 320. 1713–1716.

Eysenbach, Gunther & Christian Kohnler. 2002. How do consumers search for and appraise health information on the world wide web? Qualitative study using focus groups, usability tests, and in-depth interviews. *British Medical Journal* 324. 573–577.

Eysenbach, Gunther, John Powell, Marina Englesakis, Carlos Rizo & Anita Stern. 2004. Health related virtual communities and electronic support groups: Systematic review of the effects of online peer to peer interactions. *British Medical Journal* 328. 1166–1170.

Ferguson, Tom & Gilles Frydman. 2004. The first generation of e-patients. *British Medical Journal*, 328. 1148–1149.

Fox, Susannah & Meave Duggan. 2012. Mobile health 2012. *Pew Internet and American Life Project*. http://www.pewinternet.org/2012/11/08/mobile-health-2012/ (Accessed 14 April 2014).

Fox, Susannah & Sydney Jones. 2008. The social life of health information. *Pew Internet & American Life Project.* http://www.pewinternet.org/Reports/2009/8-The-Social-Life-of-Health-Information.aspx (Accessed 14 April 2104).

Gallant, Linda M., Cynthia Irizarry & Gary L. Kreps. 2007. User-centric hospital websites: A case for trust and personalization. *e-Service Journal* 5(2). 5–26.

Ginossar, Tamar. 2008. Online participation: A content analysis of differences in utilization of two online cancer communities by men and women, patients and family members. *Health Communication* 23. 1–12.

Ginossar, Tamar & Sara Nelson. 2010. Reducing the health and digital divides: A model for using community-based participatory research approach to e-health interventions in low-income Hispanic communities. *Journal of Computer- Mediated Communication* 15(4). 530–551.

Graham, Amanda L. & David B. Abrams. 2005. Reducing the cancer burden of lifestyle factors: Opportunities and challenges of the Internet. *Journal of Medical Internet Research* 7(3). 26. http://www.jmir.org/2005/3/e26/ (Accessed 14 April 2014).

Gustafson, David H., Michael G. Boyle, Brett R. Shaw, Andrew Isham, Fiona McTavish, Stephanie Richards & Kim Johnson. 2011. An e-health solution for people with alcohol problems. *Alcohol Research and Health* 33(4). 327.

Gustafson, David H., Thomas N. Robinson, David Ansley, Linda Adler & Patricia Flatley Brennan. 1999. Consumers and evaluation of interactive health communication applications. *American Journal of Preventive Medicine* 16(1). 23–29.

Hawkins, Robert P.m Suzanne Pingree, Timothy B. Baker, Linda J. Roberts, Brett R. Shaw, Helene McDowell, David Gustafson, Ronald C. Serlin, Lisa Dillenburg, Christopher M. Swoboda, Jeong-Yeob Han, James A. Stewart, Cindy L. Carmack, Andrew Salner, Tanya R Schlam, & Fiona McTavish. 2011. Integrating eHealth with human services for breast cancer patients. *Translational behavioral medicine* 1(1). 146–154.

Hawn, Carleen. 2009. Take two aspirin and tweet me in the morning: how Twitter, Facebook, and other social media are reshaping health care. *Health Affairs* 28(2). 361–368.

Harris, Linda M., Gary L. Kreps & Connie Dresser. 2007. Health communication technology and quality of cancer care. In Dan O'Hair, Gary L. Kreps, & Lisa Sparks (eds.), *Handbook of Communication and Cancer Care*, 59–71. Cresskill, NJ: Hampton Press

Hesse, Bradford W., Neeraj K. Arora, Ellen Burke Beckjord & Lila J. Finney Rutten. 2008. *Information support for cancer survivors. Cancer* 112(11). 2529–2540.

Hesse, Bradford W., David E. Nelson, Gary L. Kreps, Robert T. Croyle, Neeraj K. Arora & Barbara K. Rimer. 2005. Public use of health and medical information: Results of the Health Information National Trends Survey. *Archives of Internal Medicine* 165. 1–7.

Huang, Lixing, Louis-Philippe Morency & Jonathan Gratch. 2011. Virtual Rapport 2.0. In Yukiko Nakano; Michael Neff; Ana Paiva; Marilyn Walker (eds.), *Intelligent Virtual Agents*, 68–79. Berlin/Heidelberg: Springer.

Jibaja-Weiss, Maria. L., Robert J. Volk, Paul Kingery, Quentin W. Smith & J. David Holcomb. 2003. Tailored messages for breast and cervical cancer screening of low-income and minority women using medical records data. *Patient Education and Counseling* 50. 123–132.

Kreps, Gary L. 2013. Evaluating health communication interventions. In Do Kyun Kim, Arvind Singhal, and Gary L. Kreps (eds.), *Health communication: Strategies for developing global health programs*, 352–367. New York: Peter Lang Publishers.

Kreps, Gary L. 2012. Communication and palliative care: E-health interventions and pain management. In Rhonda Moore (ed.), *Handbook of pain and palliative care: Biobehavioral approaches for the life course,* 43–51. New York: Springer.

Kreps, Gary L., David Gustafson, Peter Salovey, Rosemarie Slevin Perocchia, Wayne Wilbright, Mary Anne Bright & Cathy Muha, C. 2007. The NCI Digital Divide Pilot Projects: Implications for cancer education. *Journal of Cancer Education* 22(Supplement 1). S56–S60.

Kreps, Gary L. & Linda Neuhauser. 2013. Artificial intelligence and immediacy: Designing health communication to personally engage consumers and providers. *Patient Education and Counseling* 92. 205–210.

Kreps, Gary L. & Linda Neuhauser. 2010. New directions in ehealth communication: Opportunities and challenges. *Patient Education and Counseling* 78. 329–336.

Krist, Alex & Steven Woolf. 2011. A vision for patient-centered health information systems. *Journal of the American Medical Association* 305(3). 300–301.

Krist, Alex, Steven Woolf, Stephen F. Rothemich, Robert E. Johnson, J. Eric Peele, Tine D. Cunningham, Daniel R. Longo, Ghalib A. Bello & Gary R. Matzke. 2012. Interactive preventive health record to enhance delivery of recommended care: A randomized trial. *Annals of Family Medicine* 10(4). 312–319.

Laakso, E. Liisa & Jessica J. Tandy. 2011. Use of technology as an adjunct to improve health outcomes for survivors of cancer. *Physical Therapy Reviews* 16(1). 39–45.

Lieberman, Debra. 2006. Dance games and other exergames: What the research says. Archived at the University of California at Santa Barbara, Department of Communication. http://www.comm.ucsb.edu/sites/www.comm.ucsb.edu/files/sitefiles/documents/Lieberman%20(2006)%20Dance%20Games%20and%20Other%20Exergames,%20lit%20review.pdf (Accessed 25 May 2014)

Lisetti, Christine, Ugan Yasavur, Claudia De Leon, Reza Amini, Napthali Rishe & Ubo Visser. 2012. Building an on-demand avatar-based health intervention for behavior change. In *Proceedings of the 25th International Florida Artificial Intelligence Research Society Conference*. Cambridge, MA: American Association of Artificial Intelligence Press. https://www.aaai.org/ocs/index.php/FLAIRS/FLAIRS12/paper/download/4481/4789 (Accessed 14 April 2014).

Mair, Frances S., Carl May, Catherine O'Donnell, Tracy Finch, Frank Sullivan & Elisabeth Murray, E. 2012. Factors that promote or inhibit the implementation of e-health systems: An explanatory systematic review. *Bulletin of the World Health Organization* 90(5). 357–364.

Marketing Leadership Council. 2008. Leveraging social networking sites in marketing communications. Archived at the Corporate Executive Board as an Issue Brief. http://115.112.165.74:81/AK.Dey/Collboration%20Technology/Collaboration%20Articles/Leveraging-Social-Media-Networking-Sites-in-Marketing-Communications.pdf (Accessed 25 May 2014).

Matusitz, Jonathan & Gerald-Mark Breen. 2007. Telemedicine: Its effects on health communication. *Health Communication* 21. 73–83.

Morie, Jacqueline Ford, E. Haynes, Eric Chance & D. Purohit. 2012. Virtual worlds and avatars as the frontier of telehealth care. *Studies in health technology and informatics* 181. 27–32.

Nambison, Priya, Gary L. Kreps & Stan Polit. 2013. Understanding EMR adoption in the United States: Communication and socio-cultural perspectives. *Interactive Journal of Medical Research* 2(1). e5. http://www.i-jmr.org/2013/1/e5/ (Accessed 14 April 2014).

Neuhauser, Linda. 2001. Participatory design for better interactive health communication: An interactive statewide model in the US. *Electronic Journal Of Communication La Revue Electronique De Communication* 11(3/4). http://www.cios.org/EJCPUBLIC/011/3/01134.HTML (Accessed 14 April 2014).

Neuhauser, Linda & Gary L. Kreps. 2011. Participatory design and artificial intelligence: Strategies to improve health communication for diverse audiences. In Nancy Green, Sara Rubinelli, & Donia Scott (eds.), *Artificial Intelligence and Health Communication*, 49–52. Cambridge, MA: American Association of Artificial Intelligence Press. http://www.aaai.org/ocs/index.php/SSS/SSS11/paper/viewFile/2475/2857 (Accessed 14 April 2014).

Neuhauser, Linda & Gary L. Kreps. 2010. Ehealth communication and behavior change: Promise and performance. *Social Semiotics* 20(1). 7–24.

Neuhauser, Linda & Gary L. Kreps. 2003. Rethinking communication in the E-health era. *Journal of Health Psychology* 8. 7–23.

Neuhauser, Linda, Gary L. Krep, Kathleen Morrison, Marcos Athanasoulis, Nikolai Kirienko & Deryk Van Brunt. 2013. Using design science and artificial intelligence to improve health communication: ChronologyMD case example. *Patient Education and Counseling* 92(2). 211–217.

Noar, Seth M., Nancy Grant Harrington & Rosalie Shemanski Aldrich. 2009. The role of message tailoring in the development of persuasive health communication messages. In Christina S. Beck (ed.), *Communication Yearbook* 33, 73–133. New York: Routledge.

Person, Anna K., Michela L. M. Blain, Helen Jlang, Petra W. Rasmussen & Jason E. Stout. 2011. Text messaging for enhancement of testing and treatment for tuberculosis, human immunodeficiency virus, and syphilis: a survey of attitudes toward cellular phones and healthcare. *Telemedicine and e-Health* 17(3). 189–195.

Pfeifer, Laura M. & Timothy Bickmore. 2011. Longitudinal Remote Follow-Up by Intelligent Conversational Agents for Post-Hospitalization Care. In *2011 AAAI Spring Symposium Series*. Cambridge, MA: American Association of Artificial Intelligence Press.

Piette, John D., K. C. Lun, Lincoln A, Moura Jr, Hamish S. F. Fraser, Patricia N. Mechael, John Powell, & Shariq R. Khoja. 2012. Impacts of e-health on the outcomes of care in low-and middle-income countries: where do we go from here? *Bulletin of the World Health Organization* 90(5). 365–372.

Rigby, Michael, Jan Talmon, Elske Ammenwerth, Jytte Brender, Nicolette De Keizer & Porkko Nykanen. 2009. Planning to test success-employing the Guidelines for Good Evaluation Practice in Health Informatics (GEP-HI). In Mário Macedo (ed.), *Proceedings of the IADIS International Conference E-Hea*lth 2009, 209–213. Algarve: Portugal.

Rimer, Barbara K. & B. Glassman. 1998. Tailoring communication for primary care settings. *Methods of Information in Medicine* 37. 171–177.

Rimer, Barbara K., Elisabeth J. Lyons, Kurt M. Ribisl, J. Michael Bowling, Carol E. Golin, Michael J. Forlenza & Andrea Meier. 2005. How new subscribers use cancer-related online mailing lists. *Journal of Medical Internet Research* 7(3). e32. http://www.jmir.org/2005/3/e32/ (Accessed 14 April 2013).

Rizzo, Albert A., Belinda Lange, John G. Buckwalter, Eric Forbell, Julia Kim, Kenji Sagae, Josh Williams, Barbara O. Rothbaum, JoAnn Difede, Greg Reger, Thomas Parsons, & Patrick Kenny. 2011. An intelligent virtual human system for providing healthcare information and support. In James D. Westwood, Susan W. Westwood, Li Fellander-Tsai, Randy S. Haluck, Helene M. Hoffman, Richard A. Robb, Steven Senger, Kirby G. Vosburgh (eds.), *Medicine meets virtual reality* 18, 503–509. Tacoma WAL IOS Press.

Schover, Leslie R., Andrea L. Canada, Ying Yuan, Dawen Sui, Leah Neese, Rosell Jenkins & Michelle M. Rhodes. 2012. A randomized trial of internet-based versus traditional sexual counseling for couples after localized prostate cancer treatment. *Cancer* 118(2). 500–509.

Schulz, Peter J. 2010. Potentialities and Limits of Internet Health Communication in Chronic Care: Results from a Qualitative Study. *Social Semiotics* 20(1) 57–71

Schulz Peter J., Sara Rubinelli, & Uwe Hartung. 2010. Coping with Chronic Lower Back Pain: Designing and Testing the Online Tool ONESELF. *Journal of Computer-Mediated Communication* 15(4). 625–645

Shepperd, Sasha, Deborah Chamock & Bob Gann. 1999. Helping patients access high quality health information. *British Medical Journal* 319. 764–766.

Van Bogaert, Donna, Robert Hawkins, Suzanne Pingree & David Jarrard. 2012. The Development of an eHealth Tool Suite for Prostate Cancer Patients and Their Partners. *The Journal of Supportive Oncology* 10(2). 202–208. http://www.oncologypractice.com/fileadmin/content_images/jso/PDF/v10_i05_eHealth_Tool_for_Prostate_Cancer.pdf (Accessed 25 May 2014).

Vardoulakis, Laura, Lazlo Ring, Barbara Barry, Candace Sidner & Timothy Bickmore. 2012. Designing relational agents as long term social companions for older adults. In Y. Nakano,

M. Neff, A. Paiva, & M. Walker (eds.), *Intelligent Virtual Agents* LNCS 7502, 289–302. Springer Berlin/Heidelberg.

Wen, Kuang-Yi, Gary L. Kreps, Fang Zhu & Suzanne Miller. 2010. Consumer's perceptions about and use of the Internet for personal health records and health information exchange: Analysis of the 2007 Health Information National Trends Survey. *Journal of Medical Internet Research* 12(4). 73. http://www.jmir.org/010/4/e73/ (14 April 2014).

Wen, Kuang-Yi, Fiona McTavish, Meg Wise, David Gustafson & Gary L. Kreps. 2011. From diagnosis to death: A narrative analysis of coping with breast cancer as seen through online discussion group messages. *Journal of Computer-Mediated Communication* 16(2). 331–361.

Whitten, Pamela, Gary L. Kreps & Matthew S. Eastin. 2005. Creating a framework for online cancer services research to facilitate timely and interdisciplinary applications. *Journal of Medical Internet Research* 7(3). e34. http://www.jmir.org/2005/3/e34/ (Accessed 14 April 2014).

Wright, Kevin B. & Sally B. Bell. 2003. Health-related support groups on the Internet: Linking empirical findings to social support and computer-mediated communication theory. *Journal of Health Psychology* 8(1). 39–54.

Alessandro Inversini, Zheng Xiang, and Daniel R. Fesenmaier

27 New media in travel and tourism communication: Toward a new paradigm

Abstract: This chapter discusses the advent and the impact of Information and Communication Technologies (ICT) on the travel and tourism industry. It first reviews and highlights the most important milestones in the last two decades that changed tourism communication, focusing on three main areas of development: the persuasive nature of tourism websites, social media conversations, and mobile computing. It is then argued that these technological developments have led to the transformation of tourism that is best characterized by the development of the Internet of Things (IoT) and the notion 'travelling in the network.' The chapter concludes that this ambient networking of the various tourism systems will further transform the tourism experience and therefore lead to a new paradigm for tourism communication within an increasingly ICT-support environment.

Keywords: ICT, eTourism, communication, web development, social media, mobile computing, tourism experience, Internet of Things

The development of Information and Communication Technologies (ICTs) in the last two decades has dramatically transformed the tourism industry (Inversini and Buhalis 2009; Werthner and Klein 1999). Poon (1993) predicted that a whole new system of ICTs would be rapidly diffused throughout the tourism industry and no player could escape its impact. Indeed, ICTs have had a profound impact on the operational and strategic practices of tourism organisations, and have altered the competitive landscape of enterprises and regions around the world (Fesenmaier and Xiang 2013). At the tactical level, ICTs enabled eCommerce and helped tourism organisations maximise efficiency and effectiveness whereas at the strategic level, ICTs have helped to restructure the travel and tourism value chain as well as the relationships between tourism organisations and their stakeholders (Buhalis 2003).

Modern travellers are exigent and constantly in need of information during the entire consumption process of travel (Gretzel et al. 2006). The complex array of information needed by travellers before, during, and after the tourism experience can be delivered on different devices, with different strategies, and by different stakeholders thanks to today's technologies (Inversini et al. 2009). Wise tourism managers guard different channels and use various strategies to develop the relationships with customers and to provide them with the most relevant information at the right place and at the right time. This suggests that effective communication holds the key to assisting travellers in the multifaceted pre-trip planning process, helping them making *en route* and on-site decisions, and supporting their engagement with their social circle after the trip (Jeng and Fesenmaier 2002; Gretzel et al. 2006; Inversini and Cantoni 2009).

However, the intrinsic nature of the tourism product characterised by experiential services and goods with high level of complexity poses many challenges for the industry (Gretzel and Fesenmaier 2002; Pine and Gilmore 1999). Travel and tourism is an unique industry wherein the purchased product is created at the moment when the actual experience takes place (Tresidder and Hirst 2012). The intangibility of the tourism experience gives rise to many problems for effective communications between the supply and demand. To address these challenges, many technological solutions have been proposed and tested in order to facilitate and enhance online tourism communications along with technological developments in other domains, resulting in a rich stream of literature (Fesenmaier et al. 2006). In the last two decades we have witnessed a series of significant development in online tourism communications characterised by richer content, more personalized services, higher level of interactivity, and, most importantly, highly diverse and extremely dynamic modalities of dialogues and engagement.

This chapter synthesises the literature that reflects the evolution of information and communication technologies in travel and tourism over the last two decades. Specifically, this discussion focuses attention on three important areas of ICT development including website design, social media, and mobile technology as they support a wide range of communication needs throughout the travel process. To provide an outlook for the future development, this chapter also proposes an emergent paradigm for online tourism communication.

1 New media in tourism communication: Two decades of change

Information is the lifeblood of the tourism industry, and the uses of ICTs have always played a critical role in distribution and management (O'Connor and Frew 2004; Sheldon 1997). From the development of Computer Reservation Systems (CRSs) in the 1970s to the Global Distribution Systems (GDSs) in the 1980s until the advent of the Internet in early 1990s, the tourism industry has always been confronted with the rise of new developments that posed as both opportunities and challenges (Buhalis and Law 2008; Ip et al. 2011). However, it was during this first decade of the Internet (roughly from 1991–2002) that ICTs assumed a critical importance for communication purposes. As illustrated in Figure 1, the tourism industry became one of the leaders in the use of the Internet as this new tool was enabling them to communicate easily and effectively with their existing and potential customers. Every American tourism organization had developed a website by the early 2000's, and many had gone through the evolution from a simple

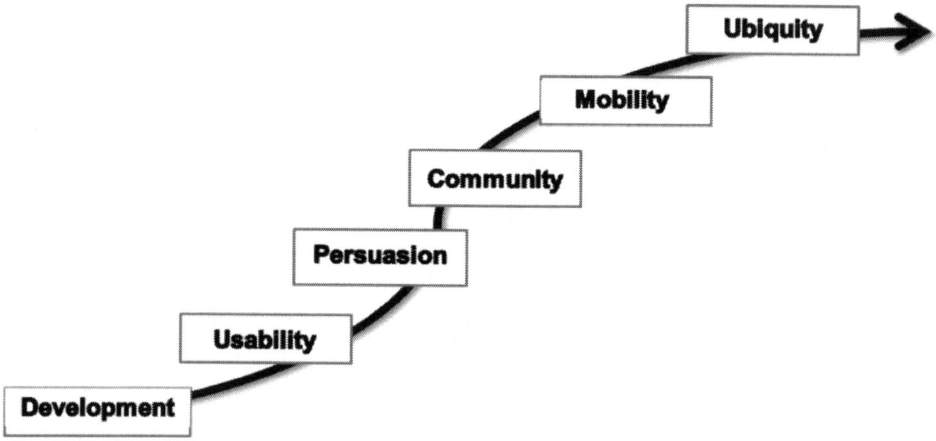

Fig. 1: Six generations of Internet development.

'electronic brochure' to highly interactive systems that supported reservations, search and even virtual tours; importantly, the website had become the primary (and in many circumstances, the only) source of contact with potential visitors (Zach et al. 2010). In retrospect, this transformation is easily understood as the use of computers already existed within the domain (i.e. the various GDSs linking travel agencies to the airlines). Also during this time, many innovative destination marketing organizations (DMOs) began to realize their new role as partners within the tourism system wherein they became 'information brokers' as they sought to develop and coordinate a range of new systems that would be used by their stakeholders (Fesenmaier and Xiang 2013; Gretzel and Fesenmaier 2002; Wang and Xiang 2012) to share relevant information.

Interestingly, the second decade (2001–present) was a time of even greater change for tourism marketing and communication where the focus of technology emerged from website development (i.e., growth in the number of websites) and usability to one of persuasion and customer empowerment, and more recently to ubiquity through mobile systems. That is, while many of the core industry changes were realized in the first decade, the rise of Web 2.0 created a new round of adaptation, which required another new and even more transformational framework for tourism marketing and communication. The first part of this second decade of the Internet saw the expansion of high-speed connections where the richness of the new interactive medium could be exploited (e.g. pictures and videos), leading eventually to the development of brochures which largely realized the interactive and personalized vision of the early developers. The most important feature of this decade is the development of new 'social' systems. TripAdvisor, for example, was launched in 2000 and fifteen years later (in 2014) it handled over 50 million inquiries monthly; Wikipedia was launched in 2001 and by 2014, it hosted over 250 million articles in 285 different languages; and Facebook, launched in 2004 grew such that it serviced over 1 billion users annually by 2013. Importantly, during about the same time mobile phone systems began to penetrate the marketplace to enable travellers to communicate more easily and from essentially anywhere on the planet.

Finally in the later part of the decade, the introduction of smartphones, mobile computing systems that incorporate a variety of technologies including communications, global positioning systems (GPS), photography and the Internet, enriched the social environment further such that it empowers users to substantially control their travel experience anytime, anywhere. Within the tourism literature, Buhalis (2003) provided the leading book which discussed in practical terms how the tourism industry could successfully navigate through this new world. More recently, Sigala, Christou, and Gretzel (2012) offer a different vision, which is situated within this new rich socially created environment whereby the "army of travellers" is now the central player within the value chain; further, Wang, Park, and Fesenmaier (2012) conclude that the increased mobility supported through smartphones substantially changes the nature of the travel experience. The Internet, together with the new interactive way of communication provided by social media and with the location based browsing, fundamentally changed the nature of travel and tourism communication and marketing (Gretzel et al. 2006).

1.1 The website: From functionality to persuasion

The Internet provided the ability to expand the customer base to cover the global population cost effectively. Large organisations such as airlines and hotel chains were able to access an international clientele and develop the tools to manage properties around the world at the touch of a button (Egger and Buhalis 2008). Smaller companies could also for the first time develop their "virtual size" and offer their services to global markets (Spencer et al. 2012). Tourism enterprises – at all levels – recognized the importance of technologies in their field and they acknowledge that a correct management of an holistic online presence is a prerequisite for success (e.g. Inversini et al. 2012).

Travel and tourism organizations rely upon the Internet to facilitate users' experience when approaching the destination in the online environment (Gretzel et al. 2006). Indeed, the cornerstone of the online tourism communication continues to be the organization's website (Baloglu and Pekcan 2006); for example, Choi, Lehto, and Oleary, (2007), as well as Wang (2008) and Zach et al. (2010), found that tourism destinations use official websites to provide users with information during the three phases of tourism goods consumption, in order to promote a coherent destination image throughout the travel experience. Similarly, hospitality businesses also put a lot of effort in creating the hotel website to be the focal point of the digital marketing and communication strategy as well as of the digital selling strategy (Phelan et al. 2011) thus fostering disintermediation (Law 2009), or in other words the possibility of selling rooms independently by third party operators. The adoption of these technologies required a lot of effort on the industry side which was not prepared to embrace the changes required by the management of a 24/7 worldwide contact point (i.e. the website) and a 24/7 worldwide sales centre (i.e. the reservations page) due to organizational problems (Card et al. 2003).

Tourism researchers drew inspiration from many other disciplines including digital communications (Hoffman and Novak 1996) and product design/marketing (Pine and Gilmore 1999) to conceptualize the key issues in online tourism communications based upon the unique characteristics of the tourism product. Early academic discussion around tourism websites (at both the destination and hospitality level) dealt with issues related to website functionalities including completeness (Choi et al. 2007), quality (Inversini and Cantoni 2009), scope (Morrison et al. 2004) and performances (Lee and Morrison 2010) using different methodologies and outcomes. This literature as well as the amount of attention given to this topic demonstrates the relevance of the subject both at academic and practical level (Ip et al. 2011). What is clear is that tourism websites, which moved from being an online brochure to a more complex communication tools, are not mere technological artefacts but also instruments that impact on management, promotion, and commercialization (Inversini et al. 2012). Moreover, they are important tools for studying the preferences of the target audiences of the destination/hotel, engaging consumers' interest and participation, and capturing key information (Wang and Fesenmaier 2006).

A key development in tourism website design evolved through seeing online communication from simply a collection of functionalities to a way of persuasion with the emphasis on change in website users' attitudes and behaviours. While this line of research has taken on perspectives from multidisciplinary fields such as human-computer interaction and digital communication, a solid grasp of the intangible, perishable, and experiential nature of the tourism product serves as the conceptual foundation for developing effective online communication strategies. Early research in this area focused attention on offering rich and interactive virtual experiences as well as community-based information exchanges for tourists looking for destination information (Cho et al. 2002; Wang et al. 2002). Gretzel and Fesenmaier (2002) argued that the inherently complex and personally-based nature of the tourism product entails the integration of the narrative logic when building a tourism information system. From a human-computer interaction perspective, Xiang and Fesenmaier (2005) further suggested that interface metaphors should serve as a critical element in website design that aims to represent the tourism product that best connects with the traveller's mental model of a tourist place. In line with the development in computer science, tourism research also placed much attention on online recommendation systems (Fesenmaier et al. 2006). For example, strategies such as case-based reasoning and collaborative filtering have been built into systems that make suggestions for tourist destinations

(Ricci et al. 2002; Ricci and Werthner 2004; Gretzel and Fesenmaier 2006). As part of this effort, Gretzel et al. (2004) tested the effectiveness of using travel personality to predict travellers' preferences of suggested tourism destinations. Recently, research on persuasive communication in online tourism also took into consideration the travel information search process to understand the impact of different channels/websites on travel planning and decision-making on the Internet (Gretzel and Fesenmaier 2006; Kim and Fesenmaier 2008; Pan et al. 2011). Perhaps, a framework that integrates various elements of persuasion such as "attract, engage, relate, and learn" epitomizes best the communication strategies that are more likely to succeed in online tourism (Gretzel et al. 2006; Wang and Fesenmaier 2006). Generally speaking, the understanding of the tourism website continues to progress from a representational tool to a communication platform that actively engages and transforms its visitors.

1.2 The power of social media: From mass broadcasting to social conversations

The emergence of web 2.0 (O'Reilly 2009) and social media has generated important impact on travel and tourism, particularly in online communication. In general, social media can be understood as Internet-based applications that encompass media impressions created by consumers, typically informed by relevant experiences, and archived or shared online for easy access by other impressionable consumers (Blackshaw and Nazzaro 2006). These Websites, in various forms of consumer generated content represented by blogs, virtual communities, product review sites, wikis, social networks, and collaborative tagging, and media sharing Websites like YouTube and Flickr, have gained substantial popularity in online travellers' use of the Internet (Xiang and Gretzel 2010). Many of these websites support consumers in posting and sharing their comments, opinions, and personal experiences, which then serve as potentially valuable information for others. This new form of online communication supports the argument by Thomas Friedman (2006) that "the world is flat" because consumers are gaining substantially more power in determining the production and distribution of information owing to the flattening of access on the Internet. Importantly, the ever growing presence of social media has changed the dynamics of online communication, i.e., from mass broadcasting to social conversations.

Research has focused attention on the socio-psychological aspects of social media use. From the early days of the commercialization of the Internet, travel related virtual communities have attracted the attention of tourism researchers because of their important role in generating online word-of-mouth. Virtual tourist communities such as LonelyPlanet and IGoUGo, where tourists can exchange opinions and experiences on topics of common interests, have been around at least since the late 1990s, and several researchers have investigated their roles and impacts in the context of travel (Tussyadiah and Fesenmaier 2008; Tussyadiah and Fesenmaier 2009; Wang and Fesenmaier 2006; Wang et al. 2002). Social media grew to include a range of new applications such as RSS-based content syndication, AJAX, tagging, wikis, ratings and evaluation systems, virtual worlds (e.g., Second Life), podcasting, blogs, and online videos. For example, consumer blogs have emerged as one of the most prominent themes in research on social media in travel and tourism. This research aims to understand the functions of blogs in creating and sharing new experiences (Pudliner 2007), its trustworthiness to online travelers (Mack et al. 2008), as well as the use of it as marketing intelligence (Pühringer and Taylor 2008; Waldhör and Rind 2008). As evidenced by the success of websites like TripAdvisor, Yelp, and Zagat, online travel-related review sites also represent a significant proportion of social media in travel and tourism (Gretzel and Yoo 2008). Multimedia sharing (i.e., video, photos, podcasting, etc), represented by Websites such as YouTube and Flickr, has attracted tourism researchers by generating interests in understanding the role of social media in transforming travel experiences (Tussyadiah and Fesenmaier 2008).

The structure of the tourism domain has changed drastically with the emergence of social media as new players in the field of online communication. Social media has given rise to issues with respect to how tourism marketers can leverage this electronic word-of-mouth in order to support their online marketing efforts. For example, Schmallegger and Carson (2008) suggested that strategies for using blogs as an information channel encompass communication, promotion, product distribution, management, and research. Many travel and tourism operations have also recognized the importance of including consumer generated content on their websites, usually in the form of edited testimonials (e.g., VisitPA.com and Sheraton.com). Adoption of social media as a marketing tool has been widespread among hospitality and tourism organizations including using it for public relations purposes (Chan and Guillet 2011; Curtis et al. 2010; Stankov 2010) as well as for prediction of market trends (Asur and Huberman 2012). However, organizations are also facing huge challenges due to the loss of control in online communication (Fotis et al. 2011; Hays et al. 2013). For example, many of them are struggling with fighting negative comments or reviews that could have devastating impact on their businesses (see, for example, http://www.bizjournals.com/boston/blog/techflash/2013/08/dirtiest-hotel-tripadvisor-lawsuit.html). Innovative organisations, on the other hand, will need to rapidly identify consumer needs and to interact with prospective clients by using comprehensive, personalised and up-to-date communication media for the design of products that satisfy tourism demand (Yu et al. 2011). Social media, thus, provide clear credentials for organizations and enable consumers to tell their side of the story with the possibility of posting online reviews about a given product/service. Social networking sites offer innovative ways to develop customer relationship management strategies and engagement that can have a direct influence a (i) company's credibility, (ii) influence, (iii) reputation and (iv) word-of-mouth advertising. It seems that social media will remain, in the foreseeable future, a significant force that drives the online conversations in travel and tourism.

1.3 Mobile computing: Engaging consumers on-the-go

Mobile communication has evolved in the last few years. The advent of the Apple iPhone followed by the spread of the so-called smartphones has revolutionized the consumer market. It was envisioned that smart phones soon would become an 'ordinary device' for consumers in order to stay in touch with friends and to browse the web (Katz 2008). into essential tools in the ordinary routines of the masses. This trend is generating an enormous impact in the tourism sector (Inversini and Violi 2013). There are indications that mobile technologies will become the next wave of innovation that drives travel and tourism. With power-efficient processors, modern operating systems, broadband Internet access, large viewing screens, user friendly interfaces, pervasive computing capabilities, as well as productivity-enhancing apps, smartphones offer a wide range of possibilities to support travellers in their use of information and communications home and away. Particularly, smartphones offer increased connectivity, communication, content consumption, and content creation. As such, mobile technologies have even been touted as the catalyst for the emergence of a new generation of modern tourists (Gretzel 2011; Gretzel and Jamal 2009) Recent research indicates that the smartphone supports tourists' needs for mobility and information/communication on-the-go and, in turn, transforms the meaning of travel (Wang et al. 2012).

There are several characteristics of mobile technology that distinguish it from the traditional, stationary technologies in terms of its support of the tourism communication. First, as discussed by Gretzel, Fesenmaier, and O'Leary (2006), tourists frequently need information support due to the uncertainty of the environment, inadequate capability in making judgments because of a lack of experience, and unexpected changes in plans. Smartphones provide nearly anywhere and anytime access to information services, and connectivity with one's social circles is taking shape, which is, in many ways, different from the time when most of information for travel is accessed on a desktop computer (Egger and Jooss 2010). As such, smartphones support the different phases in the travel process.

Second, information needs of users may change during a trip due to a possible change in location and context. As such, location-based and context aware services can then be used to support travellers on the go, particularly with the ability to suggest modification in the planned tourism experience (Angerer 2010; Lamsfus et al. 2012). The notion of context can be defined as any situation that can be used to characterize the situation of a person, place, or object (Dey and Abowd 2000). Time and location are two of the most important context parameters (Angerer 2010) but dimensions such as weather or the users' personal interests (Egger and

Jooss 2010) are crucial for the application of this technology in the tourism sector. They filter information and provide a series of choices that are relevant and tailored for the given user in the given context. Recently, context-driven services arise to help tourists on the move by leveraging not only on "physical" aspects of context such as time and location but also potentially more complex situations pertaining to the nature of the tourism experience (Lamsfus et al. 2013). As such, mobile devices, and especially the smartphone, play a central role in facilitating dynamic decision making for travellers on the move.

Third, another aspect of mobile communication is the connection between travel sphere and the traveller's everyday life. Travel can be seen as a special stage of technology use, which is connected with other settings (Gretzel 2011). As a result, tourist carries part of "himself/herself" (e.g. habits, routines, connections with friends and families) as well as her/his capability and mentality of technology use to the context of travel. Thus, use of the mobile technology consists of not only the decision-making aspects of the trip but also the information, communication, and entertainment needs resulting from the traveller's everyday life. For example, travellers use the same social network applications (e.g., Foursquare, Flickr, Instagram, Facebook, Twitter, and Skype) in both travel and everyday lives. They follow the same procedures to find direction as they do in everyday life. Tourism marketing must consider the exchange of information between a traveller and others in different physical and virtual domains.

2 Travel in the network: Toward a new paradigm for tourism communication

Technological forces of change have heavily impacted all facets of the tourism world – they have fundamentally changed how travellers experience travel, how destinations market themselves, which in turn, how the tourism industry develops and measures its success. Evidence of this restructuring is manifest in many ways, none more so that in an article by Gretzel (2012) entitled Travel in the Network: Redirected gazes, ubiquitous connections and new frontiers wherein she used, like Castells (1996), the "network" as a metaphor to describe the various systems that have fundamentally changed the travel experience; and in articles by McCabe (2012) and MacKay and Vogt (2012), who suggest that there is a huge "spillover effect" which links our daily lives, both in terms of how we behave and our use of technology, and the way we experience travel. These articles are essential in that they clearly explain the formation of new models of travel behaviour, new models for product design, and new models for research and evaluation which, in turn, establishes a new paradigm of tourism communication and marketing.

As suggested by Gretzel (2012) and others (see for example, Gretzel et al. 2006; Wang and Fesenmaier 2012), travel today differs substantially from travel fifty years ago when mass tourism began in earnest. Indeed, it appears that the change in travel follows in parallel to the "stages of change" model proposed by Contractor, Wasserman and Faust (2006), Gretzel, Yuan and Fesenmaier (2000), and Yuan, Gretzel and Fesenmaier (2005) regarding the impact of technology on organizations. In particular, they argue that the first two stages of technology use create an "expansion" in activity, but not any fundamental change in behaviour and that only in the third stage we do see real structural changes in the nature of use – that is, travel behaviour. A recent book by Turkle (20011) is mirrored by Gretzel and her colleagues (2012, 2006), MacKay and Vogt (2012), Tussyadiah (2012), Tussyadiah and Zach (2012), and Wang and Fesenmaier (2012), who suggest that there are now important structural changes in travel behaviour whereby travellers tend to 'extend' daily life into travel, become much more involved in creating/controlling the tourism experience by sharing with others, are more involved and creative as they seek authentic experiences, and today's travellers tend to adapt much better to local settings by using various forms of mobile technology.

These new notions to describe change in travel behaviour require a new paradigm of tourism communication and marketing. In particular, the metaphor of "travel in the network" suggests there are a number of new conditions for tourism marketing and communication as the result of behavioural change in travel. First, it has already been well documented that, in general, today's marketing and communication practice, from political campaigns to selling groceries, is primarily driven by information technology as the penetration rate of the Internet, for example in the United States, has reached to a maximum level of saturation (see www.pewinternet.org). Also in travel and tourism, the Internet is the predominant information source for travel planning, Within this context, the notion of "travel in the network" offers a new lens for understanding today's travellers. That is, instead of simply seeing travellers as users of technology, today's travellers can be considered an active partner in technology-supported networks, which further consist of numerous information spaces (e.g., DMO websites) and channels (e.g., OTAs and travel search engines) that support the basic information activities of the traveller. Thus, the Internet is no longer a monolithic eCommerce platform; instead, it offers countless networks and platforms vying for the traveller's attention and spending power by supporting information seeking and transactions (Xiang et al. 2008). And from a communication and marketing standpoint, creating and delivering the right message to the right person at the right moment is fundamentally important. Indeed, it may be of primary importance for marketers to focus on understanding how travellers navigate these information spaces and channels within the network structure in order to build and anticipate their upcoming travel plans and experiences.

Second, technology-supported networks are social and community-based. Indeed, tourism information on the Internet has been such from the very moment when the Internet became a public and commercial infrastructure; examples include VirtualTourist and IGoUGo whereas virtual places travellers congregate to share their experiences (Wang and Fesenmaier 2006). However, the explosive growth of Web 2.0 with a variety of tools and platforms that support consumer-generated content has further transformed the Internet into the networks for social interactions (Xiang and Gretzel 2010). Facebook, Twitter, Youtube, and Pinterest are quintessential Web 2.0 applications in that they are developed as novel ways to facilitate exchange of information and social networking. Particularly in travel and tourism, websites such as TbripAdvisor and Yelp are social spaces wherein word-of-mouth is created, distributed, and shared among peer travellers and consumers. As a result, tourism marketing and communication is no longer a practice of advertising and promotion; rather, the focus now has shifted to participating in and being part of the online conversations. Therefore, social media marketing has emerged as a new strategic area for tourism marketing with a growing interest in "listening" to the online social spaces in order to effectively manage a destination's reputation (Gretzel, et al. 2006; Inversini, Cantoni & Buhalis. 2009; Sigala et al. 2012).

Third, another important aspect of the travel in the network metaphor is that technology-supported networks are mobile, where today's cutting-edge apps offer all kinds of tools for travellers to search for information and to make decisions on-the-go thereby creating what are described as Tourism Activated Networks. It is argued that mobile technology such as smartphones (and iPads) can be used to strengthen traveller's social ties and interactions. For many people, a mobile phone is far beyond a communication tool or an accessory of daily lives, and it has become an inseparable part of his/her life or even body (Turkle 2011; Wang and Fesenmaier 2012). As such, the ubiquitous presence of these devices in people's lives potentially intensifies and encourages the participation in mobile social networking.

3 Conclusions

Two decades of ICT development has changed the fundamental nature of tourism communication. Today's tourism businesses are engaging travellers and guests with richer content and higher level of interactivity as well as through various socially constructed media spaces. Without doubt, ICT will continue to evolve and consequently drive tourism communication and marketing. The networks that surround travellers in trip planning and their mobility encompass systems that capture and generate enormous amount of consumer data, and the so-called Big Data (George, Haas, & Pentland 2014) offer numerous opportunities and pose new challenges for tourism marketing and communication offering enhanced customer intelligence, improving business processes, and, ultimately, enabling the development of new strategies for navigating an increasingly competitive environment.

References

Angerer, Fabian. 2010. Mobile kontextsensitive Dienste für die Freizeit. In Roman Egger & Mario Jooss (eds.), *mTourism: Mobile Dienste im Tourismus*, 45–58. Springer DE.

Asur, Sitaram & Bernando A. Huberman. 2010. Predicting the future with social media. *Proceedings of the International Conference on Web Intelligence and Intelligent Agent Techno*logy (WI-IAT) 1. 492–499. IEEE.

Baloglu, Seyhmus & Yakup A. Pekcan. 2006. The website design and Internet site marketing practices of upscale and luxury hotels in Turkey. *Tourism Management* 27(1). 171–176.

Blackshaw, Pete & Mike Nazzaro. 2006. Consumer-generated media (CGM) 101: Word-of-mouth in the age of the web-fortified consumer. *White Paper*. Nielsen BuzzMetrics New York. http://www.nielsen-online.com/downloads/us/buzz/nbzm_wp_CGM101.pdf (Accessed 14 April 2014).

Buhalis, Dimitios. 2003. *ETourism: information technology for strategic tourism management*. Harlow, England: Financial Times Prentice Hall.

Buhalis, Dimitrios & Rob Law. 2008. Progress in information technology and tourism management: 20 years on and 10 years after the Internet – The state of eTourism research. *Tourism Management* 29(4). 609–623.

Card, Jaclyn A., Ching-Yi Chen & Shu Tian Cole. 2003. Online Travel Products Shopping: Differences between Shoppers and Nonshoppers. *Journal of Travel Research* 42(2). 133–139.

Chan, Nga Ling & Basak Denizci Guillet. 2011. Investigation of social media marketing: how does the hotel industry in Hong Kong perform in marketing on social media websites?. *Journal of Travel & Tourism Marketing* 28(4). 345–368.

Cho, Yong-Hyun, Youcheng Wang & Daniel R. Fesenmaier. 2002. Searching for experiences: The web-based virtual tour in tourism marketing. *Journal of Travel & Tourism Marketing* 12(4). 1–17.

Choi, Soojin, Xinran Y. Lehto & Joseph T. Oleary, J. T. 2007. What does the consumer want from a DMO website? A study of US and Canadian tourists' perspectives. *International Journal of Tourism Research* 9(2). 59–72.

ComScore. 2013. comScore Reports February 2013 U.S. Mobile Subscriber Market Share. April, 4. http://www.comscore.com/ita/Insights/Press_Releases/2013/4/comScore_Reports_February_2013_US_Smartphone_Subscriber_Market_Share (Accessed 13 April 2014).

Curtis, Lindley, Carrie Edwards, Kristen L. Fraser, Sheryl Gudelsky, Jenny Holmquist, Kristin Thornton & Kaye D. Sweetser. 2010. Adoption of social media for public relations by nonprofit organizations. *Public Relations Review* 36(1). 90–92.

Dey, Anind K. & Gregory D. Abowd. 2000. CybreMinder: A Context-Aware System for Supporting Reminders. In Peter Thomas & Hans-W. Gellersen (eds.), *Handheld and Ubiquitous Computing*, 172–186. Springer Berlin Heidelberg.

Egger, Roman & Dimitrios Buhalis. 2008. *eTourism case studies: management and marketing issues*. Amsterdam; Boston; London: Butterworth-Heinemann.

Egger, Roman & Mario Jooss. 2010. *MTourism: Mobile Dienste Im Tourismus*. Springer DE.

Fesenmaier, Daniel R., Karl W. Wöber & Hannes Werthner (eds.). 2006. *Destination recommendation systems: Behavioural foundations and applications*. CABI.

Fesenmaier, Daniel R. & Zheng Xiang. 2013. Tourism Marketing from 1990–2010: Two Decades and a New Paradigm. In Scott McCabe (ed.) *The Handbook of Tourism Marketing*, 549–560. London: Routledge.

Fotis, John, Dimitrios Buhalis & Nicos Rossides. 2011. Social Media Impact on Holiday Travel Planning. *International Journal of Online Marketing* 1(4). 1–19.

Friedman, Thomas L. 2006. T*he World is Flat: A Brief History of the Twenty-First Century*. New York: Farrar, Straus and Giroux.

George, G., Haas, M. R., & Pentland, A. 2014. Big Data and Management. *Academy of Management Journal* 57(2). 321–326.

Gretzel, Ulrike. 2011. Intelligent systems in tourism: A social science perspective. Annals of Tourism Research 38(3). 757–779.

Gretzel, Ulrike & Daniel R. Fesenmaier. 2002. Building narrative logic into tourism information systems. *IEEE Intelligent Systems* 17(6). 59–61.

Gretzel, Ulrike & Daniel R. Fesenmaier. 2002. Implementing knowledge-based interfirm networks in heterogeneous B2B environments: a case study of the Illinois Tourism Network. In Karl W. Wöber, Andrew J. Frew & Martin Hitz (eds.), *Information and Communication Technology in Tourism 2002,* 39–48. Springer-Verlag Wien.

Gretzel, Ulrike & Daniel R. Fesenmaier. 2006. Persuasion in recommender systems. *International Journal of eCommerce* 11(2). 81–100.

Gretzel, Ulrike & Tazim Jamal. 2009. Conceptualizing the creative tourist class: Technology, mobility, and tourism experiences. *Tourism Analysis* 14(3). 471–481.

Gretzel, Ulrike, Daniel R. Fesenmaier & Jospeh T. O'Leary. 2006. The transformation of consumer behaviour. In Dimitrios Buhalis & Carlos Costa (eds.), *Tourism Business Frontiers: Consumers, Products and Industry*, 9–18. *Burlington, MA: Elsevier.*

Gretzel, Ulrike, Nicole Mitsche, Yeong-Hyeon Hwang & Daniel R. Fesenmaier. 2004. Tell me who you are and I will tell you where to go: use of travel personalities in destination recommendation systems. *Information Technology & Tourism* 7(1). 3–12.

Gretzel, Ulrike & Yoo, K. H. 2008. Use and impact of online travel reviews. In Peter O'Connor, Wolfram Höpken & Ulrike Gretzel (Eds.), *Information and Communication Technologies in Tourism* 2008. 35–46. New York: Springer-Verlag Wien.

Gretzel, Ulrike, Yu-Lan Yuan & Daniel R. Fesenmaier. 2000. Preparing for the New Economy: Advertising and Change in Destination Marketing Organizations, *Journal of Travel Research* 39(2). 146–156.

Hays, Stephanie, Stephen John Page & Dimitrios Buhalis. 2013. Social media as a destination marketing tool: its use by national tourism organisations. *Current Issues in Tourism* 16(3). 211–239.

Hoffman, Donna L. & Thomas P. Novak. 1996. Marketing in hypermedia computer-mediated environments: conceptual foundations. *The Journal of Marketing* 60(3). 50–68.

Inversini, Alessandro & Dimitrios Buhalis. 2009. Information Convergence in the Long Tail: The Case of Tourism Destination Information. In Wolfram Höpken, Ulrike Gretzel & Rob Law (eds.), *Information and Communication Technologies in Tourism 2009*, 381–392. Vienna: Springer Vienna.

Inversini, Alessandro & Lorenzo Cantoni. 2009. Cultural Destination Usability: The Case of Visit Bath. In Wolfram Höpken, Ulrike Gretzel & Rob Law (eds.), *Information and Communication Technologies in Tourism 2009*, 319–331. Vienna: Springer Vienna.

Inversini, Alessandro, Christoph Brülhart & Lorenzo Cantoni. 2012. Myswitzerland.com: Analysis of Online Communication and Promotion. *Information Technology & Tourism* 13(1). 39–49.

Inversini, Alessandro, Lorenzo Cantoni & Dimitrios Buhalis. 2009. Destinations' Information Competition and Web Reputation. *Information Technology & Tourism* 11(3). 221–234.

Inversini, Alessandro & Lucia Violi. 2013. Tourism Mobile Application Usability: The Case of iTicino. *International Journal of E-Services and Mobile Applications* 5(2). 54–70.

Ip, Crystal, Rob Law & Hee 'Andy' Lee. 2011. A review of website evaluation studies in the tourism and hospitality fields from 1996 to 2009. *International Journal of Tourism Research* 13(3). 234–265.

Ip, Crystal, Rosanna Leung & Rob Law. 2011. Progress and development of information and communication technologies in hospitality. *International Journal of Contemporary Hospitality Management* 23(4). 533–551.

Jeng, James & Daniel R. Fesenmaier. 2002. Conceptualizing the Travel Decision-Making Hierarchy: A Review of Recent Developments. *Tourism Analysis* 7(1). 15–32.

Katz, James E. 2008. *Handbook of Mobile Communication Studies*. The MIT Press.

Kim, Heejun & Daniel R. Fesenmaier. 2008. Persuasive design of destination web sites: An analysis of first impression. *Journal of Travel Research* 47(1). 3–13.

Lamsfus, Carlos, Aurkene Alzua-Sorzabal, David Martín & Emilio Torres. 2012. Semantic-based Context Modeling in Tourism. *Information Technology & Tourism* 13(4). 309–325.

Lamsfus, Carlos, Zheng Xiang, Aurkene Alzua-Sorzabal & David Martin. 2013. Conceptualizing context in an intelligent mobile environment for travel and tourism. In Lorenzo Cantoni & Zheng Xiang (eds.), *Information and Communication Technologies in Tourism*, 1–11. Vienna: Springer.

Law, Rob. 2009. Disintermediation of hotel reservations: The perception of different groups of online buyers in Hong Kong. *International Journal of Contemporary Hospitality Management* 21(6). 766–772.

Lee, Jungkook & Alanstair M. Morrison. 2010. A comparative study of web site performance. *Journal of Hospitality and Tourism Technology* 1(1). 50–67.

Morrison, Alanstair M., J. Stephen Taylor & A. Douglas. 2004. Website Evaluation in Tourism and Hospitality: the art is not yet stated. *Journal of Travel & Tourism Marketing* 17(2–3). 233–251.

O'Connor, Peter & Andrew J. Frew. 2004. An evaluation methodology for hotel electronic channels of distribution. *International Journal of Hospitality Management* 23(2). 179–199.

O'Reilly, Tim. 2009. *What is Web 2.0*. O'Reilly Media, Inc.

Pan, Bing, Zheng Xiang, Rob Law & Daniel R. Fesenmaier. 2011. The dynamics of search engine marketing for tourist destinations. *Journal of Travel Research* 50(4). 365–377.

Phelan, Kelly Virginia, Natasa Christodoulidou, Cary C. Countryman & Leonard J. Kistner. 2011. To book or not to book: the role of hotel web site heuristics. *Journal of Services Marketing* 25(2). 134–148.

Pine, B. Joseph & James H. Gilmore. 1999. *The Experience Economy: Work is Theatre & Every Business a Stage*. Harvard Business Press.

Poon, Auliana. 1993. *Tourism, Technology and Competitive Strategies*. CABI.

Ricci, Francesco, Bora Arslan, Nader Mirzadeh & Adriano Venturini. 2002. ITR: a case-based travel advisory system. In Susan Craw & Alun Preece (eds.), *Advances in Case-Based Reasoning*, 613–627. Springer Berlin Heidelberg.

Sheldon, Pauline J. 1997. *Tourism information technology*. CAB International.

Sigala, Marianna, Evangelos Christou & Ulrike Gretzel. 2012. *Social Media in Travel, Tourism and Hospitality: Theory, Practice and Cases*. Ashgate Publishing, Ltd.

Schmallegger, Doris & Dean Carson. (2008). Blogs in tourism: Changing approaches to information exchange. *Journal of Vacation Marketing* 14. 99–110.

Spencer, Andrew J., Dimitrios Buhalis & Miguel Moital. 2012. A hierarchical model of technology adoption for small owner-managed travel firms: An organizational decision-making and leadership perspective. *Tourism Management* 33(5). 1195–1208.

Tresidder, Richard & Craig Hirst. 2012. *Marketing in food, hospitality, tourism and events: a critical approach*. Oxford: Goodfellows Publishers.

Turkle, S. 2011. *Alone Together*. New York: Basic Books.

Tussyadiah, Iis P. 2012. 'A Concept of Location-Based Social Network Marketing', *Journal of Travel & Tourism Marketing* 29(3). 205–220.

Tussyadiah, Iis P. & Daniel R. Fesenmaier. 2008. Marketing Places Through First-Person Stories – an Analysis of Pennsylvania Roadtripper Blog. *Journal of Travel & Tourism Marketing*. 25(3–4). 299–311.

Tussyadiah, Iis. & Daniel R. Fesenmaier. 2009. Mediating tourist experiences: Access to places via shared videos, *Annals of Tourism Research* 36(1). 24–40.

Tussyadiah, Iis P. & Florian Zach. 2012. The role of geo-based technology in place experiences. *Annals of Tourism Research* 39(2). 780–800.

Yuan, Yu-Lan, Ulrike Gretzel & Daniel R. Fesenmaier. 2005. Network technology in American Convention and Visitors Bureaus. *Tourism Management* 27(2). 326–341.

Wang, Dan, Sangwon Park & Daniel R. Fesenmaier. 2012. The Role of Smartphones in Mediating the Touristic Experience. *Journal of Travel Research* 51(4). 371–387.

Wang, Dan & Zheng Xiang, Z. 2012. The New Landscape of Travel: A Comprehensive Analysis of Smartphone Apps. In Matthias Fuchs, Francesco Ricci, & Lorenzo Cantoni (eds.), *Information and Communication Technologies in Tourism*, 308–319. Springer Vienna.

Wang, Youcheng. 2008. Web-based destination marketing systems: assessing the critical factors for management and implementation. *International Journal of Tourism Research* 10(1). 55–70.

Wang, Youcheng, Quaehee Yu & Daniel R. Fesenmaier. 2002. Defining the virtual tourist community: implications for tourism marketing. *Tourism management* 23(4). 407–417.

Wang, Youcheng & Daniel R. Fesenmaier. 2006. Identifying the Success Factors of Web-Based Marketing Strategy: An Investigation of Convention and Visitors Bureaus in the United States. *Journal of Travel Research* 44(3). 239–249.

Want, Roy. 2009. When Cell Phones Become Computers. *IEEE Pervasive Computing* 8(2). 2–5. https://www.computer.org/csdl/mags/pc/2009/02/mpc2009020002.html (Accessed 14 April 2014).

Werthner, Hannes & Stefan Klein. 1999. *Information Technology and Tourism: A Challenging Relationship*. Springer-Verlag Wien.

Werthner, Hannes & Francesco Ricci. 2004. E-commerce and tourism. *Communications of the ACM* 47(12). 101–105.

Xiang, Zheng & Daniel R. Fesenmaier. 2005. An analysis of two search engine interface metaphors for trip planning. *Information Technology & Tourism* 7(2). 103–117.

Xiang, Zheng & Ulrike Gretzel. 2010. Role of social media in online travel information search. *Tourism Management* 31(2). 179–188.

Xiang, Zheng, Karl Wöber & Daniel R. Fesenmaier, D. R. 2008. Representation of the Online Tourism Domain in Search Engines. *Journal of Travel Research* 47(2). 137–150.

Yu, Bei, Miao Chen & Linchi Kwok. 2011. Toward predicting popularity of social marketing messages. In John Salerno, Shanchieh Jay Yang, Dana Nau, Sun-Ki Chai (eds.), *Social Computing, Behavioral-Cultural Modeling and Prediction*, 317–324. Springer Berlin Heidelberg.

Zach, Florian, Ulrike Gretzel & Zheng Xiang. 2010. Innovation in the Web Marketing Programs of American Convention and Visitor Bureaus. *Information Technology & Tourism* 12(1). 47–63.

John V. Pavlik

28 Journalism: From delivering information to engaging citizen dialogue

Abstract: Journalism has long served as an independent source of information about current events and matters of public importance. Technological change, economic upheaval and innovation have spurred dramatic changes in the news enterprise both in the U.S. and around the world. This chapter provides an overview of the traditions and principles of journalism and places them in the evolving context of a digital, mobile and networked world where citizen reporters operate alongside and sometimes collaboratively with professionally educated journalists.

Keywords: journalism, news, digital, mobile, networked, revenue models, innovation, ethics, diversity

1 Defining journalism

Journalism is an evolving notion. At its core, journalism fulfills a surveillance function in society, acting as roving and independent eyes and ears for the public, offering opinion as well. Journalism encompasses a spectrum of activities and content related to the reporting of information about matters of public importance in society, especially as they involve new and emerging events, developments and processes in society, both domestically and internationally. It delivers what is often referred to as news, a product generally defined in terms of information reported and packaged for the public about current events of public importance or interest. Journalists usually determine what is of public importance by considering the social, economic or environmental impact of events, processes or decisions, whether local, national or global.

Emerging digital technology and changing media economics have fostered an evolution in notions of journalism as well as the functions and processes involved in journalism. In particular, the rise of the Internet, broadband and mobile media have ushered in both a new dynamic to news flows as well as transformed the economics of news media organizations and industry. What some are calling citizen journalism has profoundly changed the nature of journalism and the use of online and mobile media to access news have impacted the traditional news media in profound ways. These changes have also upended the finances and business model of traditional media, as relatively new digital entrants such as Google, Facebook and Apple have taken control of much of the media marketplace. These changes will be discussed in detail later in this chapter.

2 An historical overview of journalism

Journalism has been an important part of society for many centuries. As long ago as 59BC Julius Caesar published the first public *Acta Diurna*, or daily acts, to share information about legal proceedings and trial outcomes, later adding notices about other noteworthy developments, such as prominent marriages or deaths. In the 17th century, the first newspapers were published, including in 1690 *Publick Occurrences* both Foreign and Domestic, the first in what is today the United States. Spurred by new printing technology, the early 19th century saw the development of mass communication in the form of the penny press, or daily newspapers published for a large audience and available for a penny. Leading the way here was James Gordon Bennett, who founded the New York Herald in 1835. Bennett not only innovated a new approach to daily news reporting but his approach was quickly imitated by other publishers. This marked the beginning of modern journalism, as both a new form of news reporting emerged as did a new funding or business model to support it. The news reported focused on developments of widespread public interest and was written or packaged for a general audience. The funding model featured an expanded use of advertising for a mass public, with newspaper readers paying only a token amount (a penny) for the paper. Previously, newspapers in the colonial era featured news particularly focused on matters of interest to the business and political elite, such as shipping news, and were relatively expensive, costing a dollar or more a copy. Circulations were relatively limited and content often had an editorial perspective. Newspapers published during this period were often called the partisan press and received much of their funding support from monetary donations from the political parties they supported.

Technological and economic developments in the mid- to late-19th century enabled further developments in newspaper journalism. Among the most notable developments was the invention of the electric telegraph. This fostered unparalleled opportunities for speed in reporting the news, but associated costs led innovative newspaper publishers to join forces in creating the Associated Press, which would share the cost of the telegraph and share the news.

The turn of the century also brought the newspaper circulation wars, predominantly fought by publishers Joseph Pulitzer and William Randolph Hearst. Vying for marketplace dominance and the majority of readers, Pulitzer and Hearst employed innovative techniques of their own to build huge circulations for their newspapers in the cities, especially New York, St. Louis and San Francisco. Their newspapers featured a variety of new techniques and formats in reporting the news. Among these were celebrity columnists, extensive sports coverage, personal interviews, and sensationalized news. The sensationalized news coverage was often called "tabloidism" as a reflection of the printed newspaper format, or "yellow journalism," as a byproduct of the cartoon "The Yellow Kid" featured in Pulitzer's New York World newspaper. The newspaper war became inflamed during the Spanish-American War, when sensationalized coverage grew hot. At one point, when the atmosphere for war seemed to cool, Hearst demonstrated his commitment to beefed up war coverage. Hearst had dispatched famed artist Frederic Remington to Cuba to report on the situation. In 1897 Remington cabled Hearst that "there will be no war." Hearst cabled back: "You furnish the pictures and I'll furnish the war." (O'Toole 1984).

Investigative journalism, such as that produced by Nellie Bly (Bly 1887) who famously reported in "Ten Days In The Mad-House" on patient abuse in Bellevue Mental Hospital, also emerged during this period as a part of mainstream reporting. Much of the most influential reporting during the first half of the 20th century was dubbed muckraking journalism. Muckrakers such as Ida Tarbell and Upton Sinclair investigated on the abuses of government and big business (Tarbell 2003; Sinclair 1905). From this era came the notion that the best journalism "comforts the afflicted, and afflicts the comfortable" (Dunne 1898). Although some muckrakers worked for newspapers, muckraking was most often found in magazines, which for several reasons were better suited to this long-form journalism. Magazines were published less often than daily newspapers, and this enabled reporters to spend more time digging up muck. Magazines also lent themselves to the longer form reporting that muckrakers often needed to tell their complex stories.

3 Principles of journalism

Through these decades of change, important principles of news emerged, as well. These principles established a foundation for much of the journalism that dominated the emergent media of most of the 20th century, including that of electronic journalism, or the radio and television news that came along with the invention of those media in the first half and middle part of the 20th century.

A core principle of journalism or news reporting that emerged in the late 19th century is the notion of objectivity (Lippmann 1931). Objectivity in the news essentially implies that the events or issues reported about should not reflect the personal bias of the journalist or the organizational bias of either the news media or society at large. That is, what is reported should be a neutral observation of events or processes. This idea is still a central principle of contemporary journalism, although it has some significant nuance and criticism. Among the questions surrounding the notion of objectivity is whether any human being is truly capable of complete impartiality, and in this sense, many argue that rather than objectivity journalists should focus on reporting the news fairly and reveal whatever bias they might bring to their reporting (Hallin and Mancini 2004). Moreover, some suggest that the very commercial nature of the news media industry, at least as practiced in much of the world, gives it an inherent commercial bias (Schiller 1981). Further, research suggests that for many journalists and news organizations it is virtually impossible to report outside the constraints of their parent society, and its essential political perspective (Schudson 1978). Consequently, when reporting on vital matters such as war, news media often report in a manner favorable to their home country, particularly in the early stages of the conflict.

The Project for Excellence in Journalism (PEJ) has identified nine principles that provide a foundation for the contemporary practice of journalism, especially in the West and developed world (Pew Research Journalism Project 2014). Core among these is the notion that a journalist's first obligation is to the truth. Although this notion is generally accepted, some suggest that it is impossible to determine what the truth is, and that there may be differing perspectives on the truth (Pew Research Journalism Project 2014).

A second principle outlined by PEJ is loyalty to citizens. Yet, even this basic notion is also somewhat problematic. Who constitutes "citizens"? Are refugees part of the citizenry, even though they are not legally qualified as citizens or holders of passports of the country in which they may reside? Perhaps it is more useful to refine this notion to assert journalists have a loyalty to serve the public. This would imply all individuals are relevant parts of the audience journalists serve, and also differentiates the purpose of journalism as serving the public good, and not serving the interests of government or other institutions.

A third core PEJ principle is the verification of information before publication. Verification is vital to maximizing the potential for accuracy in reporting. But it is also problematic. Should information be reported only once it can be verified? In the days gone by when daily newspapers dominated the journalism mix, such verification was routine and established a standard to which all journalists could aspire. But in the 21st century when social media such as the blogosphere, Twitter and Facebook all might resonate with real-time, often unverified information, what is the role of more mainstream journalism? How long can a reporter wait to enter the news conversation?

At the foundation of most U.S. journalism and much of the West is the notion of a free press or news media. This means that what is reported enjoys freedom from government or other institutional censorship. This does not mean the news media or those who report or edit them do not engage in self-censorship. Nor does it mean total freedom from government rules and regulations. In the U.S., the First Amendment to the U.S. Constitution states that Congress shall make no law abridging freedom of speech or press. Yet, there are a variety of federal and state rules, regulations and laws that can somehow limit press freedom. For instance, the landmark U.S. Supreme Court case established that in the U.S. there can be no prior restraint on publication, with certain key exceptions, most notably during war when what is to be published would result in a clear and present danger to national security or certain types of pornography, such as child pornography. Yet, there are strong, legal post-publication penalties for certain publication activities, such as libel (written word as in print media) or slander (spoken word, as on radio or television or online via video reporting). Reporters are also subject to arrest or even imprisonment, even in the U.S., for doing things such as protecting the identity of a confidential informant. Although many states have shield laws to protect journalists and their reporting methods, there is no federal shield law. Consequently, in the U.S. there have been many cases of reporters, including those reporting in the blogosphere, who have been jailed for extended periods of time by federal prosecutors and judges in an attempt to force them to reveal their sources. Journalists are often reluctant to reveal the identity of their confidential sources for several reasons, including keeping the trust of a source who has been promised anonymity in a controversial situation (perhaps out of fear of reprisal or prosecution when a whistle blower), and to protect the reputation of the reporter who may hope to obtain the cooperation of future sources otherwise reluctant to provide information possibly critical of others.

In 1947, a commission chaired by educator Robert Maynard Hutchins issued a report titled *A Free and Responsible Press* (Hutchins 1947). The Hutchins Commission, as it has come to be known, offered via the report a critique of the state of the news media in the U.S. and a prescription for the press. The Commission found the media to be lacking in a number of areas, including the diversity of staffing of the newsroom. The Commission recommended a number of changes in journalism, including increasing the diversity of news organization staffs (Hutchins 1947). The Commission concluded that because the press enjoys Constitutional guarantees of freedom, it carries with it a moral duty to act in a responsible fashion (Hutchins 1947: 138). This has been interpreted by many to provide a compelling basis for reporting in a fair, if not balanced manner, both in print and electronic and online or digital media (Nerone 1995). Moreover, the Commission emphasized that news organizations should maintain a clear distinction between news that is reported in a factual manner and opinion (Hutchins 1947: vi). Further, journalistic organizations should strive to present a representative picture of society and avoid stereotyping, especially of minority groups. The Commission also argued that government should extend press freedom to all media of news delivery (Hutchins 1947: 15).

Another long-held principle in at least U.S. newspaper journalism has been the separation of the business and editorial or news side of the journalism enterprise. Journalism leaders have held that separating the business and news gathering/reporting sides of the enterprise would shield journalists from the potential bias of commercial forces. For instance, it has been argued that separating the two sides of the newspaper endeavor would eliminate or at least minimize the potential for advertising-related issues to influence coverage decisions or how stories about source of advertising might be covered (Washington Post Standards and Ethics 2013). This is a compelling argument, but in the digital age, this line of separation is sometimes blurring as the financial footing of the journalistic enterprise has become often uncertain. In some cases in the early days of web journalism, for instance, web developers might work on both the editorial side of a newspaper's online presence as well as the advertising side.

4 Ethics

Providing the undergirding for all journalism has been the notion of ethics. Various news organizations have developed their own codes of ethical behavior but most widely known and followed is the code of ethics developed by the Society of Professional Journalists (SPJ 2014). This code outlines four key principles. First, journalists should "Seek truth and Report it." Second, they should "minimize harm," including to those who are sources or the subject of news. Third, journalists should "act independently" by avoiding conflicts of interest and the like. And, fourth, journalists should "be accountable" to each other and especially to the public.

5 Gathering the news

How stories are covered or reported is a matter of enduring importance in journalism. Reporters draw upon different types of sources in gathering the news. One of the most important are interviews with persons involved in the story in various ways, such as the subject of an investigation or an eye-witness to a newsworthy event. Often, reporters will interview expert sources to provide particularly well-informed or credible perspective. Reporters also often rely on everyday persons to provide a lay-person perspective on the news.

Documents serve as a second principle source for reporters. Especially relevant are public records (i.e., digital records kept by government). Sometimes, it is necessary for reporters to file what are called FOIA requests to access these records. FOIA refers to the federal Freedom of Information Act, although there are also a number of states that have their own versions of such legislation. Sometimes information is redacted from these records to protect private personal information or information deemed vital to national security. Sometimes agencies will redact information under the guise of protecting national security, when in fact it is merely politically embarrassing. In these cases, it is necessary for the reporter to take additional steps to gain access to the redacted information, sometimes even filing a law suit.

6 Producing the news

Once the reporter has gathered the essential facts, it is necessary to organize and present the information into what in journalism parlance is called the story. It is not necessarily a story as in a children's tale, but it may have many parallels. Journalists use a variety of media to tell the story, including words, written or spoken, audio (sometimes brief recordings of a source, called an actuality), video and interactive content (e.g., a clickable map). The information included in a news story typically features answers to what journalists call the Five W's. These are Who, What, When, Where and Why. Stories often emphasize the answers in this sequence, as well, with Who and What being the most important, especially in breaking news. Why is often relegated to follow-up stories (after the story is first reported, possibly the next day), partly because it may be more difficult to determine and complicated, but also because those in the news business tend to assume it is of less interest than the basic facts. But, in the age of hyper-news and social media, the who and what are often known immediately and hardly news worthy after the first few minutes a story breaks. The more complex information about why takes on added importance, and can help separate quality, professional, human-generated reporting from what can be created easily in the social network and even by computer. A number of working systems have been developed by collaborations between computer scientists and journalists (e.g., Narrative Science in Chicago, born out of research at Northwestern University) in which a computer uses basic data to write a news story. Clients paying for such services in 2013 include the Big Ten Sports Network. Such stories are not limited to the domain of sports, and financial reporting is another area of successful computer-generated reporting.

7 Framing the news

Research suggests that news stories are usually framed in some fashion to put them into understandable context (Tewksbury and Scheufele 2009). Framing the news means locating conceptually a single story into a larger topical domain, to help make sense of an event or series of events. For instance, a story about a bombing at a building will likely be placed into a "terrorism" frame, at least until any other interpretation of the event can be established through reporting.

Framing the news is often shaped by the type of journalism being practiced, or the type of news organization publishing or otherwise distributing the news. We can identify at least six main types of news organizations. These are Commercial, Alternative, Public Journalism, Ethnic, Hyperlocal, and Citizen Journalism ((Tewksbury and Scheufele 2009). Although each of these types of journalism shares certain basic qualities, such as a commitment to report the truth, accurately and within an ethical context, there are important differences between or among them. Commercial journalism is the most prevalent form, and includes print, broadcast and online channels, but research suggests here there is a tendency to lean toward reporting news in a way compatible with larger, mainstream audiences, as well as advertising supported content (Tewksbury and Scheufele 2009). For instance, such news channels tend to have little in the way of international news, and in the past two decades have dramatically reduced their foreign bureaus, on the assumption that mainstream audiences would have little interest in international news. Local television news perhaps best exemplifies commercial journalism, while a national newspaper such as *The New York Times* illustrates the highest quality commercial journalism.

Alternative journalism provides a contrasting approach, and when in print is often published on a weekly basis. Enterprise stories, or those featuring reporting about original story ideas linked to processes and trends, and not necessarily tied to events as is often the case with commercial journalism. *The Village Voice* is an example of a a leading alternative news weekly published in New York City.

Public journalism, based on a non-profit model, often features investigative, longer-form stories that may take a journalist or team of journalists a relatively long time to report or gather the facts. What's reported in public journalism emphasizes what editors feel is important to the public sphere, and not tailored to public appetites for celebrity news and other "soft" rather than hard news issues such as political affairs. An award-winning public journalism digital offering is ProPublica.org

Ethnic journalism features a broad array of news operations, but is defined by the targeted nature of coverage. Each ethnic news operation is focused on a particular ethnic community, such as Haitians, Indians, Chinese or Native Americans. A publication, in print, online or broadcast, offers content, often in a language other than English. Spanish-language ethnic media are perhaps the most widely available in the U.S.

Hyperlocal news is an emergent form of news reporting that focuses on the highly local, even block by block or neighborhood reporting. Invented by cable news operation News12, hyperlocal news is now common online and via mobile, including AOL's Patch, a series of community-based local news sites featuring local news often produced by citizen journalists and coordinated or curated by professional editors. Citizen journalists, or networked individuals who use mobile and social media to report news, are increasingly common in the Internet age. Of varying quality, citizen reporting can often provide on-the-ground video or other eye-witness accounts not available otherwise. For instance, citizen reporters using mobile devices provided via YouTube or other channels much of the video of the 2011 protests in Cairo's Tahrir Square during the early months of the Arab Spring in the Middle East (Baladi 2013).

8 Creating news: Covering events vs. pseudo events

Although journalism is generally thought of as involving journalists gathering information and telling stories about events or processes that occur spontaneously, research often suggests something very different (Pew Research Center Journalism Project 2010). Research shows that as many as three out of four stories reported in the news media is either generated by or features information originated in a planned event, or what historian Daniel Boorstin (1962) calls the "pseudo event". These "false" events are often part of a public relations effort to frame the news or to generate publicity. They are not necessarily an unethical manipulation of the news media or the public, and can be helpful in providing nuanced understanding of important news. For instance, an organization responsible for a natural disaster may hold a press conference as a means of disseminating information in a timely and complete fashion. On the other hand, newsworthy events may be planned as a means to generate free publicity for an individual or organization, on the belief that being covered in the news is a more credible form of public communication than paid advertising or promotion delivered via the organization's own media, such as a web site, blog or twitter feed.

Consider the case of the award-winning 2012 motion picture, Argo. Argo told the story of how the Central Intelligence Agency (CIA) planned a fictitious motion picture, a science fiction picture called *Argo*, as a cover to provide a means of aiding the escape of 6 persons caught up in the 1979 hostage crisis in Iran. A critical element in the success of the plan was to stage a pseudo event featuring actors in costume to announce the new production, using news media coverage of the *Argo* pseudo event to lend the fictitious movie credibility in the eyes of Iranian officials ... and it worked.

9 Distributing the news

A news event in which the facts have been gathered and packaged as a story must then be distributed to the public. Distribution historically involved either publication in print media (e.g., newspapers, magazines) or broadcasting (e.g., via radio or television). In the networked, digital age, the system for distributing news is much more complex, involving not only traditional channels but also via a variety of networked and digital devices, including online and mobile, as well as sharing via social networks such as Facebook, Twitter and the like. Moreover, various news aggregators, most notably Google, and other digital-first news organizations such as AOL combine both original reporting with news aggregation (e.g., collecting news produced by others and packaged for easy online or mobile consumption).

10 Technology and journalism in the digital world

Digital, networked technology (e.g., the Internet, wireless communication) has dramatically influenced journalism of the 21st century. In addition to the influences of technology on the types of journalism practiced and means of news delivery, emerging technologies are reshaping how individuals access the news. Mobile devices connected wirelessly to communication networks have grown dramatically in their use in the U.S. and around the world. Around the world, persons both young and old, rich and poor, widely use mobile devices for communication and increasingly to access news (and at least occasionally serve as citizen reporters). News media are increasingly responding to this shift in public communication patterns, and have widely created mobile apps as well as online presences as a means of packaging and delivering the news. The most innovative news organizations and journalists are also utilizing these digital media to create more interactive, multi-media forms of news stories. Location-based content such as augmented reality, tactile news interfaces (using touch screens to navigate a story) and big-data driven news graphics are among the new and emerging storytelling news formats. Social media integrated into news reports is increasingly common. Almost continuously updated 24/7 news is common, for better or worse, as competition and public appetites for the latest developments in breaking news grows, yet fact-checking suffers in times of lean, hyper-rapid news reporting (McLean 2014).

11 The business of journalism

As technology has driven a sea-change in journalistic practice and consumption, it has simultaneously spawned an overhaul of the financial underpinnings of most news organizations, whether commercial or non-profit, large or small, online or off. Advertising, the financial engine of much journalism since the advent of Benjamin Day's inaugural penny press newspaper in 1835, has experienced a seismic shift since the late 1990s (Pew Research Center 2013). By 2013, the majority of advertising dollars have fled from the world of newspapers and other traditional news media, which had held a monopoly or oligopoly on reaching consumers in most communities for nearly two centuries. Consequently, anyone seeking to reach consumers via advertising bought space or time in the media of mass communication. The dramatic growth in online and mobile media in the digital age has created a system where those seeking to sell goods and services can employ far more efficient, targeted digital tools such as Google, Facebook and Twitter to reach potential customers. Advertisers have shifted their ad dollars to the digital domain. News media, which are using digital media themselves to more efficiently reach their audiences, have faced a world where despite declining revenues (and parallel drops in audiences) have far outpaced falling costs. Instead, news media business leaders have moved with increasing urgency to develop new business or funding models based less on advertising and more on a diverse array of revenue generating techniques. Foremost among these is the introduction of the digital metered paywall. This is a business model in which consumers get limited access to a news source digitally (e.g., online or via a mobile device) but once a certain parameter is reached (e.g., 10 stories a month is the limit for *The New York Times* digital), they must pay. Evidence is increasingly indicating a willingness by consumers to pay for news, especially on mobile devices, on which many millions now do much or most of their news reading, viewing, listening and interacting (Phys.org 2014).

Consolidation is the keyword in U.S. news media ownership for nearly a century. Since World War II, the total number of daily newspapers has declined by more than a quarter. In 1940 there were 1,878 daily newspapers in the United States (Editor and Publisher Yearbook data 2007). By 2007 there were just 1,387, a decline of 491 (26 % decrease), leaving most communities with a single independent daily newspaper (Edmonds et al. 2012). Newspaper chains and broadcast groups today dominate the news enterprise. Large news operations buy others, whether smaller or not. To illustrate, in 2006 The McClatchy Company, then the eighth-largest chain in the United States, paid $ 4.5 billion to acquire Knight Ridder, then the second-largest newspaper chain (Hallinan and Berman 2006). The age of media convergence has accelerated this trend so that in 2013 very few news media online or off operate independently, at least for long, and few cities are served by truly competing local news sources. The primary exception is the hyperlocal online and mobile news operation, such as Dohanews.com in Doha, Qatar. The implications for democracy, at least on the local level, are potentially profound. Competing local newspapers had for more than a century provided the independent information an informed electorate needed to make enlightened decisions in most communities.

Technology is also making it possible for lean, efficient virtual organizations to emerge and produce high quality news at little cost. One example comes in the form of Inside Climate News (www.insideclimatenews.org). Inside Climate News has a reported annual budget of $ 550,000, most of which pays for staff, and the remainder mainly for travel and Internet services (Brian 2013). Yet, this highly lean and low-cost news organization founded in only 2008 won the 2013 Pulitzer Prize for national reporting. Its coverage of a major oil spill in Michigan involved a seven-month investigation by a news staff of three geographically distributed reporters working collaboratively online. *The New York Times* reported, "'We're a virtual organization,' said the publisher of the six-year-old Web site, David Sassoon, from his office in New York. So the celebration took place in a telephone conference call; whatever Champagne flowed, flowed in separate locations.'" (Stelter 2013: B4). The three reporters who worked on the story were each in a different city. "Elizabeth McGowan was in Washington, Lisa Song was in Boston and David Hasemyer was in New York (Stelter 2013: B4)."

Supporting the non-profit Inside Climate News is a revenue model built on individual donations from the public and charitable foundations.

12 Global perspectives

Although this chapter has focused on the state of journalism in the U.S., most of these patterns and history characterize news media in much of the world. The Internet has helped usher in an era of global interconnectedness in news and provided news consumers with unprecedented access to news from almost all parts of the globe. Citizen reporters increasingly provide a vital contribution to the news reported internationally, especially where international news bureaus have been in steep staffing declines.

13 Concluding reflections

Much has changed in journalism during recent decades, but not always for the better. Few communities are served by independent daily newpapers, and news staffs have shrunk considerably. Between 2008 and 2011, a quarter of all U.S. news media employees had been laid off or accepted buyouts (Auletta 2011). Yet, recent data suggest the employment situation is beginning to improve. Since the Hutchins Commission in 1947 challenged news media to mirror in their staffs the percentages of women and minorities in the general population, the percentage of minority employees has risen slightly, to just more than 13 percent, although this is still below the percentage in society at large (United States Demographic Profile 2014). Digital delivery of news has provided the public with access to ubiquitous on-demand news. News formats have diversified and are featuring an increasing array of multimedia and interactive stories. Will youth turn to this more fluid and dynamic news? Will the electorate of the 21st century be more engaged and informed? Will news media find a viable funding model to sustain quality journalism? These are just a few of the key questions facing journalism in the years to come. Public engagement in news may increase with the rise of mobile media and the increasing use of social media as some 5 billion persons come online around the world (Schmidt and Cohen 2013). Whether it leads to better democratic processes or increased understanding of matters of public importance remains to be seen.

References

Auletta, Ken. 2011. You've got news. Can Tim Armstrong save AOL? The New Yorker, January 24. http://www.newyorker.com/reporting/2011/01/24/110124fa_fact_auletta (Accessed 18 February 2013).

Bly, Nellie. 1887. Ten Days in a Mad-House. New York: Ian L. Munro. Published with "Miscellaneous Sketches: Trying to be a Servant" and "Nellie Bly as a White Slave".

Baladi, Lara. 2013. When Seeing is Belonging: The Photography of Tahrir Square. http://creativetimereports.org/2013/09/16/lara-baladi-photography-of-tahrir-square/ (Accessed 15 December 2014).

Boorstin, Daniel J. 1962. *The Image: A Guide to Pseudo-Events in America*. New York: Knopf Doubleday Publishing Group.

Dunne, Finley Peter. 1898. *Mr. Dooley in Peace and in War*. Boston: Small, Maynard & Company. vii–xiii.

Editor and Publisher Yearbook data. 2007. Number of U.S. Daily Newspapers, 5-Year Increments. *Pew Research Journalism Project*. http://www.journalism.org/node/1134 (Accessed 18 February 2013).

Edmonds, Rick, Emily Guskin, Tom Rosenstiel & Amy Mitchell. 2012. Newspapers: By the Numbers. *The State of News Media 2012. An annual report on American Journalism*. http://stateofthemedia.org/2012/newspapers-building-digital-revenues-proves-painfully-slow/newspapers-by-the-numbers/ (Accessed 18 February 2013).

Hallin, D. and P. Mancini. 2004. Comparing Media Systems: Three Models of Media and Politics. Cambridge: Cambridge University Press.

Hallinan, Joseph T and Denis K. Berman, 14 March 2006. "McClatchy Agrees to Acquire Knight Ridder for $ 4.5 Billion." *The Wall Street Journal*. http://www.wsj.com/articles/SB114224828072596508 (Accessed 15 December 2014).

Hutchins, Robert. 1947. A Free and Responsible Press. https://archive.org/details/freeandresponsib029216mbp (Accessed 15 December 2014).

Lippmann, Walter. 1931. Two Revolutions in the American Press. Yale Review. 433–41.

McLean, Bethany. 26 March 2014. "State of Journalism: The Lost Art of Fact Checking." https://www.linkedin.com/pulse/20140326144532–56143-state-of-journalism-the-lost-art-of-fact-checking (Accessed 15 December 2014).

Nerone, John C. 1995. *Last Rights: Revisiting Four Theories of the Press*, University of Illinois Press, On Social Responsibility, 77–100. Reprinted in McQuail's *Reader in Mass Communication Theory*, John C. Nerone, "Social Responsibility Theory," Ch. 15.

O'Toole, G. J. A. 1984. The Spanish War: An American Epic – 1898. W. W. Norton & Company: New York.

Pew Research Center's Project for Excellence in Journalism (2013). "The State of the News Media 2013." http://www.stateofthemedia.org/2013/digital-as-mobile-grows-rapidly-the-pressures-on-news-intensify/ (Accessed 15 December 2014).
Pew Research Journalism Project. 2014. "Principles of Journalism. http://www.journalism.org/resources/principles-of-journalism/ (Accessed 15 December 2014).
Pew Research Journalism Project. 2010. "How News Happens." http://www.journalism.org/2010/01/11/how-news-happens/ (Accessed 15 December 2014).
Phys.org. 5 August 2014. "Study Shows Consumers Ready to Pay for Online News." http://phys.org/news/2014-08-consumers-online-news.html (Accessed 15 December 2014).
Schiller, Dan. 1981. Objectivity and the News: The Public and the Rise o Commercial Journalism. Philadelphia, PA: University of Pennsylvania Press.
Schudson, Michael. 1978. Discovering the News: A Social History o American Newspapers. New York: Basic Books.
Schmidt, Eric and Jared Cohen. 2013. *The New Digital Age: Reshaping the Future of People, Nations and Business.* New York: Alfred E. Knopf.
Sinclair, Upton. 1905. *The Jungle: The Uncensored Original Edition.* Tucson, AZ: See Sharp Press. vi. ISBN 1884365302.
Society of Professional Journalists. 2014. SPJ Code of Ethics. http://www.spj.org/ethicscode.asp (Accessed 15 December 2014).
Stelter, Brian. 2013. A Pulitzer Prize, but without a newsroom to put it in. The New York Times, April 16: B4. http://www.nytimes.com/2013/04/17/business/media/insideclimate-news-hopes-to-build-on-pulitzer.html?_r=0 (Accessed 15 December 2015).
Tarbell, Ida M. *All in the Day's Work*, Urbana and Chicago: University of Illinois Press, 2003.
Tewksbury, David & Dietram Scheufele. 2009. News framing theory and research. In Jennings Bryant & Mary Beth Oliver (eds.), *Media effects: Advances in theory and research* (3rd edn.), 17–33. New York: Routledge.
United States Demographic Profile. 2014. http://www.indexmundi.com/united_states/demographics_profile.html (Accessed 15 December 2014).
Washington Post Standards and Ethics. http://asne.org/content.asp?pl=236 & sl=19 & contentid=335 (Accessed 15 December 2014).

Further readings

Boczkowski, Pablo. 2005. *Digitizing the News: Innovation in Online Newspapers.* MIT Press.
Downie, Leonard Jr., Robert G. Kaiser. 2002. *News About News: American Journalism in Peril.* Knopf.
Kovach, Bill & Tom Rosenstiel. 2001. *The Elements of Journalism: What Newspeople Should Know and the Public Should Expect.* Three Rivers Press.
Rosales, Rey. 2006. The Elements of Online Journalism. iUniverse.
Quinn, Stephe. 2005. *Convergent Journalism: The Fundamentals of Multimedia Reporting.* Peter Lang Publishers.

Stephen M. Griffin

29 Libraries in the digital age: Technologies, innovation, shared resources and new responsibilities

Abstract: This chapter will examine the rapidly expanding role of libraries in an unprecedented time of change driven by digital information, communication technologies and linked knowledge infrastructures. The focus will be on academic libraries, although many of the topics discussed will apply to public libraries and memory institutions more generally. Digital scholarship, characterized by computation and data-intensive tasks, has sparked transformative inquiry into new topic domains and altered the intellectual boundaries of many disciplines. As a result, scholarly workflows are becoming more complex and yet more distinctive. Comprehensive reporting of digital scholarship requires new models of scholarly communication that cannot be based on print media. Open access to publications and datasets is now a common condition for funded research. Libraries as knowledge institutions are at a unique and opportune time for examining new services and resources that support digital scholarship across the disciplinary spectrum, while also developing innovative practices that serve patrons from the broader population. The future of many libraries is likely to be defined by decisions made and actions taken in the very near term. The wisdom informing these decisions can best be acquired by continuing to participate in a rich global dialog and moving forward in a spirit of collaboration and cooperation.

Keywords: libraries, digital scholarship, digital libraries, linked data, semantic web, interoperability, open access, scholarly communication

Now well into the 3rd decade of the emergence and explosive growth and use of digital information, the debate on how libraries, museums, archives and other knowledge institutions should adapt and positively respond to the new information environment remains animated and passionate. The inherent unpredictable nature of technological change and use are such that planning discussions proceed in a volatile context – one that engages many actors in pursuit of multiple and often conflicting objectives.

Expert groups commissioned by research and policy organizations have produced an abundance of reports that attempt to identify core principles and best practices for creating, managing and using digital information. Common to the findings is that digital information has greatest value when it becomes part of a dynamic, growing and globally linked knowledge infrastructure – one that expands gracefully in scale, functionality and reach, resulting in a complex, adaptive ecology of interdependent resources in which enriching one serves to add value to others and to the whole.

A broad definition of a knowledge infrastructure will be used in this document and include systems and standards, digital content and the broad range of user services related to the information lifecycle. The reflexive relationships that exist between the development of infrastructure technologies and human interests are palpable and continuous.

The impact of the Internet as a new medium and conduit for knowledge has been such that long-established roles of libraries and other memory institutions as primary centers for collecting, organizing and providing knowledge has in many ways been eclipsed. Digital information can be acquired and managed in a decentralized way. Institutional and subject-specific repositories are just two examples. Also affecting libraries is the changing nature of scholarly practices and workflows. Research and scholarship involving large-scale, diverse datasets and computationally intensive tasks have multi-stage and complex workflows. Internet-based open data, technologies and tools are expanding the problem space of scholarship across disciplines and stimulating interdisciplinary collaboration. Considerable efforts are being devoted to creating new analytical methods and means for communicating research output.

Libraries and other memory institutions must adapt to the new realities associated with scholarly practices wed to digital resources that have no permanent home, but rather are constantly circulating within a global, network-centric knowledge environment. Often overlooked is the fact that the versatile, pliable and mercurial properties of digital information that make it so potent for transformative scholarly work also presents serious challenges for data stewardship. Digital information can be represented in multiple forms, be modified with ease and replicated and disseminated across continents with just a few keystrokes. Without a detailed record of data analysis steps associated with a research project, reproduction and validation of results may not be possible.

Accompanying data-centered approaches to inquiry have been cultural shifts in the scholarly community that challenge long-standing assumptions that underpin the structure and mores of academic institutions, as well as to call into question the efficacy and fairness of traditional models of scholarly communication. Many scholars are now demanding new models of scholarly communication that capture a comprehensive record of workflows and accommodate the complex and accretive nature of digital scholarship. Scholars are also calling for greater authority in the publication of their works and rights management.

Many memory institutions have redefined their missions and begun broad programs of change, often in concert with other organizations. Partnerships both distribute common tasks and ensure coherent agreement on fundamental principles for the creation, use and stewardship of digital content. Aggregation of resources across institutions must be a primary goal. To attain levels of interoperability in which distributed repositories, tools and other services appear and function as unified resources requires agreement on definitions, descriptions, representations, and structures of individual information objects and creating repositories that conform to these specifications. The result will be a dramatic increase in functionality and scholarly productivity. The World Wide Web is increasingly seen as the infrastructure for advancing interoperability at the level of individual information objects through semantic web and linked data standards and principles. In this way it changes from a web of linked documents to a web of linked data.

This paper will explore needs and opportunities for libraries in support of learning and scholarship in the coming years as more digital collections, tools and services are integrated into libraries operations. Libraries will assume new responsibilities, including actively supporting domain scholars in their research and establishing new models and mechanisms for scholarly communication. While academic libraries will be the focus of this paper, much of what is discussed also applies to public libraries, archives and museums.

Primary goals for libraries include undertaking the following activities at a scale and pace allowed for by available resources and to collaborate closely with other institutions performing similar activities:

- Convert unique analog holdings and special collections to digital form; establish a web presence that allows discovery, exploration and examination of holdings
- Link information objects within local collections and to those of other institutions at the data level
- Increase access to all types of digital content; develop and acquire data analytic toolsets for all types of media
- Maintain a strong focus on data stewardship across all stages of the information lifecycle
- Engage and assist local researchers and scholars in capturing and organizing scholarly workflow information for repurposing and reuse
- Participate in international dialogs on needs and opportunities
- Reduce journal subscription expenditures; redirect funds to high priority areas

The above goals imply that libraries assume a new role as publishers by: a) exposing their digital holdings and resources to all; b) implementing mechanisms by which data from the collections can be easily accessed through the internet; c) actively engage in the production of open access journals, monographs, datasets, executable documents and other forms of creative and educational digital works. Libraries as publishers will occupy a prominent space in the hierarchy of internet based information – one recognized by scholars and students as providing trustworthy, authoritative and carefully managed digital as well as analog materials. Libraries and other memory institutions have much to gain by engaging the public in conversion of their analog collections and enhancement of digital repositories, as many tasks cannot be automated. *Crowdsourcing* as a means to transcribe, tag and annotate specific types of content has proven to be an economic and credible alternative to outsourcing these tasks to commercial vendors.

Several key themes will appear through out this paper and serve as the organizing principles for topics addressed. The themes are interoperability, scalability, collaboration and context. Interoperability as noted above allows individual collections and services to function as a unified resource. Scalability allows repositories and collections of digital objects to increase in size without degradation in performance or functionality. Collaboration refers to people and institutions working together toward common ends. Context implies structuring and enhancing digital resources in ways that clarifies or extends meaning and usability.

1 Digital information and digital scholarship

In the mid-20th century, few people could imagine that information represented in its most elemental form – combinations of two values, commonly expressed as 0s and 1s – could be transformed by large-scale computing and communications networks into representations of processes and phenomena of astonishing scale, complexity and precision. That this revolution occurred so quickly, over the period of several decades, is one of historic significance whose consequences continue to unfold. Radical change associated with new information and communications technologies coupled with the creation of massive amounts of digital data have sparked transformative change in methods for discovery and exploration of new problem domains. Contemporary research and scholarship across the disciplinary spectrum is increasing characterized and dependent upon the use of large-scale internet-based data sets and computationally intensive tasks. Scientists are using vast amounts of data to explore galaxies, measure changes in natural systems, create genetic profiles of living things and study the changing behaviors and mores of societies and individuals in an increasing populated and fragile physical world. Similarly, humanists are using new information types, methodologies and tools to transform and expand their scholarly endeavors. Examples include the creation and use of digital representations of material culture by historians, introducing spatial and temporal indexed data into the study of literature and history and creating unique information visualizations to recognize, probe and analyze complex relationships (Griffin and Gradmann 2013). The open nature of the internet-based resources has resulted in changes in the research culture that has proven to transform scholarly practices as well. Rich dialog, shared and open access to resources and an emphasis on transparency pervades contemporary digital scholarship. Figure 1 shows samples of information visualizations from research projects in the sciences and the humanities.

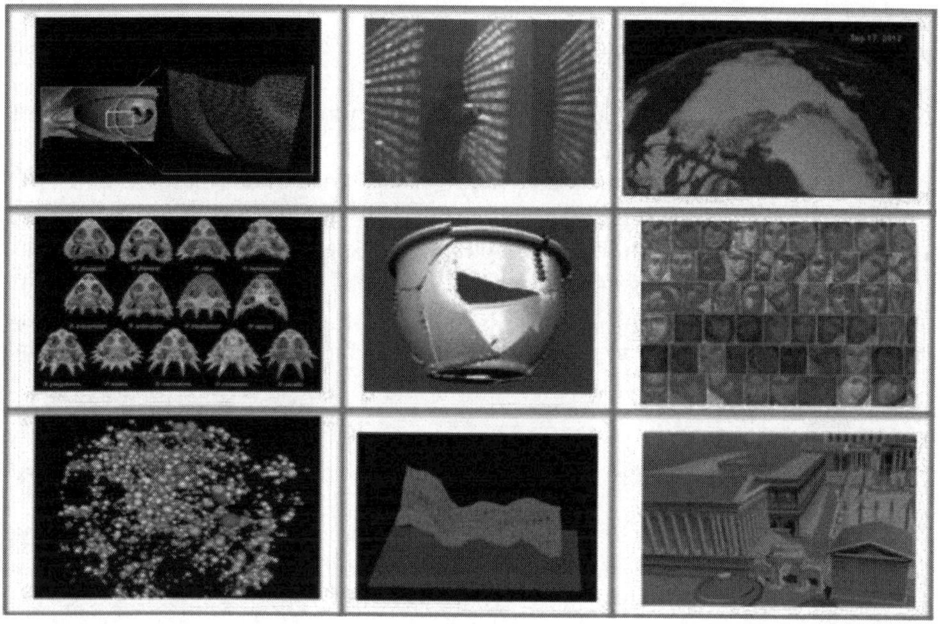

Fig. 1: Examples of Digital Scholarship.

For the purposes of this discussion, data is organized into three broad categories associated with three types of research and scholarly work and include: a) computational or eScience research; b) data-driven research; and, c) data-intensive research and scholarship. The categories roughly demarcate theoretical sciences, experimental sciences and the humanities and arts.

1.1 eScience

eScience as used here connotes large-scale simulations of phenomena too large or small, too fast or slow, too complex or otherwise impossible to explore in a research laboratory setting. Examples are simulating cosmic forces, probing the internal structures of atomic particles, constructing environmental systems models, modeling molecular structures and predicting long-term changes in the natural systems. Data consist of solution sets to mathematical models computed over broad parameter spaces. The complexity and fidelity of the computed model is directly dependent upon processor speed, primary memory and computational algorithms. The eScience community is notably cohesive. International collaboration among researchers and facilities is firmly established – in part because of the high cost of the computational infrastructure. A strong culture of open science encourages researchers to post results and data in open access repositories soon after work is complete.

Examples:
1. Netherlands eScience Center: Understanding the effects of climate change (www.esciencecenter.nl/projects)
2. Los Alamos National Laboratory Science and Innovation Research: multidisciplinary simulations (www.lanl.gov:science-innovation:index.php)
3. San Diego Supercomputer Center (SDSC): "M8" earthquake simulation (http://phys.org/news/2010-11-m8-earthquake-simulation-quake.html)
4. RIKEN Advanced Institute for Computational Science campus in Kobe, Japan: Simulation of largest general neuronal network to date (http://www.riken.jp/en/pr/press/2013/20130802_1/)
5. Pittsburgh Supercomputing Center: Multiscale modeling of biological systems (http://www.mmbios.org)

1.2 Data-driven research

Data-driven research refers to analysis of very large experimental and observational data generated by high throughput digital devices or data collected through human directed efforts. As in eScience, massive data sets can be produced. Particle accelerators can produce petabytes of data each day. Environmental sensors and telescope arrays produce similar high volumes of data. Of special value are data sets collected over decades or even centuries. These give scientists and other scholars a "long view" of change in natural and social systems. Social scientists collect data from social networks to study the habits, preferences and information flow patterns across populations. The development of new tools for web science, data science, data analytics and information visualization has proven to be essential to advancing data-driven research. Some sites are using "crowdsourcing" to enrich content. One example is the BBC World Service allowing members of the general public to annotate and add details to its archive of 45 years of radio programs (Ferne 2013).

Examples:
1. Sloan Digital Sky Survey: The Sloan Digital Sky Survey has obtained and makes freely available multi-color images covering more than a quarter of the sky. The survey has also created 3-dimensional maps containing more than 930,000 galaxies and more than 120,000 quasars (www.sdss.org)
2. Inter-university Consortium for Political and Social Research (IPCSR): International consortium of more than 700 academic institutions providing data, resources and tools, and training for the social science research community (www.icpsr.umich.edu/icpsrweb/landing.jsp)
3. Large Hadron collider: The accelerator at CERN produces and processes raw data at a petabyte per second (home.web.cern.ch/topics/large-hadron-collider)
4. Truthy: Research project that focuses on Twitter communication about politics, social movements and news; the gallery of social networking visualizations gives a number of examples visualizing hundreds of thousands of tweets (http://truthy.indiana.edu/)

5. Oyez Project: Audio and text transcripts of more than 7000 Supreme Court hearings (www.oyez.org)

eScience and Data-driven research are often considered part of a larger rubric of "big data" or "data science at scale". Many of the applications of "big data" operate at the extremes of scale and complexity in modeling and predictive analytics. Assessment of solution sets is key to quantifying uncertainties, validating results and ensuring data integrity. Data stewardship of large-scale data sets has not been sufficiently addressed and it is urgent that this be done.

1.3 A new burden of evidence: eScience and data-driven research

Defining features of scientific research include repeatability and reproducibility. Repeatability refers to the ability to duplicate an experiment under the same conditions and obtain the same result. Reproducibility refers to the ability for others to reproduce the work in different environments and obtain the same result, thus setting the stage for extending its use and value. These requirements hold for all theoretical and empirical research areas. Replication of results using proven methodologies confirms the veracity of the research process and outcomes.

Reporting the results of computation and data-driven research poses unmanageable challenges for traditional publication models. In order to reproduce reported results means having a comprehensive record of the research process and workflow including:
- Process records: algorithms, annotated source code, software pipelines and versioning, datasets and transformations, storage formats and protocols, event tracing, ...
- Resource descriptions: journals, logs, tools, methods, dialog, collaborative activities and external contributions, ...
- Intermediate forms: temporary models, concept changes, recursion points, software versions, external dialogs and contributions, ...
- Workflow artifacts: transcriptions, translations, annotations, steps taken to acknowledge distribution of effort, attribution and credit, ...

These steps are necessary if the goal of fully transparent research publications, widely endorsed by many, is to be attained. New automated methods are being developed to capture and publish a full description of the computational processes. Notable efforts include:
- VCR – Verifiable Computational Research (http://vcr.stanford.edu/)
- VisTrails (http://www.vistrails.org/index.php/Main_Page)
- Share (http://is.ieis.tue.nl/staff/pvgorp/share/)

The examples embody the concept of executable documents in which data assets associated with each stage of the workflow are created as executable components.

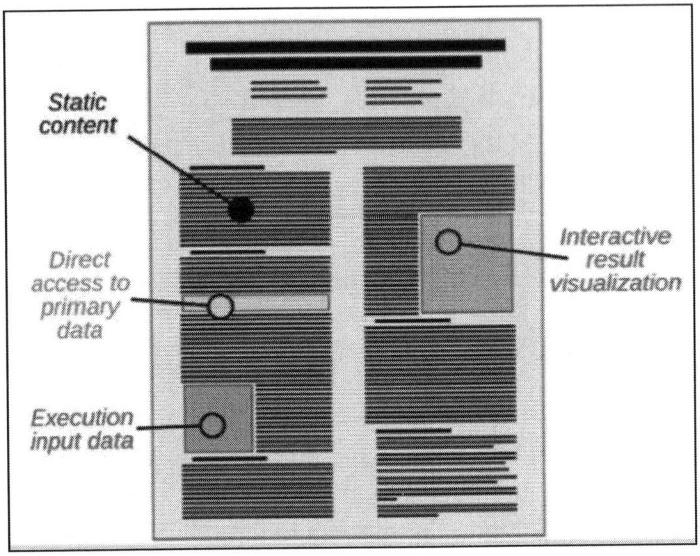

Fig. 2: Example of Executable Document Model (Nowakowski et al. 2011).

This approach allows staged publication, validation and reproducibility, while at the same time providing important provenance information. Figure 2 shows one form of a executable document that can more fully express the scholarly work on which it is based.

1.4 Data-intensive scholarship and research

Methods developed for eScience and data-driven research are of limited use to humanists. For the sciences, data as captured and recorded is often ready for use in experiments. For humanists considerable effort must be invested into the gathering and preparation of data to bring it to the form necessary to begin the tasks of establishing factual relationships, substantiating premises and building strong arguments leading to deeper understanding. When factual evidence can be situated into a larger historic context the boundaries of a study expands. Scholars in the "digital humanities" use new tools, often modified versions from the sciences, to assist in the discovery, exploration and visual representation of objects, relationships and environments.

Data-intensive research is taken here to refer to a broad range of scholarly activities characterized by use of many types of information objects derived from a variety of mediums. Source data may be located in distant geographical locations. Data-intensive scholarship can proceed through an exceptionally large number of scholarly workflow stages. For the discussion here, data will be categorized into three broad types.

A) Hybrid datasets: a combination of text, still images, moving images, illustrations, maps, audio, music, video, and other mixed and multi-media data entities. Data may be repurposed – collected for one application area, but later used in other problem domains. One example is the integration of spatial and temporal data to better understanding long-term changes of human settlement patterns and changing populations and ranges of flora and fauna. For "deep readers" of history and literature alike, stories and descriptive material can be overlaid onto maps and timelines such that the movements of characters and details of the physical environments at the time can lead to new, vivid immersive experiences full of interpretive possibilities and richness. New signal processing techniques can be applied to digital audio recordings of music ensembles, profiling individual instruments, and thereby identifying its unique sound in the overall performance.

Examples:

1. The Electronic Atlas Initiative (www.ecai.org)
2. Atlas of the Historical Geography of the United States (http://dsl.richmond.edu/historicalatlas/)
3. British Museum Semantic Web Collection (http://collection.britishmuseum.org/)
4. ECHO – Cultural Heritage Online (http://echo.mpiwg-berlin.mpg.de/home)
5. Oyez Project (www.oyez.org)
6. BBC World Service Radio Archive (http://worldservice.prototyping.bbc.co.uk/)
7. World History Center (www.worldhistory.pitt.edu)

B) Digital representations of physical entities and spaces: Digital facsimiles, or surrogates, of significant cultural and historical artifacts permit shared, distant, and non-invasive study. Ancient scripts, objects and places can be visually enhanced, rendered, reconstructed and presented in the form in which they were created. Figure 3 gives one example. Lost text can be recovered. Broken objects can be reassembled. Paintings can be analyzed according to layers of pigments, brush strokes and knowledge of chemical change over extended time periods. Archaeological sites can be restored, virtually, using 2D and 3D imagery supplemented by the contextual descriptions from historians and travelers in earlier times. Valuable cultural works of art can be digitized and visually enhanced as in Figure 3. All of these approaches allow detailed study not otherwise possible. Digital models are easily shared and aggregated into large collections. They can also be printed in 3D form, producing physical replicas at desired levels of scale and resolution. Of the examples below, several are pioneering efforts that spawned or otherwise led to transformative methods for humanistic research.

Examples:

1. Digital Hadrian's Villa Project (vwhl.clas.virginia.edu/villa/)
2. Venetus A Illiad Manuscript (chs.harvard.edu/wa/pageR?tn=ArticleWrapper&bdc=12&mn=1560)
3. Digital Michelangelo Project (graphics.stanford.edu/projects/mich/)

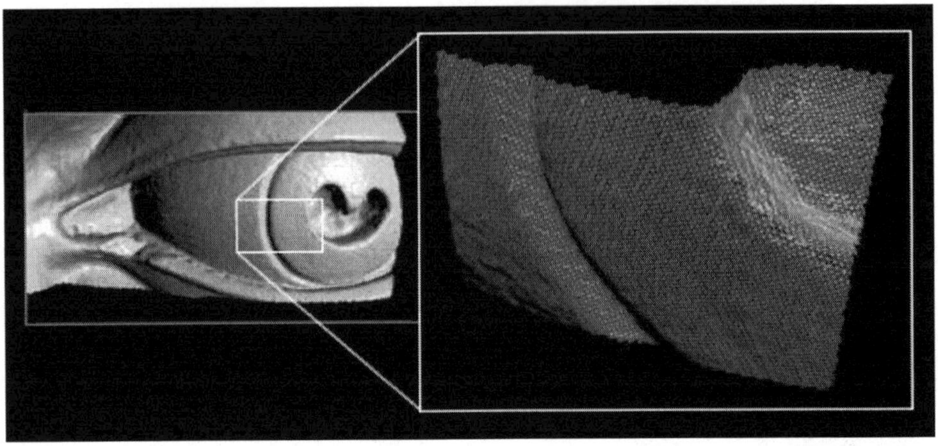

Fig. 3: Rendering of Michelangelo's David Left Eye Accurate to .25 mm, Marc Levoy, Stanford University.

4. The Archimedes Palimpsest (www.archimedespalimpsest.org)
5. Cuneiform Digital Library Initiative (cdli.ucla.edu)
6. Digital Morphology Library (www.digimorph.org/index.phtml)

C) Digitization of special collections and most valuable holdings: Some of the most valuable data to humanists, as well as researchers studying processes of social, cultural and technological change are found in library and museum collections in the form of unstructured text, periodicals and ephemera, notebooks, correspondence, maps, photographs, film, sketches, field notebooks, music, maps, specimens and artifacts, measurements and observations. These are primary sources for many scholars. Many projects and organizations are doing this.

Examples:

1. The European Library (www.theeuropeanlibrary.org:tel4) and Europeana (www.europeana.eu)
2. Library of Congress Map Collection (memory.loc.gov/ammem/gmdhtml/gmdhome.html)
3. Digitized Medieval Manuscripts Maps (DMMmaps) (digitizedmedievalmanuscripts.org)
4. Polonsky Foundation Digitization Project (bav.bodleian.ox.ac.uk/)
5. Integrating Digital Papyrology (papyri.info/)
6. French Revolution Digital Archive (frda.stanford.edu/en/about)

Traditional humanities scholarship, with a focus on textual studies and deeply-rooted analytical processes is being energized by new forms of knowledge, modeled in digital environments and containing complex digital objects, medium specific and designed for computational manipulation. Tools for understanding the course and benefits of interpretive choices are often derived from other disciplines. Spatial humanities is one example and electronic cultural atlases another. The roster of topics for conferences such as *Computer Applications and Quantitative Methods in Archaeology* (caa-international.org) reveals a potent blend of traditional and new computational techniques applicable to archaeological and related research: remote sensing and imaging; augmented reality; pattern and feature recognition; and, computing technologies uniquely designed for archaeological study. To explore the course of human expression prior to writing requires aggregating material held by many institutions and the involvement of scholars from multiple domains. The Cuneiform Digital Library Initiative is an outstanding example.

Yet, scholars in the humanities and arts have a relatively small amount of digital source data available to them. A vast portion of the historic human record is still found in the analog collections of memory institutions. The situation is often one of great fragmentation and neglect – many types of artifacts in different locations on fragile and deteriorating media. A very small portion has been converted to usable digital form. The pace of conversion is limited due to lack of resources and prolonged decision-making processes associated with technologies, media formatting, access policies and most importantly, staffing. The current lack of coherence in global collections impedes humanists' efforts to acquire factual descriptions of historic events and circumstances and to cogently examine conditions and events from within causal frameworks.

The digitization of printed materials has been ongoing for more than a decade. Projects such as Google Books, HathiTrust and Open Library have digitized and indexed textual materials on a massive scale – the combined collections of these exceed 20 million volumes. The Internet Archive (www.archive.org) has archived nearly 4 billion web pages. Memento (www.mementoweb.org), developed by researchers at Los Alamos National Laboratory and Old Dominion University allows users to recover lost web pages from the past and follow the development of web-based resources. Researchers in text mining, text analytics, predictive modeling and other domain specialties have found special value in large-scale examination of text corpora. Large digital text corpora have spawned new research domains. One such domain is "culturomics", that focuses on linguistic and cultural phenomenon as manifest in changing usage of the English language. One study on trends in "mood" words used the Google Ngram database that comprises approximately 5.2 million books and 361 billion words in English. The evidence for "joy" and "sadness" over the period 1900–2000 is shown in Figure 4. Further study showed differences between variants of English associated with different countries. As might be expected, there is considerable controversy regarding the use of such methodologies and the meaningfulness of the results obtained (Acerbi et al. 2013).

Optical character recognition alone can exclude valuable information and is limited to modern languages and typefaces. Page imaging is required to capture

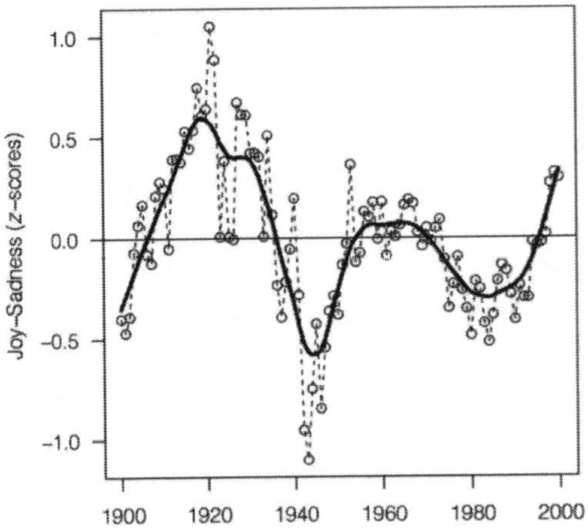

Fig. 4: The Expression of Emotion in 20th Century Books (Acerbi et al. 2013).

handwritten documents, inscriptions and markings such as annotations, scholia, tables and illustrations as well as other traces of use and visual clues of considerable scholarly value. Page imaging adds considerable cost and storage overhead – a single page may require multiple images to recover markings only visible under light of specific wavelengths. Newer approaches involve fluorescence of the chemical remnants of the writing fluid used, a phenomenon that might only occur when the document is subject to high-energy particle beams from atomic accelerators. Image management and retrieval software tools are complex and involve special technical expertise. Optical music recognition (OMR) exists, but is not designed to handle large volumes of data.

Figure 5 shows a priority scheme for an academic library whose mission is to serve the entire academic community (Griffin 2013).

There is little need for libraries to provide data storage and software support for eScience and Data-driven research. National and disciplinary data centers have taken on this responsibility and for the sciences are relatively well funded. Currently there are international efforts such as the Research Data Alliance (RDA), www.rd-alliance.org, and the Global Research Data Infrastructures Initiative (GRDI2020), www.grdi2020.eu, dedicated to open sharing of data across national boundaries. National centers for disciplinary research in the physical sciences, geosciences and biological sciences have been in place for decades (e.g. ncar.ucar.edu/). Libraries should maintain current reference and descriptive materials relevant to ongoing work and provide toolkits for users who are interested in analyzing the datasets, but even this may be beyond many libraries means.

A high priority is given to conversion of analog holdings into digital form, although this may be costly and time consuming. Often, a library is known for its special collections that contain unique and singular assets. To make these internet-accessible is an exceptional service to external communities of scholars and per-

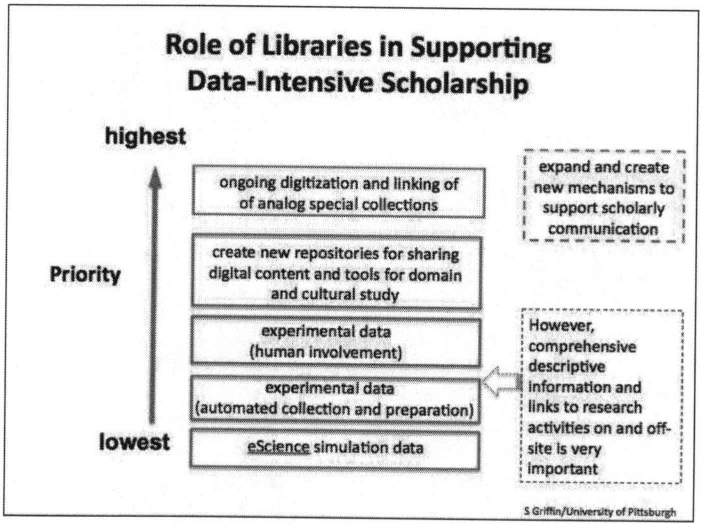

Fig. 5: Priorities for Libraries.

mits the development of larger, more complete and more sophisticated aggregate collections that in turn, allow new lines of inquiry and increased scope of studies. Conversion into digital form does not imply a devaluation of the analog collection. Both should be maintained for the long term. It has been the case that new technologies extract additional information of value from analog materials on a regular basis.

1.5 Data and information visualization tools

New tools are required to analyze and present vast data sets. A great amount of effort has been devoted to develop information visualization and analytical toolsets. Publications and conferences are devoted to the topic (e.g. ieeexplore.ieee.org/xpl/RecentIssue.jsp?punumber=2945). Using such methods assists in (Ware 2004):

- Facilitating the understanding of large amounts of data
- Promoting the perception of emergent properties that were not anticipated
- Highlighting problems in the quality of the data
- Promoting the understanding of large-scale and small-scale features of the data
- Hypothesis formation

Many data visualization tools for analysis and exploration are openly available on the web, as well as samples and templates for creating effective visualizations.

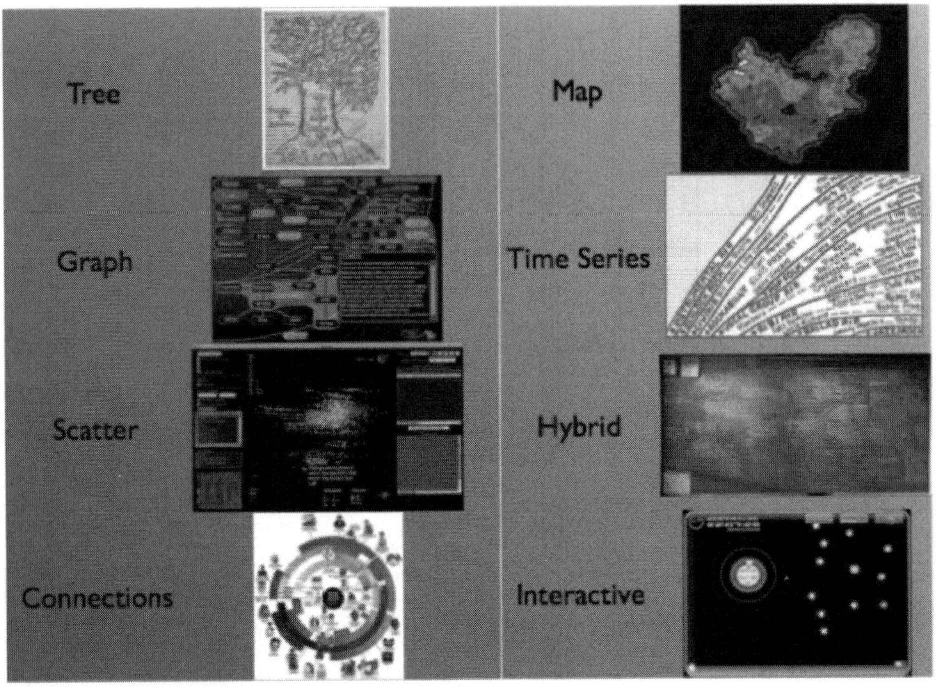

Fig. 6: Types of Visualizations.

Examples include:
- SIMILE: simile-widgets.org
- Tableau: www.tableausoftware.com
- Gephi: gephi.org/
- IBM Many Eyes: www-958.ibm.com/
- Truthy: truthy.indiana.edu/
- ARC 3D: homes.esat.kuleuven.be/~visit3d/webservice/v2/
- Places and Spaces: Mapping Science: scimaps.org/maps/map/tree_of_life_113/detail/
- Humdrum Toolkit: www.musiccog.ohio-state.edu/Humdrum/
- Synapsify: gosynapsify.com/

Most recently, registries of research tools for digital humanists and other primarily non-scientific applications' use are becoming well established. A notable example is the DiRT Directory, containing hundreds of open source tools. It can be found at Digital Research Tools: http://dirtdirectory.org/.

Among the more popular types of data visualizations for large data sets are network diagrams (relationships), bar and bubble charts (value comparison), line and stack graphs (changes over time), pie and tree maps (fractions of a whole), word tree and tag clouds (basic text analysis) and various forms of geographical maps and historic atlases. Some examples are shown in Figure 6. The widespread appeal and demonstrated value of visual analysis has resulted in new specialties within university departments. Termed "visual studies" in many, these studies offer new methods and approaches for scholars and students in many disciplines including art history, archaeology, cultural studies, film and media studies, anthropology, sociology, and the humanities.

2 Interregnum: The digital libraries initiatives and coming to terms with digital content

Much of the seminal work in developing the data types, information environments and analytical resources that support data-intensive research and digital scholarship can be linked directly to digital libraries research – past and present. Pioneering digital libraries research illuminated essential core information architectures and environments and inspired a generation of researchers to look beyond the confines of their own discipline, collaborate with others and pursue interdisciplinary projects.

The USA Interagency Digital Libraries Initiative (DLI) proceeded through 10 years of funding projects and activities that were instrumental in demonstrating the power and scope of data-centered scholarship. The phrase "digital libraries" became a metaphor for organizing discussion of the use of digital content in the internet age; one that inspired creative thinking and innovative research into the creation and use of large-scale, distributed information environments and new communication and learning practices. International collaborative programs were established with the Joint Information Systems Committee (JISC) of the United Kingdom, the Deutsche Forschungsgemeinshchaft (DFG) of Germany and the European Commission. International working groups focusing on specific topics were organized with the European Commission funded DELOS Network of Excellence. The logo for the Digital Libraries Initiative, shown in Figure 7 was a prominent feature on the web pages of projects and working groups sponsored by the DLI.

The Digital Libraries Initiative produced many of the new tools and approaches for data-intensive research, in some cases spawning entirely new research areas. Google, Inc., a result of research in the Stanford Digital Library Technologies project, was the most significant commercial success. Fedora, EPrints and the Open Archive Initiative repository projects were funded, in part, through the collaborative program with the Joint Information Systems Committee. Other new technology fields initiated or catalyzed by the DLI projects included:

- Computational photography
- Enhanced non-invasive 2D and 3D imaging technologies
- Music information retrieval and analysis

Fig. 7: Digital Libraries Initiative Logo.

- Integration of spatial and temporal indexed data into global cultural and historic atlases
- Character recognition for cursive text
- Automatic classification of very large image collections
- Computational tools for modeling, visualizing and analyzing archaeological sites
- Reconstruction of cultural artifacts from large sets of fragments
- Multi-lingual, mixed media digital libraries
- Digital libraries for children
- New approaches for provenance, curation and preservation of archival material
- ... and many others

Digital libraries research revealed a new realm of possibilities for scholars in many disciplines. It marked a transition from a disproportionate amount of attention on computational applications in the sciences to data-centered scholarship across many subject areas – including the humanities and arts. The second phase of the program provided funding to researchers from more than 30 academic departments, many engaged in interdisciplinary projects (Griffin 2005).

3 Open access and interoperability

3.1 Open access

The debate on open access to scholarly resources has had an enormous and beneficial influence on research productivity. Free and immediate access to research findings accelerates the research cycle and offers new models for forms of expression and dissemination of scholarly works. Widespread support for web-based access to research publications has broadened to include access to algorithms, code and data sets. Funders and institutions around the world have established open access policies and issued mandates for open access publishing. Open access is an essential part of the broader dialog on necessary changes in scholarly communication practices and a broader critical assessment of the disciplinary departmental structure and rewards systems in academic institutions.

A comprehensive timeline of open access milestones was created by Peter Suber, widely considered as one of the pioneers of the open access movement (OAD 2013). Stevan Harnad was an early and ardent advocate for open access and moderated the American Scientist Open Access (AmSci) Forum, begun in 1998 and hosted for 13 years (1998–2011). The Global Open Access List (mailman.ecs.soton.ac.uk/mailman/listinfo/goal), is the successor of the AmSci Forum. Together, these forums are a comprehensive and incisive record of the open access discussion.

The Directory of Open Access Journals (www.doaj.org), lists almost 10,000 journals containing more than 1,500,000 articles searchable at the article level. The Registry of Open Access Repositories (roar.eprints.org/) lists more than 3500 open access repositories.

The Registry of Open Access Repositories Mandatory Archiving Policies (ROARMAP, roarmap.eprints.org/), is a central registry updating open access mandates.

The first large-scale example of an open access repository was the physics preprints repository arXiv (arxiv.org). Today arXiv contains more than 900,000 articles from a variety of disciplines in the natural sciences. The arXiv logo is shown in Figure 8.

arXiv.org

Fig. 8: arXiv Logo

While the benefits of open access are widely acknowledged, there remains a host of practical, cultural and organizational issues impeding broad implementation. University departments' reward and recognition practices often compel faculty to work and publish according to norms established as much by political as by intellectual considerations. Tenure committees persist in rewarding faculty based on publications in paid subscription journals, even though publication in open access repositories is viewed as superior reporting by many scholars and funders. A new approach to measuring research contributions termed "altmetrics" is being explored. Altmetrics combines social web data with traditional bibliometrics data to gauge scholarly impact. A growing consensus in the research community is that change in institutional norms related to reporting research and scholarly production will add value and enhance the global research enterprise in many ways, benefitting nearly all stakeholders. Open access repositories can offer innovative interfaces to their holdings. One such access interface is shown in Figure

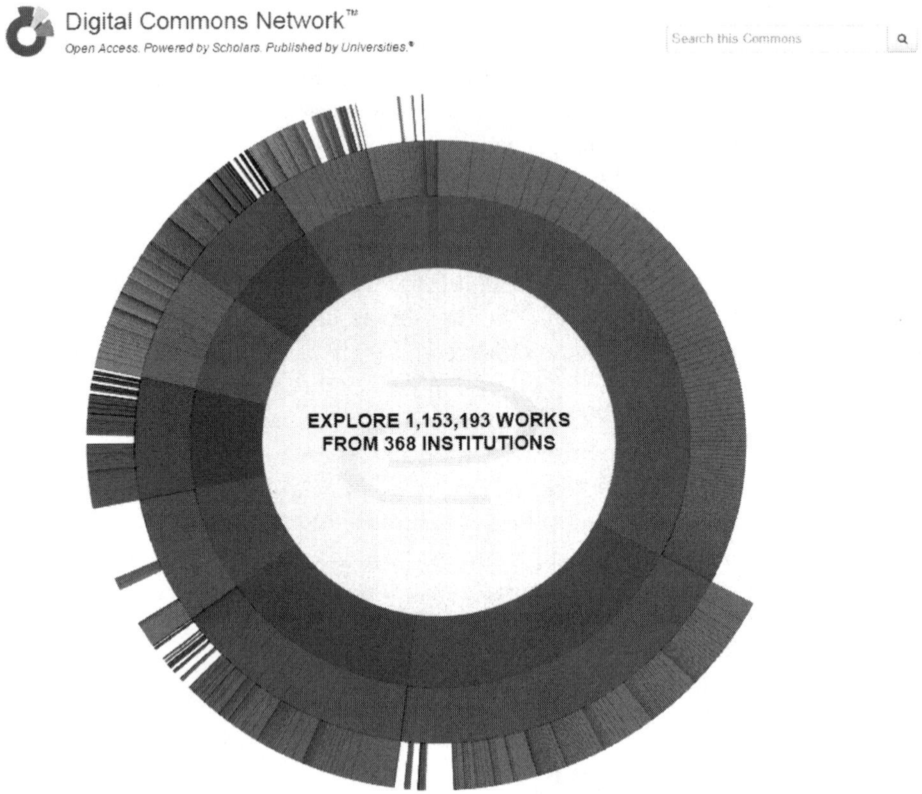

Fig. 9: Digital Commons Access Tool (http://network.bepress.com).

9 used by the Digital Commons Network and an instantiation of the Docuburst visualization tool for document content (http://vialab.science.uoit.ca/portfolio/docuburst-visualizing-document-content-using-language-structure).

Libraries stand to benefit considerably from open access activities and should support them. Libraries at large research universities pay millions of dollars per year for journal subscriptions. A single year's subscription to some of the higher priced journals would cover the purchase of several hundred monographs. The paradox for researchers is bewildering: they perform the research, write papers and help in the review manuscripts for journals, but are often required to purchase them after publication. Libraries can be proactive in advocating university-led publishing efforts, with libraries playing a key role. Costs associated with knowledge dissemination can be greatly reduced by leveraging and scaling technological and support systems already in place.

3.2 Interoperability

Interoperability across web-based repositories was a primary goal of digital libraries research. The Open Archives Initiative (www.openarchives.org) officially established in 2000, extended the scope and functionality of earlier interoperability efforts by introducing the Protocol for Metadata Harvesting (OAI-PMH). OAI-PMH compliant repositories exposed their metadata in such a way as to allow other repositories to identify and collect materials. OAI-PMH became a de facto standard for many repositories and led to federation of distributed open access institutional repositories – albeit at a high level. The primary application was for retrieving text documents. EPrints was among the earliest software for creating institutional repositories, followed quickly by Fedora and DSpace.

A follow-on development by the Open Archives research group was The Object Reuse and Exchange (ORE) standards for the description and exchange of distributed Web resources – including text, images, data, video and compound digital objects.

The Open Access Infrastructure for Research in Europe (www.openaire.eu) offers a comprehensive toolset for researchers, project coordinators and repository managers.

Managing repository infrastructure places new demands and responsibilities on institutions that provide access to and sharing of knowledge. But the benefits are enormous. Open access, interoperable repository networks are seen as essential for advancing research and digital scholarship. They reduce costs by aggregating resources; enhance research capabilities across topic domains by providing standards and methodological frameworks; and, preserve and sustain data and resources for reuse. Perhaps most importantly they create new environments in which researchers can interact, share ideas and develop collaborative efforts.

4 The semantic web and linked data

The World Wide Web today consists of billions of individual web pages containing many types of information objects and hyperlinks to connect to other web-based resources. Web sites are often portals to an extended set of sites or interfaces to larger collections of data that can be searched from within the site. But much of the content of the World Wide Web is invisible to search engines. The functionality of the web today is primarily for use by the general public and is not optimal for scholarly work. In order to identify and retrieve information from multiple web sites, it might be necessary to move stepwise through individual sites and perform the same set of operations in order to build data sets, unless common metadata standards have been adopted, in which case retrieval of *documents* from multiple repositories is possible. In its current form, the web cannot be used for advanced

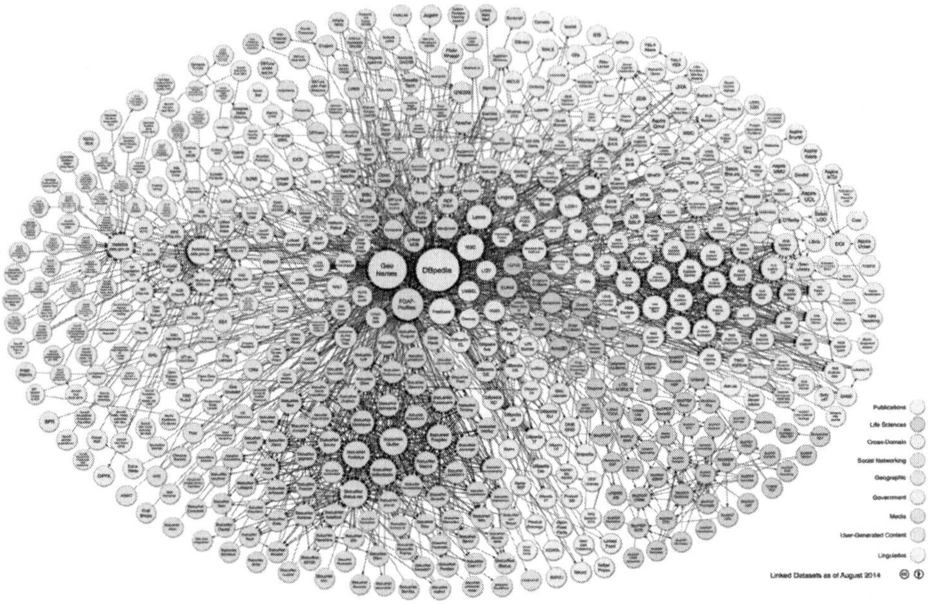

Fig. 10: Linking Open Data Cloud (Cyganiak and Anja Jentzsch 2011) http://lod-cloud.net/.

information retrieval applications, such as those regularly accomplished in database systems. Ideally, scholars should be able to query across web pages at the content, or individual data element level and create various forms of result sets at the data level.

Sir Tim Berners-Lee and collaborators put forward a vision for a "semantic web" that would greatly enhance the functionality of the World Wide Web. The Semantic Web was described in a 2001 article in Scientific American and the World Wide Web Consortium (the W3C) was established to advance the Semantic Web principles. Linking data means encoding and publishing data on the Web in conformance with Semantic Web rules and standards developed under the auspices of the World Wide Web Consortium (W3C). Semantic web and linked data technologies provide a way for performing advanced computing applications across web based content by structuring data and using reference models that allow individual information objects to be connected and continually enriched so that a myriad of relationships can be formed and data sets created. W3C categorizes data in terms of quality and accessibility. Annotating data with temporal and provenance information adds another important dimension.

Although there was initial skepticism due to the additional complexity and labor of creating interlinked data, the benefits of doing so across large numbers of repositories are enormous. The result would be a highly functional new form of web infrastructure, a "web of data" that grows in capacity, capability and richness as additional linked datasets are created, added and new applications are conceived and developed. This is now happening.

The W3C has produced the technology standards for defining the Semantic Web, including a data model, schema, taxonomy, ontology languages for data description, a distributed query language and other products for making the Semantic Web a reality. W3C has numerous working groups and hundreds of contributors (www.w3.org/2005/Incubator/lld/XGR-lld-20111025/).

Still to be explored with linked data repositories are issues of scaling. As more repositories are added, the network of potential relationships between individual elements increases immensely and immediately and can lead to circumstances in which the addition of a single datum to a repository can spawn indefinitely large networks of object relationships across multiple repositories, greatly expanding the information space associated with a given topic as well as the scale of query result sets. Very large, complex systems containing a large variety of information objects are prone to instability and unpredictable behavior. Given the large number of actors involved in developing the "web of data", research on scaling of linked data repositories is very important.

5 Scholarly communication

Scholarly communication across all disciplines is changing profoundly as research and scholarly practices become tied to the ever increasingly rich and interoperable ecology of global information and scholarly tools and resources. Internet-based communication and social media technologies foster on-going informal dialogue, interaction and collaboration among scholars within and across traditional disciplinary lines. A new culture of openness has emerged in which many scholars freely share ideas, data, resources and tools and communicate with peers throughout the scholarly research lifecycle. A growing consensus in the research community is that change in institutional norms related to reporting research and scholarly production will add value and enhance the global research enterprise in many ways, benefitting nearly all stakeholders.

Digital scholarship is radically reshaping knowledge discovery, presentation and dissemination in many subject areas. The processes and products of digital scholarship often involve new types of data analytics, information objects and heuristic representation of findings that cannot be accurately or completely described in current scholarly communication models. In addition, the final value of the research product is diminished unless it can naturally and easily link and become part of larger and often global data research infrastructures. The latter is critical in inspiring new efforts and to advance on-going work. Since open data today is globally accessible, often in real-time, the impact of research findings is potentially immediate and far-reaching[1].

These circumstances have altered the responsibilities for communication of scholarly work. Fundamental reform of scholarly communication practices and the adoption of new models for communicating the full extent and nuances of scholarly efforts, research findings and the workflows associated with creating new knowledge are necessary.

Ideally, research reporting and documentation should be dynamic and recombinant. Dynamic implies updating or revision either by manual or automated processes to add corroborating evidence and to increase its value as part of a larger communicative medium. Recombinant documents are assemblages of well-defined components, each of which contains substantive information of value in its own right. They are structured to permit reassembly by others to recreate the original or to enable new forms that incorporate additional information (Crane 2007).

Full research reporting must contain the broad mix of resources and artifacts associated with each stage of the research workflow. This might include detailed descriptive information, access to all data sets, algorithms and applications software, logs, event traces, data representations, processing protocols, linked media, other resources and executables, presentation schemes, and so on. Figure 11 gives a new model, applicable to any discipline that would meet these requirements.

This new modular workflow-based research "document" model accomplishes many goals that reflect the realities of today's network-centric, computation and data-intensive scholarship. The model:

- provides a variety of presentation forms to accommodate the different expressive forms of disciplinary domains
- facilitates automated retrieval
- gives direct access to datasets, tools and other workflow elements
- anticipates future needs for repurposing, re-use, preservation and access (curation, stewardship, provenance issues)
- is capable of aggregation at the component level with other research documents in other locations
- allows annotation, creation of relationships and indefinite versioning
- contains features to increase dissemination, diffusion and reach of new concepts and findings
- allows accurate measure of contribution and fair attribution, acknowledgement, credit and reward for those involved in carrying out the work

Full reporting of scholarly workflows offers multiple advantages in regenerating and sharing research products. Repurposing and reuse of project assets by others

1 Pittsburgh Scholarly Communication Workshop, January 2013. See www.ischool.pitt.edu/scholarlycom/.

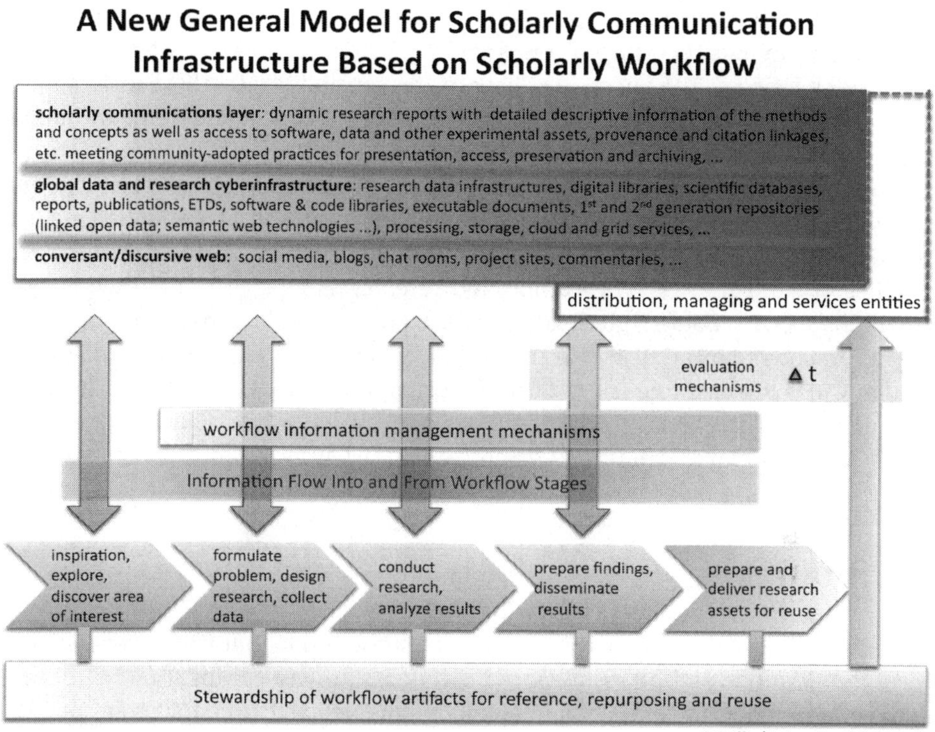

Fig. 11: New Conceptual Model for Scholarly Communication.

is made easier. This view of "publication" also proposes a process of staged release of project activities and findings. Rather than compiling and reporting after the project is complete, scholars would release details of ongoing work and assets across each workflow stage at their discretion. This allows considerable flexibility in packaging and formatting publications. Researchers would not be obliged to follow the fixed formats often required by publishing houses. Reviewers would have the advantage of disassembling the documentation to reproduce the entire or selected parts of an experiment.

Although most of the debate and criticism of traditional scholarly communication has focused on access and dissemination issues, the shortcomings are broader and deeper. Full reporting that supports review, evaluation and reproducibility of research is necessary for the growing body of *in silico* research in which the core activities of the research process might be management of data and software associated with the workflow stages.

Stewardship of digital resources is a challenging, but essential task. If modified data cannot be traced back to its original form and format, verification of results is at risk. Ideally, the process of capturing transitional forms and the operations that produced these would be automated. This has not yet been done and is unlikely to happen unless policies are established to require comprehensive documentation and reporting of research and scholarly workflows as part of the publication process. The question of how the responsibilities for these tasks should be distributed is yet to be answered and has been put to the side of the larger discussion of scholarly communication.

All of this reflects the of complexity of contemporary scholarship and research – processes and results that cannot be captured or communicated in printed form. An aggressive effort to reform scholarly communication to mirror the realities of contemporary scholarship and research is long overdue. New ad hoc communities of scholars, publishers, data managers and funding organizations are vigorously advocating for changes that exploit the many benefits of digital media. One such example is the FORCE11 community (www.force11.org). Its goals are summarized in the FORCE11 Manifesto that sets out a bold agenda for reforming scholarly communication and e-Scholarship (www.force11.org/white_paper).

6 Libraries

By providing access to the human record in all the forms that it is expressed and recorded, libraries and other memory institutions are critical institutions for personal intellectual growth, scholarship and helping people to understand a complex world – in past and present times. Much has been accomplished by the libraries working with information technology communities over the past decade to establish basic principles, standards, object representations, descriptive metadata and reference models needed to ensure interoperability, scalability, access, and long-term preservation and archiving of digital content. By structuring and organizing collections in accordance with new linked data specifications the ability of users to discover and explore resources will be enhanced. In this way individual libraries increase both the reach and value of their holdings. A prodigious and costly effort will be required and may take years to accomplish, so decisions on priorities must be made. Collaboration and coordination with other institutions is very important.

Librarians must come to common understandings of the consequences resulting from choices made in all stages of knowledge creation, conversion and management. Understanding and managing the technical roles and choices will likely be easier than creating the most productive social arrangements – balancing the skills of staff, the expectations of users and building productive alliances with peer institutions.

Yet to be fully appreciated is that information objects have layers of social nuance that only become clear when fully contextualized. In this sense information objects gain social import. Historians, classicists, literature scholars and others in the humanities depend upon context as requisite for understanding. Much of this information is still in the analog collections of memory institutions – and often far down on the list of institutional priorities.

Public libraries face many of the same issues. Those in small towns and cities often lack the resources for digitizing local collections. As a result, valuable provincial information is off-line and unavailable to distant users. These collections are important in telling a nation's story, or in tracing back contemporary social identities to the experiences of one's ancestors in the countries from which they came and the circumstances that motivated emigration. In libraries with limited Internet bandwidth, entire collections of large-scale materials (photos, audio, video) might be loaded onto tablet computers along with access tools. Patrons could then check out with entire collections containing hundreds of gigabytes of content, in addition to other materials. More libraries and memory institutions are recognizing that the general public can play a significant role in digitizing their analog holdings and enhancing digital collections, as many tasks cannot be automated. Crowdsourcing and Citizen Science are growing in popularity as the value of public participation in activities such as metadata tagging, keyword addition and annotation of collections – or in simply reporting on observations of interest to specialists. Speech recognition technologies cannot reliably transcribe all spoken word collections, particularly if cross-language translation is desired, as is the case for oral tradition recordings. Optical character recognition is not effective in converting cursive, handwritten documents such as letters, memoirs, diaries, log books, and field notebooks into machine-readable form. These collections are ubiquitous and are primary sources for many domain scholars. Careful monitoring by library staff and error correction strategies must be included in the projects. But the rewards are significant.

Librarians and the new "digital librarians" are being trained to help create the new Libraries of the future. There is a new generation of Library and Information Science graduate schools, termed "iSchools" that are producing graduates with extensive skills relevant to management, access and use of digital content (http://ischools.org).

There is also much to be said about the design of physical spaces for libraries as the digital environment evolves. This is being actively discussed and many institutions have undergone modernization and created spaces that accommodate new learning styles. Of particular note are spaces designed to encourage collaborative work and study.

Universities and libraries have not yet taken aggressive steps to benefit from all that the global knowledge infrastructure offers, or to contribute to it. Institutions of higher education often define themselves in terms of particular attributes of excellence, exclusivity and comparative measures of quality. Collaboration across universities is common at the level of individuals and even disciplinary departments; yet collaboration at the institutional level has been more difficult to achieve and competition for resources is intense. Institutional behaviors have proven to be entrenched and self-perpetuating. How to reconcile ingrained competition for academic status and resources with the necessity for working toward common ends to create a robust global knowledge infrastructure, open to all on the same terms, is still a vexing question.

The introduction of this chapter outlined goals and opportunities, as well as challenges for libraries. Throughout this chapter the themes of interoperability, scaling, collaboration and context have been stressed. Emphasis has been placed on the value of:

- Open Data
- Open Access Journals
- Open Repositories and Archives
- Open Source Software
- Open and Transparent Community Governance
- Open Scholarly and Practitioner Communities
- Open Scholarship

Libraries can take a leadership role in advancing these goals by demonstrating the power of digital content and tools for many different purposes, both traditional and new. Attention to all stages of the information lifecycle is central to achieving this. A continuing dialog with users is an important part of a library's planning processes in developing new services and resources. By working in collaboration with other institutions the many challenges and opportunities can be addressed and the crucial role of libraries in advancing the creation of knowledge and promoting the social good will be sustained.

References

Acerbi, Alberto, Vasileios Lampos, Philip Garnett, R. Alexander Bentley. 2013. The Expression of Emotions in 20th Century Books. *PLoS ONE* 8(3). e59030.

Berners-Lee, Tim, James Hendler, Ora Lassila. 2001. The Semantic Web. *Scientific American* May. 29–37.

Crane, Gregory. 2007. eCulture, Cyberinfrastructure, Virtual Research Environments and the Cultural Heritage of the Maghreb. Paper presented at Implementing a Maghreb Digital Library for Education, Science and Culture, Rabat, Morocco, January 25–27, 2007.

Ferne, Tristan. 2013. Crowdsourcing the World Service Radio Archive: an experiment from BBC R & D. *BBC Internet Blog*, Tuesday 24 September. http://www.bbc.co.uk/blogs/internet/posts/An-experiment-with-the-World-Service-Radio-Archive (Accessed 14 April 2014).

Griffin, Stephen M. 2005. Funding for Digital Libraries Research. *D-Lib Magazine* 11 (07/08). http://www.dlib.org/dlib/july05/griffin/07griffin.html (Accessed 14 April 2014).

Griffin, Stephen M. 2013. New Roles for Libraries in Supporting Data-Intensive Research and Advancing Scholarly Communication. *International Journal of Humanities and Arts Computing* 7. 59–71.

Griffin, Stephen M. & Stefan Gradmann (eds.). 2013. Call for Papers. *International Journal on Digital Libraries* [Special Issue on Digital Scholarship].

Nowakowski et al. 2011. The Collage Authoring Environment. Procedia Computer Science v4.

OAD – Open Access Directory. 2013. Timeline. http://oad.simmons.edu/oadwiki/Timeline (Accessed 14 April 2014).

Ware, Colin. 2004. Information Visualization – Perception for Design. New York: Morgan Kaufmann.

Loet Leydesdorff

30 The sciences are discursive constructs: The communication perspective as an empirical philosophy of science

Abstract: The new communication technologies have introduced a new dynamics in the sciences: in addition to the (local) context of discovery and the (global) context of justification, a third context of mediation enables scholars to reflect on the sciences as discursive constructs. As against the philosophical perspective, the communication perspective on the sciences enables us to proceed to the operationalization. One can, for example, expect the variation of scientific communications as knowledge claims in manuscripts. These knowledge claims are selectively organized in journals, and the disciplinary structures in the relations among journals can be used to visualize the latent dimensions of coding into specialisms and disciplines. The reconstructions in texts rewrite the sciences so that new words are needed or existing words (e.g., "force" in Newtonian and Aristotelian physics) are provided with new meaning. Thus, an evolutionary dynamics – restructuring from the perspective of hindsight – is added to the historical perspective following the arrow of time. From this perspective, the epistemological status of the sciences is changed, and an empirical philosophy of science can be envisaged. The different disciplines and interdisciplines can be expected to develop by exploring possible configurations of the three contexts.

Keywords: philosophy of science, discourse, knowledge, semantic maps, disciplines, interdisciplinarity

1 Introduction

In his famous Preface to the second edition of the *Critique of Pure Reason* (1787), Immanuel Kant described Galileo's crucial step in developing the modern sciences as follows:

> When Galileo caused balls, the weights of which he had himself previously determined, to roll down an inclined plane; when Torricelli made the air carry a weight which he had calculated beforehand to be equal to that of a definite column of water; or in more recent times, when Stahl changed metal into lime, and lime back into metal, by withdrawing something and then restoring it, a light broke upon all students of nature. [...] Reason, holding in one hand its principles, according to which alone concordant appearances can be admitted as equivalent to laws, and in the other hand the experiment which it has devised in conformity with these

> principles, must approach nature in order to be taught by it. [...] It is thus that the study of nature has entered on the secure path of a science, after having for so many centuries been nothing but a process of merely random groping (Kant 1787: B XII f.).

In addition to his many great discoveries, Galileo changed the philosophy of science of his time by considering the Book of Nature not as God's Revelation, but as a text open to debate. Unlike Kant, Galilei himself was thoroughly aware of this communicative nature of the new sciences. The orientation is empirical not only in terms of the subject matter, but also methodologically: the methods are also the subject of communications. Galileo presented his own arguments discursively in *Dialogo* (1632) and *Discorsi* (1638). The older form of a *disputatio* – that focused only on substance – had been used by him in earlier work (e.g., *Disputatio de coelo*, 1616) and was used also by his adversaries[1]. However, this mono-logical style was here abandoned in favor of open dialogue as the prime mode for generating, validating, and reproducing knowledge (Biagioli 2003).

Discursive knowledge – that is, proceeding to a conclusion through reasoning – is shaped in communication among reflexive participants with reference to a supra-individual horizon of meaning exchanges (Husserl 1929; 1935/1936). Kuhn (1962) called the structure in this horizon a *paradigm*, with the "Copernican turn" of the 16th century as his prime example. The organization of written knowledge in scientific literature as we know it today followed on the Scientific Revolution of the 17th century as an achievement of mainly the 18th century (Bazerman 1988). The mediation of knowledge production by scientific literature was made possible in principle by the invention of the printing press in the 15th century (Eisenstein 1979; 1983; Luhmann 1981); first in Geneva and Lausanne, as cities with free presses during the 16th century, and then in Amsterdam and Leiden in the early 17th century. The free circulation of discursive knowledge triggered the Scientific Revolution.

The dynamics of knowledge diffusion in the modern regime is different from the medieval one: whereas previously a manuscript had to be transcribed in a manual process that was prone to error, a new edition of a scholarly book can be expected to improve on previous ones by correcting errors, and old knowledge is continuously revised and updated in scholarly communications (Price 1965). Thus, one no longer proceeds from an original text (like the Bible) to an inferior copy – because of mistakes in the transcriptions – but from a previous to a next version by rewriting and correcting the contents in terms of their meaning. This process is social, and in more recent times (since the 19th century) increasingly organized purposefully (e.g., in industrial laboratories; cf. Bernal 1939; Braverman 1984; Noble 1977).

The development of *discursive* knowledge presumes the communication of meaning as analytically different from the communication of information. Some-

1 E.g., Orazio Crassi, *De tribus cometis anni MDCXVIII disputatio astronomica* (Rome, 1619).

times new words are needed (for example, when "oxygen" was invented as against "phlogiston"), but at other times the same word can be provided with a new meaning (such as "force" in Newtonian or Aristotelian physics). The theoretical framework can make a difference to both the meaning and the expected information content of the messages that are exchanged (Hesse 1984; Leydesdorff 1997; Quine 1951).

Whereas the communication of information is studied in the information sciences and in scientometrics, the communication of meaning has been central to Luhmann's (1995) attempts to make the theory of *autopoiesis* (Maturana and Varela 1980) or "self-organization" relevant for sociology. Luhmann (e.g. 1986) argued that unlike living systems, communications can become self-organizing in terms of the contents of the communication. Unlike the communication carriers, the contents and meanings of communications are immaterial and volatile, and are thus subject to a very different dynamics.

These three processes – the communication of meaning, (Shannon-type) information, and knowledge – are interwoven in layers that condition and enable one another; developments in the one subdynamic condition developments in the other two. The communication of knowledge, for example, leaves a paper trail as an expanding universe of scientific journal literature that can be made the subject of scientometric analysis (Price 1965). Meanings, however, may change from one instance to another, and the communication of uncertainty – that is, Shannon-type information – is always implied. Whereas Luhmann focused on the top-layers of codification and its dynamics, in my opinion, one must also take the underlying processes into account in developing a richer understanding of the sciences as processes of communication.

2 The communication perspective on the sciences

As against the philosophy of science, in which one distinguished traditionally between a context of discovery and a context of justification (Popper 1964 [1939]), the communication perspective enables us to distinguish three dynamic contexts: the local interactions and exchanges in the context of discovery operating bottom-up in generating knowledge claims, the validation of these knowledge claims in a globalized context of justification that can operate as a self-organizing control mechanism partly anonymously, and the mediation between these two levels in terms of manuscripts and other texts in which the arguments are organized so that specific audiences can be targeted. This last context could also be called the context of mediation.

The three contexts are intermediated, but can be distinguished analytically: whereas the context of discovery operates bottom-up, the codes emerging in the context of justification as paradigmatic operate top-down as control mechanisms, while the context of mediation mediates between these two other levels (Mulkay et al. 1983) and provides the possibility of overflowing between the disciplines as different frames (Callon 1998), as well as mediating between science and society (Gibbons et al. 1994). Implicitly referencing Herbert Simon's (1962; 1973) theory of complex systems, Lumann (1975) proposed these three levels in the dynamics of communication: interaction, organization, and self-organization. At the self-organizing level, codes of communication are reconstructed as paradigmatic frameworks that operate in scientific disciplines and specialties. These codes are not manifest as variations, but remain latent selection mechanisms that can be hypothesized (and updated!) on the basis of theoretical and empirical arguments.

The interactions first provide variation. However, the variation has to be organized (e.g, in a specific text) before it can further be selected for globalization. Using another metaphor of Herbert Simon (1973), one can consider coding as an alphabet that is used for next-order organization selectively. For example, some knowledge content can be considered as a part of chemistry, whereas a text submitted to a psychology journal is expected to contain psychological knowledge. Both the two texts and their contents will vary differently in their structures and their semantics. From this perspective, interdisciplinarity can no longer mean a mixing of the codes, but requires the careful construction of hierarchies of meaning and juxtapositions that may enrich (or confuse) the semantics when compared with mono-disciplinary perspectives.

Simon (1973) hypothesized that all systems operate with codes such as DNA or the twenty amino-acids that shape proteins. Following Parsons (1968), Luhmann (e.g. 1974) suggested that when symbolically generalized in functionally different directions, the codes of communication can drive self-organizing subsystems of society such as the sciences or the economy. The symbolic generalization spans the codes orthogonally as eigenvectors (principal components) with reference to horizons of meaning. Note that this is an analytical model (Leydesdorff 2012).

Communication systems – more than communities of people – can also be globalized, as in the case of economic exchange processes under the conditions of modern capitalism. In the Marxist tradition the dominance of exchange processes over use has been considered as "alienation," since the driving forces of social developments are then beyond control. In *Capital I*, Marx (1971 [1867]) went to great lengths to explain why the exchange of commodities through the mediation of money ("Ware-Geld-Ware" or WGW) is different from the exchange of money via the mediation of commodities ("Geld-Ware-Geld" or GWG).[2] The dynamics of money is more abstract than that of commodities – the price of a commodity can be considered as an expectation of its value – and thus this symbolic generalization of commodities in terms of prices enables us to handle greater complexity in the communication.

2 "Ware-Geld-Ware" can be translated as "Commodity-Money-Commodity" (Marx, 1995, pp. 66 ff.).

Similarly, in the sciences – albeit perhaps less frequently than in daily transactions on the market – the submission of a knowledge claim (usually in the form of a manuscript) brings the context of justification into operation within the context of discovery in which this new insight had previously been shaped. The epistemological status of the knowledge content under discussion is changed by entering into this differently organized layer of communications (Luhmann 1975). Note that the context of discovery remains always needed to provide the historical variation, whereas the context of justification functions as a next-order control mechanism that operates as a latent dimension of the communication. The two communication dynamics can operate with different frequencies, and they can be expected to disturb each other.

In summary, networks which are themselves composed of volatile communications of knowledge claims can be stabilized and then also globalized by developing next-order structures and dynamics endogenously. Unlike social networks, which find their stability in human and institutional agents, communication networks are event-based: communication is an operation that disappears as it happens (Snijders et al. 2010). Stability is always constructed (Latour 1987) and thus to be explained. The reproduction of communication structures is grounded, on the one side, in the layer of agents carrying the system; but communications, on the other side, are selected with reference to next-order structures such as codes of communication emerging recursively within the communication networks.

3 Operationalization

As against the philosophical perspective, the communication perspective on the sciences as evolving communication structures enables us to proceed to operationalization. One can, for example, expect the variation of scientific communications to be visible first as knowledge claims in manuscripts (Myers 1985; Pinch 1985). These knowledge claims are organized in journals, and the disciplinary structures in the relations among journals can be used to visualize the latent dimensions of coding into specialisms and disciplines.

Figure 1 shows, for example, the relevant citation environment for the 141 authors publishing 67 papers in *Public Understanding of Science* (PUS) in 2011. After removing the incidental citations (< 1 %), three main groups are indicated at the aggregated level (using the community-finding algorithm of Blondel et al. 2008)[3]: communication studies, science communication, and science education.

At the paper level, however, the 67 papers published in 2011 couple only one subset among these journals in terms of their references, namely the ten journals

3 Three factors explain 43.06 % of the variance; T(xyz) = +190.27 mbits.

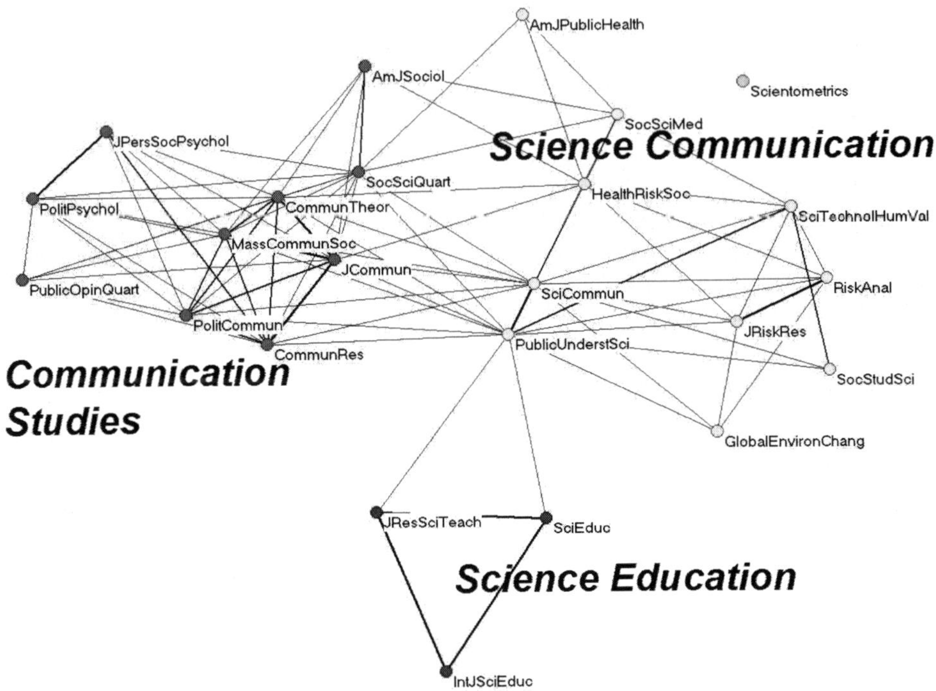

Fig. 1: Aggregated citation network among 24 journals cited in *Public Understanding of Science* in 2011 to the extent of more than 1% of all references; cosine > 0.1; Blondel et al.'s (2008) modularity used for the clustering in Pajek; $Q = 0.447$.

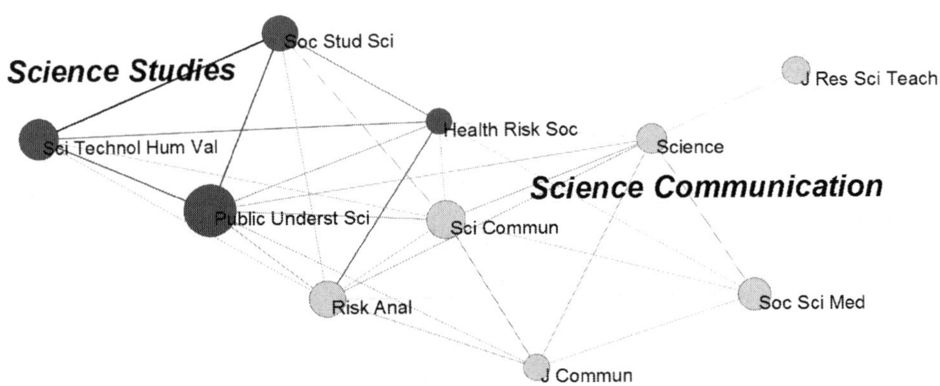

Fig. 2: Ten journals bibliographically coupled above the 1%-level in 67 papers published in *Public Understanding of Science* in 2011; cosine > 0.1; $Q = 0.1373$.

shown in Figure 2. While the 2011-volume of this journal (*PUS*) makes references at the aggregate level to a larger set, certain individual papers selectively organize a local interface in their referencing between science studies as a specialty and

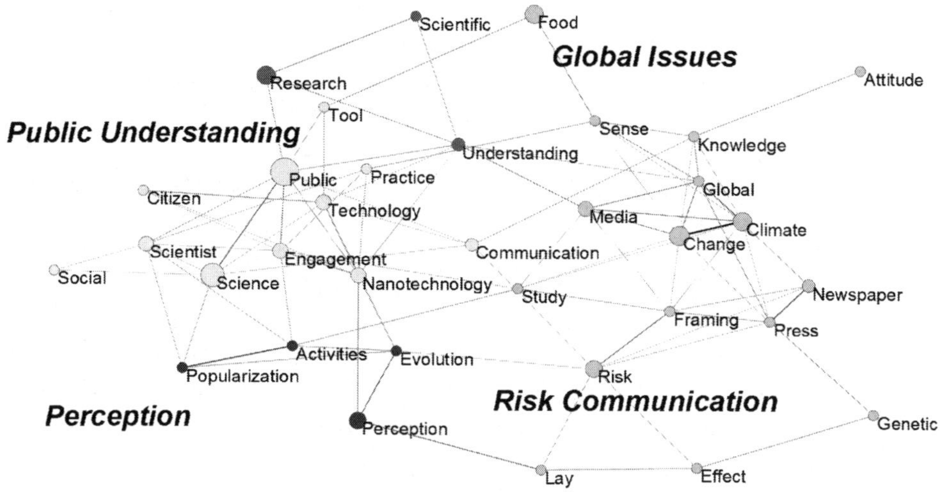

Fig. 3: Topical network indicated by 34 words occurring more than twice in the titles of 64 documents[4] published in *Public Uniderstanding of Science* during 2011; cosine > 0.2; 5 clusters indicated with $Q = 0.449$.

communication studies as a larger domain. These are the instantiations in this specific year of the latent and self-organizing structures of the disciplines involved in Figure 1.

Figure 3 adds the substantive variation in terms of the topics under study in the papers of this journal volume. One can add to the topics also a visualization of the co-author network: their institutional affiliations, national and international collaborations, etc. (Leydesdorff et al. 2008). The tools of social and semantic network analysis enable us to study the dynamics in the different dimensions of science in considerable detail and perhaps also to simulate these (Scharnhorst et al. 2012).

4 Changes in the context of mediation

The shaping of the context of mediation has followed the development of communication technologies. I mentioned above the changes which the printing press made to the development of the sciences during the 15th to the 19th centuries (Eisenstein 1979). The late 20th century, however, witnessed the emergence of the internet as a communication technology which changed scientific practices both at the level of the context of discovery and increasingly – albeit with a lag – also

4 Three more documents (Editorials) have no title words within this set of 34 words.

the context of justification. New patterns of communication can be expected to change the codes of communication reflexively (Larivière et al. 2008).

As publication databases first became more inclusive of a wider range of disciplines from the 1970s through the beginnings of the 1990s, scientists made increasingly use of them with the assistance of research librarians trained in information retrieval. During the same period, listservs and email became communication tools changing laboratory life. The very process of science became intensively embedded into literature and communication. The flows of communication and literature could be increased with orders of magnitude (Hull et al. 2008).

In the early 1990s, another communication technology emerged, as the Mosaic Internet browser was introduced, soon followed by Netscape, Firebox, Internet Explorer, Chrome, and other browsers. Windows 3 (1990) brought the graphical user environment. These technologies made web pages an increasingly important means of scholarly communication as well as the study of it.

The most recent phase has been the development of Google Scholar (2004), including the accessibility of conference proceedings, open journals, portions of books, and personal collections of .PDFs of manuscripts. In this period another key technological development has been real time communication among collaborating scientists using Skype, shared databases of references, and webinar tools. The mapping and ranking of science using scientometric tools made it possible to visualize the evolution of the sciences as concept networks (Börner 2010).

5 Conclusion

I have argued that the communication turn in the philosophy of science adds to the linguistic turn (Rorty 1962 [1967]) a sociological perspective. By analyzing the sciences as communication systems in which knowledge can be constructed discursively, one obtains a richer conceptualization: the context of mediation can be added to the context of discovery and that of justification. The study of the communicative layer opens the philosophy of science to empirical operationalization, further refinement, and/or the possible rejection of hypotheses on statistical grounds (Leydesdorff 1995).

References

Bazerman, Charles. 1988. *Shaping Written Knowledge: The Genre and Activity of the Experimental Article in Science*. Madison. WI: University of Wisconsin Press.

Bernal, John Desmond. 1939. *The Social Function of Science*. Cambridge, MA: MIT Press, 1967.

Biagioli, Mario. 2003. Stress in the Book of Nature: The Supplemental Logic of Galileo's Realism. *MLN* 118(3). 557–585.

Blondel, Vincent D., Jean-Loup Guillaume, Renaud Lambiotte & Etienne Lefebvre. 2008. Fast unfolding of communities in large networks. *Journal of Statistical Mechanics: Theory and Experiment* 8(10). 10008.

Börner, Katy. 2010. *Atlas of Science: Visualizing what we know*. Cambridge, MA: MIT Press.

Braverman, Harry. 1984. Labor and Monopoly Capital. The Degradation of Work in the Twentieth Century. New York/London: Monthly Review Press.

Callon, Michel. 1998. *The Laws of the Market*. Oxford and Malden, MA: Blackwell.

Eisenstein, Elizabeth L. 1979. *The Printing Press as an Agent of Change*. Cambridge: Cambridge University Press.

Eisenstein, Elizabeth L. 1983. *The Printing Revolution in Early Modern Europe*. Cambridge: Cambridge University Press.

Galilei, Galileo. 1632. *Dialogo sopra i due massimi sistemi del mondo*, [Dialogue concerning the two chief world systems] Berkeley: University of California Press, 1953.

Galilei, Galileo. 1638. *Discorsi. Leiden*. [Dialogue concerning the two new sciences]. Chicago, 1946.

Gibbons, Michael, Camille Limoges, Helga Nowotny, Simon Schwartzman, Peter Scott & Martin Trow. 1994. *The new production of knowledge: the dynamics of science and research in contemporary societies*. London: Sage.

Grassi, Orazio. 1619. *De tribus cometis anni MDCXVIII disputatio astronomica*. Rome.

Hesse, Mary B. 1984. *Revolutions and Reconstructions in the Philosophy of Science*. London: Harvester Press.

Hull, Duncan, Steve R. Pettifer, & Douglas B. Kell. 2008. Defrosting the digital library: bibliographic tools for the next generation web. *PLoS computational biology* 4(10). e1000204.

Husserl, Edmund. 1929. *Cartesianische Meditationen und Pariser Vorträge [Cartesian Meditations and the Paris Lectures, translated by Dorion Cairns]*. The Hague: Martinus Nijhoff, 1973. http://ia600601.us.archive.org/10/items/CartesiamMeditations/12813080-husserl-cartesian-meditations.pdf (Accessed 14 April 2014).

Husserl, Edmund. 1935/1936. *Die Krisis der Europäischen Wissenschaften und die Transzendentale Phänomenologie*. Den Haag: Martinus Nijhoff, 1962.

Kant, Immanuel. 1787. *Kritik der reinen Vernunft*. Hamburg: Felix Meiner, 1956.

Larivière, Vincent, Éric Archambault, & Yves Gingras. 2008. Long-term variations in the aging of scientific literature: From exponential growth to steady-state science (1900–2004). *Journal of the American Society for Information Science and Technology* 59(2). 288–296.

Latour, Bruno. 1987. *Science in Action*. Milton Keynes: Open University Press.

Leydesdorff, Loet. 1995. *The Challenge of Scientometrics: The development, measurement, and self-organization of scientific communications*. Leiden: DSWO Press, Leiden University. http://www.universal-publishers.com/book.php?method=ISBN & book=1581126816 (Accessed 14 April 2014).

Leydesdorff, Loet. 1997. Why Words and Co-Words Cannot Map the Development of the Sciences. *Journal of the American Society for Information Science* 48(5). 418–427.

Leydesdorff, Loet. 2012. Radical Constructivism and Radical Constructedness: Luhmann's Sociology of Semantics, Organizations, and Self-Organization. *Constructivist Foundations* 8(1). 85–92.

Leydesdorff, Loet, Thomas Schank, Andrea Scharnhorst & Wouter De Nooy. 2008. Animating the Development of *Social Networks* over Time using a Dynamic Extension of Multidimensional Scaling. *El Profesional de la Información* 17(6). 611–626.

Luhmann, Niklas. 1974. Einführende Bemerkungen zu einer Theorie symbolisch generalisierter Kommunikationsmedien. *Zeitschrift für Soziologie* 3. 236–255.

Luhmann, Niklas. 1975. Interaktion, Organisation, Gesellschaft: Anwendungen der Systemtheorie. In M. Gerhardt (ed.), *Die Zukunft der Philosophie*, 85–107. München: List.

Luhmann, Niklas. 1981. Gesellschaftsstrukturelle Bedingungen und Folgeprobleme der naturwissenschaftlich-technischen Fortschirtts. In R. Löw, P. Koslowski & P. Kreuzer (eds.), *Fortschritt ohne Maß: Eine Ortsbestimmung der wissenschaftlich-technischen Zivilisation*, 112–131. München: Piper.

Luhmann, Niklas. 1986. The autopoiesis of social systems. In F. Geyer & J. v. d. Zouwen (Eds.), *Sociocybernetic Paradoxes*, 172–192. London: Sage.

Luhmann, Niklas. 1995. *Social Systems*. Stanford, CA: Stanford University Press.

Marx, Karl. 1971 [1867]. *Das Kapital I*. Berlin: Dietz.

Maturana, Humberto R. & Francisco Varela. 1980. *Autopoiesis and Cognition: The Realization of the Living*. Boston: Reidel.

Mulkay, Michael, Jonathan Potter & Steven Yearley. 1983. Why an Analysis of Scientific Discourse is Needed. In K. D. Knorr & M. J. Mulkay (eds.), *Science Observed: Perspectives on the Social Study of Science*, 171–204. London: Sage.

Myers, Greg. 1985. Texts as knowledge claims: The social construction of two biology articles. *Social Studies of Science* 15. 593–630.

Noble, David. 1977. *America by Design: Science, Technology, and the Rise of Corporate Capitalism*. New York: Knopf.

Parsons, Talcott. 1968. Interaction: I. Social Interaction. In D. L. Sills (ed.), *The International Encyclopedia of the Social Sciences* 7, 429–441. New York: McGraw-Hill.

Pinch, Trevor J. 1985. Towards an Analysis of Scientific Observation: The Externality and Evidential Significance of Observational Reports in Physics. *Social Studies of Science* 15. 3–36.

Popper, Karl R. 1959 [1935]. *The Logic of Scientific Discovery*. London: Hutchinson.

Price, Derek J. de Solla. 1965. Networks of scientific papers. *Science* 149(3683). 510–515.

Rorty, Richard M. (ed.). 1992 [1967]. *The Linguistic Turn: Essays in Philosophical Method*. Chicago: University of Chicago Press.

Quine, Willard V. 1951. Main trends in recent philosophy: two dogmas of empiricism. *The Philosophical Review* 60(1). 20–43.

Scharnhorst, Andrea, Katy Börner & Peter Van den Besselaar (eds.). 2012. *Models of Science Dynamics: Encounters Between Complexity Theory and Information Sciences*. Heidelberg, etc.: Springer.

Simon, Herbert A. 1962. The architecture of complexity. *Proceedings of the American Philosophical Society* 106(6). 467–482.

Simon, Herbert A. 1973. The Organization of Complex Systems. In H. H. Pattee (ed.), *Hierarchy Theory: The Challenge of Complex Systems*, 1–27. New York: George Braziller Inc.

Snijders, Tom A. B., Gerhard G. Van de Bunt & Christian E. G. Steglich. 2010. Introduction to stochastic actor-based models for network dynamics. *Social Networks* 32(1). 44–60.

Biographical sketches

Daniel Arasa is Associate Professor of Digital and Strategic Communications and Vice Dean of the School of Church Communications at the Pontificia Università della Santa Croce (Rome, Italy). He has a Bachelor's Degree in Journalism from the Universitat Autònoma de Barcelona, a M.A. in Television and Radio from the Southern Methodist University, Dallas, and a Ph.D. by Santa Croce. Moreover, he is member of the Board of Directors of *Rome Reports TV*, news agency specialized in covering the Pope, the Vatican and the Catholic Church. He is author of *Church Communications Through Diocesan Websites. A Model of Analysis* (2008), and co-editor of *Religious Internet Communication. Facts, Experiences and Trends in the Catholic Church* (2010), and of *Church Communication & the Culture of Controversy* (2010).

Gabriele Balbi is Assistant Professor in Media Studies at USI – Università della Svizzera italiana (Lugano, Switzerland). At the Faculty of Communication Sciences, he is Vice and Study Dean and teaches undergraduate courses in "Media history" and "Sociology of mass communication". His main areas of research are media history and history of telecommunications. On these topics, he published four books (two on the history of the telephone, one on the history of International Telecommunication Union, and one on the history of digital media) and several articles in international journals such as: *Technology and Culture, Media History, History of Technology, International Journal of Communication, Media, Culture and Society.*

Davide Bolchini is an Associate Professor and the Department Chair of Human-Centered Computing at Indiana University School of Informatics and Computing (Indianapolis), where he teaches and conducts research in human-computer interaction. Through over 100 scholarly publications, he has made important contributions to user experience design methods, web accessibility for the blind and aural user interfaces, and requirements engineering for the web. He secured over $ 1.5 M of external research funding as PI or Co-PI from the US National Science Foundation (NSF), Swiss National Science Foundation (SNSF) and a Google Faculty Research Award.

Lorenzo Cantoni is Professor at USI – Università della Svizzera italiana (Lugano, Switzerland), Faculty of Communication Sciences, where he served as Dean in the years 2010–2014. He is currently director of the Institute for Communication Technologies and scientific director of the laboratories webatelier.net, NewMinE Lab: New Media in Education Lab, and eLab: eLearning Lab. He is chair-holder of the UNESCO chair in ICT to develop and promote sustainable tourism in World Heritage Sites, and president of IFITT – International Federation for Information Technologies in Travel and Tourism. His research interests are where communication, education and new media overlap, ranging from computer mediated communication to usability, from eLearning to eTourism, and from ICT4D to eGovernment.

Wenhong Chen is an Assistant Professor in the Department of Radio-TV-Film, Moody College of Communication, at University of Texas at Austin. Her research has been focused on social implications of digital media and communication technologies. Dr. Chen has published in top-tier journals in the field of communication and media studies, management, and sociology such as *Social Networks, Human Communication Research, Journal of the Association for Information Science and Technology (JASIST), Management & Organization Review,* and *Entrepreneurship Theory & Practice.* She co-edited the book *Networked China: Global Dynamics of Digital Media and Civic Engagement* (Routledge 2015).

Pauline Hope Cheong is Associate Professor at the Hugh Downs School of Human Communication, Arizona State University. She studies the complex interactions between communication technologies and different cultural communities around the world, including changing authority and community relations in spiritual organizations. Dr. Cheong has published more than 60 articles and books. She is the Lead Editor of *Digital Religion, Social Media and Culture: Perspectives, Practices, Futures* and *New Media and Intercultural Communication: Identity, Community and Politics.* She has received several top research book and paper awards by the National and International Communication Associations.

James Danowski has been Assistant Professor at the Annenberg School for Communication at the University of Southern California and at the University of Wisconsin-Madison, and then Associate Professor of Communication at the University of Illinois at Chicago. He is currently chair of the Communication and Technology (CAT) division of ICA, the International Communication Association, a division he founded back in 1978. Beside his research and teaching activities, focused on the history, development and nature of Information and Communication Technologies, he has also developed WORDij: a family of programs that automate semantic network analysis of big data text from social media, email, news and other sources.

Aziz Douai, Ph.D. is Associate Professor of Communication and Digital Media Studies, Faculty of Social Science and Humanities, at the University of Ontario Institute of Technology (Canada). His research focuses on the socio-political implications of new media, Arab journalism, global media and international conflict, among other areas of international communications. In addition to co-editing *New media influence on social and political change in Africa* (IGI-Global, 2013), Dr. Douai has published more than 40 book chapters and articles in leading international communication journals.

Martin J. Eppler is a Chaired Professor of media and communications management at the University of St. Gallen in Switzerland, researching knowledge communication and visualization (knowledge-communication.org). He is a director of the university's =mcm institute for media and communications management and the academic director of the International Study MBA program. He is the inventor of the software en.lets-focus.com and the collabcards.com set. He has been an advisor to organizations such as the United Nations, the European Central Bank, EY, Daimler, Swiss Re, or Tawuniya. He has been a guest professor in the US, Finland, China, and Peru.

Adriano Fabris is Full Professor in Moral Philosophy at Pisa University (Italy), where he teaches also Communication Ethics. At Pisa University he is actually director of the C.I.Co. (Interdisciplinary Center for Research in Communication). He is director of the international review "Teoria" too. From 2014 he serves as President of the "Committee for dissemination of the scientific and technological culture in Italy", established by the Italian Government. His research interests are: Moral Philosophy, Applied ethics, Communication Ethics, Roboethics, responsibility in new technologies, Philosophy of Religions, Interreligious dialogue, Religions on the web.

Dr. Daniel Fesenmaier received his Ph.D. in geography from Canada's University of Western Ontario. He is a Professor at the University of Florida and Director of the National Laboratory for Tourism & eCommerce, Eric Friedheim Tourism Institute. Dr. Fesenmaier is author of articles dealing with tourism marketing, advertising evaluation and information technology. He has co-authored a book and a monograph and co-edited five books focusing on various aspects of tourism marketing, information technology and development.

Dr. Jacquelyn Ford Morie is well-known for creating meaningful Virtual Reality experiences. Starting in 1990, she developed multi-sensory techniques for virtual reality (VR) that can predictably elicit emotional responses from participants, for example inventing a scent collar that can deliver scents to a participant in VR. She is also active in online 3D virtual worlds and has been bringing her VR techniques to these social worlds for Mindfulness and NASA astronauts. Her other research interests include space, identity and play in virtual worlds, and she has presented extensively at conferences worldwide. She has appeared in 2 films: The Mindfulness Movie (2013) and Dsknectd (2013) as herself.

Ulrike Gretzel is a Professor of Tourism in the UQ Business School, University of Queensland, Australia and a member of the Smart Tourism Research Centre at Kyung Hee University, South Korea. She received her Ph.D. in Communications from the University of Illinois at Urbana-Champaign. Her research focusses on technology use in tourism, with an emphasis on social media, both from organizational as well as consumer perspectives, and the design of intelligent systems, in particular recommender systems. Her research has been published in major tourism and eCommerce journals. She has also co-authored and co-edited several books related to technology in tourism.

Stephen M. Griffin is Visiting Professor and Mellon Cyberscholar at the University of Pittsburgh, School of Information Sciences. Before coming to the University of Pittsburgh he had a 32-year career at the National Science Foundation (NSF) in Washington, DC. He served as a Program Director in the Division of Information and Intelligent Systems (IIS) for the period 1994–2011. Prior to that period, he was engaged in program planning, development and evaluation for the Division of Advanced Scientific Computing and the Directorate for Computer and Information Science and Engineering (CISE). In 2004–2005 and again in 2010–2011 he was on special assignment to the Library of Congress, Office of Strategic Initiatives, to assist with planning the National Digital Information and Infrastructure Preservation Program and to serve as a Senior Advisor for repository and digital content development. His research interests include new models of scholarly communication, cultural heritage informatics and data-intensive scholarship.

Claire Hewson is Lecturer in Psychology at The Open University (UK). She has a wide range of publications on Internet-mediated research methods, including *Internet Research Methods*, Sage, 2003 (second edition forthcoming). She acted as Convenor and Editor for the recently published British Psychological Society guidelines on ethics in Internet-mediated research (2013). Her research interests include: online research and assessment methods, common sense understandings, particularly folk psychology, and related issues in philosophy of mind and cognitive science.

Dr. Alessandro Inversini is Senior Lecturer at Bournemouth University (UK), Faculty of Management. Alessandro holds a Ph.D. in Communication Science from USI – Università della Svizzera italiana (Lugano, Switzerland). In 2011, and was appointed as Managing Director of Ticinoinfo SA, a technology competence center for Tourism active in the field of technological innovation, ePromotion and eMarketing in tourism for the Tessin region in Switzerland. In January 2012 Alessandro was elected as Board Member of the IFITT (International Federation of IT in Travel and Tourism), where he supports eMarketing initiatives and the ICT4D scholarship.

Tomasz Janowski is the founding Head of the United Nations University Operating Unit on Policy-Driven Electronic Governance (UNU-EGOV), a member of the United Nations system, and an international research, policy analysis and leadership education organization located in Portugal. He founded and coordinates the International Conference on Theory and Practice of Electronic Governance (ICEGOV), a leading conference in the area, and is Co-Editor-in-Chief of Government Information Quarterly, a leading journal in the area. He received funding for his projects from Macao Foundation, World Bank, Microsoft, UNDP, European Commission, Commonwealth Secretariat and several national governments. His research focuses on the design and performance of technology-enabled governance mechanisms in different local and sectorial contexts, and on the transfer of such mechanisms between contexts.

Richard R. John is a Professor of history and communications at the Columbia Journalism School, Columbia University, where he teaches in Columbia's Ph.D. program in communications. His publications include *Spreading the News: The American Postal System from Franklin to Morse* (Cambridge, Mass.: Harvard University Press, 1995) and *Network Nation: Inventing American Telecommunications* (Cambridge, Mass.: Belknap Press of Harvard University Press, 2010). *Spreading the News* received the Allan Nevins Prize from the Society of American Historians; *Network Nation* received the Ralph Gomory Prize from the Business History Conference. His current research interests include the history of global communications networks.

Gary L. Kreps is a university distinguished Professor and Director of the Center for Health and Risk Communication (http://chrc.gmu.edu) at George Mason University (Fairfax, Virginia). He studies health and risk communication, with a major focus on reducing health inequities. He publishes widely (more than 400 articles, books, and chapters). He has been funded by numerous federal agencies, foundations, health systems, foreign governments, and corporations. He coordinates the Fairfax County Health Literacy Initiative community collaborative that develops culturally-sensitive health information programs for at-risk populations (http://chrc.gmu.edu/fchli/), and co-directs the Global Advocacy Leadership Academy (GALA) for promoting effective consumer health advocacy (www.advocacyleader.org).

Joanna Kulesza, Ph.D., is an Assistant Professor of international public law at the Faculty of Law and Administration, University of Lodz (Poland). She specializes in international Internet law. She is the author of five monographs and over fifty peer-reviewed publications on international and Internet law. She has been a Visiting Professor with the Oxford Internet Institute, Norwegian Research Center for Computers and Law, Institut für Informations-, Telekommunikations- und Medienrecht, Westfälische Wilhelms Universität Münster and Justus-Liebig-Universität Gießen. She was a Post-Doctoral Researcher at the University of Cambridge and Ludwig-Maximilians-Universität München, a scholar of the Robert Bosch Stiftung, Polish Ministry of Foreign Affairs and the Foundation for Polish Science. She worked for the European Parliament, Polish Ministry of Foreign Affairs and the Council of Europe. She currently is the Membership Committee Chair of the Global Internet Governance Academic Network (GigaNet). Her main research interests cover human rights protection in cyberspace, international cooperation on Internet governance and the future of World Wide Web.

Rita M. Lauria, JD, Ph.D. is Collegiate Professor, University of Maryland; Counselor of Law, Metalaw®.US; and Research Associate, Media Interface and Network Design (M.I.N.D.) Labs. Rita works transdisciplinarily across fields to distill fundamental principles applicable to constructing a conceptual framework for the understanding of *Virtuality.* She is interested in the quality of consciousness in relation to electric sensibility. The goal of intelligence augmentation drives Rita's interests as she focuses upon the paradigm shift occurring because of the pervasive use of metamedia that generate virtualities. Pragmatically, her work explores psychological, philosophical, and aesthetic considerations for the design and use of interactive, computational media. Previously, she served as Advisor for Communications to the Secretary of Transportation, Communications, and Infrastructure of the Federated States of Micronesia, where she witnessed first-hand the diffusive and transformative nature of new media upon the indigenous people of the Pacific. Rita's early *The Law and Regulation of International Space Communication* still stands as a foundational reference work in international telecommunications law and policy.

Loet Leydesdorff (Ph.D. Sociology, M.A. Philosophy, and M.Sc. Biochemistry) is Professor at the Amsterdam School of Communications Research (ASCoR) of the University of Amsterdam (Netherlands). He is Honorary Professor of the Science and Technology Policy Research Unit (SPRU) of the University of Sussex, visiting professor of the Institute of Scientific and Technical Information of China (ISTIC) in Beijing, Guest Professor at Zhejiang University in Hangzhou, and Visiting Professor at the School of Management, Birkbeck, University of London. He has published extensively in systems theory, social network analysis, scientometrics, and the sociology of innovation (see at www.leydesdorff.net/list.htm). With Henry Etzkowitz, he initiated a series of workshops, conferences, and special issues about the Triple Helix of University-Industry-Government Relations. He received the Derek de Solla Price Award for Scientometrics and Informetrics in 2003 and held "The City of Lausanne" Honor Chair at the School of Economics, Université de Lausanne (Switzerland), in 2005. In 2007, he was Vice-President of the 8th International Conference on *Computing Anticipatory Systems* (CASYS'07, Liège). In 2014, Thomson Reuters listed him as a highly-cited author (http://highlycited.com).

Rich Ling (Ph.D, University of Colorado, sociology) is the Shaw Foundation Professor of Media Technology, at Nanyang Technological University, Singapore. He also has connections with Telenor Research and has an adjunct position at the University of Michigan. Professor Ling has studied the social consequences of mobile communication for the past two decades. He has written *The mobile connection* (Morgan Kaufmann, 2004), *New Tech, New Ties* (MIT, 2008) and most recently *Taken for grantedness* (MIT, 2012). He is a founding Co-Editor of *Mobile media and communication* (Sage) and the Oxford University Press series *Studies in Mobile Communication.*

Dr. Anne Linke is a Research Associate at the Chair of Communication Management and Public Relations at the University of Leipzig (Germany). She graduated in communication and media studies and sociology at the University of Leipzig and Universitat de Valencia (Spain). She got practical experience in the fields of public relations, controlling, strategic management and corporate communication. From 2010 until 2013 she received a Ph.D. scholarship from Fink & Fuchs Public Relations AG to research the effects of social media on the process of communication management. Her research interests are online public relations, social media, communication management, innovation communication, international communication.

Tom McCourt is an Associate Professor of communication at Fordham University. He received his Ph.D. in radio/TV/film from the University of Texas. He is the author of *Digital Music Wars: Ownership and Control of the Celestial Jukebos* (with Patrick Burkart) and *Conflicting Communication Interests in America: The Case of National Public Radio*. He recently co-produced a documentary film, *Drop City*, with Joan Grossman (www.dropcitydoc.com). His research interests include the history of media, communication technology, music and cultural studies, with particular focus on the relationship of technology to the public sphere.

Brett Oppegaard is an Assistant Professor at University of Hawaii, in the School of Communications, within the College of Social Sciences, at the state's flagship Manoa campus. He was the individual recipient of the regional and national 2012 George and Helen Hartzog Award for his research into mobile app development and media delivery systems within the National Park Service, as well as the national 2013 John Wesley Powell Prize winner for outstanding achievement in the field of historical displays. His research projects often include collaborations with the U. S. National Park Service, including recent work at America's first national park, Yellowstone.

Alejandro Pardo is Associate Professor at the Department of Film, TV & Digital Media, School of Communication, University of Navarra (Spain). Graduated in Production for Film & TV (UCLA-Ext), Entertainment & Media Management (UCLA Anderson School of Management) and Audiovisual Management (Media Business School, Madrid). Visiting Researcher at the UCLA School of Theater, Film & TV, as well as Guest Scholar at the DeSantis Business and Economics Center (Florida Atlantic University). He has written extensively (books, book chapters and journal articles) about the audiovisual industry. Member of the European Media Management Association (EMMA), the Society of Cinema and Media Studies (SCMS), and the European Network for Cinema and Media Studies (NECS). He is currently researching on the impact of digitization on the film industry.

John V. Pavlik is a Professor of journalism and media studies at Rutgers, the State University of New Jersey (USA). He is a member of the Advisory Board of the Global Communication Research Institute at Shanghai Jiao Tong University (China). Pavlik has written widely on the impact of new technology on journalism, media and society. His books include *Converging Media, Media in the Digital Age, Journalism and New Media* and *The People's Right to Know*. He is co-developer of the Situated Documentary, a form of location-based storytelling using the emerging mobile and wearable technology known as Augmented Reality.

Prof. Francesc Pedró is the Chief of Education Policy at UNESCO, whose education sector he joined in 2010. His multinational team carries out national policy reviews and comparative international research in areas such as the monitoring and assessment of education, policies to support school leadership and emerging issues in governance in developing countries, as well as on technology in education. Prior to this he worked at the OECD Centre for Educational Research and Innovation (CERI). In his capacity as senior policy analyst, he was responsible for a number of projects including the *Teaching and Learning International Survey* (TALIS) and the *New Millennium Learners Project*. He was also in charge of the *reviews of educational R&D* in OECD countries and of a project on *Systemic Innovation in Education*. His most recent publications are: *Are the New Millennium Learners Making the Grade? Technology Use and Educational Performance in PISA* (2010), *Inspired by Technology, Driven by Pedagogy* (2010), *Tecnología y educación: lo que funciona y porqué* (2011), *Connected Minds. Technology and Today's Learners* (2012), and *Le numérique: une chance pour l'école* (2013).

Constance Elise Porter is Visiting Assistant Professor of Marketing at the Jones Graduate School of Business, Rice University. Her research sits at the intersection of marketing, management, communications and information systems and is focused on relational issues such as trust and engagement within virtual communities and other technological domains. Her research has appeared in academic journals that span various disciplinary boundaries including *Management Science, California Management Review, Journal of Management Information Systems, Journal of Computer-Mediated Communication, Journal of Business Research, Industrial Marketing Management* and *Journal of Marketing Theory and Practice.*

Emanuele Rapetti graduated in Education at the Catholic University of the Sacred Heart in Milan (Italy) with a Master Thesis on Media Education. He holds a Ph.D. in Communication Sciences at USI – Università della Svizzera italiana (Lugano, Switzerland); his research work was entitled "LoDE – Learners of Digital Era. From the analysis of the current debate to an empirical study emphasizing the learners' perspective". He collaborates/ed with several academical institutions in Switzerland and Italy (Università della Svizzera italiana, University of Applied Sciences and Arts of Southern Switzerland, Università Cattolica del Sacro Cuore di Milano, MIP Politecnico di Milano, University of Genova). His research interests are pedagogy, digital pedagogy, education, eLearning, educational evaluation, and research methodology.

Patricia M. Reeves, Ph.D. is a Professor of Social Work at The University of Georgia. She led the development of a dual degree program encompassing a Masters of Social Work and a Masters of Public Health at The University of Georgia. She has won numerous teaching awards. Among other courses, she teaches qualitative research methods, human development, and most recently animal-assisted social work. She has served as a qualitative research methodologist on multi-year research projects focused on diverse problems such as the coping skills of individuals living with HIV/AIDS, teen dating violence, and the needs of severely burn-injured children.

Thomas C. Reeves, Ph.D. is Professor Emeritus of Learning, Design, and Technology in the College of Education at The University of Georgia. He was a Fulbright Lecturer in Peru and has been an invited speaker in the USA and more than 30 other countries. His books include *Interactive Learning Systems Evaluation* (with John Hedberg), a *Guide to Authentic E-Learning* (with Jan Herrington and Ron Oliver), *Conducting Educational Design Research* (with Susan McKenney), and the forthcoming *MOOCs and Open Education Around the World* (with Curt Bonk, Mimi Lee, and Thomas Reynolds). His research interests include educational technology in developing countries, design-based research, and evaluation.

Marko Siitonen works as a University Lecturer at the Department of Communication, University of Jyväskylä (Finland). His Ph.D. dissertation from 2007 looked at social interaction in online multiplayer communities. His research interests include game studies, especially player-to-player interaction, virtual teaming in working life, and newsgames. Currently, he is the editor of the Finnish Yearbook of Speech Communication, and a member of a research consortium looking at the "ludic turn" in industrialized societies.

Stefano Tardini, after a degree in Classical Philology, in 2002 got a Ph.D. in language sciences, with a thesis on the linguistic and semiotic features of virtual communities. Currently he is the Managing Director of eLab: eLearning Lab at USI – Università della Svizzera italiana (Lugano, Switzerland). His research interests include ICT mediated communication, eLearning, (online) communities and social networks. He is developing his research in three interrelated directions: an approach to ICT mediated communication from a socio-historical perspective; an analysis of the impact of the introduction of eLearning activities and tools into given communities; an elaboration of a semiotic approach to online communities and social networks. He is author and co-author of several international publications.

Professor Tim Unwin was appointed Secretary General of the Commonwealth Telecommunications Organisation (CTO) in September 2011, and he also leads the CTO's activities on ICTs for people with disabilities and ICTs in education, focusing particularly on skills development and entrepreneurship. He serves on the ITU's m-Powering Development Advisory Board and as a Member of the UK Department for International Development's Digital Advisory Panel. He was Chair of the Commonwealth Scholarship Commission from 2009–2014, and a Commissioner from 2004. His continuing academic roles include his role as UNESCO Chair in ICT4D and Emeritus Professor of Geography at Royal Holloway, University of London, as well as Honorary Professor at Lanzhou University in China. He has written or edited 15 books and more than 200 academic papers and chapters, many of which focus on the use of technology in development practices. His book, *Information and Communication Technologies for Development*, was published by Cambridge University Press in 2009.

Rolf Wigand is Maulden-Entergy Chair and Distinguished Professor of Information Science and Business Information Systems, University of Arkansas at Little Rock. He is Past Director of the Graduate Program in Information Management at Syracuse University. Current NSF research grants continue with standards development research in several industries; the analysis of virtual organizations, networks in disaster settings and the analysis of online blogs and social media as instruments in collective action. Wigand authored eight books and over 320 articles, chapters, and monographs. His research has appeared in such journals as *MIS Quarterly, Sloan Management Review, Journal of Information Technology* , *Journal of Communication*, and the *Journal of Applied Psychology.*

Brian Winston is the Lincoln Professor at the University of Lincoln (UK) and is a Guest Professor at Beijing Normal University. A prize-winning documentary scriptwriter and journalist, he has written 15 academic books and has held decanal posts in universities in Britain and America. Winston's current main areas of research interest are documentary film, freedom of expression and media technology theory and history. In the later area, his concept of a 'law of suppression of radical potential' and 'supervening social necessity' (as outlined in the prize-winning Media Technology and Society, 1998) have been widely noted.

Kevin B. Wright (Ph.D., University of Oklahoma) is a Professor in the Department of Communication at George Mason University. His main research interests include new technologies and health information-seeking, social support and health outcomes, and stress and burnout among healthcare workers. In addition, his research examines the links between conflict and stress as well as work/life balance issues within healthcare organizations.

Dr. Zheng Xiang holds a Ph.D. in business administration with a concentration in tourism management from Temple University. He is currently an Assistant Professor in the Department of Hospitality and Tourism Management at Virginia Polytechnic Institute and State University. His research interests include travel information search, social media marketing, and business analytics for the tourism and hospitality industries. He is a Board Member of International Federation for IT and Travel & Tourism (IFITT) and Editorial Board Member of several international journals including *Journal of Business Research, Journal of Travel Research*, and *Journal of Information Technology and Tourism.*

Subject index